MODELING and ANALYSIS of DYNAMIC SYSTEMS

MODELING and ANALYSIS of DYNAMIC SYSTEMS

Ramin S. Esfandiari
Bei Lu

CRC Press
Taylor & Francis Group
Boca Raton London New York

CRC Press is an imprint of the
Taylor & Francis Group, an **informa** business

MATLAB® is a trademark of The MathWorks, Inc. and is used with permission. The MathWorks does not warrant the accuracy of the text or exercises in this book. This book's use or discussion of MATLAB® software or related products does not constitute endorsement or sponsorship by The MathWorks of a particular pedagogical approach or particular use of the MATLAB® software.

CRC Press
Taylor & Francis Group
6000 Broken Sound Parkway NW, Suite 300
Boca Raton, FL 33487-2742

© 2010 by Taylor and Francis Group, LLC
CRC Press is an imprint of Taylor & Francis Group, an Informa business

No claim to original U.S. Government works

Printed in the United States of America on acid-free paper
10 9 8 7 6 5 4 3 2 1

International Standard Book Number: 978-1-4398-0845-0 (Hardback)

This book contains information obtained from authentic and highly regarded sources. Reasonable efforts have been made to publish reliable data and information, but the author and publisher cannot assume responsibility for the validity of all materials or the consequences of their use. The authors and publishers have attempted to trace the copyright holders of all material reproduced in this publication and apologize to copyright holders if permission to publish in this form has not been obtained. If any copyright material has not been acknowledged please write and let us know so we may rectify in any future reprint.

Except as permitted under U.S. Copyright Law, no part of this book may be reprinted, reproduced, transmitted, or utilized in any form by any electronic, mechanical, or other means, now known or hereafter invented, including photocopying, microfilming, and recording, or in any information storage or retrieval system, without written permission from the publishers.

For permission to photocopy or use material electronically from this work, please access www.copyright.com (http://www.copyright.com/) or contact the Copyright Clearance Center, Inc. (CCC), 222 Rosewood Drive, Danvers, MA 01923, 978-750-8400. CCC is a not-for-profit organization that provides licenses and registration for a variety of users. For organizations that have been granted a photocopy license by the CCC, a separate system of payment has been arranged.

Trademark Notice: Product or corporate names may be trademarks or registered trademarks, and are used only for identification and explanation without intent to infringe.

Library of Congress Cataloging-in-Publication Data

Esfandiari, Ramin S.
   Modeling and analysis of dynamic systems / authors, Ramin S. Esfandiari, Bei Lu.
       p. cm.
   "A CRC title."
   Includes bibliographical references and index.
   ISBN 978-1-4398-0845-0 (hardcover : alk. paper)
   1. Mathematical models. 2. Dynamics. 3. System theory. I. Lu, Bei, 1974- II. Title.

TA342.E88 2010
621.01'5118--dc22   2009053433

Visit the Taylor & Francis Web site at
http://www.taylorandfrancis.com

and the CRC Press Web site at
http://www.crcpress.com

To my wife Haleh, my sisters Mandana and Roxana, and my parents to whom I owe it all.

Ramin Esfandiari

To my husband Qifu and my parents.

Bei Lu

Contents

Acknowledgments xiii
Preface xv
Authors xvii

Chapter 1 Introduction to MATLAB® and Simulink® 1

    1.1   MATLAB Command Window and Command Prompt 1
    1.2   User-Defined Functions 1
    1.3   Defining and Evaluating Functions 3
    1.4   Iterative Calculations 4
    1.5   Matrices and Vectors 4
    1.6   Differentiation and Integration 6
    1.7   Plotting in MATLAB 7
           1.7.1   Plotting Data Points 7
           1.7.2   Plotting Analytical Expressions 9
    1.8   Simulink 10
           1.8.1   Block Library 10
           1.8.2   Building a New Model 12
           1.8.3   Simulation 13

Chapter 2 Complex Analysis, Differential Equations, and Laplace Transformation 19

    2.1   Complex Analysis 19
           2.1.1   Complex Numbers in Rectangular Form 19
                 2.1.1.1   Magnitude 20
                 2.1.1.2   Complex Conjugate 21
           2.1.2   Complex Numbers in Polar Form 22
                 2.1.2.1   Complex Algebra Using Polar Form 23
                 2.1.2.2   Integer Powers of Complex Numbers 25
                 2.1.2.3   Roots of Complex Numbers 25
           2.1.3   Complex Variables and Functions 25
    2.2   Differential Equations 27
           2.2.1   Linear, First-Order Differential Equations 27
           2.2.2   Second-Order Differential Equations with Constant Coefficients 28
                 2.2.2.1   Homogeneous Solution 28
                 2.2.2.2   Particular Solution 29
    2.3   Laplace Transformation 31
           2.3.1   Linearity of Laplace and Inverse Laplace Transforms 32
           2.3.2   Differentiation and Integration of Laplace Transforms 32
           2.3.3   Special Functions 35
                 2.3.3.1   Unit-Step Function 35
                 2.3.3.2   Unit-Ramp Function 37
                 2.3.3.3   Unit-Pulse Function 38
                 2.3.3.4   Unit-Impulse (Dirac Delta) Function 38

| | | 2.3.3.5 | The Relation between Unit-Impulse and Unit-Step Functions .... 39 |
| | | 2.3.3.6 | Periodic Functions .... 39 |

- 2.3.4 Laplace Transforms of Derivatives and Integrals................................ 40
- 2.3.5 Inverse Laplace Transformation 41
  - 2.3.5.1 Partial Fractions Method 41
  - 2.3.5.2 ◆ Performing Partial Fractions in MATLAB 44
  - 2.3.5.3 Convolution Method 45
  - 2.3.5.4 Solving Initial-Value Problems 46
- 2.3.6 Final-Value Theorem and Initial-Value Theorem 47
- 2.4 Summary 50

Chapter 3 Matrix Analysis 55

- 3.1 Matrices 55
  - 3.1.1 Special Matrices 57
  - 3.1.2 Elementary Row Operations 57
  - 3.1.3 Determinant of a Matrix 58
    - 3.1.3.1 Properties of a Determinant 59
  - 3.1.4 Block Diagonal and Block Triangular Matrices 60
  - 3.1.5 Cramer's Rule 61
    - 3.1.5.1 Homogeneous Systems 63
  - 3.1.6 Inverse of a Matrix 63
    - 3.1.6.1 Adjoint Matrix 63
    - 3.1.6.2 Properties of Inverse 65
- 3.2 Matrix Eigenvalue Problem 68
  - 3.2.1 Solving the Eigenvalue Problem 68
  - 3.2.2 Eigenvalue Properties of Special Matrices 69
    - 3.2.2.1 Algebraic Multiplicity and Geometric Multiplicity 70
    - 3.2.2.2 Generalized Eigenvectors 71
  - 3.2.3 Similarity Transformations 72
    - 3.2.3.1 Matrix Diagonalization 72
    - 3.2.3.2 Defective Matrices 73
- 3.3 Summary 74

Chapter 4 System Model Representation 79

- 4.1 Configuration Form 79
  - 4.1.1 Second-Order Matrix Form 80
- 4.2 State-Space Form 82
  - 4.2.1 State Variables, State-Variable Equations, and State Equation 82
    - 4.2.1.1 State-Variable Equations 83
    - 4.2.1.2 State Equation 83
  - 4.2.2 Output Equation, State-Space Form 85
    - 4.2.2.1 Output Equation 85
    - 4.2.2.2 State-Space Form 86
  - 4.2.3 Decoupling the State Equation 88
- 4.3 I/O Equation, Transfer Function 91
  - 4.3.1 I/O Equations from the System Model 92
  - 4.3.2 Transfer Functions from the System Model 93

	4.4	Relations between State-Space Form, I/O Equation, and Transfer Function 95
		4.4.1 I/O Equation to State-Space Form 96
		4.4.1.1 Controller Canonical Form ◆ 97
		4.4.2 State-Space Form to Transfer Function 99
	4.5	Block Diagram Representation 103
		4.5.1 Block Diagram Operations 103
		4.5.1.1 Summing Junction 103
		4.5.1.2 Series Combinations of Blocks 103
		4.5.1.3 Parallel Combinations of Blocks 104
		4.5.1.4 Integration 105
		4.5.1.5 Closed-Loop Systems 105
		4.5.2 Block Diagram Reduction Techniques 107
		4.5.2.1 Moving a Branch Point 107
		4.5.2.2 Moving a Summing Junction 107
		4.5.2.3 Mason's Rule 109
		4.5.3 Block Diagram Construction from a System Model 111
	4.6	Linearization 116
		4.6.1 Linearization of a Nonlinear Element 116
		4.6.1.1 Functions of Two Variables 117
		4.6.2 Linearization of a Nonlinear Model 117
		4.6.2.1 Operating Point 118
		4.6.2.2 Linearization Procedure 118
		4.6.2.3 Small-Angle Linearization 120
	4.7	Summary 122

Chapter 5 Mechanical Systems 127

	5.1	Mechanical Elements 127
		5.1.1 Mass Elements 127
		5.1.2 Spring Elements 129
		5.1.3 Damper Elements 130
		5.1.4 Equivalence 131
	5.2	Translational Systems 138
		5.2.1 Newton's Second Law 138
		5.2.2 Free-Body Diagram 139
		5.2.3 Static Equilibrium Position and Coordinate Reference 141
		5.2.4 D'Alembert's Principle 145
		5.2.5 Massless Junctions 146
	5.3	Rotational Systems 152
		5.3.1 General Moment Equation 153
		5.3.2 Modeling of Rigid Bodies in Plane Motion 154
		5.3.3 Mass Moment of Inertia 156
		5.3.4 Pure Rolling Motion 160
	5.4	Mixed Systems: Translational and Rotational 167
		5.4.1 Force and Moment Equations 167
		5.4.2 Energy Method 171
	5.5	Gear–Train Systems 178
	5.6	Summary 182

Chapter 6 Electrical, Electronic, and Electromechanical Systems 189

- 6.1 Electrical Elements 189
  - 6.1.1 Resistors 190
  - 6.1.2 Inductors 192
  - 6.1.3 Capacitors 192
- 6.2 Electric Circuits 196
  - 6.2.1 Kirchhoff's Voltage Law 196
  - 6.2.2 Kirchhoff's Current Law 198
  - 6.2.3 Node Method 201
  - 6.2.4 Loop Method 202
  - 6.2.5 State Variables of Circuits 204
- 6.3 Operational Amplifiers 209
- 6.4 Electromechanical Systems 214
  - 6.4.1 Elemental Relations of Electromechanical Systems 215
  - 6.4.2 Armature-Controlled Motors 217
  - 6.4.3 Field-Controlled Motors 222
- 6.5 Impedance Methods 225
  - 6.5.1 Impedances of Electric Elements 225
  - 6.5.2 Series and Parallel Impedances 225
  - 6.5.3 Mechanical Impedances 228
- 6.6 Summary 229

Chapter 7 Fluid and Thermal Systems 235

- 7.1 Pneumatic Systems 235
  - 7.1.1 Ideal Gases 235
  - 7.1.2 Pneumatic Capacitance 236
  - 7.1.3 Modeling of Pneumatic Systems 237
- 7.2 Liquid-Level Systems 241
  - 7.2.1 Hydraulic Capacitance 241
  - 7.2.2 Hydraulic Resistance 243
  - 7.2.3 Modeling of Liquid-Level Systems 244
- 7.3 Thermal Systems 253
  - 7.3.1 First Law of Thermodynamics 253
  - 7.3.2 Thermal Capacitance 254
  - 7.3.3 Thermal Resistance 255
  - 7.3.4 Modeling of Heat Transfer Systems 258
- 7.4 Summary 265

Chapter 8 System Response 269

- 8.1 Transient Response of First-Order Systems 270
  - 8.1.1 Free Response of First-Order Systems 270
  - 8.1.2 Step Response of First-Order Systems 271
  - 8.1.3 Ramp Response of First-Order Systems 272
    - 8.1.3.1 Steady-State Error 273
- 8.2 Transient Response of Second-Order Systems 274
  - 8.2.1 Free Response of Second-Order Systems 275
    - 8.2.1.1 Initial Response in MATLAB® 276
  - 8.2.2 Impulse Response of Second-Order Systems 278
    - 8.2.2.1 Impulse Response in MATLAB 278

Contents

	8.2.3	Step Response of Second-Order Systems 279
		8.2.3.1 ◆ Step Response in MATLAB 280
	8.2.4	◆ Response to Arbitrary Inputs 281
8.3	Frequency Response 287
	8.3.1	Frequency Response of Stable, Linear Systems 288
		8.3.1.1 Frequency Response of First-Order Systems 289
		8.3.1.2 Frequency Response of Second-Order Systems 290
	8.3.2	Bode Diagram 291
		8.3.2.1 ◆ Plotting Bode Diagrams in MATLAB 292
		8.3.2.2 Bode Diagram of First-Order Systems 292
		8.3.2.3 Bode Diagram of Second-Order Systems 295
8.4	Solving the State Equation 300
	8.4.1	Formal Solution of the State Equation 300
		8.4.1.1 Matrix Exponential 300
		8.4.1.2 ◆ Formal Solution in MATLAB 302
	8.4.2	Solution of the State Equation via Laplace Transformation 303
	8.4.3	Solution of the State Equation via State-Transition Matrix 304
8.5	Response of Nonlinear Systems 305
	8.5.1	RK4 Method 306
		8.5.1.1 ◆ Running RK4 in MATLAB 307
8.6	Summary 310

Chapter 9 Introduction to Vibrations 313

9.1	Free Vibration 313
	9.1.1	Logarithmic Decrement 313
	9.1.2	Coulomb Damping 316
9.2	Forced Vibration 320
	9.2.1	Half-Power Bandwidth 321
	9.2.2	Rotating Unbalance 323
	9.2.3	Harmonic Base Excitation 325
9.3	Vibration Suppressions 329
	9.3.1	Vibration Isolators 329
	9.3.2	Vibration Absorbers 331
9.4	Modal Analysis 336
	9.4.1	Eigenvalue Problem 336
	9.4.2	Orthogonality of Modes 340
	9.4.3	Response to Initial Excitations 342
	9.4.4	Response to Harmonic Excitations 344
9.5	Vibration Measurement and Analysis 347
	9.5.1	Vibration Measurement 348
	9.5.2	System Identification 349
9.6	Summary 352

Chapter 10 Introduction to Feedback Control Systems 359

10.1	Basic Concepts and Terminologies 359
10.2	Stability and Performance 362
	10.2.1	Stability of Linear Time-Invariant Systems 362
	10.2.2	Time-Domain Performance Specifications 365
	10.2.3	Frequency-Domain Performance Specifications 368
10.3	Benefits of Feedback Control 370

   10.3.1 Stabilization 370
   10.3.2 Disturbance Rejection... 373
   10.3.3 Reference Tracking.. 375
   10.3.4 Sensitivity to Parameter Variations 377
  10.4 Proportional–Integral–Derivative Control... 380
   10.4.1 Proportional Control 380
   10.4.2 Proportional–Integral Control.. 382
   10.4.3 PID Control 383
   10.4.4 Ziegler–Nichols Tuning of PID Controllers 385
  10.5 Root Locus.. 389
   10.5.1 Root Locus of a Basic Feedback System... 389
   10.5.2 Analysis Using Root Locus 394
   10.5.3 Control Design Using Root Locus... 396
  10.6 Bode Plot 401
   10.6.1 Bode Plot of a Basic Feedback System.. 401
   10.6.2 Analysis Using Bode Plot 407
   10.6.3 Control Design Using Bode Plot 409
  10.7 Full-State Feedback 414
   10.7.1 Analysis of State-Space Equations 414
   10.7.2 Control Design for Full-State Feedback 417
  10.8 Summary... 421

Bibliography 425

Appendix A 427

Appendix B 429

Index 431

Acknowledgments

The authors express their deep gratitude to Jonathan Plant (Senior Editor, Mechanical, Aerospace, Nuclear and Energy Engineering) and Amber Donley (Project Coordinator, Editorial Project Development) at Taylor & Francis/CRC Press for their assistance during various stages of the development of this book.

Preface

The principal goal of this book is to provide the reader with a thorough knowledge of mathematical modeling and analysis of dynamic systems. MATLAB® and Simulink® are introduced at the outset and are utilized throughout the book to perform symbolic, graphical, numerical, and simulation tasks. The textbook, written at the junior level, meticulously covers techniques for modeling dynamic systems, methods of response analysis, and an introduction to vibration and control systems.

The book comprises 10 chapters and two appendices. Chapter 1 methodically introduces MATLAB and Simulink to the reader. The essential mathematical background is covered in Chapter 2 (complex analysis, differential equations, and Laplace transformation) and Chapter 3 (matrix analysis). Different forms of system model representation (state-space form, transfer function, input–output equation, block diagram, etc.) as well as linearization are discussed in Chapter 4. Each topic is also handled using MATLAB. Block diagrams are constructed and analyzed by using Simulink.

Chapter 5 treats translational, rotational, and mixed mechanical systems. The free-body-diagram approach is greatly emphasized in the derivation of the systems' equations of motion. Electrical and electromechanical systems are covered in Chapter 6. Also included are operational amplifiers and impedance methods. Chapter 7 discusses pneumatic, liquid-level, and thermal systems.

Chapter 8 deals with the time-domain and frequency-domain analysis of dynamic systems. Time-domain analysis entails the transient response of first-, second-, and higher-order systems. The sinusoidal transfer function (frequency response function) is introduced and utilized in obtaining the system's frequency response, as well as the Bode diagram. An analytical solution of the state equation is also included in this chapter. MATLAB and Simulink play significant roles in determining and simulating system response, and are used throughout the chapter.

Chapter 9 presents an introduction to vibrations and includes free and forced vibrations of single- and multiple-degree-of-freedom systems, vibration suppression including vibration isolators and absorbers, modal analysis, and vibration testing. Also included are some applications of vibrations: logarithmic decrement for experimental determination of the damping ratio, rotating unbalance, and harmonic base excitation.

Chapter 10 gives an introduction to control systems analysis and design in the time and frequency domains. Basic concepts and terminology are presented first, followed by stability analysis, types of control, root locus analysis, Bode plot, and full-state feedback. All these analytical techniques are implemented using MATLAB and Simulink.

APPENDICES

Appendix A gives a summary of systems of units and conversion tables. Appendix B contains useful formulas such as trigonometric identities and integrals.

EXAMPLES AND EXERCISES

Each covered topic is followed by at least one example for better comprehension of the subject matter at hand. More complex topics are accompanied by multiple, painstakingly worked-out examples. Each section of each chapter is followed by several exercises so that the reader can immediately apply the ideas just learned. End-of-chapter review exercises help in learning how a combination of different ideas can be used to analyze a problem.

CHAPTER SUMMARIES

Chapter summaries provide succinct reviews of the key aspects of each chapter.

INSTRUCTOR'S SOLUTIONS MANUAL

A solutions manual, featuring complete solution details of all exercises, is prepared by the authors and will be available to instructors adopting the book.

An ample portion of the material in this book has been rigorously class tested over the past several years. And the valuable remarks and suggestions made by students have greatly contributed to making this book as complete and user-friendly as possible.

MATLAB® and Simulink® are registered trademarks of The MathWorks, Inc. For product information, please contact:

The MathWorks, Inc.
3 Apple Hill Drive
Natick, MA 01760-2098, USA
Tel: 508 647 7000
Fax: 508-647-7001
E-mail: info@mathworks.com
Web: www.mathworks.com

Ramin Esfandiari
Bei Lu

Authors

Dr. Ramin Esfandiari is a professor of Mechanical and Aerospace Engineering at California State University, Long Beach (CSULB), where he has served as a faculty member since 1989. He received his BS in Mechanical Engineering, and MA and PhD in Applied Mathematics (Optimal Control), all from the University of California, Santa Barbara. He has authored several refereed research papers in high-quality engineering and scientific journals such as *Journal of Optimization Theory and Applications*, *Journal of Sound and Vibration*, *Optimal Control Applications and Methods*, and *ASME Journal of Applied Mechanics*. Dr. Esfandiari is the author of *Applied Mathematics for Engineers*, 4th edition (Atlantis, 2007), *Matrix Analysis and Numerical Methods for Engineers* (Atlantis, 2007), and *MATLAB Manual for Advanced Engineering Mathematics* (Atlantis, 2007). He is one of the select few contributing authors for the latest edition of *Mechanical Engineering Handbook* (Springer-Verlag, 2009) and coauthor (with Dr. H. V. Vu) of *Dynamic Systems: Modeling and Analysis* (McGraw-Hill, 1997). Professor Esfandiari is the recipient of numerous teaching and research awards, including Meritorious Performance and Professional Promise Award, TRW Excellence in Teaching and Scholarship Award, and the Distinguished Faculty Teaching Award.

Dr. Bei Lu is an assistant professor of Mechanical and Aerospace Engineering at California State University, Long Beach (CSULB). She received her BS and MS degrees in Power and Mechanical Engineering from Shanghai Jiaotong University, China, in 1996 and 1999, respectively, and her PhD in Mechanical Engineering from North Carolina State University in 2004. Subsequently, she worked in the Department of Mechanical and Aerospace Engineering at North Carolina State University as a research associate for five months. Her main research interests include robust control, linear parameter-varying control of nonlinear systems, and application of advanced control and optimization techniques to aerospace, mechanical, and electromechanical engineering problems. She has published over 20 conference papers and journal articles. She is a member of American Institute of Aeronautics and Astronautics (AIAA), American Society of Mechanical Engineers (ASME), and Institute of Electrical and Electronic Engineers (IEEE).

1 Introduction to MATLAB® and Simulink®

This chapter introduces some fundamental features of MATLAB® and Simulink®. These include the description and application of several commonly used built-in functions (commands) in MATLAB and the basics of building block diagrams for the purpose of simulation of dynamic systems using Simulink. MATLAB and Simulink are integrated throughout the book, and most of the functions utilized in the upcoming chapters depend on the basic ones presented here.

1.1 MATLAB COMMAND WINDOW AND COMMAND PROMPT

Once a MATLAB session is opened, commands can be entered at the MATLAB command prompt ">>" (Figure 1.1). For example,

```
>> sqrt(3) + 1
ans =
    2.7321
```

The outcome of a calculation can be stored under a variable name:

```
>> c = cos(pi/4)
c =
    0.7071
```

The result may be suppressed (not displayed) by using a semicolon at the end of the statement:

```
>> c = cos(pi/4);
```

Commands such as `sqrt` (square root) and `cos` (cosine of an angle in radians) are MATLAB built-in functions. Each of these functions is accompanied by a brief but sufficient description through the `help` command.

```
>> help sqrt
 SQRT Square root.
    SQRT(X) is the square root of the elements of X. Complex
    results are produced if X is not positive.
    See also sqrtm, realsqrt, hypot.
    Overloaded methods:
       sym/sqrt
    Reference page in Help browser
       doc sqrt
```

1.2 USER-DEFINED FUNCTIONS

User-defined M file functions and scripts may be created, saved, and edited in MATLAB using the `edit` command. For example, suppose we want to create a function (say, `eqline`) that returns

FIGURE 1.1 Screen capture of a MATLAB session.

the equation of a line passing through a point with a given slope. The function can be saved in a folder on the MATLAB path or in the current directory. The current directory can be viewed and/or changed using the drop-down menu at the top of the MATLAB command window. Once the current directory has been properly selected, type

```
>> edit eqline
```

This will open a new window where the function is constructed by typing the following code:

```
function y = eqline(x0, y0, m)
%
% EQLINE finds the equation of a line with slope m going through
% a given point (x0, y0).

syms x     % Declare x as a symbolic variable
y = y0 + m*(x-x0);
```

To execute this successfully, the current directory must be where the function was saved. And the three input arguments (x0, y0, m) must be supplied. For example, the equation of the line going through the point $(-1, 1)$ with slope 2 can be found as

```
>> y = eqline(-1, 1, 2)
y =
3+2*x
```

Introduction to MATLAB® and Simulink®

1.3 DEFINING AND EVALUATING FUNCTIONS

The built-in function `inline` is ideal for defining and evaluating functions of one or more variables. For instance, consider $h(t) = t^2$ which is a function of a single variable t and hence must be defined as such:

```
>> h = inline('t^2','t')
h =
     Inline function:
     h(t) = t^2
```

For the case of one independent variable, as above, dependence on t may simply be omitted:

```
>> h = inline('t^2')
h =
     Inline function:
     h(t) = t^2
```

For functions of two or more independent variables, for example, $g = y \sin x$, the desired order of the variables must be specified. Otherwise, MATLAB will list them in alphabetical order.

```
>> g = inline('y*sin(x)')
g =
     Inline function:
     g(x,y) = y*sin(x)
```

If $g(y, x)$ is desired, then

```
>> g = inline('y*sin(x)','y','x')
g =
     Inline function:
     g(y,x) = y*sin(x)
```

To evaluate g when $x = \pi/3$ and $y = 1.1$,

```
>> g(1.1,pi/3)
ans =
    0.9526
```

Another way to define and evaluate a function is as follows. Let us consider $g = y \sin x$ once again. Then, function g is defined symbolically in one of two ways:

```
>> g = 'y*sin(x)'
g =
y*sin(x)
```

or

```
>> g = sym('y*sin(x)')
g =
y*sin(x)
```

Evaluation of a function can be achieved by using the subs command:

```
>> help subs
   Utilities for obsolete MUTOOLS commands.
subs is both a directory and a function.
SUBS Symbolic substitution.
   SUBS(S) replaces all the variables in the symbolic expression S with
   values obtained from the calling function, or the MATLAB workspace.
   SUBS(S,NEW) replaces the free symbolic variable in S with NEW.
   SUBS(S,OLD,NEW) replaces OLD with NEW in the symbolic expression S.
```

For instance, to evaluate g when $x = \pi/3$ and $y = 1.1$,

```
>> x = pi/3; y = 1.1;
>> subs(g)
ans =
    0.9526    % Agrees with the earlier result
```

1.4 ITERATIVE CALCULATIONS

Iterations involve statements that are repeated a specific or indefinite number of times. For a specific number of repeated statements, the command for is used. As an example, to generate the sequence $\{\frac{1}{2}, \frac{1}{4}, \frac{1}{8}, \frac{1}{16}, \frac{1}{32}\}$, we proceed as follows.

```
>> for n=1:5,
x(n)=1/(2^n);
end
>> x
x =
    0.5000 0.2500 0.1250 0.0625 0.0313
```

Situations entailing an indefinite number of repeated statements can be handled in several ways. One way is to use the command for in conjunction with the command if, as follows. The code below will generate elements of the sequence $x_n = 1/n$ ($n = 1, 2, 3, \ldots$), terminating the process as soon as $|x_n - x_{n-1}| < 0.1$ is met.

```
>> x(1) = 1;        % Define the first element
>> for n = 2:20,    % Maximum number of iterations is 20
x(n) = 1/n;         % Generate subsequent elements
if abs(x(n) - x(n-1)) < 0.1, break; end    % Terminating condition
end
>> x
x =
    1.0000 0.5000 0.3333 0.2500
```

Note that not all 20 values of the index n were used here because the terminating condition was satisfied when $n = 4$.

1.5 MATRICES AND VECTORS

A matrix can be created by using brackets that enclose all elements of the matrix, rows being separated by a semicolon.

```
>> A = [-1 1 2;2 0 -1;0 1 5]      % 3-by-3 matrix A
A =
    -1    1    2
     2    0   -1
     0    1    5
```

An entry can be accessed by using the row and column number of the location of that entry.

```
>> A(3,2)   % Entry at the intersection of the 3rd row and 2nd column
ans =
     1
```

An entire row or an entire column of a matrix is accessed via a colon operator.

```
>> Row2 = A(2,:)     % Second row of A and any column
Row2 =
     2    0   -1
>> Col3 = A(:,3)     % Third column of A and any row
Col3 =
     2
    -1
     5
```

To replace an entire column of matrix A by a given vector v, we proceed as follows.

```
>> v = [-4;1;2]    % Define the given vector v
v =
    -4
     1
     2
>> Anew = A;    % Pre-allocate matrix Anew
>> Anew(:,2) = v    % Assign vector v to the second column of Anew
Anew =
    -1   -4    2
     2    1   -1
     0    2    5    % Same as A except for the 2nd column which is v
```

The $m \times n$ zero matrix is created by using zeros(m,n); for instance, the 2×3 zero matrix:

```
>> A = zeros(2,3)
A =
     0    0    0
     0    0    0
```

This is a common way of preallocating memory for a matrix. Now, any entry can be altered while others remain unchanged.

```
>> A(2,2) = -1; A(1,3) = 2;
>> A
A =
     0    0    2
     0   -1    0
```

Suppose we wish to construct a 4 × 4 matrix whose diagonal elements are all −1s, entries directly above the diagonal are 2s, entries directly below are −2s, and all other entries are zeros.

```
>> for m = 1:4,
for n = 1:4,
if m == n,
A(m,n) = -1;
elseif m-n == 1
A(m,n) = -2;
elseif n-m == 1
A(m,n) = 2;
end
end
end
>> A
A =
    -1     2     0     0
    -2    -1     2     0
     0    -2    -1     2
     0     0    -2    -1
```

1.6 DIFFERENTIATION AND INTEGRATION

Differentiation of a function with respect to an independent variable is done through the "diff" command. Suppose $x = t^2 - \sin t$ and we are interested in dx/dt. We first need to define the function symbolically. As mentioned in Section 1.3, this can be done in one of two ways:

```
>> x = sym('t^2-sin(t)')
x =
t^2-sin(t)
```

or

```
>> x = 't^2-sin(t)'
x =
t^2-sin(t)
```

Then, dx/dt is calculated as

```
>> dxdt = diff(x)
dxdt =
2*t-cos(t)
```

To evaluate dxdt when t=1, we use the subs command as before.

```
>> eval(subs(dxdt,'1'))
ans =
    1.4597
```

The second derivative d^2x/dt^2 can be obtained as

```
>> diff(x,2)
ans =
2+sin(t)
```

Introduction to MATLAB® and Simulink®

Now consider a function of two variables, say, $g = t^2 + a\sin s^2$, where a is a constant. The partial derivatives of g with respect to its independent variables t and s are found as follows.

```
>> syms a      % Define a symbolically
>> g = 't^2+a*sin(s^2)';    % Define function g
>> diff(g,'t')    % Differentiate with respect to t
ans =
2*t
>> diff(g,'s')    % Differentiate with respect to s
ans =
2*a*cos(s^2)*s
```

Indefinite and definite integrals are calculated in MATLAB via the "int" command.

```
INT      Integrate.
   INT(S) is the indefinite integral of S with respect to its symbolic
     variable as defined by FINDSYM. S is a SYM (matrix or scalar).
     If S is a constant, the integral is with respect to 'x'.
   INT(S,v) is the indefinite integral of S with respect to v. v is a
     scalar SYM.
   INT(S,a,b) is the definite integral of S with respect to
     its symbolic variable from a to b. a and b are each double or
     symbolic scalars.
   INT(S,v,a,b) is the definite integral of S with respect to v from
     a to b.
```

For instance, the indefinite integral $\int \left(t + \frac{1}{2}\cos t\right) dt$ is calculated as

```
>> syms t
>> int(t+1/2*cos(t))
ans =
1/2*t^2+1/2*sin(t)
```

But the definite integral $\int_0^1 (a + 2t) \, dt$, where a is a constant, is evaluated as

```
>> syms a t
>> int(a+2*t,t,0,1)
ans =
a+1
```

1.7 PLOTTING IN MATLAB

1.7.1 PLOTTING DATA POINTS

Plotting a vector of values versus another vector of values is done by using the plot command.

```
>> help plot
 PLOT    Linear plot.
   PLOT(X,Y) plots vector Y versus vector X. If X or Y is a matrix,
    then the vector is plotted versus the rows or columns of the matrix,
    whichever line up. If X is a scalar and Y is a vector, disconnected
    line objects are created and plotted as discrete points vertically
    at X.
```

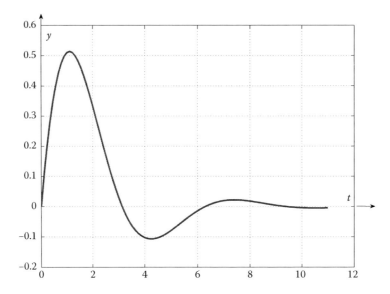

FIGURE 1.2 Single plot.

For instance, to plot $y = e^{-t/2} \sin t$ versus $t \in [0, 7\pi/2]$ using 100 points, we proceed as follows (Figure 1.2).

```
>> t = linspace(0,7*pi/2);   % Generate 100 (default) equally-spaced points
>> y = exp(-t/2).*sin(t);    % Evaluate the function at these points
>> plot(t,y)     % Generate Figure 1.2
>> grid on       % Add grid
```

Because t is an array, we have used ".*" to allow for element-by-element multiplication. The plot command can also be used to generate multiple plots. For example, to plot $x_1 = e^{-t/3} \sin 2t$ and $x_2 = e^{-t/2} \sin t$ versus $0 \leq t \leq 10$, the code below may be executed.

```
>> t = linspace(0,10);
>> x1 = exp(-t/3).*sin(2*t);
>> x2 = exp(-t/2).*sin(t);
>> plot(t,x1,t,x2)    % Figure 1.3
>> grid on
```

The built-in command subplot is designed to create multiple figures in tiled positions.

```
>> help subplot
 SUBPLOT Create axes in tiled positions.
    H = SUBPLOT(m,n,p), or SUBPLOT(mnp), breaks the Figure window into an
    m-by-n matrix of small axes, selects the p-th axes for the current
    plot, and returns the axis handle. The axes are counted along the
    top row of the Figure window, then the second row, etc.
```

Let $u(x,t) = e^{-x} \sin(t+x)$, where $0 \leq x \leq 5$. To plot $u(x,t)$ versus x for $t = 0, 1, 2, 3$, we execute the following code.

```
>> x = linspace(0,5);
>> t = 0:1:3;
>> for i = 1:4,
```

Introduction to MATLAB® and Simulink®

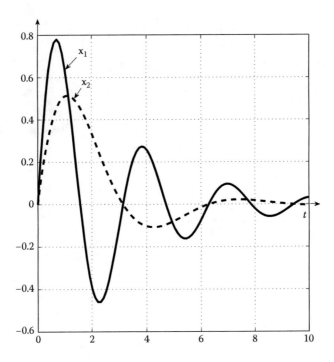

FIGURE 1.3 Multiple plots.

```
for j = 1:100,
u(j,i) = exp(-x(j))*sin(t(i)+x(j));
end
end
>> subplot(2,2,1), plot(x,u(:,1)), grid on    % Initiate Figure 1.4
>> title('t = 0')
>> subplot(2,2,2), plot(x,u(:,2)), grid on
>> title('t = 1')
>> subplot(2,2,3), plot(x,u(:,3)), grid on
>> title('t = 2')
>> subplot(2,2,4), plot(x,u(:,4)), grid on
>> title('t = 3')      % Complete Figure 1.4
```

1.7.2 PLOTTING ANALYTICAL EXPRESSIONS

An alternative method is to use the `ezplot` command, which plots the function without requiring data generation.

```
>> syms t
>> y = exp(-t/2)*sin(t);
>> ezplot(y,[0,7*pi/2])    % Also generates Figure 1.2
```

Multiple plots can also be generated by using `ezplot`. Reconsidering an earlier example (which led to Figure 1.3), we plot the two functions $x_1 = e^{-t/3}\sin 2t$ and $x_2 = e^{-t/2}\sin t$ versus $0 \leq t \leq 10$ as follows (Figure 1.4).

```
>> syms t
>> x1 = exp(-t/3)*sin(2*t);
```

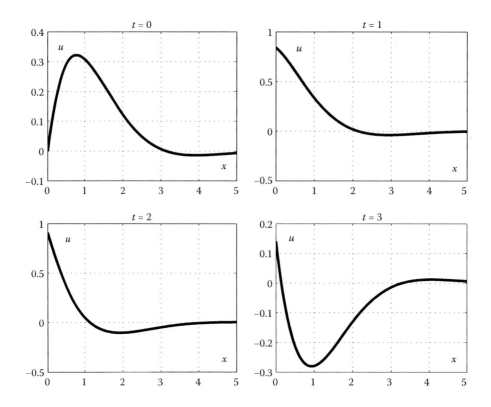

FIGURE 1.4 Subplot.

```
>> x2 = exp(-t/2)*sin(t);
>> ezplot(x1,[0,10])      % Initiate plot
>> hold on
>> ezplot(x2,[0,10])      % Complete plot
>> axis([0 10 -0.6 0.8])  % Set the limits to resemble Figure 1.3
>> grid on
```

Note that the limits for the horizontal and vertical axes have been reset so that the resulting graph matches that in Figure 1.3.

1.8 SIMULINK

Simulink is a powerful software package widely used in academia and industry for the modeling, analysis, and simulation of dynamic systems. In the modeling phase, models are built as block diagrams (see Section 4.5) via a graphical user interface (GUI). Once a model is built, it can be simulated with the aid of Simulink menus or by entering commands in MATLAB's command window. One of the greatest advantages of Simulink is the fact that it allows us to analyze and simulate the more realistic nonlinear models rather than the idealized linear ones.

1.8.1 BLOCK LIBRARY

Typing Simulink at the MATLAB command prompt opens a new window labeled Simulink Library Browser (Figure 1.5), which includes a complete block library of sinks, sources, components, and connectors. Clicking on any of these categories reveals the list of blocks it contains. For instance, clicking on Commonly Used Blocks results in the window shown in Figure 1.6.

Introduction to MATLAB® and Simulink®   11

FIGURE 1.5 Screen capture of Simulink library.

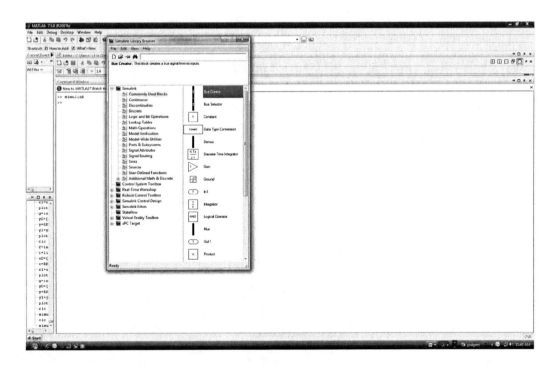

FIGURE 1.6 Screen capture of Commonly Used Blocks.

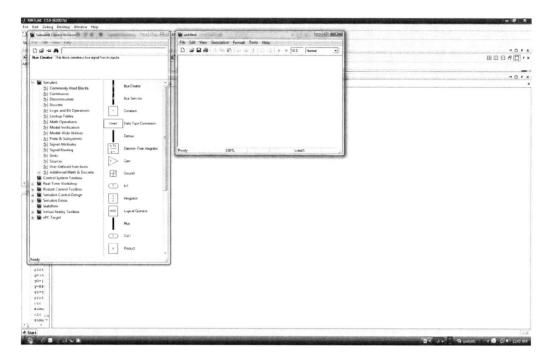

FIGURE 1.7 Screen capture of a new model window.

1.8.2 Building a New Model

To create a new model, select **Model** from the **New** submenu of the Simulink library window's **File** menu or simply press the **New Model** button on the Library Browser's toolbar. This will open a new model window (Figure 1.7). Suppose we want to build a model that integrates a step signal and displays the result, along with the step signal itself. The block diagram of this model will resemble Figure 1.8. To create this model, we will need to copy (drag and drop) the following blocks into our new model:

- The Step signal (from Sources library)
- The Integrator block (from Continuous library)
- The Mux block (from Signal Routing library)
- The Scope block (from Sinks library)

This results in what is shown in Figure 1.9. Note that double-clicking on each block reveals more detailed properties for that block. The ">" symbol pointing out of a block is an output port. The symbol ">" pointing to a block is an input port. A signal travels through a connecting line out

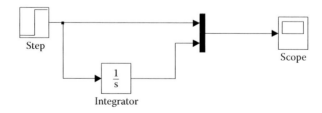

FIGURE 1.8 New model.

Introduction to MATLAB® and Simulink®

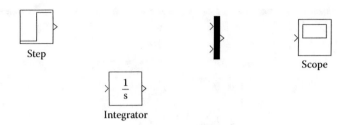

FIGURE 1.9 Basic blocks involved in a new model.

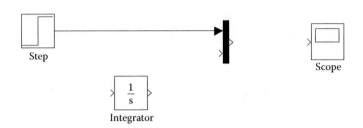

FIGURE 1.10 Connecting line.

of an output port of a block and into an input port of another block. The port symbols disappear as soon as the blocks are connected. Let us now connect the blocks. First connect the Step block to the top input port of the Mux block. Place the pointer over the output port of the Step block and note that the cursor shape changes to cross hairs. Hold down the mouse button and move the cursor to the top input port of the Mux block, and then release the mouse button. If the connecting line is not straight, simply drag the Step block up or down until the ports are lined up (Figure 1.10). The signal going from the Step block to the Mux block must also go through the integrator. This may be done by means of a branch line. Place the pointer on the connecting line between the Step and Mux blocks. Press and hold down the **Ctrl** key. Press the mouse button and drag the pointer to the Integrator block's input port, and then release the mouse button. This results in Figure 1.11. Finally connect from the output port of the Integrator to the bottom input port of the Mux block, and draw a connecting line from the output port of the Mux to the input port of the Scope. The completed block diagram will look like the original Figure 1.8.

1.8.3 SIMULATION

The simulation parameters can be set by choosing **Configuration Parameters** from the **Simulation** menu. Note that the default stop time is 10.0. Next, choose **Start** from the **Simulation** menu.

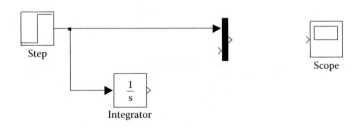

FIGURE 1.11 Branch line.

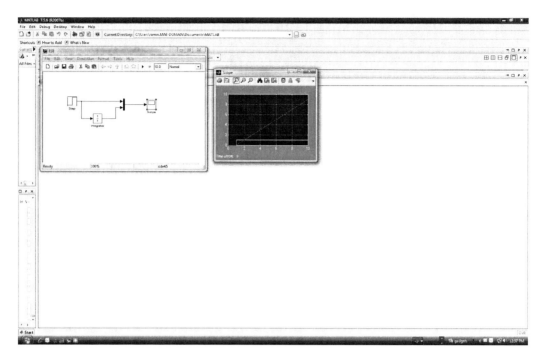

FIGURE 1.12 Simulation.

Double-click on the Scope block and then choose the binocular option (autoscale) to see the step signal as well as its integral (Figure 1.12). Obviously, the simulation stops when the stop time of 10.0 is reached. The simulation thus obtained cannot be copied and pasted into a document and is only for observation while in a MATLAB session. To have access to the actual output data, the following needs to be done. Select the Out1 block from the Commonly Used Blocks library and copy it onto the existing model. Then draw the appropriate branch line to obtain the completed diagram shown in Figure 1.13. Once again, run the simulation. As a result, the time vector is automatically saved in tout, while the output is saved in yout. Next at the MATLAB command prompt, type

```
>> plot(tout,yout)
```

This yields Figure 1.14. Note that this is the same result as observed in the simulation.

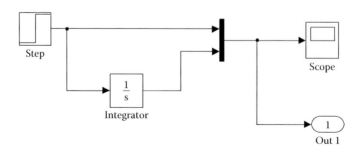

FIGURE 1.13 Storing output data.

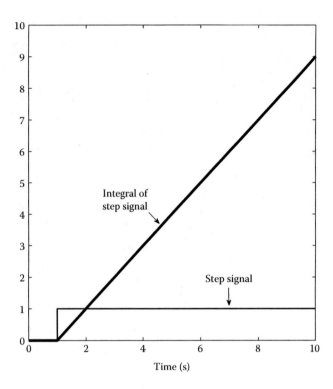

FIGURE 1.14 Output data.

REVIEW PROBLEMS

1. Create a function called `circ_area` to compute the area A of a circle of radius r. Then, execute the function to calculate the area of a circle of radius 1.45.
2. Create a function called `eval_func` that has inputs $f(x)$, defined as an inline function, and constants a and b, where $a < b$. It finds the midpoint m of $[a,b]$ and then computes

$$Q = f(a) + 2f(m) + f(b).$$

   The sole output of the function should be Q. Execute `eval_func` for $f(x) = \sin 2x$, $a = \pi/3$, $b = \pi$.
3. Write a function called `laplace_eval` with the following task. Its inputs are $f(x,y)$, defined symbolically, and constants a and b. The function first finds $f_{xx} + f_{yy}$ and then evaluates the result at $x = a$, $y = b$. Call the output Q. Execute the function for the case of $f = x^2 \cos y - 1/y$, $a = 0$, $b = 1$. Note that $f_{xx} = \partial^2 f/\partial x^2$.
4. Write a function called `mid_point` with the following task. The inputs are constants a, b (such that $b > a$) and tolerance e. The function first calculates the midpoint of $[a,b]$, called m_1. Next it finds the midpoint of $[a,m_1]$, called m_2. Then the midpoint of $[a,m_2]$, called m_3, and so on. The process terminates as soon as $|m_k - m_{k-1}| < e$. Use a maximum of 20 iterations. The function output will be the entire sequence (array) $m_1, m_2, \ldots$. Execute it for the case of $a = 1$, $b = 8$, and $e = 10^{-2}$ (in MATLAB, written as 1e-2).
5. Evaluate $\int_0^t e^{t-x} \sin x \, dx$.
6. Evaluate $\int_0^\infty (\sin x/x) \, dx$.

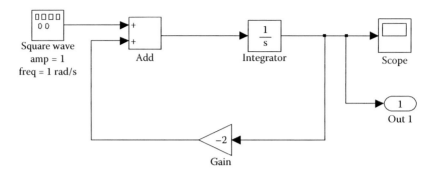

FIGURE 1.15 Problem 13.

7. Write a code to generate

$$\mathbf{A} = \begin{bmatrix} 2 & 1 & -1 & 0 & 0 \\ 0 & 2 & 1 & -1 & 0 \\ 0 & 0 & 2 & 1 & -1 \\ 0 & 0 & 0 & 2 & 1 \\ 0 & 0 & 0 & 0 & 2 \end{bmatrix}.$$

8. Write a code to generate

$$\mathbf{B} = \begin{bmatrix} 1 & 0 & 2 & 0 & 0 \\ 0 & 2 & 0 & 2 & 0 \\ -1 & 0 & 3 & 0 & 2 \\ 0 & -1 & 0 & 4 & 0 \\ 0 & 0 & -1 & 0 & 5 \end{bmatrix}.$$

9. Plot the two functions $x_1(t) = \frac{2}{\sqrt{3}} e^{-t/2} \sin\left(\frac{\sqrt{3}}{2} t\right)$ and $x_2(t) = t e^{-t}$ versus $0 \le t \le 10$ s in the same graph. Adjust the limits of the vertical axis to -0.2 and 0.7. Add grid and label.

10. Plot $w(x,t) = \sin\left(\frac{1}{2}\pi x\right) \cos\left(\frac{\sqrt{5}}{2}\pi t\right)$ versus $0 \le x \le 4$ for $t = 0.1, 0.5, 1, 2$ in a 2×2 tile and add title.

11. Consider

$$\int_0^1 e^{-x/3}\, dx.$$

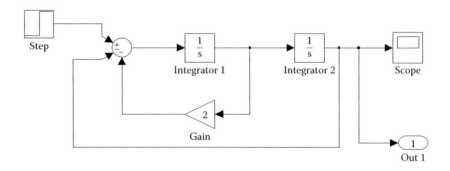

FIGURE 1.16 Problem 14.

a. Use the `quad` function to evaluate the integral. Note that `quad` requires the integrand to be defined as an inline function.
b. Use the `int` command to evaluate the integral.
12. Plot $\int_0^t e^{-x/2} dx$ versus $0 \le t \le 10$.
13. Create the Simulink model shown in Figure 1.15. Double-clicking on each block allows you to explore its properties. Choose the signal generator as a square wave with an amplitude of 1 and a frequency of 1 rad/s. To flip the gain block (Commonly Used Blocks), right click on it; then go to format and choose flip block. Perform simulation and generate a figure that can be imported into a document.
14. Repeat Problem 13 for the model shown in Figure 1.16. Note that the input signal is a unit-step now.

2 Complex Analysis, Differential Equations, and Laplace Transformation

This chapter covers complex analysis, differential equations, and Laplace transformation, fundamental tools that facilitate the understanding process of various ideas, and implementation of many techniques, involved in the analysis of dynamic systems. Complex analysis refers to the study of complex numbers, variables, and functions. Ordinary differential equations arise in situations involving the rate of change of a function with respect to its only independent variable. Differential equations of various orders—with constant coefficients—are generally difficult to solve. To that end, Laplace transformation is used to transform the data from the time domain to the s domain, where equations are algebraic and hence easier to treat. The solution of the differential equation is ultimately obtained when information is transformed back to the time domain.

2.1 COMPLEX ANALYSIS

Complex analysis comprises complex numbers, variables, and functions, which occur in a wide range of areas in dynamic systems analysis from frequency response to natural frequencies.

2.1.1 COMPLEX NUMBERS IN RECTANGULAR FORM

A complex number z in rectangular form is expressed as

$$z = x + jy, \quad j = \sqrt{-1} = \text{imaginary number}, \tag{2.1}$$

where x and y are real numbers, known as the real and imaginary parts of z, respectively, and denoted by $x = \text{Re}\{z\}$, $y = \text{Im}\{z\}$. For example, if $z = -3 + j$, then $x = \text{Re}\{z\} = -3$ and $y = \text{Im}\{z\} = 1$. A complex number with zero real part is known as pure imaginary, for example, $z = 2j$. Two complex numbers are equal if and only if their respective real and imaginary parts are equal. Addition of complex numbers is performed component-wise, that is, if $z_1 = x_1 + jy_1$ and $z_2 = x_2 + jy_2$, then

$$\begin{aligned} z_1 + z_2 &= (x_1 + jy_1) + (x_2 + jy_2) \\ &= (x_1 + x_2) + j(y_1 + y_2). \end{aligned}$$

Multiplication of two complex numbers is performed in the same way as two binomials with the provision that $j^2 = -1$, that is,

$$\begin{aligned} z_1 z_2 &= (x_1 + jy_1)(x_2 + jy_2) = x_1 x_2 + jy_1 x_2 + jx_1 y_2 + j^2 y_1 y_2 \\ &= (x_1 x_2 - y_1 y_2) + j(x_1 y_2 + x_2 y_1). \end{aligned}$$

Complex numbers have a two-dimensional character because they consist of a real part and an imaginary part. Therefore, they may be represented geometrically as points in a Cartesian coordinate system, known as the complex plane. The x-axis of the complex plane is the real axis, and its y-axis is the imaginary axis (Figure 2.1). Since $z = x + jy$ is uniquely identified by an ordered pair (x, y) of real numbers, it can be represented as a position vector in the complex plane, with initial point 0 and

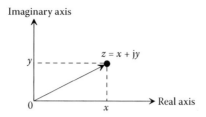

FIGURE 2.1 Geometry of a complex number.

terminal point $z = x + jy$. The concept of vector addition hence applies to the addition of complex numbers. For example, consider $z_1 = 3 + j$ and $z_2 = -2 + 4j$ as in Figure 2.2. It is then clear that the sum $z_1 + z_2 = 1 + 5j$ agrees precisely with the resultant of the position vectors of z_1 and z_2.

2.1.1.1 Magnitude

The magnitude (or modulus) of a complex number $z = x + jy$ is denoted by $|z|$ and defined as

$$|z| = \sqrt{x^2 + y^2}. \tag{2.2}$$

Referring to Figure 2.1, the magnitude of z is simply the distance from the origin to z. If z is a real number ($z = x$), it is located on the real axis and its magnitude is equal to its absolute value. If z is pure imaginary ($z = jy$), it is on the imaginary axis and $|z| = |y|$. The distance between two complex numbers z_1 and z_2 is provided by $|z_1 - z_2|$. Addition of complex numbers follows the triangle inequality rule

$$|z_1 + z_2| \leq |z_1| + |z_2|. \tag{2.3}$$

Example 2.1: Magnitude

Let $z_1 = -1 - 2j$, $z_2 = 3 - 4j$.

  a. Calculate the distance between z_1 and z_2, and verify the triangle inequality.
  b. Perform Part (a) in MATLAB®.

Solution

  a. $z_1 - z_2 = -4 + 2j$ and $|z_1 - z_2| = \sqrt{(-4)^2 + 2^2} = \sqrt{20}$. The triangle inequality, Equation 2.3, can also be verified for the problem at hand as follows. First

$$|z_1| = |-1 - 2j| = \sqrt{5}, \quad |z_2| = |3 - 4j| = 5, \quad |z_1 + z_2| = |2 - 6j| = \sqrt{40}.$$

Subsequently, it is observed that $\sqrt{40} \leq \sqrt{5} + 5$.

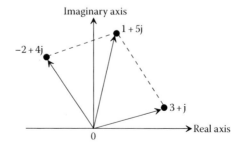

FIGURE 2.2 Addition of complex numbers.

b.
```
>> z1 = -1-2*j; z2 = 3-4*j;
% MATLAB recognizes both i and j as the imaginary number
>> abs(z1 - z2)
ans =
    4.4721    % sqrt(20)
>> abs(z1 + z2)
ans =
    6.3246    % sqrt(40)
>> abs(z1) + abs(z2)
ans =
    7.2361    % Triangle inequality is verified
```

2.1.1.2 Complex Conjugate

The complex conjugate of $z = x + jy$ is denoted by $\bar{z}$ and defined as $\bar{z} = x - jy$. It is readily seen that the product of a complex number ($z \neq 0$) and its conjugate is a positive, real number, equal to the square of the magnitude of z,

$$z\bar{z} = (x + jy)(x - jy) = x^2 + y^2 = |z|^2. \tag{2.4}$$

Geometrically, $\bar{z}$ is the reflection of z through the real axis. Conjugates play an important role in the treatment of division of complex numbers. Let us consider z_1/z_2, where $z_1 = x_1 + jy_1$ and $z_2 = x_2 + jy_2$ ($z_2 \neq 0$). Multiplication of the numerator and the denominator by the conjugate of the denominator ($\bar{z}_2 = x_2 - jy_2$) results in

$$\frac{x_1 + jy_1}{x_2 + jy_2} = \frac{(x_1 + jy_1)(x_2 - jy_2)}{(x_2 + jy_2)(x_2 - jy_2)} \quad \text{Using Equation 2.4 in the denominator} \quad = \frac{(x_1 x_2 + y_1 y_2) + j(y_1 x_2 - y_2 x_1)}{x_2^2 + y_2^2}$$

$$= \frac{x_1 x_2 + y_1 y_2}{x_2^2 + y_2^2} + j\frac{y_1 x_2 - y_2 x_1}{x_2^2 + y_2^2}$$

Note that the resulting complex number has been expressed in standard rectangular form.

Example 2.2: Conjugation

a. Perform $(3 + 2j)/(1 - 3j)$ and express the result in rectangular form.
b. Repeat (a) in MATLAB.

SOLUTION

a. $\dfrac{3 + 2j}{1 - 3j} = \dfrac{(3 + 2j)\overline{(1 - 3j)}}{(1 - 3j)\overline{(1 - 3j)}} = \dfrac{(3 + 2j)(1 + 3j)}{(1 - 3j)(1 + 3j)} = \dfrac{-3 + 11j}{10} = \dfrac{-3}{10} + \dfrac{11}{10}j.$

b.
Following the exact steps as in the solution above, and using the `conj` command, we find

```
>> (3 + 2*j)*conj(1 - 3*j)/((1 - 3*j)*conj(1 - 3*j))
ans =
    -0.3000 + 1.1000i    % Note that MATLAB returns i instead of j
```

Alternatively, we can let MATLAB perform the division directly.

```
>> (3 + 2*j)/(1 - 3*j)
ans =
    -0.3000 + 1.1000i    % Both results agree with (a)
```

2.1.2 Complex Numbers in Polar Form

The standard rectangular form $z = x + jy$ turns out to be very inefficient in many situations; for instance, a simple calculation such as $(3-2j)^5$. To remedy this, the polar form of a complex number is utilized. As its name suggests, the polar form uses polar coordinates to represent a complex number in the complex plane. The location of any point $z = x + jy$ in the complex plane can be determined by a radial coordinate r and an angular coordinate θ. The relationship between the rectangular and polar coordinates is given by (Figure 2.3)

$$x = r\cos\theta, \quad y = r\sin\theta.$$

We first introduce Euler's formula

$$e^{j\theta} = \cos\theta + j\sin\theta. \tag{2.5}$$

Consequently,

$$z = x + jy = \underbrace{r\cos\theta}_{x} + j\underbrace{r\sin\theta}_{y} = r\underbrace{(\cos\theta + j\sin\theta)}_{e^{j\theta}} = re^{j\theta}. \tag{2.6}$$

The result, $z = re^{j\theta}$, is known as the polar form of z. Here, r is the magnitude of z given by

$$r = |z| = \sqrt{(\text{Re}\{z\})^2 + (\text{Im}\{z\})^2} = \sqrt{x^2 + y^2}$$

and the phase (argument) of z is determined by

$$\theta = \arg z = \tan^{-1}\left(\frac{\text{Im}\{z\}}{\text{Re}\{z\}}\right) = \tan^{-1}\left(\frac{y}{x}\right). \tag{2.7}$$

The angle θ is measured from the positive real axis and, by convention, is regarded as positive in the counterclockwise direction. It is measured in radians (rad) and is determined in terms of integer multiples of 2π. The specific value of θ that lies in the interval $(-\pi, \pi]$ is called the principal value of arg z and is denoted by Arg z. In engineering applications, it is also common to express the polar form as

$$z = r\angle\theta,$$

where $\angle$ denotes the angle.

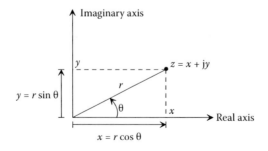

FIGURE 2.3 Connection between polar and rectangular coordinates.

Example 2.3: Polar Form

a. Express $z = 1 - 2j$ in polar form.
b. Repeat in MATLAB.

SOLUTION

a. The location of z will facilitate the phase calculation. By Equation 2.7,

$$\theta = \tan^{-1}\left(\frac{-2}{1}\right) = -1.1071 \text{ rad} \cong -63.43°.$$

This completely agrees with what we observe in Figure 2.4. Noting that $r = \sqrt{5}$, we have

$$z = 1 - 2j = \sqrt{5}e^{-1.1071j}.$$

b.
```
>> z = 1 - 2*j;
>> mag = abs(z)        % Magnitude
mag =
      2.2361
>> theta = atan(imag(z)/real(z))    % Phase
theta =
      -1.1071
>> mag*exp(j*theta)    % Polar form
ans =
      1.0000 - 2.0000i   % Agrees with the original z
```

2.1.2.1 Complex Algebra Using Polar Form

Working with the polar form substantially simplifies complex algebra. Consider two complex numbers, in their respective polar forms, $z_1 = r_1 e^{j\theta_1}$ and $z_2 = r_2 e^{j\theta_2}$. Then,

$$z_1 z_2 = r_1 r_2 e^{j(\theta_1 + \theta_2)} \quad \text{or, alternatively,} \quad r_1 r_2 \angle (\theta_1 + \theta_2).$$

This means that the magnitude and phase of the product $z_1 z_2$ are

$$|z_1 z_2| = r_1 r_2 = |z_1||z_2| \quad \text{and} \quad \arg(z_1 z_2) = \theta_1 + \theta_2 = \arg(z_1) + \arg(z_2)$$

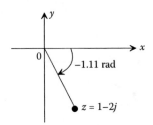

FIGURE 2.4 Example 2.3.

Similarly, in the case of division of complex numbers,

$$\frac{z_1}{z_2} = \frac{r_1}{r_2} e^{j(\theta_1 - \theta_2)} \quad \text{or} \quad \frac{r_1}{r_2} \angle(\theta_1 - \theta_2).$$

Therefore, the magnitude and phase of z_1/z_2 are

$$\left|\frac{z_1}{z_2}\right| = \frac{r_1}{r_2} = \frac{|z_1|}{|z_2|} \quad \text{and} \quad \arg\left(\frac{z_1}{z_2}\right) = \theta_1 - \theta_2 = \arg(z_1) - \arg(z_2).$$

Example 2.4: Division Using Polar Form

Express the result in polar form:

$$\frac{-2+j}{1-j}.$$

Solution

Noting that the numerator and the denominator are in the second and fourth quadrants, respectively, their polar forms are obtained as

$$-2+j = \sqrt{5}\, e^{2.6780j}; \quad 1-j = \sqrt{2}\, e^{-(\pi/4)j}.$$

As a result,

$$\frac{-2+j}{1-j} = \frac{\sqrt{5}\, e^{2.6780j}}{\sqrt{2}\, e^{-(\pi/4)j}} = \sqrt{\frac{5}{2}}\, e^{3.4634j}.$$

This may be verified as follows:

$$\frac{(-2+j)(1+j)}{(1-j)(1+j)} = \frac{-3-j}{2} = -\frac{3}{2} - \frac{1}{2}j \xrightarrow{\text{Polar form third quadrant}} \sqrt{\frac{5}{2}}\, e^{3.4634j} \xrightarrow{\text{Equivalently}} \sqrt{\frac{5}{2}}\, e^{-2.8198j}.$$

Given $z = re^{j\theta}$, its conjugate is obtained as

$$\bar{z} = x - jy = r\cos\theta - j(r\sin\theta) = r(\cos\theta - j\sin\theta) = r[\cos(-\theta) + j\sin(-\theta)] \xrightarrow{\text{Euler's formula}} re^{-j\theta}.$$

This makes sense geometrically, since a complex number and its conjugate are reflections of one another through the real axis. Hence, they are equidistant from the origin, that is, $|z| = |\bar{z}| = r$, and the phase of one is the negative of the phase of the other, that is, $\arg(z) = -\arg(\bar{z})$ (Figure 2.5). The magnitude property of complex conjugation, Equation 2.4, can now be confirmed as

$$z\bar{z} = (re^{j\theta})(re^{-j\theta}) = r^2 = |z|^2.$$

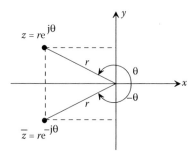

FIGURE 2.5 A complex number and its conjugate.

2.1.2.2 Integer Powers of Complex Numbers

As mentioned at the outset of this section, one area that demonstrates the efficacy of the polar form is in dealing with expressions in the form z^n for integer n. The idea is simple and shown below:

$$z^n \stackrel{\text{Polar form}}{=} (re^{j\theta})^n = r^n e^{jn\theta} \stackrel{\text{Euler's formula}}{=} r^n(\cos n\theta + j\sin n\theta)$$
$$\stackrel{\text{Rectangular form}}{=} r^n \cos n\theta + jr^n \sin n\theta. \qquad (2.8)$$

Example 2.5: Integer Power

Simplify $(1 - 2j)^4$.

Solution

Using the result of Example 2.3, and following Equation 2.8, we have

$$(1 - 2j)^4 = (\sqrt{5}e^{-1.1071j})^4 = 25e^{-4.4284j} \stackrel{\text{Equivalently}}{=} 25e^{1.8548j} = -7 + 24j.$$

2.1.2.3 Roots of Complex Numbers

In real calculus, if a is a real number then $\sqrt[n]{a}$ has a single value. On the contrary, given a complex number $z \neq 0$ and a positive integer n, then the nth root of z, written as $\sqrt[n]{z}$, is *multivalued*. In fact, there are n different values of $\sqrt[n]{z}$ corresponding to each value of $z \neq 0$. Given $z = re^{j\theta}$, it can readily be shown that

$$\sqrt[n]{z} = \sqrt[n]{r}\left(\cos\frac{\theta + 2k\pi}{n} + j\sin\frac{\theta + 2k\pi}{n}\right), \quad k = 0, 1, \ldots, n-1. \qquad (2.9)$$

Geometrically, these n values lie on a circle centered at the origin and a radius of $\sqrt[n]{r}$ and are the n vertices of an n-sided regular polygon.

Example 2.6: Third Roots of 1

Calculate all three values of $\sqrt[3]{1}$.

Solution

$z = 1$ is located on the positive real axis; hence it has a phase of 0 and is expressed as $1 = 1e^{j0}$. Using Equation 2.9 with $n = 3$, $r = 1$, and $\theta = 0$, we find

$$\sqrt[3]{1} = \sqrt[3]{1}\left(\cos\frac{2k\pi}{3} + j\sin\frac{2k\pi}{3}\right), \quad k = 0, 1, 2 \Rightarrow \begin{array}{l} \cos 0 + j\sin 0 = 1 \\ \cos(2\pi/3) + j\sin(2\pi/3) = -\dfrac{1}{2} + j\dfrac{\sqrt{3}}{2} \\ \cos(4\pi/3) + j\sin(4\pi/3) = -\dfrac{1}{2} - j\dfrac{\sqrt{3}}{2} \end{array}$$

Note that these values lie on the unit circle (each has a magnitude of 1) and are the vertices of a three-sided regular polygon (equilateral triangle).

2.1.3 Complex Variables and Functions

If x or y or both vary, then $z = x + jy$ is called a complex variable. The Laplace variable (Section 2.3) is a recognized example of a complex variable. A complex function F defined on a set S is a rule that assigns a complex number w to each $z \in S$. The notation is $w = F(z)$ and the set S is the domain of the definition of F. For instance, the domain of the function $F(z) = z/(z-1)$ is any region that does not contain the point $z = 1$. Because z assumes different values from the set S, it is clearly a complex variable. Transfer functions (Section 4.4) and frequency response functions (Section 8.3) are examples of complex functions that arise in system dynamics.

PROBLEM SET 2.1

In each case,

    a. Perform z_1/z_2 and express the result in rectangular form.

    b. Verify that $|z_1/z_2| = |z_1|/|z_2|$.

    c. Repeat in MATLAB.

1. $\dfrac{2+j}{1-2j}$

2. $\dfrac{-3-j}{2j}$

3. $\dfrac{-3j}{2+3j}$

4. $\dfrac{4}{-4+3j}$

Express each complex number in its polar form.

5. $-1 + \dfrac{1}{2}j$

6. $1 - \dfrac{3}{2}j$

7. $3 + j\sqrt{3}$

8. $-\sqrt{3} - 3j$

Perform using the polar form and express the result in rectangular form.

9. $\dfrac{1 + (2/3)j}{-(1/3) + j}$

10. $\dfrac{3 - j\sqrt{3}}{\sqrt{3} + 3j}$

11. $\dfrac{3 - 5j}{2j}$

12. $\dfrac{-3j}{-1 + j}$

13. $\left(\dfrac{1}{3} - j\right)^4$

14. $(-2 - j)^5$

15. $\dfrac{(1 + 3j)^3}{(-1 + 2j)^2}$

16. $\dfrac{-100j}{(1 + 4j)^3}$

In each case, find all possible values.

17. $(1+j)^{1/4}$
18. $(-1+j)^{1/3}$
19. $(\sqrt{3}-3j)^{1/2}$
20. $(2j)^{1/2}$

2.2 DIFFERENTIAL EQUATIONS

Differential equations are divided into two general categories: *ordinary differential equations* (ODEs) and *partial differential equations* (PDEs). An equation involving an unknown function and one or more of its derivatives is called a differential equation. When there is only one independent variable, the equation is called an ODE. For example, $\dot{x} + 4x = e^{2t}$ is an ODE involving the unknown function $x(t)$, its first derivative $\dot{x} = dx/dt$, as well as a given function e^{2t}. Similarly, $t\ddot{x} - x\dot{x} = \sin t$ is an ODE relating $x(t)$ and its first and second derivatives with respect to t, as well as the function $\sin t$. Note that while dealing with dynamic system models, the independent variable is time t. If the unknown function is a function of more than one independent variable, the equation is referred to as a PDE. The derivative of the highest order of the unknown function $x(t)$ with respect to t is the order of the ODE. For instance, $\dot{x} + 4x = e^{2t}$ is of order 1 and $t\ddot{x} - x\dot{x} = \sin t$ is of order 2. Consider an nth-order ODE in the form

$$a_n x^{(n)} + a_{n-1} x^{(n-1)} + \cdots + a_1 \dot{x} + a_0 x = F(t), \quad (2.10)$$

where $x = x(t)$ and $x^{(n)} = d^n x/dt^n$. If all coefficients $a_0, a_1, \ldots, a_n$ are either constants or functions of the independent variable t, then the ODE is linear. Otherwise, the ODE is nonlinear. If $F(t) \equiv 0$, the ODE is called homogeneous. Otherwise, it is nonhomogeneous. Therefore $\dot{x} + 4x = e^{2t}$ is linear, while $t\ddot{x} - x\dot{x} = \sin t$ is nonlinear, and both are nonhomogeneous.

2.2.1 LINEAR, FIRST-ORDER DIFFERENTIAL EQUATIONS

In accordance with Equation 2.10, linear, first-order ODEs are represented as

$$a_1 \dot{x} + a_0 x = F(t) \quad \xRightarrow[\text{and rewrite as}]{\text{Divide by } a_1} \quad \dot{x} + g(t)x = f(t) \quad (2.11)$$

with a general solution in the form

$$x(t) = e^{-h}\left[\int e^h f(t)\,dt + c\right], \quad \text{where } h = \int g(t)\,dt, \quad (2.12)$$

and c is a constant. In order to derive a particular solution, an initial condition must be specified. Assuming that the initial time is $t = 0$, the initial condition for x refers to the value of x immediately before the initial time and is denoted by $x(0^-)$. On the other hand, the initial value of x is its value immediately after the initial time and expressed as $x(0^+)$. Although the initial condition and the initial value of a quantity are almost always the same, there are rare instances occurring in dynamic systems where they are different (see Section 2.3). In the meantime, we simply assume the two are the same and denoted by $x(0)$. That said, a first-order initial-value problem (IVP) is described as

$$\dot{x} + g(t)x = f(t), \quad x(0) = x_0.$$

Example 2.7: IVP

a. Solve $2\dot{x} + 3x = e^{-t}, \quad x(0) = 0$.
b. ◢ Perform (a) in MATLAB.

Solution

a. Rewrite the ODE as $\dot{x} + (3/2)x = (1/2)e^{-t}$ to resemble the standard form of Equation 2.11. Noting that $g = 3/2$, we find $h = (3/2)t$ and by Equation 2.12,

$$x(t) = e^{-3t/2}\left[\int e^{3t/2}\frac{1}{2}e^{-t}\,dt + c\right] = e^{-3t/2}[e^{t/2} + c] = e^{-t} + ce^{-3t/2}.$$

Applying the initial condition yields $c = -1$ and hence $x(t) = e^{-t} - e^{-3t/2}$.

b.
```
>> x = dsolve('2*Dx + 3*x = exp(-t)','x(0) = 0')
x =
exp(-t)-exp(-3/2*t)
```

2.2.2 Second-Order Differential Equations with Constant Coefficients

A specific class of linear, second-order ODEs that arises quite often in dynamic system models is in the form

$$\underbrace{a_2\ddot{x} + a_1\dot{x} + a_0 x}_{\text{Constant coefficients}} = F(t) \stackrel{\text{Rewrite}}{\Rightarrow} \ddot{x} + b\dot{x} + cx = f(t). \tag{2.13}$$

This is normally accompanied by a set of two initial conditions $x(0)$ and $\dot{x}(0)$. A general solution $x(t)$ of Equation 2.13 is a superposition of the homogeneous solution $x_h(t)$—also known as the complementary solution $x_c(t)$—and the particular solution $x_p(t)$, that is, $x = x_h + x_p$.

2.2.2.1 Homogeneous Solution

To find $x_h(t)$ we solve the homogeneous equation

$$\ddot{x} + b\dot{x} + cx = 0$$

as follows. Assume $x = e^{\lambda t}$, with λ to be determined, and substitute to obtain

$$[\lambda^2 + b\lambda + c]e^{\lambda t} = 0 \stackrel{e^{\lambda t} \neq 0}{\Rightarrow} \underbrace{\lambda^2 + b\lambda + c = 0}_{\text{Characteristic equation}} \stackrel{\text{Solve}}{\Rightarrow} \begin{matrix}\lambda_1\\\lambda_2\end{matrix}.$$

The characteristic values λ_1 and λ_2 ultimately establish the two linearly independent solutions that form $x_h(t)$.

Case (1) $\lambda_1 \neq \lambda_2$ real
The two independent solutions are $e^{\lambda_1 t}$ and $e^{\lambda_2 t}$, and a linear combination of the two yields

$$x_h(t) = c_1 e^{\lambda_1 t} + c_2 e^{\lambda_2 t}, \quad c_1, c_2 = \text{const.}$$

Case (2) $\lambda_1 = \lambda_2 = \lambda$
The independent solutions are $e^{\lambda t}$ and $te^{\lambda t}$ so that

$$x_h(t) = (c_1 + c_2 t)e^{\lambda t}, \quad c_1, c_2 = \text{const.}$$

Case (3) $\lambda_2 = \bar{\lambda}_1$ complex conjugates
Letting $\lambda_1 = \alpha + j\beta$, the two independent solutions are $e^{\alpha t}\cos\beta t$ and $e^{\alpha t}\sin\beta t$, leading to

$$x_h(t) = e^{\alpha t}(c_1 \cos\beta t + c_2 \sin\beta t), \quad c_1, c_2 = \text{const.}$$

TABLE 2.1
Selection of Particular Solution; Method of Undetermined Coefficients

Term in $f(t)$	Recommended $x_p(t)$
$A_n t^n + \cdots + A_1 t + A_0$	$K_n t^n + \cdots + K_1 t + K_0$
Ae^{at}	Ke^{at}
$A \sin \omega t$ or $A \cos \omega t$	$K_1 \cos \omega t + K_2 \sin \omega t$
$Ae^{\sigma t} \sin \omega t$ or $Ae^{\sigma t} \cos \omega t$	$e^{\sigma t}(K_1 \cos \omega t + K_2 \sin \omega t)$

2.2.2.2 Particular Solution

The particular solution $x_p(t)$ of Equation 2.13 is obtained using the method of undetermined coefficients. The method is limited in its applications only to cases where $f(t)$ is a polynomial, an exponential, sinusoidal, or some combinations of them. Table 2.1 contains the recommended x_p for different scenarios involving $f(t)$. These are subject to modification in some special cases as follows. If x_p contains a term that coincides with a solution of the homogeneous equation, and that solution corresponds to a nonrepeated characteristic value, then the recommended x_p must be multiplied by t. If the characteristic value is repeated—as in Case (2) above—then multiply x_p by t^2.

Example 2.8: Second-Order ODE

a. Solve $\ddot{x} + 3\dot{x} + 2x = e^{-t}$, $x(0) = 1$, $\dot{x}(0) = 0$.
b. Repeat in MATLAB.

Solution

a. The characteristic equation $\lambda^2 + 3\lambda + 2 = 0$ yields $\lambda = -1, -2$ so that $x_h = c_1 e^{-t} + c_2 e^{-2t}$. Based on Table 2.1, we have $x_p = Ke^{-t}$. However, e^{-t} coincides with the independent solution in x_h associated with $\lambda = -1$. Therefore, x_p is modified to $x_p = Kte^{-t}$. Inserting into the original ODE and equating the two sides, we find $K = 1$ so that $x_p = te^{-t}$. This means that a general solution is $x = c_1 e^{-t} + c_2 e^{-2t} + te^{-t}$. Finally, application of the initial conditions results in $c_1 = 1$, $c_2 = 0$; hence $x = (1+t)e^{-t}$.

b.
```
>> x = dsolve('D2x + 3*Dx + 2*x = exp(-t)','x(0) = 1,Dx(0) = 0')
x =
   (t+1)*exp(-t)
```

Example 2.9: Second-Order ODE

a. Solve $\ddot{x} + \omega^2 x = 0$, $x(0) = 0$, $\dot{x}(0) = 1$.
b. Perform (a) in MATLAB.

Solution

a. The characteristic equation $\lambda^2 + \omega^2 = 0$ has roots $\lambda = \pm j\omega$ so that $x(t) = c_1 \cos \omega t + c_2 \sin \omega t$. Applying the initial conditions leads to $c_1 = 0, c_2 = 1/\omega$; hence $x(t) = \sin \omega t / \omega$.

b.
```
>> syms w
>> x = dsolve('D2x + w^2*x = 0','x(0) = 0, Dx(0) = 1')
x =
sin(w*t)/w
```

2.2.2.2.1 Expressing $A\cos\omega t + B\sin\omega t$ as $D\sin(\omega t + \phi)$

We often come across expressions in the form $A\cos\omega t + B\sin\omega t$, where the sine and cosine waves have the same frequency ω. In these situations, we generally prefer to replace the expression with a single trigonometric term such as $D\sin(\omega t + \phi)$, where D is the amplitude and ϕ is the phase shift. The details are as follows. Using a trigonometric expansion,

$$D\sin(\omega t + \phi) = D\sin\omega t \cos\phi + D\cos\omega t \sin\phi.$$

Comparing with $A\cos\omega t + B\sin\omega t$, we find

$$\begin{matrix} D\sin\phi = A \\ D\cos\phi = B \end{matrix} \quad \text{Divide the two equations} \Rightarrow \quad \tan\phi = \frac{A}{B}.$$

Next, construct a right triangle in which angle ϕ satisfies $\tan\phi = A/B$ (Figure 2.6). Then,

$$\sin\phi = \frac{A}{\sqrt{A^2 + B^2}}, \quad \cos\phi = \frac{B}{\sqrt{A^2 + B^2}}.$$

Noting that $\sin\phi = A/D$ and $\cos\phi = B/D$, we find $D = \sqrt{A^2 + B^2}$. In summary,

$$A\cos\omega t + B\sin\omega t = D\sin(\omega t + \phi), \quad D = \sqrt{A^2 + B^2}, \quad \phi = \tan^{-1}\frac{A}{B}. \tag{2.14}$$

Example 2.10: Amplitude, Phase Shift

Express $2\sin t - \cos t$ as $D\sin(\omega t + \phi)$.

SOLUTION

Comparing with the general form, $A = -1$ and $B = 2$. But $D\sin\phi = A < 0$, $D\cos\phi = B > 0$, and $D = \sqrt{5} > 0$. Thus ϕ is in the fourth quadrant and $\phi = \tan^{-1}(-1/2) = -26.5651° = -0.4636\,\text{rad}$ agrees with the location of ϕ. In conclusion,

$$2\sin t - \cos t = \sqrt{5}\sin(t - 0.4636).$$

PROBLEM SET 2.2

In each case,

    a. Solve the IVP.
    b. Confirm the result in MATLAB.

1. $\dot{x} + x = \sin t, \quad x(0) = -\dfrac{1}{2}$

2. $\dfrac{1}{2}\dot{x} + x = 0, \quad x(0) = \dfrac{1}{2}$

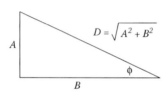

FIGURE 2.6 Phase shift angle.

3. $\dot{x} + tx = t, \quad x(0) = 0$
4. $\dot{y} + y = \cos t, \quad y(0) = 1$
5. $\ddot{x} + 2\dot{x} + 2x = 0, \quad x(0) = 0, \dot{x}(0) = 1$
6. $\ddot{x} + 2\dot{x} + x = e^{-2t}, \quad x(0) = 1, \dot{x}(0) = -1$
7. $\ddot{x} + 3\dot{x} = 10\sin t, \quad x(0) = 0, \dot{x}(0) = -1$
8. $\ddot{x} + x = 3\sin 2t, \quad x(0) = 1, \dot{x}(0) = 0$
9. $\ddot{u} + 4\dot{u} + 3u = 2e^{-t}, \quad u(0) = 1, \dot{u}(0) = 0$
10. $2\ddot{y} + 3\dot{y} + y = 0, \quad y(0) = 0, \dot{y}(0) = \frac{1}{2}$

Write each expression as $D\sin(\omega t + \phi)$, where D is amplitude and ϕ is phase angle.

11. $\cos t + 2\sin t$
12. $\cos 2t - 3\sin 2t$
13. $-\cos 2t - \frac{1}{2}\sin 2t$
14. $\sin \omega t - \frac{1}{3}\cos \omega t$

2.3 LAPLACE TRANSFORMATION

In Section 2.2, we learned how to solve ODEs with constant coefficients entirely in the time domain. This was achieved by using the method of undetermined coefficients, which has serious limitations. To remedy this, we introduce a systematic approach to solve such ODEs in a more expedient manner. The primary advantage gained here is that the arbitrary constants in the general solution need not be found separately. The idea is simple: in order to solve an ODE subject to initial conditions (i.e., IVP), transform the data to the s domain, in which the transformed problem is an algebraic one and thus much easier to handle. This algebraic problem is then treated properly, and the data are ultimately transformed back to the time domain to find the solution of the original problem. If a function $x(t)$ is defined for all $t \geq 0$, its Laplace transform is defined by

$$X(s) \stackrel{\text{Notation}}{=} \mathcal{L}\{x(t)\} \stackrel{\text{Definition}}{=} \int_0^\infty e^{-st} x(t)\, dt \qquad (2.15)$$

provided that the integral exists. The complex variable s is the Laplace variable and $\mathcal{L}$ is the Laplace transform operator. It is common practice to denote a time-dependent function by a lowercase letter such as $x(t)$ and its Laplace transform by the same letter in capital, $X(s)$. Transforming the data back to the time domain is done through inverse Laplace transformation $\mathcal{L}^{-1}$ as (Figure 2.7)

$$x(t) = \mathcal{L}^{-1}\{X(s)\}.$$

Laplace transforms of several functions are listed in Table 2.2 and will be referred to frequently.

Example 2.11: Laplace Transform

a. If $f(t) = e^{-at}$ $(a = \text{const} > 0)$ for $t \geq 0$, determine $\mathcal{L}\{f(t)\}$.
b. Repeat in MATLAB.

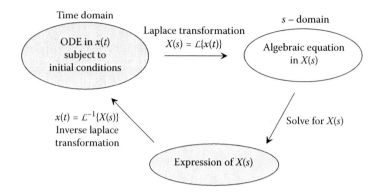

FIGURE 2.7 Operations involved in the Laplace transformation method.

SOLUTION

a. $F(s) = \mathcal{L}\{e^{-at}\} = \int_0^\infty e^{-st} e^{-at} dt = \int_0^\infty e^{-(s+a)t} dt = \left.\dfrac{e^{-(s+a)t}}{-(s+a)}\right|_{t=0}^\infty = \dfrac{1}{s+a}$ for $s + a > 0$.

b.
```
>> syms a t
>> F = laplace(exp(-a*t))
F =
1/(s+a)
```

2.3.1 LINEARITY OF LAPLACE AND INVERSE LAPLACE TRANSFORMS

The Laplace transform operator $\mathcal{L}$ is linear, that is, if the Laplace transforms of functions $x_1(t)$ and $x_2(t)$ exist, and a_1 and a_2 are constant scalars, then

$$\mathcal{L}\{a_1 x_1(t) + a_2 x_2(t)\} = \int_0^\infty e^{-st}[a_1 x_1(t) + a_2 x_2(t)] dt = a_1 \int_0^\infty e^{-st} x_1(t) dt + a_2 \int_0^\infty e^{-st} x_2(t) dt$$

$$= a_1 \mathcal{L}\{x_1(t)\} + a_2 \mathcal{L}\{x_2(t)\} = a_1 X_1(s) + a_2 X_2(s).$$

To establish the linearity of $\mathcal{L}^{-1}$, take the inverse Laplace transforms of the expressions on the far left and far right of the equation above to obtain

$$a_1 x_1(t) + a_2 x_2(t) = \mathcal{L}^{-1}\{a_1 X_1(s) + a_2 X_2(s)\}.$$

Noting that $x_1(t) = \mathcal{L}^{-1}\{X_1(s)\}$ and $x_2(t) = \mathcal{L}^{-1}\{X_2(s)\}$, the result follows.

2.3.2 DIFFERENTIATION AND INTEGRATION OF LAPLACE TRANSFORMS

We now consider two specific types of situations: (i) $\mathcal{L}\{tg(t)\}$ and (ii) $\mathcal{L}\{g(t)/t\}$. In both cases, we assume that $G(s) = \mathcal{L}\{g(t)\}$ either is known directly from Table 2.2 or can be determined by other means. Either way, once $G(s)$ is available, the two transforms labeled (i) and (ii) will be obtained in terms of the derivative and integral of $G(s)$, respectively. Let us first make the following definition. If a transform function is in the form $G(s) = N(s)/D(s)$, then any value of s for which $D(s) = 0$ is called a pole of $G(s)$. A pole with a *multiplicity* (number of occurrences) of one is known as a simple pole.

TABLE 2.2
Laplace Transform Pairs

No.	$f(t)$	$F(s)$
1	Unit impulse $\delta(t)$	1
2	1, Unit step $u_s(t)$	$\dfrac{1}{s}$
3	t, Unit ramp $u_r(t)$	$\dfrac{1}{s^2}$
4	$\delta(t-a)$	e^{-as}
5	$u(t-a)$	$\dfrac{e^{-as}}{s}$
6	$t^{n-1}, \quad n=1,2,\ldots$	$\dfrac{(n-1)!}{s^n}$
7	$t^{a-1}, \quad a>0$	$\dfrac{\Gamma(a)}{s^a}$
8	e^{-at}	$\dfrac{1}{s+a}$
9	te^{-at}	$\dfrac{1}{(s+a)^2}$
10	$t^n e^{-at}, \quad n=1,2,\ldots$	$\dfrac{n!}{(s+a)^{n+1}}$
11	$\dfrac{1}{b-a}(e^{-at}-e^{-bt}), \, a\neq b$	$\dfrac{1}{(s+a)(s+b)}$
12	$\dfrac{1}{a-b}(ae^{-at}-be^{-bt}), \, a\neq b$	$\dfrac{s}{(s+a)(s+b)}$
13	$\dfrac{1}{ab}\left[1+\dfrac{1}{a-b}(be^{-at}-ae^{-bt})\right]$	$\dfrac{1}{s(s+a)(s+b)}$
14	$\dfrac{1}{a^2}(-1+at+e^{-at})$	$\dfrac{1}{s^2(s+a)}$
15	$\dfrac{1}{a^2}(1-e^{-at}-ate^{-at})$	$\dfrac{1}{s(s+a)^2}$
16	$\sin\omega t$	$\dfrac{\omega}{s^2+\omega^2}$
17	$\cos\omega t$	$\dfrac{s}{s^2+\omega^2}$
18	$e^{-\sigma t}\sin\omega t$	$\dfrac{\omega}{(s+\sigma)^2+\omega^2}$
19	$e^{-\sigma t}\cos\omega t$	$\dfrac{s+\sigma}{(s+\sigma)^2+\omega^2}$
20	$1-\cos\omega t$	$\dfrac{\omega^2}{s(s^2+\omega^2)}$
21	$\omega t-\sin\omega t$	$\dfrac{\omega^3}{s^2(s^2+\omega^2)}$
22	$t\cos\omega t$	$\dfrac{s^2-\omega^2}{(s^2+\omega^2)^2}$
23	$\dfrac{1}{2\omega}t\sin\omega t$	$\dfrac{s}{(s^2+\omega^2)^2}$
24	$\dfrac{1}{2\omega^3}(\sin\omega t-\omega t\cos\omega t)$	$\dfrac{1}{(s^2+\omega^2)^2}$
25	$\dfrac{1}{2\omega}(\sin\omega t+\omega t\cos\omega t)$	$\dfrac{s^2}{(s^2+\omega^2)^2}$

continued

TABLE 2.2 (continued)
Laplace Transform Pairs

No.	$f(t)$	$F(s)$
26	$\dfrac{1}{\omega_2^2 - \omega_1^2}\left[\dfrac{1}{\omega_2}\sin\omega_2 t - \dfrac{1}{\omega_1}\sin\omega_1 t\right],\ \omega_1^2 \neq \omega_2^2$	$\dfrac{1}{(s^2+\omega_1^2)(s^2+\omega_2^2)}$
27	$\dfrac{1}{\omega_2^2 - \omega_1^2}(\cos\omega_1 t - \cos\omega_2 t),\ \omega_1^2 \neq \omega_2^2$	$\dfrac{s}{(s^2+\omega_1^2)(s^2+\omega_2^2)}$
28	$\sinh at$	$\dfrac{a}{s^2-a^2}$
29	$\cosh at$	$\dfrac{s}{s^2-a^2}$
30	$\dfrac{1}{a^2-b^2}\left[\dfrac{1}{a}\sinh at - \dfrac{1}{b}\sinh bt\right],\ a \neq b$	$\dfrac{1}{(s^2-a^2)(s^2-b^2)}$
31	$\dfrac{1}{a^2-b^2}[\cosh at - \cosh bt],\ a \neq b$	$\dfrac{s}{(s^2-a^2)(s^2-b^2)}$
32	$\dfrac{1}{3a^2}\left[e^{-at} + 2e^{(1/2)at}\sin\left(\dfrac{\sqrt{3}}{2}at - \dfrac{1}{6}\pi\right)\right]$	$\dfrac{1}{s^3+a^3}$
33	$\dfrac{1}{3a}\left[-e^{-at} + 2e^{(1/2)at}\sin\left(\dfrac{\sqrt{3}}{2}at + \dfrac{1}{6}\pi\right)\right]$	$\dfrac{s}{s^3+a^3}$
34	$\dfrac{1}{3a^2}\left[e^{at} - 2e^{-(1/2)at}\sin\left(\dfrac{\sqrt{3}}{2}at + \dfrac{1}{6}\pi\right)\right]$	$\dfrac{1}{s^3-a^3}$
35	$\dfrac{1}{3a}\left[e^{-at} + 2e^{-(1/2)at}\sin\left(\dfrac{\sqrt{3}}{2}at - \dfrac{1}{6}\pi\right)\right]$	$\dfrac{s}{s^3-a^3}$
36	$\dfrac{1}{4a^3}[\cosh at \sin at - \sinh at \cos at]$	$\dfrac{1}{s^4+4a^4}$
37	$\dfrac{1}{2a^2}\sinh at \sin at$	$\dfrac{s}{s^4+4a^4}$
38	$\dfrac{1}{2a^3}(\sinh at - \sin at)$	$\dfrac{1}{s^4-a^4}$
39	$\dfrac{1}{2a^2}(\cosh at - \cos at)$	$\dfrac{s}{s^4-a^4}$

Differentiation of Laplace Transforms: If $G(s) = \mathcal{L}\{g(t)\}$ exists, then at any point except at the poles of $G(s)$, we have

$$\mathcal{L}\{tg(t)\} = -\frac{dG(s)}{ds}. \qquad (2.16)$$

In the general case,

$$\mathcal{L}\{t^n g(t)\} = (-1)^n \frac{d^n G(s)}{ds^n}.$$

Example 2.12: Derivative of Laplace Transforms

a. Find $\mathcal{L}\{te^{-t}\}$.
b. Repeat in MATLAB.

Solution

a. Comparison with Equation 2.16 reveals $g(t) = e^{-t}$ so that $G(s) = 1/(s+1)$. Then,

$$\mathcal{L}\{te^{-t}\} = -\frac{d}{ds}\left(\frac{1}{s+1}\right) = \left(\frac{1}{s+1}\right)^2.$$

b.
```
>> syms t
>> laplace(t*exp(-t))
ans =
1/(1+s)^2
```

Integration of Laplace Transforms: If $\mathcal{L}\{g(t)/t\}$ exists and the order of integration can be interchanged, then

$$\mathcal{L}\left\{\frac{g(t)}{t}\right\} = \int_s^\infty G(\sigma)\,d\sigma. \tag{2.17}$$

Example 2.13: Integral of Laplace Transforms

a. Find $\mathcal{L}\{\sin\omega t/t\}$.
b. Repeat in MATLAB.

Solution

a. Comparing with Equation 2.17, we find $g(t) = \sin\omega t$ so that $G(s) = \omega/(s^2 + \omega^2)$. Consequently,

$$\mathcal{L}\left\{\frac{\sin\omega t}{t}\right\} = \int_s^\infty \frac{\omega}{\sigma^2 + \omega^2}\,d\sigma = \int_s^\infty \frac{1}{1+(\sigma/\omega)^2}\frac{d\sigma}{\omega} = \left[\tan^{-1}\frac{\sigma}{\omega}\right]_{\sigma=s}^\infty$$

$$= \frac{\pi}{2} - \tan^{-1}\frac{s}{\omega} = \tan^{-1}\frac{\omega}{s}.$$

b.
```
>> syms w t
>> laplace(sin(w*t)/t)
ans =
atan(w/s)
```

2.3.3 Special Functions

Much can be learned about a dynamic system's characteristics by studying how it responds to specific external disturbances. Mathematical models of external disturbances are represented by special functions. In this section we introduce the step, ramp, pulse, and impulse functions, as well as their Laplace transforms.

2.3.3.1 Unit-Step Function

The unit-step function (Figure 2.8) is analytically defined as

$$u(t) = \begin{cases} 1 & \text{if } t > 0, \\ 0 & \text{if } t < 0, \\ \text{undefined (finite)} & \text{if } t = 0. \end{cases}$$

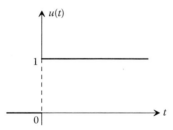

FIGURE 2.8 Unit-step occurring at $t = 0$.

Physically, this may be realized as a constant signal (of magnitude 1) suddenly applied to the system at time $t = 0$. Using the definition of the Laplace transform, Equation 2.15, we find

$$\mathcal{L}\{u(t)\} \stackrel{\text{Notation}}{=} U(s) = \int_0^\infty e^{-st} u(t)\, dt = \int_0^\infty e^{-st}\, dt = \frac{1}{s}.$$

In general, when the magnitude is A, the signal is a step function, denoted by $Au(t)$ and

$$\mathcal{L}\{Au(t)\} = \int_0^\infty e^{-st} A\, dt = \frac{A}{s}.$$

When the unit-step occurs at some time $a \neq 0$ (Figure 2.9), it is denoted by $u(t - a)$, and

$$u(t - a) = \begin{cases} 1 & \text{if } t > a, \\ 0 & \text{if } t < a, \\ \text{undefined (finite)} & \text{if } t = a. \end{cases}$$

In order to find the Laplace transform of $u(t - a)$, we first need to discuss the shifted translation of a given function (Figure 2.10). Note that $f(t - a)u(t - a) = 0$ for $t < a$.

Shift on t-axis: If $F(s) = \mathcal{L}\{f(t)\}$ exists and $a \geq 0$, then

$$\mathcal{L}\{f(t - a)u(t - a)\} = e^{-as} F(s). \tag{2.18}$$

Comparing $\mathcal{L}\{u(t - a)\}$ to the left side of Equation 2.18, we find $f(t - a) = 1$, which in turn implies $f(t) = 1$ and hence $F(s) = 1/s$. Therefore,

$$\mathcal{L}\{u(t - a)\} = \frac{e^{-as}}{s}.$$

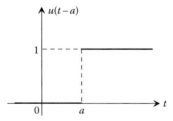

FIGURE 2.9 Unit-step occurring at $t = a$.

Complex Analysis, Differential Equations, and Laplace Transformation

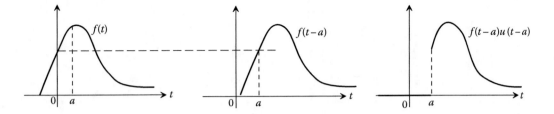

FIGURE 2.10 Effect of step function on shifted translation.

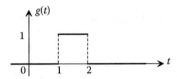

FIGURE 2.11 Example 2.14.

Example 2.14: Shift on *t*-Axis

Consider the function $g(t)$ defined in Figure 2.11.

a. Express $g(t)$ in terms of unit-step functions.
b. Find $G(s)$ using the shift on the *t*-axis, Equation 2.18.
c. Confirm the result in MATLAB.

Solution

a. Following the definition of unit-step function, it is readily seen that

$$g(t) = u(t-1) - u(t-2).$$

b. Applying Equation 2.18 to each term yields

$$G(s) = \frac{e^{-s}}{s} - \frac{e^{-2s}}{s} = \frac{e^{-s} - e^{-2s}}{s}.$$

c.
```
>> syms t
>> g = heaviside(t-1) - heaviside(t-2);    % Heaviside is the unit-step
>> G = laplace(g)
G =
(exp(-s)-exp(-2*s))/s
```

2.3.3.2 Unit-Ramp Function

The unit-ramp function (Figure 2.12) is analytically defined as

$$u_r(t) = \begin{cases} t & \text{if } t \geq 0, \\ 0 & \text{if } t < 0. \end{cases}$$

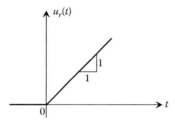

FIGURE 2.12 Unit-ramp.

Physically, this models a signal that changes linearly with a unit rate. By Equation 2.15,

$$\mathcal{L}\{u_r(t)\} = U_r(s) = \int_0^\infty t e^{-st}\,dt = \left\{t\frac{e^{-st}}{-s}\right\}_{t=0}^\infty - \int_0^\infty \frac{e^{-st}}{-s}\,dt = \left[\frac{e^{-st}}{-s^2}\right]_{t=0}^\infty = \frac{1}{s^2}.$$

In general, when the rate is A, the signal is called a ramp function, denoted by $Au_r(t)$ and

$$\mathcal{L}\{Au_r(t)\} = \frac{A}{s^2}.$$

2.3.3.3 Unit-Pulse Function

The unit-pulse function (Figure 2.13a) is defined as

$$u_p(t) = \begin{cases} 1/t_1 & \text{if } 0 < t < t_1, \\ 0 & \text{if } t < 0,\ t > t_1. \end{cases}$$

The word "unit" refers to the area of unity that the signal occupies. Then,

$$\mathcal{L}\{u_p(t)\} = U_p(s) = \int_0^{t_1} \frac{1}{t_1} e^{-st}\,dt = \frac{1 - e^{-st_1}}{st_1}.$$

2.3.3.4 Unit-Impulse (Dirac Delta) Function

In the unit-pulse function of Figure 2.13a let $t_1 \to 0$ (Figure 2.13b). In the limit, the rectangular-shaped signal occupies a region with an infinitesimally small width and a very large height (Figure 2.13c). The area, however, remains unity throughout the process. This limiting signal is known as the unit-impulse (or Dirac delta) function, denoted by $\delta(t)$. In general, if the area is A, it is called an impulse, denoted by $A\delta(t)$. An impulse with zero duration and infinite magnitude is

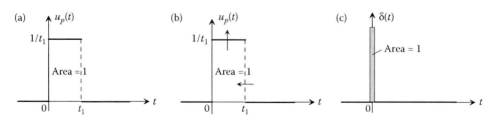

FIGURE 2.13 (a) Unit-pulse, (b) $t_1 \to 0$, and (c) unit-impulse.

mathematical fabrication and does not occur physically. However, if an external disturbance (such as an applied force, voltage, torque, etc.) is a pulse with very large magnitude and applied for a very short period of time, then it can be approximated as an impulse. Since $\delta(t)$ is the limit of $u_p(t)$ as $t_1 \to 0$, we have

$$\mathcal{L}\{\delta(t)\} \stackrel{\text{Notation}}{=} \Delta(s) = \lim_{t_1 \to 0}\left\{\frac{1-e^{-st_1}}{st_1}\right\} \stackrel{\text{L'Hôpital's rule}}{=} \lim_{t_1 \to 0}\left\{\frac{se^{-st_1}}{s}\right\} = 1.$$

If the unit-impulse occurs at $t = a$, it is represented by $\delta(t-a)$ and has the properties

$$\delta(t-a) = \begin{cases} 0 & \text{if } t \neq a \\ \infty & \text{if } t = a \end{cases}$$

and

$$\int_{-\infty}^{\infty} \delta(t-a)\,dt = 1.$$

It can also be shown that

$$\mathcal{L}\{\delta(t-a)\} = e^{-as}.$$

2.3.3.5 The Relation between Unit-Impulse and Unit-Step Functions

Consider $u(t-a)$ of Figure 2.9. Note that $t = a$ is a point of discontinuity of $u(t-a)$. The unit-impulse signal $\delta(t-a)$ can then be regarded as the derivative of $u(t-a)$ at this point of discontinuity, that is,

$$\frac{d}{dt}u(t-a) = \delta(t-a).$$

Thus, the idea of the impulse function allows us to differentiate time-varying functions with discontinuities.

2.3.3.6 Periodic Functions

Physical systems are often subjected to external disturbances that exhibit repeated behavior over long periods of time. In order to determine a system's response to this type of input, its Laplace transform must be identified properly. A function $f(t)$ is periodic with period $P > 0$ if it is defined for all $t > 0$, and $f(t+P) = f(t)$ for all $t > 0$. We assume that $f(t)$ is also piecewise continuous, as basically all signals of physical interest are. By Equation 2.15,

$$F(s) = \int_0^{\infty} e^{-st}f(t)\,dt = \int_0^{P} e^{-st}f(t)\,dt + \int_P^{2P} e^{-st}f(t)\,dt + \cdots = \sum_{k=0}^{\infty}\left\{\int_{kP}^{(k+1)P} e^{-st}f(t)\,dt\right\}.$$

To make the lower and upper limits of the integral independent of P, we introduce the dummy variable $\tau = t - kP$. Then,

$$F(s) = \sum_{k=0}^{\infty}\left\{\int_0^{P} e^{-s(\tau+kP)}f(\tau)\,d\tau\right\} = \sum_{k=0}^{\infty} e^{-skP}\left\{\int_0^{P} e^{-s\tau}f(\tau)\,d\tau\right\}.$$

We note that the summation only affects the exponential term because the integral term is independent of the summation index k. But the exponential term is a geometric series so that

$$\sum_{k=0}^{\infty} e^{-skP} = \sum_{k=0}^{\infty} (e^{-sP})^k \stackrel{\text{Geometric series}}{=} \frac{1}{1-e^{-sP}}.$$

Using this information in the expression of $F(s)$, we find

$$F(s) = \left\{\int_0^P e^{-s\tau} f(\tau)\,d\tau\right\} \sum_{k=0}^{\infty} e^{-skP} = \left\{\int_0^P e^{-s\tau} f(\tau)\,d\tau\right\} \frac{1}{1-e^{-sP}}.$$

Therefore, the Laplace transform of the periodic function $f(t)$ is determined as

$$F(s) = \frac{1}{1-e^{-Ps}} \int_0^P e^{-st} f(t)\,dt. \qquad (2.19)$$

Example 2.15: Periodic Signal

Find the Laplace transform of the periodic function $f(t)$ defined in Figure 2.14.

Solution

It is clear that the period is $P = 2$. Evaluating the integral in Equation 2.19,

$$\int_0^2 e^{-st} f(t)\,dt \underset{f(t)=0 \text{ for } 1<t<2}{\overset{f(t)=1 \text{ for } 0<t<1}{=}} \int_0^1 e^{-st}\,dt = \frac{1-e^{-s}}{s}.$$

Then, by Equation 2.19,

$$F(s) = \frac{1-e^{-s}}{s(1-e^{-2s})} \quad \overset{1-e^{-2s}=1-(e^{-s})^2=(1-e^{-s})(1+e^{-s})}{\Rightarrow} \quad F(s) = \frac{1}{s(1+e^{-s})}.$$

2.3.4 Laplace Transforms of Derivatives and Integrals

Dynamic systems are generally modeled by differential equations of various orders. On other occasions, the system may be described by an equation that contains not only derivatives but also integrals; for instance, certain electrical circuits. Solving such equations using Laplace transformation requires a knowledge of the Laplace transforms of derivatives and integrals.

Laplace Transforms of Derivatives: The Laplace transforms of the first and second derivatives of a function $x(t)$ are defined by

$$\mathcal{L}\{\dot{x}(t)\} = sX(s) - x(0),$$
$$\mathcal{L}\{\ddot{x}(t)\} = s^2 X(s) - sx(0) - \dot{x}(0). \qquad (2.20)$$

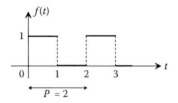

FIGURE 2.14 Periodic signal.

Laplace Transforms of Integrals: The Laplace transform of the integral of a function $x(t)$ is

$$\mathcal{L}\left\{\int_0^t x(t)\,dt\right\} = \frac{1}{s}X(s). \tag{2.21}$$

Alternatively,

$$\mathcal{L}^{-1}\left\{\frac{1}{s}X(s)\right\} = \int_0^t x(t)\,dt. \tag{2.22}$$

Example 2.16: Inverse Laplace

a. Find $\mathcal{L}^{-1}\left\{\dfrac{1}{s(s+2)}\right\}$.

b. ◀ Confirm in MATLAB.

Solution

a. Comparing with the left side of Equation 2.22, $X(s) = 1/(s+2)$ so that $x(t) = e^{-2t}$. Then,

$$\mathcal{L}^{-1}\left\{\frac{1}{s(s+2)}\right\} = \int_0^t e^{-2t}\,dt = \frac{1}{2}(1 - e^{-2t}).$$

b. ◀
```
>> syms s
>> ilaplace(1/s/(s+2))
ans =
1/2-1/2*exp(-2*t)
% Alternatively, performing the integration above yields the same result.
>> int(exp(-2*t),0,t)
ans =
1/2-1/2*exp(-2*t)     % Result confirmed
```

2.3.5 Inverse Laplace Transformation

As indicated in Figure 2.7, the final step in solving an IVP involves inverse Laplace transform $\mathcal{L}^{-1}\{X(s)\}$. This can be done in many ways, but the simplest is either direct use of the table of Laplace transforms (Table 2.2), if possible, or the partial fractions method.

2.3.5.1 Partial Fractions Method

In system dynamics, the transform function $X(s)$ is normally in the form

$$X(s) = \frac{N(s)}{D(s)} = \frac{\text{Polynomial of degree } m}{\text{Polynomial of degree } n}, \quad m < n. \tag{2.23}$$

The idea is simple: express $X(s)$ as a suitable sum of fractions, find the inverse Laplace transform of each fraction, and ultimately—by the linearity of $\mathcal{L}^{-1}$—the sum of the resulting time functions yields $x(t)$. How these partial fractions are formed depends on the nature of the poles of $X(s)$, that is, the roots of $D(s)$, which can be real or complex. For instance, the roots of $D(s) = s(s+1)(s+2)$ are $0, -1, -2$, all real and distinct. But the roots of $D(s) = s(s^2 + 2s + 2)$ are $0, -1 \pm j$, real and complex conjugates. Any second-degree polynomial such as $s^2 + 2s + 2$ with complex roots is called an irreducible polynomial and should not be decomposed into linear factors.

Case (1) Linear factor $s - p_i$

If p_i is a simple pole of $X(s)$, then $D(s)$ contains the factor $s - p_i$. This factor is associated with a fraction in the form of

$$\frac{A}{s - p_i},$$

where A is a constant called a residue and is to be determined appropriately.

Example 2.17: Inverse Laplace; Linear Factor

a. Find $\mathcal{L}^{-1}\{X(s)\}$, where $X(s) = \dfrac{s+3}{(s+1)(s+2)}$.

b. Verify in MATLAB.

Solution

a. $D(s)$ contains two linear factors, meaning poles of $X(s)$ are real and distinct; they are $-1, -2$. Each linear term is associated with a simple fraction as mentioned above. Thus,

$$X(s) = \frac{s+3}{(s+1)(s+2)} = \frac{A_1}{s+1} + \frac{A_2}{s+2} \quad (A_1, A_2 = \text{const})$$

$$= \frac{A_1(s+2) + A_2(s+1)}{(s+1)(s+2)} \xrightarrow{\text{Collect like terms}} \frac{(A_1 + A_2)s + 2A_1 + A_2}{(s+1)(s+2)}.$$

The denominators of the original and the final fractions are identical (by design), so we force their respective numerators to be identical, that is,

$$s + 3 \equiv (A_1 + A_2)s + 2A_1 + A_2.$$

But this identity holds only if the coefficients of like powers of s on both sides are the same. So, we have

$$\begin{array}{l}\text{Coefficient of s:} \quad 1 = A_1 + A_2 \\ \text{Constant term:} \quad 3 = 2A_1 + A_2\end{array} \xrightarrow{\text{Solve}} \begin{array}{l} A_1 = 2, \\ A_2 = -1. \end{array}$$

Insert the two residues into the partial fractions, and perform term-by-term inverse Laplace transformation, to obtain

$$X(s) = \frac{2}{s+1} - \frac{1}{s+2} \xrightarrow{\mathcal{L}^{-1}} x(t) \xlongequal{\text{Linearity}} 2\mathcal{L}^{-1}\left\{\frac{1}{s+1}\right\} - \mathcal{L}^{-1}\left\{\frac{1}{s+2}\right\} = 2e^{-t} - e^{-2t}.$$

b.
```
>> syms s
>> x = ilaplace((s+3)/(s+1)/(s+2))
x =
2*exp(-t)-exp(-2*t)
```

Case (2) Repeated linear factor $(s - p_i)^k$

If p_i is a pole of $X(s)$ with multiplicity k, then $D(s)$ contains the factor $(s - p_i)^k$. This factor is then associated with partial fractions

$$\frac{A_k}{(s-p_i)^k} + \frac{A_{k-1}}{(s-p_i)^{k-1}} + \cdots + \frac{A_2}{(s-p_i)^2} + \frac{A_1}{s-p_i},$$

where the residues $A_k, \ldots, A_1$ are determined as in Case (1).

Complex Analysis, Differential Equations, and Laplace Transformation

Case (3) Irreducible polynomial $s^2 + as + b$

Each irreducible polynomial is associated with a single fraction in the form of

$$\frac{Bs + C}{s^2 + as + b}$$

with constants B and C to be determined. Before taking the inverse Laplace transform, complete the square in the irreducible polynomial, that is, $s^2 + as + b = (s + \sigma)^2 + \omega^2$. For example, $s^2 + 2s + 2 = (s + 1)^2 + 1^2$. Then, at some point, we need to determine

$$\mathcal{L}^{-1}\left\{\frac{Bs + C}{(s + \sigma)^2 + \omega^2}\right\}.$$

The key is to split the fraction in terms of the two expressions

$$\frac{\omega}{(s + \sigma)^2 + \omega^2}, \qquad \frac{s + \sigma}{(s + \sigma)^2 + \omega^2}$$

so that we can ultimately use the relations (see Table 2.2)

$$\mathcal{L}^{-1}\left\{\frac{\omega}{(s + \sigma)^2 + \omega^2}\right\} = e^{-\sigma t} \sin \omega t, \qquad \mathcal{L}^{-1}\left\{\frac{s + \sigma}{(s + \sigma)^2 + \omega^2}\right\} = e^{-\sigma t} \cos \omega t.$$

Example 2.18: Real and Complex Poles

Find $\mathcal{L}^{-1}\{X(s)\}$, where $X(s) = \dfrac{2}{(s + 3)(s^2 + 2s + 5)}$.

Solution

The term $s^2 + 2s + 5 = (s + 1)^2 + 2^2$ is irreducible, while $s + 3$ is a linear factor. Therefore,

$$X(s) = \frac{2}{(s + 3)(s^2 + 2s + 5)} = \frac{A}{s + 3} + \frac{Bs + C}{s^2 + 2s + 5}$$

$$= \frac{A(s^2 + 2s + 5) + (Bs + C)(s + 3)}{(s + 3)(s^2 + 2s + 5)} \overset{\text{Collect like terms}}{=} \frac{(A + B)s^2 + (2A + 3B + C)s + 5A + 3C}{(s + 3)(s^2 + 2s + 5)}.$$

Proceeding as before,

$$A + B = 0 \qquad A = \frac{1}{4},$$
$$2A + 3B + C = 0 \quad \Rightarrow \quad B = -\frac{1}{4},$$
$$5A + 3C = 2 \qquad C = \frac{1}{4}.$$

Substituting into the partial fractions, we arrive at

$$X(s) = \frac{1}{4}\left[\frac{1}{s + 3} - \frac{s - 1}{s^2 + 2s + 5}\right] = \frac{1}{4}\left[\frac{1}{s + 3} - \frac{s + 1}{(s + 1)^2 + 2^2} + \frac{2}{(s + 1)^2 + 2^2}\right].$$

Term-by-term Laplace inversion yields $x(t) = \frac{1}{4}(e^{-3t} - e^{-t}\cos 2t + e^{-t}\sin 2t)$.

Case (4) Repeated irreducible polynomial $(s^2 + as + b)^k$

Each factor $(s^2 + as + b)^k$ in $D(s)$ is associated with partial fractions

$$\frac{B_k s + C_k}{(s^2 + as + b)^k} + \cdots + \frac{B_2 s + C_2}{(s^2 + as + b)^2} + \frac{B_1 s + C_1}{s^2 + as + b}.$$

2.3.5.2 ♦ Performing Partial Fractions in MATLAB

As we saw in Example 2.17, the command `ilaplace` returns the inverse Laplace transform but not the actual partial fractions. For that purpose, the `residue` command needs to be used instead. This command is concerned with the partial-fraction expansion of the ratio of two polynomials $B(s)/A(s)$, where, unlike the restriction cited in Equation 2.23, the degree of $B(s)$ can be higher than that of $A(s)$. When $\deg B(s) \geq \deg A(s)$, polynomial division results in

$$\frac{B(s)}{A(s)} = K(s) + \frac{R(1)}{s - P(1)} + \frac{R(2)}{s - P(2)} + \cdots + \frac{R(n)}{s - P(n)}, \qquad (2.24)$$

where $K(s)$ is called the direct term and $R(1), \ldots, R(n)$ are the residues. Note that the assumption here is that there are no multiple poles. For further information, type `help residue` at the MATLAB command prompt. The structure `[R, P, K] = residue(B, A)` indicates that there are two input arguments (B, A) and three outputs [R, P, K]. Here B and A are vectors of the coefficients of the numerator and denominator, respectively. The residues are returned in the column vector R, the pole locations in the column vector P, and the direct terms in the row vector K.

Example 2.19 Revisiting Example 2.17 ♦

Find the partial-fraction expansion for $X(s) = \dfrac{s+3}{(s+1)(s+2)}$.

SOLUTION

```
>> A = [1 3 2];    % denominator s^2+3*s+2
>> B = [1 3];      % numerator s+3
>> [R,P,K] = residue(B,A)
R =
    -1
     2    % residues are -1 and 2
P =
    -2
    -1    % poles (corresponding to residues) are -2 and -1
K =
    []    % empty set because deg B(s)<deg A(s)
```

Following the standard form in Equation 2.24, we have

$$X(s) = \frac{-1}{s-(-2)} + \frac{2}{s-(-1)} = \frac{-1}{s+2} + \frac{2}{s+1},$$

which completely agrees with the result of Example 2.17.

Example 2.20: Revisiting Example 2.18 ♦

Find the partial-fraction expansion for $X(s) = \dfrac{2}{(s+3)(s^2 + 2s + 5)}$.

SOLUTION

```
>> B = 2;
>> A = conv([1 3],[1 2 5])    % ''conv'' gives the product
A =
     1     5    11    15
>> [R,P,K] = residue(B,A)
```

```
R =
    0.2500
   -0.1250 - 0.1250i
   -0.1250 + 0.1250i

P =
   -3.0000
   -1.0000 + 2.0000i
   -1.0000 - 2.0000i

K =
   []
```

Based on these findings, the partial fractions are formed as

$$X(s) = \frac{2}{(s+3)(s^2+2s+5)} = \frac{(1/4)}{s-(-3)} + \frac{-(1/8)-(1/8)j}{s-(-1+2j)} + \frac{-(1/8)+(1/8)j}{s-(-1-2j)}.$$

Inspection confirms that the above is equivalent to the expansion obtained in Example 2.18.

2.3.5.3 Convolution Method

In addition to partial-fraction expansion, there is a second method of practical importance known as the convolution method. In systems analysis, the problem of determining the time history of a function often boils down to finding $\mathcal{L}^{-1}\{F(s)\}$, where $F(s) = G(s)H(s)$ and $g(t)$ and $h(t)$ are available. Then,

$$\mathcal{L}^{-1}\{F(s)\} = f(t) = \underbrace{(g*h)(t)}_{\text{convolution of } g \text{ and } h} = \int_0^t g(\tau)h(t-\tau)\,d\tau \stackrel{\text{Symmetry}}{=} \int_0^t h(\tau)g(t-\tau)\,d\tau = (h*g)(t).$$

To prove this, it suffices to show

$$\mathcal{L}\left\{\int_0^t g(\tau)h(t-\tau)\,d\tau\right\} = G(s)H(s).$$

First, using the fact that $u(t-\tau) = 0$ for all $\tau > t$, the left side is rewritten as

$$\mathcal{L}\left\{\int_0^t g(\tau)h(t-\tau)\,d\tau\right\} = \mathcal{L}\left\{\int_0^\infty g(\tau)h(t-\tau)u(t-\tau)\,d\tau\right\}.$$

Next, using Equation 2.15 and switching the order of integration, the above reduces to

$$\mathcal{L}\left\{\int_0^\infty g(\tau)h(t-\tau)u(t-\tau)\,d\tau\right\} = \int_0^\infty e^{-st}\left[\int_0^\infty g(\tau)h(t-\tau)u(t-\tau)\,d\tau\right]dt$$

$$= \int_0^\infty e^{-st}h(t-\tau)u(t-\tau)\,dt \int_0^\infty g(\tau)\,d\tau.$$

This is allowed because the Laplace transforms of g and h exist. Once again, since $u(t-\tau) = 0$ for all $\tau > t$, the first integral can be rewritten to give

$$\mathcal{L}\left\{\int_0^\infty g(\tau)h(t-\tau)u(t-\tau)\,d\tau\right\} = \int_\tau^\infty e^{-st}h(t-\tau)\,dt \int_0^\infty g(\tau)\,d\tau.$$

Introducing the change of variables $\xi = t - \tau$, the above becomes

$$\mathcal{L}\left\{\int_0^\infty g(\tau)h(t-\tau)u(t-\tau)\,d\tau\right\} = \int_0^\infty e^{-s(\xi+\tau)}h(\xi)\,d\xi \int_0^\infty g(\tau)\,d\tau$$

$$= \underbrace{\int_0^\infty e^{-s\xi}h(\xi)\,d\xi}_{H(s)} \underbrace{\int_0^\infty e^{-s\tau}g(\tau)\,d\tau}_{G(s)} = H(s)G(s).$$

Example 2.21: Convolution

Find $\mathcal{L}^{-1}\left\{\dfrac{1}{s^2(s+1)}\right\}$.

SOLUTION

Let

$$F(s) = \frac{1}{s^2(s+1)} = G(s)H(s)$$

with $G(s) = \dfrac{1}{s^2}$, $H(s) = \dfrac{1}{s+1}$. Noting that $g(t) = t$ and $h(t) = e^{-t}$,

$$f(t) = (g * h)(t) = \int_0^t \tau e^{-(t-\tau)}\,d\tau \overset{\text{Integration by parts}}{=} \left[\tau e^{-(t-\tau)}\right]_{\tau=0}^t - \int_0^t e^{-(t-\tau)}\,d\tau = t - 1 + e^{-t}.$$

2.3.5.4 Solving Initial-Value Problems

We now have the necessary tools to complete the procedure depicted in Figure 2.7. This will be illustrated by the following examples.

Example 2.22: IVP

a. Solve $\ddot{x} + 9x = u(t)$, $x(0) = 0$, $\dot{x}(0) = 0$, where $u(t)$ is the unit-step function.
b. Repeat in MATLAB.

SOLUTION

a. Taking the Laplace transform of the ODE, using Equation 2.20, and accounting for the zero initial conditions, we arrive at

$$(s^2 + 9)X(s) = \frac{1}{s} \Rightarrow X(s) = \frac{1}{s(s^2+9)} \overset{\text{Partial fractions}}{=} \frac{A}{s} + \frac{Bs+C}{s^2+9} = \frac{(A+B)s^2 + Cs + 9A}{s(s^2+9)}.$$

Equating the coefficients of like powers of s on both sides yields $A = \frac{1}{9}$, $B = -\frac{1}{9}$, and $C = 0$. Then

$$X(s) = \frac{1}{s(s^2+9)} = \frac{(1/9)}{s} + \frac{-(1/9)s}{s^2+9} = \frac{1}{9}\left[\frac{1}{s} - \frac{s}{s^2+9}\right].$$

Inverse Laplace transformation gives $x(t) = \frac{1}{9}(1 - \cos 3t)$, $t \geq 0$. Of course, the convolution method would have led to the same result (Verify).

b.
```
>> x = dsolve('D2x + 9*x = heaviside(t)','x(0) = 0, Dx(0) = 0')
x =
-1/9*heaviside(t)*(-1+cos(3*t))
```
Noting that `heaviside` represents the unit-step, the result agrees with that in Part (a).

Example 2.23: IVP

Solve
$$\ddot{x} + 3\dot{x} + 2x = 0, \quad x(0) = 0, \; \dot{x}(0) = 1.$$

SOLUTION

Taking the Laplace transform,
$$s^2 X(s) - sx(0) - \dot{x}(0) + 3[sX(s) - x(0)] + 2X(s) = 0.$$

Using the initial conditions and solving for $X(s)$ yields
$$X(s) = \frac{1}{(s+1)(s+2)} \xrightarrow[\text{or convolution}]{\mathcal{L}^{-1} \text{ using partial fractions}} x(t) = e^{-t} - e^{-2t}.$$

2.3.6 FINAL-VALUE THEOREM AND INITIAL-VALUE THEOREM

Suppose that a function $x(t)$ attains a finite limit as $t \to \infty$, that is, it settles down after a sufficiently long time. This finite, limiting value is the steady-state value (or final value) of $x(t)$, denoted by x_{ss}. However, there are situations where the time history $x(t)$ is not available, but instead $X(s)$ is. The final-value theorem (FVT) allows us to find x_{ss}, if it exists, by directly using $X(s)$ without knowledge of $x(t)$.

Final-value theorem (FVT): Suppose $X(s)$ has no poles in the right half-plane (RHP) or on the imaginary axis, except possibly a simple pole at the origin. Then, $x(t)$ has a definite steady-state value and it is given by

$$x_{ss} = \lim_{s \to 0} \{sX(s)\}. \tag{2.25}$$

It is important to use FVT only when applicable. Otherwise, it can lead to misleading results.

Example 2.24: FVT

a. Find x_{ss} if $X(s) = \dfrac{s+1}{s(s^2 + 2s + 2)}$.

b. Confirm in MATLAB.

SOLUTION

a. The poles of $X(s)$ are at 0 and $-1 \pm j$. The conjugate pair lies in the left half-plane, and 0 is a simple pole (at the origin), all permitted by the FVT. Therefore,

$$x_{ss} = \lim_{s \to 0} \{sX(s)\} = \lim_{s \to 0} \left\{ \frac{s+1}{s^2 + 2s + 2} \right\} = \frac{1}{2}.$$

b.
```
>> syms s
>> X = (s+1)/s/(s^2+2*s+2);
>> xss = limit(s*X,s,0)
xss =
1/2
```

```
% Re-confirm by inspecting the limit of x(t) as t goes to infinity
>> xt = ilaplace(X)
xt =
1/2+1/2*(-cos(t)+sin(t))*exp(-t)    % time history x(t)
>> syms t
>> xss = limit(xt,t,inf)
xss =
1/2
```

Example 2.25: FVT Not Applicable

Let $X(s) = 1/(s^2 + 4)$ so that its poles are at $\pm 2j$, on the imaginary axis, not permitted by the FVT. Therefore, FVT is not applicable and should not be applied! If it were to be applied,

$$x_{ss} = \lim_{s \to 0}\{sX(s)\} = \lim_{s \to 0}\left\{\frac{s}{s^2 + 4}\right\} = 0,$$

which is clearly not true. To explain this, we first find $x(t) = \mathcal{L}^{-1}\{X(s)\} = \frac{1}{2}\sin 2t$. Then, it is clear that $\lim_{t \to \infty} x(t)$ does not exist, since $x(t)$ is oscillatory and there is no steady-state value.

Initial-Value Theorem (IVT): If $\lim_{s \to \infty}\{sX(s)\}$ exists, then the initial value (see Section 2.2) of $x(t)$ is given by

$$x(0^+) = \lim_{s \to \infty}\{sX(s)\}. \tag{2.26}$$

Note that in the case of IVT, the poles of $X(s)$ are not limited to specific regions in the complex plane as they were with FVT.

Example: 2.26 IVT

If $X(s) = \dfrac{s^2 + 3}{s^2(2s + 1)}$, find $x(0^+)$.

Solution

By Equation 2.26,

$$x(0^+) = \lim_{s \to \infty}\{sX(s)\} = \lim_{s \to \infty}\left\{\frac{s^2 + 3}{s(2s + 1)}\right\} = \frac{1}{2}.$$

Example 2.27: Initial Condition ≠ Initial Value

Consider $\ddot{x} + \dot{x} + 2x = \delta(t)$, $x(0^-) = 0$, $\dot{x}(0^-) = 0$, where $\delta(t)$ denotes the unit-impulse. Recall that 0^- refers to the time immediately prior to $t = 0$ and $x(0^-)$ is the initial condition of x. Determine the initial values of x and $\dot{x}$, that is, $x(0^+)$ and $\dot{x}(0^+)$.

Solution

Taking the Laplace transform of the ODE and using the zero initial conditions, we have

$$X(s) = \frac{1}{s^2 + s + 2}.$$

By IVT,

$$x(0^+) = \lim_{s \to \infty}\{sX(s)\} = \lim_{s \to \infty}\left\{\frac{s}{s^2 + s + 2}\right\} = 0.$$

Thus $x(0^+) = x(0^-)$. To evaluate $\dot{x}(0^+)$, we apply the IVT while x is replaced with $\dot{x}$:

$$\dot{x}(0^+) = \lim_{s \to \infty} \{s\mathcal{L}\{\dot{x}\}\} = \lim_{s \to \infty} \{s[sX(s)]\} = \lim_{s \to \infty} \left\{ \frac{s^2}{s^2 + s + 2} \right\} = 1.$$

Therefore, $\dot{x}(0^+) \neq \dot{x}(0^-)$. This clearly shows that when impulsive forces are present in the system, initial values and initial conditions are indeed different.

PROBLEM SET 2.3

In each case,

a. Find the Laplace transform of the given function. Use Table 2.2 when applicable.
b. Confirm the result in MATLAB.

1. e^{-at+b}, $a, b = \text{const}$
2. e^{jat}, $a = \text{const}$, $j = \sqrt{-1}$. Express the result in rectangular form of a complex number.
3. $\sin(\omega t - \phi)$, $\omega, \phi = \text{const}$
4. $\cos(\omega t + \phi)$, $\omega, \phi = \text{const}$

In each case,

a. Express the signal in terms of unit-step functions.
b. Find the Laplace transform of the expression in (a) using the shift on the t-axis.

5. $g(t) = \begin{cases} 1 & \text{if } 0 < t < 1, \\ -1 & \text{if } 1 < t < 2, \\ 0 & \text{otherwise.} \end{cases}$

6. $g(t) = \begin{cases} t & \text{if } 0 < t < 1, \\ 0 & \text{otherwise.} \end{cases}$

7. $h(t) = \begin{cases} 0 & \text{if } t < 0, \\ t & \text{if } 0 < t < 1, \\ 1 & \text{if } t > 1. \end{cases}$

8. $h(t) = \begin{cases} 0 & \text{if } t < 0, \\ 1-t & \text{if } 0 < t < 1, \\ 0 & \text{if } t > 1. \end{cases}$

Find the Laplace transform of each periodic function whose definition in one period is given.

9. $f(t) = \begin{cases} 1 & \text{if } 0 < t < 1, \\ -1 & \text{if } 1 < t < 2. \end{cases}$

10. $f(t) = t$ if $0 < t < 1$.

11. $g(t) = \begin{cases} t & \text{if } 0 < t < 1, \\ 1 & \text{if } 1 < t < 2. \end{cases}$

12. $g(t) = \begin{cases} -t & \text{if } 0 < t < 1, \\ 2-t & \text{if } 1 < t < 2. \end{cases}$

In each case,

a. Find the inverse Laplace transform using the partial-fraction expansion method.
b. Repeat in MATLAB.

13. $\dfrac{3s+2}{s(s+1)}$

14. $\dfrac{3s^2+s+5}{(s^2+1)(s+2)}$

15. $\dfrac{s+5}{s(s^2+2s+5)}$

16. $\dfrac{4s+5}{s^2(s^2+4s+5)}$

17. $\dfrac{s^2+s-2}{(s+3)(s+1)^2}$

18. $\dfrac{s^2+s-1}{(s+3)(s^2+2s+2)}$

In each case,

a. Solve the IVP.
b. Confirm the result in MATLAB.

19. $\ddot{x}+2\dot{x}+2x=0, \quad x(0)=1, \; \dot{x}(0)=-3$

20. $\ddot{x}+2\dot{x}+2x=1, \quad x(0)=1, \; \dot{x}(0)=0$

21. $2\ddot{x}+\dot{x}=e^{-t}, \quad x(0)=0, \; \dot{x}(0)=\dfrac{1}{2}$

22. $\ddot{x}+4x=\sin t, \quad x(0)=1, \; \dot{x}(0)=0$

23. $\ddot{y}+3\dot{y}+2y=e^{-t}, \quad y(0)=1, \; \dot{y}(0)=-1$

24. $\ddot{y}+3\dot{y}=9t, \quad y(0)=0, \; \dot{y}(0)=1$

In each case, decide whether the FVT is applicable, and if so, determine x_{ss}.

25. $X(s)=\dfrac{1}{s(s+2)}$

26. $X(s)=\dfrac{2s+3}{(s+3)(s^2+4s+5)}$

27. $X(s)=\dfrac{s+4}{s^2(s+1)(s+2)}$

28. $X(s)=\dfrac{s+3}{(s+1)^2(s+2)}$

In each case, evaluate $x(0^+)$ using the IVT.

29. $X(s)=\dfrac{s^2+1}{s(2s^2+s+1)}$

30. $X(s)=\dfrac{3s+2}{(s+1)(s+2)^2}$

2.4 SUMMARY

A complex number z in rectangular form is $z = x + jy$, where x and y are the real and imaginary parts of z, respectively. The magnitude of z is $|z| = \sqrt{x^2 + y^2}$. The distance between two complex numbers z_1 and z_2 is $|z_1 - z_2|$. The complex conjugate of z, denoted by $\bar{z}$, is defined as $\bar{z} = x - jy$ with the property that $z\bar{z} = x^2 + y^2 = |z|^2$. The polar form of z is $z = re^{j\theta}$ where $r = |z|$ and θ is measured from the positive real axis, and regarded as positive in the counterclockwise sense. Given a complex number $z = re^{j\theta} \neq 0$, and a positive integer n, the nth root of z is multivalued and defined by

$$\sqrt[n]{z} = \sqrt[n]{r}\left(\cos\frac{\theta + 2k\pi}{n} + j\sin\frac{\theta + 2k\pi}{n}\right), \quad k = 0, 1, \ldots, n-1.$$

An nth-order ODE

$$a_n x^{(n)} + a_{n-1} x^{(n-1)} + \cdots + a_1 \dot{x} + a_0 x = F(t)$$

is linear if coefficients $a_0, a_1, \ldots, a_n$ are either constants or functions of t. If $F(t) \equiv 0$, the ODE is homogeneous. Otherwise, it is nonhomogeneous. A linear, first-order ODE

$$\dot{x} + g(t)x = f(t)$$

has a general solution

$$x(t) = e^{-h}\left[\int e^h f(t)\, dt + c\right], \quad h = \int g(t)\, dt, \; c = \text{const.}$$

If a function $x(t)$ is defined for all $t \geq 0$, its Laplace transform is defined by

$$X(s) \stackrel{\text{Notation}}{=} \mathcal{L}\{x(t)\} \stackrel{\text{Definition}}{=} \int_0^\infty e^{-st} x(t)\, dt$$

provided that the integral exists. If $F(s) = \mathcal{L}\{f(t)\}$ exists and $a \geq 0$, then

$$\mathcal{L}\{f(t-a)u(t-a)\} = e^{-as} F(s).$$

If $f(t)$ is periodic with period P, then

$$F(s) = \frac{1}{1 - e^{-Ps}} \int_0^P e^{-st} f(t)\, dt.$$

The Laplace transforms of the first and second derivatives of a function $x(t)$ are defined by

$$\mathcal{L}\{\dot{x}(t)\} = sX(s) - x(0),$$
$$\mathcal{L}\{\ddot{x}(t)\} = s^2 X(s) - sx(0) - \dot{x}(0).$$

The Laplace transform of the integral of a function $x(t)$ is

$$\mathcal{L}\left\{\int_0^t x(t)\, dt\right\} = \frac{1}{s} X(s).$$

Inverse Laplace transformation can be accomplished by either partial-fraction expansion or the convolution method, which states

$$\mathcal{L}^{-1}\{G(s)H(s)\} = \underbrace{(g * h)(t)}_{\text{convolution of } g \text{ and } h} = \int_0^t g(\tau) h(t-\tau)\, d\tau \stackrel{\text{symmetry}}{=} \int_0^t h(\tau) g(t-\tau)\, d\tau = (h * g)(t).$$

If $X(s)$ has no poles in the RHP or on the imaginary axis, except possibly a simple pole at the origin, then $x(t)$ has a definite steady-state value and it is given by

$$x_{ss} = \lim_{s \to 0}\{sX(s)\} \quad \text{FVT.}$$

If $\lim_{s \to \infty}\{sX(s)\}$ exists, then

$$x(0^+) = \lim_{s \to \infty}\{sX(s)\} \quad \text{IVT.}$$

REVIEW PROBLEMS

1. Perform the complex algebra and express the result in rectangular form
$$\frac{2-j}{3+2j} \cdot \frac{1+2j}{2-3j}.$$

2. Repeat Problem 1 for
$$\frac{4+j}{2j(3-2j)}.$$

3. Perform $(2-3j)^3$ using polar form and express the result in rectangular form.

4. Repeat Problem 3 for
$$\frac{(1+2j)^3}{(-2+j)^4}.$$

5. Find all possible values of $\left(-1-\frac{1}{2}j\right)^{1/3}$.

6. Solve $3\dot{y} + 2y = t$, $y(0) = \frac{1}{4}$.

7. Using the method of undetermined coefficients solve
$$\ddot{x} + 3\dot{x} = 10e^{-3t}, \quad x(0) = 1, \quad \dot{x}(0) = -\frac{10}{3}.$$

8. a. Solve
$$4\ddot{x} + 4\dot{x} + 5x = -1 + 5t, \quad x(0) = -1, \quad \dot{x}(0) = 1.$$
   b. Repeat in MATLAB.

9. Find amplitude D and phase ϕ so that $\cos t - 2\sin t = D\cos(t+\phi)$.

10. Find $\mathcal{L}\left\{e^{(2-3j)t}\right\}$.

11. Consider the signal $g(t)$ described in Figure 2.15.
    a. Express $g(t)$ in terms of unit-step functions.
    b. Using the shift on the t-axis find $G(s)$.

12. Repeat Problem 11 for $h(t)$ in Figure 2.16.

13. a. Using partial-fraction expansion, find
$$\mathcal{L}^{-1}\left\{\frac{2s^2+1}{s^2(4s^2+1)}\right\}.$$
   b. Confirm the result in MATLAB.

14. Repeat Problem 13 for
$$\mathcal{L}^{-1}\left\{\frac{\frac{1}{2}s^2+s-2}{(s+1)(s^2+4)}\right\}.$$

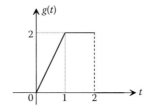

FIGURE 2.15 Problem 11.

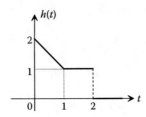

FIGURE 2.16 Problem 12.

15. a. Solve
$$\ddot{x} + x = u(t-1), \quad x(0) = 0, \; \dot{x}(0) = 0.$$

   (*Hint:* In the final step, use the shift on the *t*-axis to transfer data from the *s* domain to the time domain.)

   b. Confirm the result in MATLAB.

16. a. Solve
$$\ddot{x} + 3\dot{x} = 0, \quad x(0) = 0, \; \dot{x}(0) = 1.$$

   b. Repeat in MATLAB.

17. Let
$$X(s) = \frac{1}{s(s+1)^2}.$$

   a. Using the FVT, if applicable, evaluate x_{ss}.

   b. Confirm the result of Part (a) by evaluating $\lim_{t \to \infty} x(t)$.

18. Using the IVT, evaluate $x(0^+)$ if
$$X(s) = \frac{s}{2(s^2 + s + 1)}.$$

3 Matrix Analysis

Fundamentals of linear algebra, specifically, vectors and matrices, and their relation to linear systems of algebraic and differential equations are presented in this chapter. Methods of linear algebra are mainly useful in the treatment of systems of equations that are heavily coupled, that is, when a large number of equations in the system involve many of the unknown variables. In these cases, techniques such as direct substitution and elimination are no longer suitable due to their lack of computational efficiency. We focus on algebraic systems first, and then extend the ideas to systems of differential equations and the matrix eigenvalue problem.

3.1 MATRICES

An n-dimensional vector v is an ordered set of n scalars, written as $v = (v_1, v_2, \ldots, v_n)$, where each v_i ($i = 1, 2, \ldots, n$) is called a component of vector v. Brackets and braces will be used interchangeably to represent vectors. A matrix is a collection of numbers (real or complex), or possibly functions, arranged in a rectangular array and enclosed by square brackets. Each of the elements in a matrix is called an entry of the matrix. The horizontal and vertical lines are referred to as rows and columns of the matrix, respectively. The number of rows and columns of a matrix determine the size of that matrix. If a matrix $\mathbf{A}$ has m rows and n columns, then it is said to be of size $m \times n$. A matrix is square if the number of its rows and columns are the same, and rectangular if different. We denote matrices by boldfaced capital letters, such as $\mathbf{A}$. The abbreviated form of an $m \times n$ matrix is

$$\mathbf{A} = [a_{ij}]_{m \times n}, \quad i = 1, 2, \ldots, m, \ j = 1, 2, \ldots, n,$$

where a_{ij} is known as the (i, j) entry of $\mathbf{A}$, located at the intersection of the ith row and the jth column of $\mathbf{A}$ so that a_{12}, for instance, occupies the entry at which the first row and the second column meet. If $\mathbf{A}$ is a square matrix ($m = n$), the elements $a_{11}, a_{22}, \ldots, a_{nn}$ are the diagonal entries of $\mathbf{A}$. These diagonal elements form the so-called main diagonal of $\mathbf{A}$. Two matrices $\mathbf{A} = [a_{ij}]$ and $\mathbf{B} = [b_{ij}]$ are equal if they have the same size and the same entries in the respective locations. If some rows or columns (or possibly both) of $\mathbf{A}$ are deleted, a submatrix of $\mathbf{A}$ is generated. The sum of $\mathbf{A} = [a_{ij}]_{m \times n}$ and $\mathbf{B} = [b_{ij}]_{m \times n}$ is $\mathbf{C} = [c_{ij}]_{m \times n} = [a_{ij} + b_{ij}]_{m \times n}$. The product of a scalar k and matrix $\mathbf{A} = [a_{ij}]_{m \times n}$ is $k\mathbf{A} = [ka_{ij}]_{m \times n}$. Consider $\mathbf{A} = [a_{ij}]_{m \times n}$ and $\mathbf{B} = [b_{ij}]_{n \times p}$ so that the number of columns of $\mathbf{A}$ is equal to the number of rows of $\mathbf{B}$. Then, their product $\mathbf{C} = \mathbf{AB}$ is $m \times p$ whose entries are obtained as

$$c_{ij} = \sum_{k=1}^{n} a_{ik} b_{kj}, \quad i = 1, 2, \ldots, m, \ j = 1, 2, \ldots, p.$$

This is schematically shown in Figure 3.1. If the number of columns of $\mathbf{A}$ does not match the number of rows of $\mathbf{B}$, the product is undefined. If the product is defined, then the (i, j) entry of $\mathbf{C}$ is simply the dot (inner) product of the ith row of $\mathbf{A}$ and the jth column of $\mathbf{B}$.

Given an $m \times n$ matrix $\mathbf{A}$, its transpose, denoted by $\mathbf{A}^T$, is an $n \times m$ matrix with the property that its first row is the first column of $\mathbf{A}$, its second row is the second column of $\mathbf{A}$, and so on. Given that all matrix operations are valid,

$$(\mathbf{A} + \mathbf{B})^T = \mathbf{A}^T + \mathbf{B}^T,$$
$$(k\mathbf{A})^T = k\mathbf{A}^T, \text{ scalar } k,$$
$$(\mathbf{AB})^T = \mathbf{B}^T \mathbf{A}^T.$$

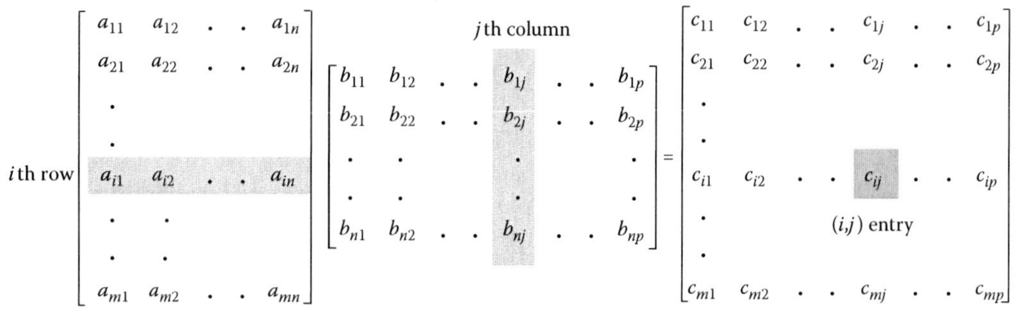

$$\begin{bmatrix} a_{11} & a_{12} & \cdots & a_{1n} \\ a_{21} & a_{22} & \cdots & a_{2n} \\ \vdots & & & \vdots \\ a_{i1} & a_{i2} & \cdots & a_{in} \\ \vdots & & & \vdots \\ a_{m1} & a_{m2} & \cdots & a_{mn} \end{bmatrix} \begin{bmatrix} b_{11} & b_{12} & \cdots & b_{1j} & \cdots & b_{1p} \\ b_{21} & b_{22} & \cdots & b_{2j} & \cdots & b_{2p} \\ \vdots & & & \vdots & & \vdots \\ b_{n1} & b_{n2} & \cdots & b_{nj} & \cdots & b_{np} \end{bmatrix} = \begin{bmatrix} c_{11} & c_{12} & \cdots & c_{1j} & \cdots & c_{1p} \\ c_{21} & c_{22} & \cdots & c_{2j} & \cdots & c_{2p} \\ \vdots & & & & & \vdots \\ c_{i1} & c_{i2} & \cdots & c_{ij} & \cdots & c_{ip} \\ \vdots & & & & & \vdots \\ c_{m1} & c_{m2} & \cdots & c_{mj} & \cdots & c_{mp} \end{bmatrix}$$

FIGURE 3.1 Construction of the matrix product **C=AB**.

Example 3.1: Transpose

Consider
$$\mathbf{A} = \begin{bmatrix} -2 & 1 \\ -1 & 3 \end{bmatrix}, \quad \mathbf{B} = \begin{bmatrix} 1 & 0 \\ 4 & -2 \end{bmatrix}.$$

a. Verify $(\mathbf{AB})^T = \mathbf{B}^T \mathbf{A}^T$.
b. ◢ Confirm the result in MATLAB®.

Solution

a. Working on the left side,

$$\mathbf{AB} = \begin{bmatrix} -2 & 1 \\ -1 & 3 \end{bmatrix} \begin{bmatrix} 1 & 0 \\ 4 & -2 \end{bmatrix} = \begin{bmatrix} 2 & -2 \\ 11 & -6 \end{bmatrix} \xRightarrow{\text{Transpose}} (\mathbf{AB})^T = \begin{bmatrix} 2 & 11 \\ -2 & -6 \end{bmatrix}.$$

Performing the operations on the right side,

$$\mathbf{B}^T \mathbf{A}^T = \begin{bmatrix} 1 & 4 \\ 0 & -2 \end{bmatrix} \begin{bmatrix} -2 & -1 \\ 1 & 3 \end{bmatrix} = \begin{bmatrix} 2 & 11 \\ -2 & -6 \end{bmatrix}.$$

b. ◢
```
>> A = [-2 1;-1 3];
>> B = [1 0;4 -2];
>> At = A.'     % A.' is the non-conjugate transpose
At =
    -2   -1
     1    3
>> Bt = B.'
Bt =
     1    4
     0   -2
>> left = (A*B).'
left =
     2   11
    -2   -6
>> right = Bt*At
right =
     2   11
    -2   -6
```

Matrix Analysis

3.1.1 Special Matrices

A square matrix $\mathbf{A}$ is symmetric if $\mathbf{A}^T = \mathbf{A}$ and skew-symmetric if $\mathbf{A}^T = -\mathbf{A}$. A square matrix $\mathbf{A}_{n \times n} = [a_{ij}]$ is upper-triangular if $a_{ij} = 0$ for all $i > j$, that is, all elements below the main diagonal are zeros, lower-triangular if $a_{ij} = 0$ for all $i < j$, that is, all elements above the main diagonal are zeros, and diagonal if $a_{ij} = 0$ for all $i \neq j$. In the upper- and lower-triangular matrices, the diagonal elements may all be zeros. However, in a diagonal matrix, at least one diagonal entry must be nonzero. The $n \times n$ identity matrix, denoted by $\mathbf{I}$, is a diagonal matrix whose diagonal entries are all equal to 1.

3.1.2 Elementary Row Operations

In matrix analysis, we often encounter matrices that do not appear in any special form such as triangular or diagonal. Such matrices are commonly known as full matrices. However, it would be desirable if a full matrix could be transformed (or reduced) to a special matrix since they are so much easier to work with. This is done through the use of elementary row operations (EROs). There are three types of EROs as listed below.

(ERO$_1$)—Multiply a row by a nonzero constant.
(ERO$_2$)—Interchange any two rows.
(ERO$_3$)—Multiply the ith row by a scalar $\alpha \neq 0$, add the result to the jth row, and then replace the jth row with the result. The ith row is called the pivot row.

These EROs are applied to a matrix $\mathbf{A}$ to transform it to an upper-triangular matrix. In the final step of this process, the form of matrix $\mathbf{A}$ is called the row-echelon form of $\mathbf{A}$, denoted REF($\mathbf{A}$). It is important to realize that the original matrix and any subsequent one generated by an ERO are completely different matrices and that the row-echelon form of a matrix is not unique. The number of nonzero rows in REF($\mathbf{A}$) is called the rank of matrix $\mathbf{A}$, denoted rank($\mathbf{A}$). Let us consider the general case of an $m \times n$ matrix $\mathbf{A}$. Then rank($\mathbf{A}$) can at most be equal to m or n, whichever is smaller. For instance, the rank of a 3×4 matrix can at most be 3.

Example 3.2: Rank via REF

For the matrix given below,
  a. Find the row-echelon form and rank.
  b. Repeat in MATLAB.

$$\mathbf{A} = \begin{bmatrix} -1 & 1 & 2 \\ 3 & -1 & 1 \\ -1 & 3 & 4 \end{bmatrix}.$$

Solution

a. We first use the first row to generate zeros below the (1, 1) entry (Figure 3.2). This means that the first row is the pivot row and both operations are ERO$_3$.

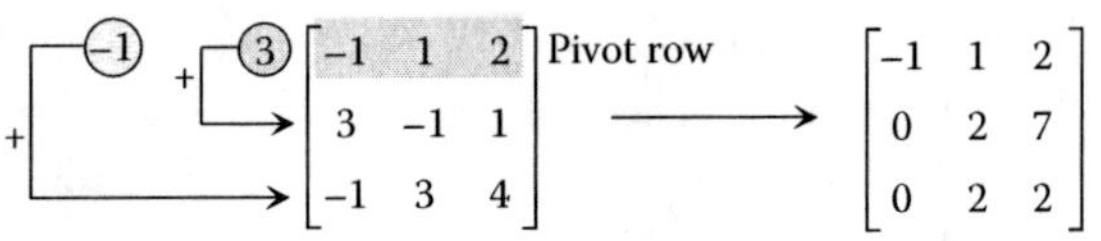

$$\begin{bmatrix} -1 & 1 & 2 \\ 3 & -1 & 1 \\ -1 & 3 & 4 \end{bmatrix} \text{Pivot row} \longrightarrow \begin{bmatrix} -1 & 1 & 2 \\ 0 & 2 & 7 \\ 0 & 2 & 2 \end{bmatrix}$$

FIGURE 3.2 First step in generating the REF.

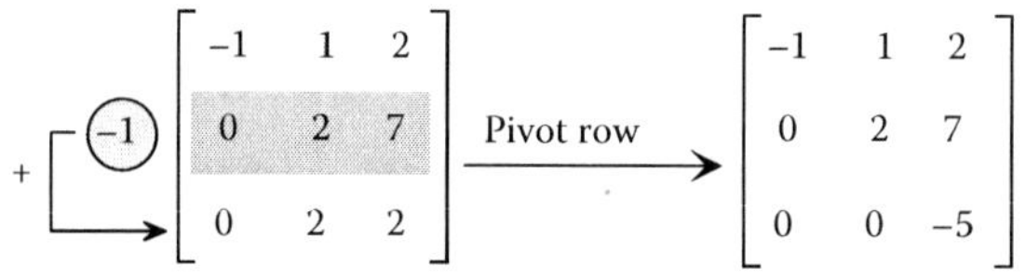

$$\begin{array}{c}+\\\end{array}\boxed{\begin{array}{c}\boxed{-1}\\\\\end{array}}\begin{bmatrix}-1 & 1 & 2\\ 0 & 2 & 7\\ 0 & 2 & 2\end{bmatrix}\xrightarrow{\text{Pivot row}}\begin{bmatrix}-1 & 1 & 2\\ 0 & 2 & 7\\ 0 & 0 & -5\end{bmatrix}$$

FIGURE 3.3 Second step in Example 3.2.

Next, in the newly generated matrix, we use the second row (pivot row) to create zeros below the (2, 2) entry (Figure 3.3).

The resulting matrix is indeed upper-triangular, as intended, and hence represents the row-echelon form of **A**. Therefore,

$$\text{REF}(\mathbf{A}) = \begin{bmatrix} -1 & 1 & 2 \\ 0 & 2 & 7 \\ 0 & 0 & -5 \end{bmatrix}.$$

Since there are three nonzero rows in REF(**A**), we conclude that rank(**A**) = 3.

b.

```
>> A = [-1 1 2;3 -1 1;-1 3 4];
```

The `rref` command in MATLAB produces the *reduced* row-echelon form by trying to create ones in the diagonal entries and zeros elsewhere for a square matrix. For nonsquare matrices, the outcome will contain either an additional row or an additional column, depending on the size of the matrix.

```
>> rref(A)
ans =
     1   0   0
     0   1   0
     0   0   1
% As expected, the result differs from REF(A).
>> rank(A)
ans =
     3   % This agrees with the earlier finding.
```

3.1.3 Determinant of a Matrix

The determinant of a square matrix $\mathbf{A} = [a_{ij}]_{n \times n}$ is a real scalar denoted by $|\mathbf{A}|$ or $\det(\mathbf{A})$. The most trivial case is $\mathbf{A} = [a_{11}]_{1 \times 1}$, for which the determinant is simply $|\mathbf{A}| = a_{11}$. For $n \geq 2$, the determinant may be calculated using any row—with preference given to the row with the most number of zeros—as

$$|\mathbf{A}| = \sum_{k=1}^{n} a_{ik}(-1)^{i+k} M_{ik}, \quad i = 1, 2, \ldots, n. \tag{3.1}$$

In Equation 3.1, M_{ik} is called the minor of the entry a_{ik}, defined as the determinant of the $(n-1) \times (n-1)$ submatrix of **A** obtained by deleting the ith row and the kth column of **A**. The quantity $(-1)^{i+k} M_{ik}$ is the cofactor of a_{ik} and is denoted by C_{ik}. Also note that $(-1)^{i+k}$ is responsible for whether a term is multiplied by $+1$ or -1. A square matrix with a nonzero determinant is called nonsingular. Otherwise, it is called singular.

Matrix Analysis

Example 3.3: Determinant

a. Calculate the determinant of

$$A = \begin{bmatrix} -2 & 1 & 1 & 3 \\ 1 & 2 & 4 & -1 \\ 3 & -1 & 2 & 4 \\ 1 & 5 & 6 & -2 \end{bmatrix}.$$

b. Repeat in MATLAB.

SOLUTION

a. We will use the first row since it contains the smallest (in magnitude) entries.

$$\begin{vmatrix} -2 & 1 & 1 & 3 \\ 1 & 2 & 4 & -1 \\ 3 & -1 & 2 & 4 \\ 1 & 5 & 6 & -2 \end{vmatrix} \stackrel{\text{Using the first row}}{=} -2\begin{vmatrix} 2 & 4 & -1 \\ -1 & 2 & 4 \\ 5 & 6 & -2 \end{vmatrix} - \begin{vmatrix} 1 & 4 & -1 \\ 3 & 2 & 4 \\ 1 & 6 & -2 \end{vmatrix}$$

$$+ \begin{vmatrix} 1 & 2 & -1 \\ 3 & -1 & 4 \\ 1 & 5 & -2 \end{vmatrix} - 3\begin{vmatrix} 1 & 2 & 4 \\ 3 & -1 & 2 \\ 1 & 5 & 6 \end{vmatrix}.$$

Each of the four 3×3 determinants is computed via Equation 3.1, resulting in

$$|A| = -2(32) - (-4) + (-14) - 3(16) = -122.$$

b.
```
>> A=[-2 1 1 3;1 2 4 -1;3 -1 2 4;1 5 6 -2];
>> det(A)
ans =
    -122
```

3.1.3.1 Properties of Determinant

- A square matrix and its transpose have the same determinant, that is, $|A| = |A^T|$.
- The determinant of diagonal, upper-triangular, and lower-triangular matrices is the product of the diagonal entries.
- If an entire row (or column) of a square matrix A is zero, then $|A| = 0$.
- $|AB| = |A| |B|$.
- A matrix with any number of linearly dependent rows (or columns) is singular.
- If A is $n \times n$ and k is a scalar, then $|kA| = k^n |A|$.
- EROs affect the determinant as follows: (1) ERO$_1$ causes the determinant to be multiplied by the constant involved, (2) ERO$_2$ causes the determinant to be negated, and (3) ERO$_3$ does not alter the determinant.

Earlier in this chapter we defined the rank of a matrix as the number of nonzero rows in its REF. With the knowledge of the determinant, we define the rank of any matrix A as the size of the largest nonsingular submatrix of A. In other words, rank$(A) = r$ if there exists an $r \times r$ submatrix of A with nonzero determinant and any other $p \times p$ (with $p > r$) submatrix of A is singular. If matrix A is square $(n \times n)$ and $|A| \neq 0$, then rank$(A) = n$. If $|A| = 0$, then A has linearly dependent rows and hence rank$(A) < n$.

Example 3.4: Rank via Determinant

a. Find the rank of
$$A = \begin{bmatrix} 2 & 3 & 4 \\ 0 & -1 & 2 \\ 1 & 1 & 3 \\ 1 & 0 & 5 \end{bmatrix}.$$

b. ◆ Confirm in MATLAB.

Solution

a. Since **A** is 4×3, rank(**A**) cannot exceed 3. We first inspect all possible 3×3 submatrices of **A**.

$$\begin{vmatrix} 2 & 3 & 4 \\ 0 & -1 & 2 \\ 1 & 1 & 3 \end{vmatrix} = 0, \quad \begin{vmatrix} 2 & 3 & 4 \\ 0 & -1 & 2 \\ 1 & 0 & 5 \end{vmatrix} = 0, \quad \begin{vmatrix} 2 & 3 & 4 \\ 1 & 1 & 3 \\ 1 & 0 & 5 \end{vmatrix} = 0, \quad \begin{vmatrix} 0 & -1 & 2 \\ 1 & 1 & 3 \\ 1 & 0 & 5 \end{vmatrix} = 0.$$

Therefore rank(**A**) < 3 so that it may be either 2 or 1. If we find any nonsingular 2×2 submatrix of **A**, we conclude that the rank is 2. That said,

$$\begin{vmatrix} 2 & 3 \\ 0 & -1 \end{vmatrix} = -2 \neq 0 \quad \Rightarrow \quad \text{rank}(\mathbf{A}) = 2.$$

b. ◆
```
>> A = [2 3 4;0 -1 2;1 1 3;1 0 5];
>> rref(A)
ans =
     1   0    5
     0   1   -2
     0   0    0
     0   0    0        % See remarks in the solution of Example 3.2
>> rank(A)
ans =
     2
```

3.1.4 Block Diagonal and Block Triangular Matrices

A block diagonal matrix is defined as a square matrix partitioned such that its diagonal elements are square matrices, while all other elements are zeros (Figure 3.4a). Similarly, we define a block triangular matrix as a square matrix partitioned in such a way that its diagonal elements are square blocks, while all entries either above or below this main block diagonal are zeros (Figure 3.4b and c). It is interesting to note that many properties of these special block matrices basically generalize those of diagonal and triangular matrices. In particular, the determinant of each of these matrices is equal to the product of the individual determinants of the blocks along the main diagonal. Consequently, a block diagonal (or triangular) matrix is singular if and only if one of the blocks along the main diagonal is singular. We also mention that the rank of a block triangular matrix is at least equal to the sum of the ranks of the individual diagonal blocks.

Matrix Analysis

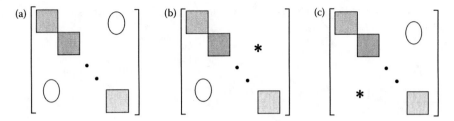

FIGURE 3.4 (a) Block diagonal matrix, (b) block upper-triangular matrix, and (c) block lower-triangular matrix.

Example 3.5: Block Diagonal Matrix

Evaluate det(**A**) and rank(**A**) where

$$\mathbf{A} = \begin{bmatrix} 1 & 3 & 0 & 0 \\ -2 & 4 & 0 & 0 \\ 0 & 0 & 5 & 0 \\ 0 & 0 & 0 & 2 \end{bmatrix}.$$

Solution

This matrix can be thought of as a block diagonal matrix in various ways, two of which are listed in Figure 3.5. In Figure 3.5a, the 2×2 block in the upper-left corner has a determinant of 10 and a rank of 2, while the other two blocks (single entries) are 5 and 2, each with a rank of 1. Therefore, $|\mathbf{A}| = (10)(5)(2) = 100$ and rank(**A**) $= 2 + 1 + 1 = 4$. In Figure 3.5b, the two 2×2 blocks each have a determinant of 10 and a rank of 2; thus the determinant of the matrix is $(10)(10) = 100$ and its rank is 4, as expected.

3.1.5 Cramer's Rule

Consider a linear system of n equations in n unknowns, $x_1, x_2, \ldots, x_n$, in the general form of

$$\begin{cases} a_{11}x_1 + a_{12}x_2 + \cdots + a_{1n}x_n = b_1 \\ a_{21}x_1 + a_{22}x_2 + \cdots + a_{2n}x_n = b_2 \\ \quad \vdots \\ a_{n1}x_1 + a_{n2}x_2 + \cdots + a_{nn}x_n = b_n \end{cases} \quad \text{or in matrix form} \quad \mathbf{A}\mathbf{x} = \mathbf{b}. \tag{3.2}$$

If the coefficient matrix **A** is nonsingular, then each unknown x_i is uniquely determined by

$$x_i = \frac{\Delta_i}{\Delta}, \quad i = 1, 2, \ldots, n, \tag{3.3}$$

where $\Delta = |\mathbf{A}| \neq 0$ and Δ_i is obtained by replacing the ith column in Δ by the vector **b** (Figure 3.6).

FIGURE 3.5 Example 3.5. (a) Block matrix with three blocks and (b) block matrix with two blocks.

Example 3.6: Cramer's Rule

a. Solve the following system using Cramer's rule.
b. Solve in MATLAB using the "`solve`" command.

$$\begin{bmatrix} 2 & 3 & -1 \\ -1 & 2 & 1 \\ 1 & -3 & -2 \end{bmatrix} \begin{Bmatrix} x_1 \\ x_2 \\ x_3 \end{Bmatrix} = \begin{Bmatrix} 1 \\ 8 \\ -13 \end{Bmatrix}.$$

Solution

a. The determinants involved are Δ, Δ_1, Δ_2, and Δ_3. The determinant of the coefficient matrix is

$$\Delta = \begin{vmatrix} 2 & 3 & -1 \\ -1 & 2 & 1 \\ 1 & -3 & -2 \end{vmatrix} = -6.$$

Since $\Delta \neq 0$, we proceed by evaluating Δ_1, Δ_2, and Δ_3 as

$$\Delta_1 = \begin{vmatrix} 1 & 3 & -1 \\ 8 & 2 & 1 \\ -13 & -3 & -2 \end{vmatrix} = 6, \quad \Delta_2 = \begin{vmatrix} 2 & 1 & -1 \\ -1 & 8 & 1 \\ 1 & -13 & -2 \end{vmatrix} = -12, \quad \Delta_3 = \begin{vmatrix} 2 & 3 & 1 \\ -1 & 2 & 8 \\ 1 & -3 & -13 \end{vmatrix} = -18.$$

Subsequently,

$$x_1 = \frac{\Delta_1}{\Delta} = \frac{6}{-6} = -1, \quad x_2 = \frac{\Delta_2}{\Delta} = \frac{-12}{-6} = 2, \quad x_3 = \frac{\Delta_3}{\Delta} = \frac{-18}{-6} = 3.$$

FIGURE 3.6 Matrices involved in Cramer's rule.

b.
```
>> [x1,x2,x3] =solve('2*x1+3*x2-x3=1','-x1+2*x2+x3=8','x1
-3*x2-2*x3=-13')
x1 =
-1
x2 =
2
x3 =
3
```

3.1.5.1 Homogeneous Systems

Consider the homogeneous system $\mathbf{Ax} = \mathbf{0}$ with $\Delta = |\mathbf{A}| \neq 0$. Because $\mathbf{b} = \mathbf{0}$, each Δ_i contains a zero column, making $\Delta_i = 0$ for $i = 1, 2, \ldots, n$. Consequently, the only solution is the trivial solution, that is, $\mathbf{x} = \mathbf{0}$. In order for $\mathbf{Ax} = \mathbf{0}$ to have a nontrivial solution, the coefficient matrix must be singular, that is, $|\mathbf{A}| = 0$. This indicates that the equations are linearly dependent, causing at least one free variable to be present, and thereby infinitely many solutions are generated.

3.1.6 Inverse of a Matrix

The inverse of a square matrix $\mathbf{A}_{n \times n}$ is denoted by $\mathbf{A}^{-1}$ with the property $\mathbf{A}\mathbf{A}^{-1} = \mathbf{I} = \mathbf{A}^{-1}\mathbf{A}$, where $\mathbf{I}$ is the $n \times n$ identity matrix. The inverse of $\mathbf{A}$ exists only if $|\mathbf{A}| \neq 0$ and is obtained using the adjoint matrix of $\mathbf{A}$, denoted by adj($\mathbf{A}$).

3.1.6.1 Adjoint Matrix

If $\mathbf{A} = [a_{ij}]_{n \times n}$, then the adjoint of $\mathbf{A}$ is defined as

$$\text{adj}(\mathbf{A}) = \begin{bmatrix} (-1)^{1+1}M_{11} & (-1)^{2+1}M_{21} & \cdots & (-1)^{n+1}M_{n1} \\ (-1)^{1+2}M_{12} & (-1)^{2+2}M_{22} & \cdots & (-1)^{n+2}M_{n2} \\ \cdot & \cdot & & \cdot \\ \cdot & \cdot & & \cdot \\ (-1)^{1+n}M_{1n} & (-1)^{2+n}M_{2n} & \cdots & (-1)^{n+n}M_{nn} \end{bmatrix} = \begin{bmatrix} C_{11} & C_{21} & \cdots & C_{n1} \\ C_{12} & C_{22} & \cdots & C_{n2} \\ \cdot & \cdot & & \cdot \\ \cdot & \cdot & & \cdot \\ C_{1n} & C_{2n} & \cdots & C_{nn} \end{bmatrix},$$
(3.4)

where M_{ij} is the minor of a_{ij} and $C_{ij} = (-1)^{i+j}M_{ij}$ is the cofactor of a_{ij}. Note that each minor M_{ij} (or cofactor C_{ij}) occupies the (j, i) position in the adjoint matrix, the opposite of what one would normally expect. Then,

$$\mathbf{A}^{-1} = \frac{1}{|\mathbf{A}|} \text{adj}(\mathbf{A}).$$
(3.5)

Example 3.7: Inverse

a. Find the inverse of $\mathbf{A} = \begin{bmatrix} 3 & 1 & 0 \\ 1 & -1 & 2 \\ 1 & 1 & 1 \end{bmatrix}$.

b. Confirm in MATLAB.

Solution

a. We first calculate $|\mathbf{A}| = -8$. Next, minors and cofactors are computed as

$$M_{11} = \begin{vmatrix} -1 & 2 \\ 1 & 1 \end{vmatrix} = -3, \quad C_{11} = (-1)^{1+1} M_{11} = -3, \quad M_{12} = \begin{vmatrix} 1 & 2 \\ 1 & 1 \end{vmatrix} = -1, \quad C_{12} = 1$$

$$M_{13} = \begin{vmatrix} 1 & -1 \\ 1 & 1 \end{vmatrix} = 2, \quad C_{13} = 2, \quad M_{21} = \begin{vmatrix} 1 & 0 \\ 1 & 1 \end{vmatrix} = 1, \quad C_{21} = -1, \quad M_{22} = \begin{vmatrix} 3 & 0 \\ 1 & 1 \end{vmatrix} = 3$$

$$C_{22} = 3, \quad M_{23} = \begin{vmatrix} 3 & 1 \\ 1 & 1 \end{vmatrix} = 2, \quad C_{23} = -2, \quad M_{31} = \begin{vmatrix} 1 & 0 \\ -1 & 2 \end{vmatrix} = 2, \quad C_{31} = 2$$

$$M_{32} = \begin{vmatrix} 3 & 0 \\ 1 & 2 \end{vmatrix} = 6, \quad C_{32} = -6, \quad M_{33} = \begin{vmatrix} 3 & 1 \\ 1 & -1 \end{vmatrix} = -4, \quad C_{33} = -4.$$

By Equations 3.4 and 3.5, we find

$$\text{adj}(\mathbf{A}) = \begin{bmatrix} C_{11} & C_{21} & C_{31} \\ C_{12} & C_{22} & C_{32} \\ C_{13} & C_{23} & C_{33} \end{bmatrix} = \begin{bmatrix} -3 & -1 & 2 \\ 1 & 3 & -6 \\ 2 & -2 & -4 \end{bmatrix}$$

$$\Rightarrow \mathbf{A}^{-1} = \frac{1}{-8} \begin{bmatrix} -3 & -1 & 2 \\ 1 & 3 & -6 \\ 2 & -2 & -4 \end{bmatrix} = \begin{bmatrix} \frac{3}{8} & \frac{1}{8} & -\frac{1}{4} \\ -\frac{1}{8} & -\frac{3}{8} & \frac{3}{4} \\ -\frac{1}{4} & \frac{1}{4} & \frac{1}{2} \end{bmatrix}.$$

b.
```
>> A = [3 1 0;1 -1 2;1 1 1];
>> inv(A)
ans =
    0.3750    0.1250   -0.2500
   -0.1250   -0.3750    0.7500
   -0.2500    0.2500    0.5000
```

Example 3.8: Inverse of a Symbolic Matrix

a. Determine $(s\mathbf{I} - \mathbf{A})^{-1}$, where s is the Laplace variable, $\mathbf{I}$ is the 2×2 identity matrix, and

$$\mathbf{A} = \begin{bmatrix} 0 & 1 \\ -4 & -5 \end{bmatrix}.$$

b. Repeat in MATLAB.

Solution

a. First

$$s\mathbf{I} - \mathbf{A} = \begin{bmatrix} s & 0 \\ 0 & s \end{bmatrix} - \begin{bmatrix} 0 & 1 \\ -4 & -5 \end{bmatrix} = \begin{bmatrix} s & -1 \\ 4 & s+5 \end{bmatrix}$$

$$\Rightarrow |s\mathbf{I} - \mathbf{A}| = s^2 + 5s + 4 = (s+1)(s+4).$$

Matrix Analysis

Then by Equation 3.5,

$$(s\mathbf{I} - \mathbf{A})^{-1} = \frac{\text{adj}(s\mathbf{I} - \mathbf{A})}{|s\mathbf{I} - \mathbf{A}|} = \frac{1}{(s+1)(s+4)}\begin{bmatrix} s+5 & 1 \\ -4 & s \end{bmatrix} = \begin{bmatrix} \dfrac{s+5}{(s+1)(s+4)} & \dfrac{1}{(s+1)(s+4)} \\ \dfrac{-4}{(s+1)(s+4)} & \dfrac{s}{(s+1)(s+4)} \end{bmatrix}.$$

b.
```
>> syms s
>> A = [0 1;-4 -5];
>> inv(s*eye(2) - A)
ans =
  [ (s+5)/(s^2+5*s+4), 1/(s^2+5*s+4)]
  [-4/(s^2+5*s+4), s/(s^2+5*s+4)]
```

Example 3.9: System of Equations

Solve the system described in Example 3.6 using the inverse of the coefficient matrix.

Solution

```
>> A = [2 3 -1;-1 2 1;1 -3 -2];
>> b = [1;8;-13];
>> x = A\b    % backslash is preferred to inv(A)*b
x =
   -1.0000
    2.0000
    3.0000
```

3.1.6.2 Properties of Inverse

- $(\mathbf{A}^{-1})^{-1} = \mathbf{A}$
- $(\mathbf{A}^T)^{-1} = (\mathbf{A}^{-1})^T$
- $(\mathbf{A}^p)^{-1} = (\mathbf{A}^{-1})^p$, p is a positive integer
- $\det(\mathbf{A}^{-1}) = 1/\det(\mathbf{A})$
- $(\mathbf{AB})^{-1} = \mathbf{B}^{-1}\mathbf{A}^{-1}$
- The inverse of a symmetric matrix is symmetric

- $$\begin{bmatrix} a_{11} & 0 & 0 \\ & a_{22} & \\ 0 & & 0 \\ & \cdot & \\ 0 & 0 & a_{nn} \end{bmatrix}^{-1} = \begin{bmatrix} 1/a_{11} & 0 & 0 \\ & 1/a_{22} & \\ 0 & & 0 \\ & \cdot & \\ 0 & 0 & 1/a_{nn} \end{bmatrix}$$

PROBLEM SET 3.1

In Problems 1 through 6, perform the indicated operations, if defined, and identify those that are undefined.

$$\mathbf{A} = \begin{bmatrix} 1 & -1 & 2 \\ 0 & -2 & 3 \\ 2 & 1 & 4 \end{bmatrix}, \quad \mathbf{B} = \begin{bmatrix} 1 & 0 \\ 0 & -1 \\ 0 & 1 \end{bmatrix}, \quad \mathbf{a} = \begin{bmatrix} 3 \\ 1 \\ 0 \end{bmatrix}, \quad \mathbf{b} = \begin{bmatrix} 2 \\ -1 \end{bmatrix}.$$

1. $\mathbf{B}^T\mathbf{a}$, $\mathbf{Aa}$, $\mathbf{Ab}$
2. $\mathbf{a}^T\mathbf{A}$, $\mathbf{b}^T\mathbf{B}^T$
3. $\mathbf{Bb}$, $\mathbf{A}^T\mathbf{a}$, $\mathbf{A}^T\mathbf{B}$
4. $\mathbf{B}^T\mathbf{A}$, $\mathbf{BA}$, $\mathbf{ab}^T$
5. $\mathbf{Aa}+\mathbf{Bb}$, $(\mathbf{Aa})^T + \mathbf{b}^T\mathbf{B}^T$
6. $(\mathbf{AB})^T$, $\mathbf{B}^T\mathbf{A}^T$, $\mathbf{A}^T\mathbf{B}^T$

For each given matrix,

(a) Find the row-echelon form and use it to determine the rank.
(b) ◀ Repeat in MATLAB.

7. $\mathbf{A} = \begin{bmatrix} 3 & -1 \\ 1 & 4 \end{bmatrix}$

8. $\mathbf{A} = \begin{bmatrix} 2 & -1 & 1 \\ -1 & 3 & 2 \\ 4 & -2 & 0 \end{bmatrix}$

9. $\mathbf{A} = \begin{bmatrix} 1 & -1 & 1 \\ 4 & 1 & -2 \\ 1 & 4 & -5 \end{bmatrix}$

10. $\mathbf{A} = \begin{bmatrix} -3 & 1 & -2 \\ -1 & 2 & 1 \\ 2 & -1 & 1 \end{bmatrix}$

11. $\mathbf{A} = \begin{bmatrix} 4 & -2 & 0 & 1 \\ 1 & -1 & -1 & 0 \\ 0 & 2 & 4 & 1 \\ 1 & 1 & 3 & 1 \end{bmatrix}$

12. $\mathbf{A} = \begin{bmatrix} -1 & 1 & 2 & 0 \\ 3 & 2 & -1 & 1 \\ 2 & 3 & 1 & 3 \\ 1 & 1 & 3 & 3 \end{bmatrix}$

Evaluate each determinant.

13. $\begin{vmatrix} 1 & 2 & 3 & 5 \\ -2 & 2 & 1 & 4 \\ 3 & 1 & 0 & -1 \\ 2 & 1 & -3 & 1 \end{vmatrix}$

14. $\begin{vmatrix} a & -1 & a & 2 \\ 0 & a+1 & 0 & 1 \\ 0 & 0 & 3 & -2 \\ 0 & 0 & 0 & 2a \end{vmatrix}$, a is a parameter

15. $\begin{vmatrix} s+2 & 1 & -1 \\ 0 & s+1 & 2 \\ -2 & 2 & s \end{vmatrix}$, s is a parameter,

16. $\begin{vmatrix} -1 & 2 & 3 & 5 \\ 0 & 0 & 1 & 4 \\ 0 & 1 & 5 & -1 \\ 0 & 0 & 0 & 1 \end{vmatrix}$

Solve each system of equations

a. Using Cramer's rule.
b. ◢ Using the "`solve`" command in MATLAB.

17. $\begin{bmatrix} 2 & 0 & -1 \\ 1 & 1 & -2 \\ 3 & -1 & 2 \end{bmatrix} \begin{Bmatrix} x_1 \\ x_2 \\ x_3 \end{Bmatrix} = \begin{Bmatrix} 3 \\ 3 \\ 1 \end{Bmatrix}$

18. $\begin{bmatrix} 3 & -1 & 1 \\ 2 & 0 & -2 \\ 1 & -2 & 3 \end{bmatrix} \begin{Bmatrix} x_1 \\ x_2 \\ x_3 \end{Bmatrix} = \begin{Bmatrix} -5 \\ -10 \\ 3 \end{Bmatrix}$

19. $\begin{cases} (s+1)X_1 - X_2 = 0 \\ -2X_1 + (s+2)X_2 = 1/s, \end{cases}$ s is a parameter.

20. $\begin{cases} x + 2y - 3z = 3 \\ -2x + y = 5 \\ x + 4y - 2z = 4 \end{cases}$

Find the inverse of each given matrix.

21. $\mathbf{A} = \begin{bmatrix} 4 & 0 & 1 \\ 0 & -3 & 2 \\ 1 & 2 & 1 \end{bmatrix}$

22. $\mathbf{A} = \begin{bmatrix} 2 & 0 & -3 \\ 0 & 1 & 2 \\ 0 & 0 & 5 \end{bmatrix}$ Comment on the structure of $\mathbf{A}^{-1}$.

23. $\mathbf{A} = \begin{bmatrix} -1 & 0 & 0 \\ 0 & 5 & 0 \\ 4 & 3 & -2 \end{bmatrix}$ Comment on the structure of $\mathbf{A}^{-1}$.

24. $\mathbf{A} = \begin{bmatrix} s & 0 & -1 \\ 0 & s+1 & 2 \\ 1 & 0 & s+2 \end{bmatrix}$, s is a parameter

25. $\mathbf{A} = \begin{bmatrix} a+1 & 0 & 0 \\ 0 & 2a & 0 \\ 0 & 0 & 2(a+1) \end{bmatrix}$, a is a parameter $\neq 0, -1$

26. $\mathbf{A} = \begin{bmatrix} 0 & 1 & 0 \\ 0 & 0 & 1 \\ -1 & -2 & -1 \end{bmatrix}$

Solve each system of equations $\mathbf{A}\mathbf{x} = \mathbf{b}$

a. Using the inverse of the coefficient matrix.
b. Using the "\" command in MATLAB.

27. $\begin{bmatrix} -1 & 2 & -1 \\ 3 & 0 & 2 \\ 2 & 1 & -2 \end{bmatrix} \mathbf{x} = \begin{bmatrix} 0 \\ a \\ 4a \end{bmatrix}$, a is a parameter

28. $\begin{bmatrix} 2 & 4 & -1 \\ 1 & 3 & 1 \\ -1 & 2 & 3 \end{bmatrix} \mathbf{x} = \begin{bmatrix} 0 \\ 2 \\ 1 \end{bmatrix}$

29. $\mathbf{A} = \begin{bmatrix} m & 1 & -2 \\ -1 & 2m & 1 \\ 0 & 1 & 3m \end{bmatrix}$, $\mathbf{b} = \begin{bmatrix} m \\ 4m \\ 3m+2 \end{bmatrix}$, m is a parameter $\neq 0$

30. $\mathbf{A} = \begin{bmatrix} 2 & 1 & -1 \\ 0 & 3 & 1 \\ 0 & 0 & 4 \end{bmatrix}$, $\mathbf{b} = \begin{bmatrix} 1 \\ 2a+3 \\ 8a \end{bmatrix}$, a is a parameter

3.2 MATRIX EIGENVALUE PROBLEM

Let $\mathbf{A}$ be an $n \times n$ matrix, $\mathbf{v}$ a nonzero $n \times 1$ vector, and λ a number (complex in general). Then, the eigenvalue problem for matrix $\mathbf{A}$ is formulated as

$$\mathbf{A}\mathbf{v} = \lambda \mathbf{v}, \quad \mathbf{v} \neq \mathbf{0}. \tag{3.6}$$

A number λ for which Equation 3.6 has a nontrivial solution ($\mathbf{v} \neq \mathbf{0}_{n \times 1}$) is called an eigenvalue or characteristic value of $\mathbf{A}$. The corresponding solution $\mathbf{v} \neq \mathbf{0}_{n \times 1}$ of Equation 3.6 is the eigenvector or characteristic vector of $\mathbf{A}$ corresponding to λ. The set of all eigenvalues of $\mathbf{A}$ is denoted by $\lambda(\mathbf{A})$ and known as the spectrum of $\mathbf{A}$.

3.2.1 Solving the Eigenvalue Problem

Rewrite Equation 3.6 as

$$\mathbf{A}\mathbf{v} - \lambda \mathbf{v} = \mathbf{0}_{n \times 1}.$$

Note that every single term here is an $n \times 1$ vector. On the left-hand side, both terms contain vector $\mathbf{v}$. While in the second term λ and $\mathbf{v}$ commute, the same is not true with $\mathbf{A}$ and $\mathbf{v}$ in the first term. Therefore, we can only factor out $\mathbf{v}$ from the right to obtain

$$(\mathbf{A} - \lambda \mathbf{I})\mathbf{v} = \mathbf{0}. \tag{3.7}$$

The identity matrix $\mathbf{I}$ has been inserted so that the two terms in parentheses are compatible. Equation 3.7 has a nontrivial solution ($\mathbf{v} \neq \mathbf{0}$) if and only if the coefficient matrix $\mathbf{A} - \lambda \mathbf{I}$ is singular (see earlier remarks on homogeneous systems under Cramer's rule). Thus

$$|\mathbf{A} - \lambda \mathbf{I}| = 0.$$

This is called the characteristic equation of $\mathbf{A}$. The determinant $|\mathbf{A} - \lambda \mathbf{I}|$ is an nth-degree polynomial in λ and is known as the characteristic polynomial of $\mathbf{A}$ whose roots are precisely the eigenvalues of $\mathbf{A}$. For each eigenvalue we find the corresponding eigenvector by solving Equation 3.7. Since $\mathbf{A} - \lambda \mathbf{I}$ is singular, it has at least one row dependent on the other rows, which means the row-echelon form of $\mathbf{A} - \lambda \mathbf{I}$ will have at least one zero row. Therefore, for each fixed λ, Equation 3.7 has infinitely many solutions. A basis of solutions will then represent all eigenvectors associated with each λ.

3.2.2 Eigenvalue Properties of Special Matrices

- Eigenvalues of upper- and lower-triangular and diagonal matrices are the entries along the main diagonal of the matrix.
- All eigenvalues of a symmetric matrix are real.
- Eigenvalues of a skew-symmetric matrix are either zero or pure imaginary.
- A matrix $\mathbf{A}$ is orthogonal if $\mathbf{A}^T = \mathbf{A}^{-1}$. All eigenvalues of an orthogonal matrix have absolute values of 1.
- The eigenvalues of block diagonal and block triangular matrices are the eigenvalues of block matrices along the diagonal.

Example 3.10: Eigenvalue Problem

a. Find all eigenvalues and eigenvectors of

$$\mathbf{A} = \begin{bmatrix} 1 & 0 & 1 \\ 0 & 1 & 0 \\ 1 & 0 & 1 \end{bmatrix}.$$

b. Repeat in MATLAB.

Solution

a. The characteristic equation yields

$$|\mathbf{A} - \lambda \mathbf{I}| = \begin{vmatrix} 1-\lambda & 0 & 1 \\ 0 & 1-\lambda & 0 \\ 1 & 0 & 1-\lambda \end{vmatrix} = (1-\lambda)^3 - (1-\lambda)$$

$$= \lambda(\lambda - 1)(\lambda - 2) = 0 \Rightarrow \lambda_{1,2,3} = 0, 1, 2.$$

For $\lambda_1 = 0$, we solve

$$(\mathbf{A} - \lambda_1 \mathbf{I})v_1 = \mathbf{0} \Rightarrow \mathbf{A}v_1 = \mathbf{0} \Rightarrow \begin{bmatrix} 1 & 0 & 1 \\ 0 & 1 & 0 \\ 1 & 0 & 1 \end{bmatrix} \begin{bmatrix} a \\ b \\ c \end{bmatrix} = \begin{bmatrix} 0 \\ 0 \\ 0 \end{bmatrix} \xRightarrow[\text{row operations}]{\text{Elementary}} \begin{bmatrix} 1 & 0 & 1 \\ 0 & 1 & 0 \\ 0 & 0 & 0 \end{bmatrix} \begin{bmatrix} a \\ b \\ c \end{bmatrix} = \begin{bmatrix} 0 \\ 0 \\ 0 \end{bmatrix}.$$

The second row gives $b = 0$. The first row yields $a + c = 0$. The last row of zeros suggests that a free variable exists, which can be either a or c. Choosing $a = 1$ results in $c = -1$ and

$$v_1 = \begin{bmatrix} 1 \\ 0 \\ -1 \end{bmatrix}.$$

For $\lambda_2 = 1$, we solve

$$(\mathbf{A} - \lambda_2 \mathbf{I})v_2 = \mathbf{0} \Rightarrow (\mathbf{A} - \mathbf{I})v_2 = \mathbf{0} \Rightarrow \begin{bmatrix} 0 & 0 & 1 \\ 0 & 0 & 0 \\ 1 & 0 & 0 \end{bmatrix} \begin{bmatrix} a \\ b \\ c \end{bmatrix} = \begin{bmatrix} 0 \\ 0 \\ 0 \end{bmatrix} \Rightarrow \begin{array}{l} c = 0 \\ b = \text{free} = 1. \\ a = 0 \end{array}$$

Therefore

$$v_2 = \begin{bmatrix} 0 \\ 1 \\ 0 \end{bmatrix}.$$

For $\lambda_3 = 2$, we solve

$$(\mathbf{A} - \lambda_3\mathbf{I})\mathbf{v}_3 = \mathbf{0} \Rightarrow (\mathbf{A} - 2\mathbf{I})\mathbf{v}_3 = \mathbf{0} \Rightarrow \begin{bmatrix} -1 & 0 & 1 \\ 0 & -1 & 0 \\ 1 & 0 & -1 \end{bmatrix} \begin{bmatrix} a \\ b \\ c \end{bmatrix} = \begin{bmatrix} 0 \\ 0 \\ 0 \end{bmatrix}$$

$$\overset{\text{EROs}}{\Rightarrow} \begin{bmatrix} -1 & 0 & 1 \\ 0 & -1 & 0 \\ 0 & 0 & 0 \end{bmatrix} \begin{bmatrix} a \\ b \\ c \end{bmatrix} = \begin{bmatrix} 0 \\ 0 \\ 0 \end{bmatrix}.$$

The second row gives $b = 0$ while the first row yields $a = c$. The last row suggests a free variable, which can be either a or c. Choosing $a = 1$ results in $c = 1$, and

$$\mathbf{v}_3 = \begin{bmatrix} 1 \\ 0 \\ 1 \end{bmatrix}.$$

b.
```
>> A = [1 0 1;0 1 0;1 0 1];
```

In order to find the eigenvalues as well as the eigenvectors, we use the command "`eig`" in the form of `[V,D] = eig(A)` so that matrix V contains the eigenvectors (normalized as unit vectors) in its columns and diagonal matrix D has the eigenvalues along its main diagonal. The eigenvalue in the (1, 1) entry of D corresponds to the eigenvector in the first column of V, and so on.

```
>> [V,D] = eig(A)
V =
    0.7071         0    0.7071
         0   -1.0000         0
   -0.7071         0    0.7071
D =
    0    0    0
    0    1    0
    0    0    2
```

As mentioned above, the eigenvectors are normalized to unit vectors by dividing the vector by its length (norm). For instance, eigenvector $\mathbf{v}_1$ in Part (a) has length $\sqrt{2}$ and

$$\frac{1}{\sqrt{2}}\mathbf{v}_1 = \frac{1}{\sqrt{2}} \begin{bmatrix} 1 \\ 0 \\ -1 \end{bmatrix} = \begin{bmatrix} 0.7071 \\ 0 \\ -0.7071 \end{bmatrix}.$$

This is returned by the "`eig`" command in MATLAB. Also note that $\mathbf{v}_2$ returned by "`eig`" matches the one in Part (a)—except for a negative multiple—because it is a unit vector to begin with. This, of course, is not a concern because they both have the same basis.

3.2.2.1 Algebraic Multiplicity and Geometric Multiplicity

The algebraic multiplicity (AM) of an eigenvalue is the number of times it occurs. Its geometric multiplicity (GM) is the number of linearly independent eigenvectors associated with it. For instance, in Example 3.10 each of the three eigenvalues has an AM of 1 since each occurs only once, and a GM of 1 because there is only one independent eigenvector for each. In general, we have GM $\leq$ AM. Therefore, any eigenvalue with an AM of 1 automatically has a GM of 1.

Example 3.11: AM and GM

Find all eigenvalues and eigenvectors of

$$A = \begin{bmatrix} 1 & 0 & -3 \\ 0 & 1 & 2 \\ 0 & 0 & -2 \end{bmatrix}.$$

SOLUTION

Since A is upper-triangular, $\lambda(A) = 1, 1, -2$ so that $\lambda = 1$ has an AM of 2 and $\lambda = -2$ has an AM of 1. For $\lambda = 1$, we solve

$$(A - I)v = 0 \Rightarrow \begin{bmatrix} 0 & 0 & -3 \\ 0 & 0 & 2 \\ 0 & 0 & -3 \end{bmatrix} \begin{bmatrix} a \\ b \\ c \end{bmatrix} = \begin{bmatrix} 0 \\ 0 \\ 0 \end{bmatrix} \xrightarrow[\text{row operations}]{\text{Elementary}} \begin{bmatrix} 0 & 0 & 1 \\ 0 & 0 & 0 \\ 0 & 0 & 0 \end{bmatrix} \begin{bmatrix} a \\ b \\ c \end{bmatrix} = \begin{bmatrix} 0 \\ 0 \\ 0 \end{bmatrix}.$$

The first row gives $c = 0$. Two zero rows indicate two free variables (a and b here) so that two linearly independent eigenvectors can be obtained. Letting $a = 1$, $b = 0$ and $a = 0$, $b = 1$ yields

$$\begin{bmatrix} 1 \\ 0 \\ 0 \end{bmatrix}, \quad \begin{bmatrix} 0 \\ 1 \\ 0 \end{bmatrix}.$$

Therefore, $\lambda = 1$ has a GM of 2. For $\lambda = -2$, the only independent eigenvector is obtained as

$$\begin{bmatrix} 3 \\ -2 \\ 3 \end{bmatrix}.$$

Thus $\lambda = -2$ has a GM of 1.

3.2.2.2 Generalized Eigenvectors

Suppose that the AM of a certain eigenvalue λ is m so that m corresponding eigenvectors are expected for λ. However, if the GM of λ is $k < m$, then only k linearly independent eigenvectors will be generated for λ. This implies that there are $m - k$ missing eigenvectors. These are referred to as generalized eigenvectors and can be found using a systematic approach (see Ref. [6]). Any matrix with a generalized eigenvector is called defective. The following example illustrates how defective matrices can be handled in MATLAB.

Example 3.12: Generalized Eigenvectors

Find the eigenvalues and eigenvectors of

$$A = \begin{bmatrix} 1 & -1 & -3 \\ 0 & 4 & 0 \\ -3 & 1 & 1 \end{bmatrix}.$$

Solution

```
>> A = [1 -1 -3;0 4 0;-3 1 1];
>> [V,D] = eig(A)
V =
    0.7071    0.7071    0.7071
         0         0    0.0000
   -0.7071    0.7071   -0.7071
D =
    4.0000         0         0
         0   -2.0000         0
         0         0    4.0000
```

Since the first and third columns of V are the same eigenvector, we conclude that there is a generalized eigenvector. In order to find this missing eigenvector, we switch from "`eig`" to "`jordan`," which is designed for this purpose.

```
>> [V,D] = jordan(A)
V =
    0.1667   -1.0000    0.1667
         0         0    1.0000
    0.1667    1.0000   -0.1667
D =
    -2     0     0
     0     4     1
     0     0     4
```

Note that matrix D is no longer diagonal and is known as a Jordan matrix. The eigenvalues of **A** are $-2, 4, 4$. The first column of V contains the eigenvector for $\lambda = -2$. The next two columns correspond to $\lambda = 4$, with the last one being a generalized eigenvector.

3.2.3 Similarity Transformations

Consider an $n \times n$ matrix **A** and a nonsingular $n \times n$ matrix **S**, and suppose

$$\mathbf{S}^{-1}\mathbf{A}\mathbf{S} = \mathbf{B}. \tag{3.8}$$

We say **A** has been transformed into **B** through a similarity transformation. One of the main features here is that **A** and **B** have the same eigenvalues. Similarity transformations are used to transform a matrix into a diagonal matrix. And eigenvectors play a central role in that process.

3.2.3.1 Matrix Diagonalization

Suppose $\mathbf{A}_{n \times n}$ has eigenvalues $\lambda_1, \ldots, \lambda_n$ and linearly independent eigenvectors $v_1, \ldots, v_n$ with no generalized eigenvectors. Form the modal matrix $\mathbf{V}_{n \times n} = [v_1 \cdots v_n]$, which is guaranteed to be nonsingular because its columns are linearly independent (see Section 3.1.3.1). Then,

$$\mathbf{V}^{-1}\mathbf{A}\mathbf{V} = \mathbf{D} = \begin{bmatrix} \lambda_1 & 0 & 0 \\ & \cdot & 0 \\ 0 & \cdot & \\ 0 & 0 & \lambda_n \end{bmatrix}. \tag{3.9}$$

Matrix **A** has clearly been transformed into a diagonal matrix **D** by a similarity transformation.

3.2.3.2 Defective Matrices

Suppose $\mathbf{A}_{n\times n}$ has eigenvalues $\lambda_1,\ldots,\lambda_n$ and linearly independent eigenvectors $\boldsymbol{v}_1,\ldots,\boldsymbol{v}_n$ including at least one generalized eigenvector. Again, the modal matrix $\mathbf{V}_{n\times n} = [\boldsymbol{v}_1 \cdots \boldsymbol{v}_n]$ is guaranteed to be nonsingular because its columns are linearly independent, and

$$\mathbf{V}^{-1}\mathbf{AV} = \mathbf{D}, \tag{3.10}$$

where $\mathbf{D}$ is not diagonal and is called a Jordan matrix (see Example 3.12).

Example 3.13: Diagonalization

In Example 3.10 we found $\lambda(\mathbf{A}) = 0,1,2$ and no generalized eigenvectors. The modal matrix then transforms $\mathbf{A}$ into the diagonal matrix $\mathbf{D}$ as in Equation 3.9.

```
>>  V\A*V      % Perform V⁻¹AV
ans =
      0    0    0
      0    1    0
      0    0    2     % Diagonal matrix D comprised of the eigenvalues of A
```

Example 3.14: Jordan Matrix

In Example 3.12 we found $\lambda(\mathbf{A}) = -2,4,4$ and one generalized eigenvector. The modal matrix transforms $\mathbf{A}$ into a Jordan matrix $\mathbf{D}$ as in Equation 3.10.

```
>>  V\A*V
ans =
     -2    0    0
      0    4    1
      0    0    4     % Jordan matrix
```

PROBLEM SET 3.2

For each given matrix,

a. Find the eigenvalues and eigenvectors.
b. Repeat in MATLAB.

1. $\mathbf{A} = \begin{bmatrix} 1 & 3 \\ -4 & -6 \end{bmatrix}$

2. $\mathbf{A} = \begin{bmatrix} 0 & 4 \\ 1 & 3 \end{bmatrix}$

3. $\mathbf{A} = \begin{bmatrix} 0 & a \\ a & 0 \end{bmatrix}$, a is a parameter

4. $\mathbf{A} = \begin{bmatrix} 1 & 0 \\ -2 & -3 \end{bmatrix}$

5. $\mathbf{A} = \begin{bmatrix} 1 & 1 & 0 \\ 2 & 2 & 0 \\ 0 & 0 & 1 \end{bmatrix}$

6. $\mathbf{A} = \begin{bmatrix} 1 & 3 & 1 \\ 0 & 2 & -3 \\ 0 & 0 & -1 \end{bmatrix}$

7. $\mathbf{A} = \begin{bmatrix} 1 & 0 & 0 \\ 1 & 2 & 0 \\ 2 & 3 & 4 \end{bmatrix}$

8. $\mathbf{A} = \begin{bmatrix} 1 & 0 & 0 \\ 0 & -1 & -3 \\ 0 & 4 & 6 \end{bmatrix}$

For each matrix $\mathbf{A}$, find the eigenvalues, eigenvectors, AM, and GM of each eigenvalue, and decide whether $\mathbf{A}$ is defective or not. Using an appropriate modal matrix, transform $\mathbf{A}$ into either a diagonal or a Jordan matrix, whichever is applicable.

9. $\mathbf{A} = \begin{bmatrix} 1 & 3 & 1 \\ 0 & 2 & -3 \\ 0 & 0 & 1 \end{bmatrix}$

10. $\mathbf{A} = \begin{bmatrix} 1 & 0 & 0 \\ 0 & 1 & 1 \\ 5 & 0 & -2 \end{bmatrix}$

11. $\mathbf{A} = \begin{bmatrix} 0 & 1 & 0 \\ 0 & 0 & 1 \\ -1 & -1 & -1 \end{bmatrix}$

12. $\mathbf{A} = \begin{bmatrix} -2 & -1 & -1 \\ 3 & 2 & 1 \\ 1 & 1 & 0 \end{bmatrix}$

13. $\mathbf{A} = \begin{bmatrix} 3 & 2 & 1 \\ 0 & 2 & 0 \\ 0 & 0 & 2 \end{bmatrix}$

14. $\mathbf{A} = \begin{bmatrix} 3 & 0 & 0 \\ 0 & 0 & -1 \\ 0 & 1 & 0 \end{bmatrix}$

3.3 SUMMARY

A matrix is a collection of elements arranged in a rectangular array and enclosed by square brackets. Matrix $\mathbf{A}_{m \times n}$ has m rows and n columns, and is said to be of size $m \times n$. The abbreviated form of an $m \times n$ matrix is

$$\mathbf{A} = [a_{ij}]_{m \times n}, \quad i = 1, 2, \ldots, m, \; j = 1, 2, \ldots, n.$$

Matrix addition is performed entry-wise. If k is scalar, then $k\mathbf{A} = [ka_{ij}]_{m \times n}$. If $\mathbf{A} = [a_{ij}]_{m \times n}$ and $\mathbf{B} = [b_{ij}]_{n \times p}$, then $\mathbf{C} = \mathbf{AB}$ is $m \times p$ whose entries are obtained as

$$c_{ij} = \sum_{k=1}^{n} a_{ik} b_{kj}, \quad i = 1, 2, \ldots, m, \; j = 1, 2, \ldots, p.$$

Matrix Analysis

The transpose of $\mathbf{A}_{m \times n}$, denoted by $\mathbf{A}^T$, is an $n \times m$ matrix whose rows are the columns of $\mathbf{A}$. $\mathbf{A}_{n \times n} = [a_{ij}]$ is symmetric if $\mathbf{A}^T = \mathbf{A}$, upper-triangular if $a_{ij} = 0$ for all $i > j$, lower-triangular if $a_{ij} = 0$ for all $i < j$, and diagonal if $a_{ij} = 0$ for all $i \neq j$. Matrix transformations may be achieved by using EROs:

(ERO$_1$)—Multiply a row by a nonzero constant.
(ERO$_2$)—Interchange any two rows.
(ERO$_3$)—Multiply the ith row by a scalar $\alpha \neq 0$, add the result to the jth row, and then replace the jth row with the result. The ith row is called the pivot row.

The rank of $\mathbf{A}$ is the number of nonzero rows in the row-echelon form of $\mathbf{A}$. The determinant of $\mathbf{A}_{n \times n}$ is a real scalar, calculated as

$$|\mathbf{A}| = \sum_{k=1}^{n} a_{ik}(-1)^{i+k} M_{ik}, \quad i = 1, 2, \ldots, n,$$

where M_{ik} is the minor of a_{ik} and $(-1)^{i+k} M_{ik}$ is the cofactor of a_{ik}. The adjoint matrix of $\mathbf{A}$ is defined as

$$\mathrm{adj}(\mathbf{A}) = \begin{bmatrix} (-1)^{1+1} M_{11} & (-1)^{2+1} M_{21} & \cdot & \cdot & (-1)^{n+1} M_{n1} \\ (-1)^{1+2} M_{12} & (-1)^{2+2} M_{22} & \cdot & \cdot & (-1)^{n+2} M_{n2} \\ \cdot & \cdot & & & \cdot \\ \cdot & \cdot & & & \cdot \\ (-1)^{1+n} M_{1n} & (-1)^{2+n} M_{2n} & \cdot & \cdot & (-1)^{n+n} M_{nn} \end{bmatrix}.$$

Then, the inverse of matrix $\mathbf{A}$ is obtained as

$$\mathbf{A}^{-1} = \frac{1}{|\mathbf{A}|} \mathrm{adj}(\mathbf{A}).$$

The eigenvalue problem for matrix $\mathbf{A}$ is formulated as

$$\mathbf{A}_{n \times n} \mathbf{v}_{n \times 1} = \lambda_{1 \times 1} \mathbf{v}_{n \times 1}, \quad \mathbf{v}_{n \times 1} \neq \mathbf{0}_{n \times 1},$$

where λ is an eigenvalue of $\mathbf{A}$ and $\mathbf{v}$ is the corresponding eigenvector. The eigenvalues are the roots of the characteristic equation $|\mathbf{A} - \lambda \mathbf{I}| = 0$. The AM of an eigenvalue is the number of times it occurs. Its GM is the number of linearly independent eigenvectors associated with it. In general, GM $\leq$ AM. A matrix with at least one eigenvalue for which GM $<$ AM is called defective. Any nondefective matrix $\mathbf{A}$ can be diagonalized via a similarity transformation $\mathbf{V}^{-1} \mathbf{A} \mathbf{V} = \mathbf{D}$, where $\mathbf{V}_{n \times n} = [\mathbf{v}_1 \cdots \mathbf{v}_n]$ is the modal matrix of $\mathbf{A}$, and $\mathbf{D}$ is diagonal comprised of the eigenvalues of $\mathbf{A}$. For defective matrices, matrix $\mathbf{D}$ is almost diagonal and is known as the Jordan matrix.

REVIEW PROBLEMS

1. Prove that if rank $(\mathbf{A}_{n \times n}) < n$, then $\mathbf{A}^{-1}$ does not exist.
2. Find the rank of

$$\mathbf{A} = \begin{bmatrix} -2 & 1 & 0 & -3 \\ 4 & 9 & -2 & 7 \\ 3 & 4 & -1 & 5 \end{bmatrix}.$$

3. Consider

$$\mathbf{A} = \begin{bmatrix} 1 & -2 & -1 & 0 \\ -2 & 3 & 4 & 1 \\ 3 & 5 & 0 & -2 \\ 2 & 6 & a & -1 \end{bmatrix}, \quad a \text{ is a parameter.}$$

Determine a such that rank$(\mathbf{A}) = 3$.

4. Find α such that the homogeneous system below has a nontrivial ($\mathbf{x} \neq \mathbf{0}$) solution:

$$\begin{bmatrix} 2 & 1 & -1 \\ -2 & 3 & \alpha \\ 2 & 5 & 2 \end{bmatrix} \mathbf{x} = \mathbf{0}_{3 \times 1}.$$

5. Find the inverse of

$$\mathbf{R} = \begin{bmatrix} \cos\theta & \sin\theta & 0 \\ -\sin\theta & \cos\theta & 0 \\ 0 & 0 & 1 \end{bmatrix}.$$

6. Solve

$$\begin{bmatrix} 2 & 0 & 2 & 1 \\ 0 & 1 & 3 & -2 \\ 2 & 1 & 4 & 3 \\ 2 & -1 & -1 & 4 \end{bmatrix} \mathbf{x} = \begin{bmatrix} 1 \\ -5 \\ 1 \\ 7 \end{bmatrix}.$$

   a. Using the inverse of the coefficient matrix.
   b. ◢ Using the "\" operator in MATLAB.

7. ◢ Solve in MATLAB using Cramer's rule:

$$\begin{bmatrix} 1 & 0 & -1 & 2 \\ 2 & 1 & 0 & -2 \\ 0 & 3 & 1 & -1 \\ -1 & 0 & 2 & 3 \end{bmatrix} \mathbf{x} = \begin{bmatrix} -2 \\ 4 \\ 2 \\ -2 \end{bmatrix}.$$

8. ◢ Repeat Problem 7 for

$$\begin{bmatrix} -1 & 1 & 0 & 3 \\ 0 & 2 & 1 & -1 \\ 1 & 0 & 3 & 1 \\ -1 & 1 & 2 & 0 \end{bmatrix} \mathbf{x} = \begin{bmatrix} 2 \\ -2 \\ 4 \\ 1 \end{bmatrix}.$$

9. a. Find the eigenvalues and eigenvectors of

$$\mathbf{A} = \begin{bmatrix} 1 & -2 & 0 \\ 0 & 3 & 0 \\ 1 & -1 & 2 \end{bmatrix}.$$

   b. ◢ Perform Part (a) in MATLAB.

10. Repeat Problem 9 for

$$\mathbf{A} = \begin{bmatrix} 2 & 0 & 0 \\ 0 & -2 & 0 \\ 0 & 3 & 1 \end{bmatrix}.$$

11. Find the eigenvalues and the AM and GM of each. Then decide whether the matrix is defective.

$$\mathbf{A} = \begin{bmatrix} 1 & 0 & 0 & 0 \\ -2 & 3 & 0 & 0 \\ 1 & 0 & -1 & 1 \\ 0 & 2 & 0 & -1 \end{bmatrix}.$$

12. Repeat Problem 11 for
$$A = \begin{bmatrix} 2 & -1 & 0 & 0 \\ 0 & 4 & 0 & 0 \\ 0 & 0 & 1 & 0 \\ 0 & 0 & 3 & 2 \end{bmatrix}.$$

13. Prove that any matrix with distinct eigenvalues is nondefective.
14. Suppose **A** has been transformed into **B** through the similarity transformation $S^{-1}AS = B$. Prove that the eigenvalues of **A** and **B** are the same. (*Hint*: Show that if λ is an eigenvalue of **B**, it is also an eigenvalue of **A**.)
15. Let
$$A = \begin{bmatrix} 1 & 4 & -1 & 2 \\ 0 & 3 & 1 & 4 \\ 0 & 0 & 1 & 0 \\ 0 & 0 & 2 & -3 \end{bmatrix}.$$

Find the modal matrix and use it to transform **A** into a diagonal or a Jordan matrix.

16. Repeat Problem 15 for
$$A = \begin{bmatrix} 2 & 0 & 0 & 0 \\ -3 & 4 & 0 & 0 \\ 2 & 1 & -4 & 5 \\ 1 & -1 & 0 & -2 \end{bmatrix}.$$

4 System Model Representation

This chapter discusses various forms involved in the representation of mathematical models of dynamic systems. The actual techniques to derive these models will be presented in Chapters 5 through 7. The main topics covered in this chapter include the configuration form, state-space form, input–output (I/O) equation, transfer function, block diagram, and how one form may be obtained from the other. All of these will be discussed on the premise that the dynamic system is linear. Linearization of nonlinear systems is covered in the final section of this chapter.

4.1 CONFIGURATION FORM

A set of coordinates that describes the motion of a system completely is known as a set of generalized coordinates. This set is not unique so that more than one set of coordinates can be chosen for this purpose. But the number of coordinates remains the same regardless of the set selected for a specific system. If there are n generalized coordinates, they are usually denoted by $q_1, q_2, \ldots, q_n$. Suppose a dynamic system model is described by

$$\begin{cases} \ddot{q}_1 = f_1(q_1, q_2, \ldots, q_n, \dot{q}_1, \dot{q}_2, \ldots, \dot{q}_n, t) \\ \ddot{q}_2 = f_2(q_1, q_2, \ldots, q_n, \dot{q}_1, \dot{q}_2, \ldots, \dot{q}_n, t) \\ \qquad \vdots \\ \ddot{q}_n = f_n(q_1, q_2, \ldots, q_n, \dot{q}_1, \dot{q}_2, \ldots, \dot{q}_n, t) \end{cases}, \qquad (4.1)$$

where $\dot{q}_1, \dot{q}_2, \ldots, \dot{q}_n$ are the generalized velocities and $f_1, f_2, \ldots, f_n$ are the algebraic functions of $q_i, \dot{q}_i$ ($i = 1, 2, \ldots, n$) and time t, known as the generalized forces. Assuming initial time is $t = 0$, Equation 4.1 subjected to initial generalized coordinates $q_1(0), q_2(0), \ldots, q_n(0)$ and initial generalized velocities $\dot{q}_1(0), \dot{q}_2(0), \ldots, \dot{q}_n(0)$, is called the configuration form.

Example 4.1: Configuration Form

The mechanical system shown in Figure 4.1 consists of blocks of mass m_1 and m_2, springs with stiffness coefficients k_1 and k_2, a damper with coefficient of viscous damping c, and force $f(t)$ applied to block m_2. The equations of motion are derived as (Chapter 5)

$$\begin{cases} m_1 \ddot{x}_1 + c\dot{x}_1 + k_1 x_1 - k_2(x_2 - x_1) = 0, \\ m_2 \ddot{x}_2 + k_2(x_2 - x_1) = f(t), \end{cases}$$

where x_1, x_2 are the displacements of the blocks and $\dot{x}_1, \dot{x}_2$ are the respective velocities. The system is subjected to initial conditions $x_1(0) = x_{10}$, $x_2(0) = x_{20}$, $\dot{x}_1(0) = \dot{x}_{10}$, and $\dot{x}_2(0) = \dot{x}_{20}$. Obtain the configuration form.

Solution

A comparison with Equation 4.1 reveals $n = 2$ so that there are two generalized coordinates for this system: $q_1 = x_1$ and $q_2 = x_2$. Simplifying and rewriting the equations of motion to resemble

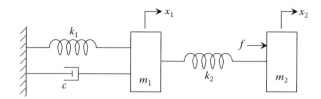

FIGURE 4.1 Mechanical system.

the form in Equation 4.1 yields

$$\begin{cases} \ddot{x}_1 = \dfrac{1}{m_1}[-c\dot{x}_1 - (k_1 + k_2)x_1 + k_2 x_2] = f_1(x_1, x_2, \dot{x}_1, \dot{x}_2, t), \\ \ddot{x}_2 = \dfrac{1}{m_2}[-k_2 x_2 + k_2 x_1 + f(t)] = f_2(x_1, x_2, \dot{x}_1, \dot{x}_2, t), \end{cases}$$

where f_1 and f_2 are the generalized forces. These together with the four initial conditions constitute the configuration form.

4.1.1 Second-Order Matrix Form

Mathematical models of dynamic systems that are governed by n-dimensional systems of second-order differential equations can conveniently be expressed as

$$\mathbf{M}\ddot{\mathbf{x}} + \mathbf{C}\dot{\mathbf{x}} + \mathbf{K}\mathbf{x} = \mathbf{f}, \qquad (4.2)$$

which is commonly known as the standard, second-order matrix form. Here

- $\mathbf{x}_{n \times 1}$ = configuration vector, $\mathbf{f}_{n \times 1}$ = vector of external forces
- $\mathbf{M}_{n \times n}$ = mass matrix, $\mathbf{C}_{n \times n}$ = damping matrix, $\mathbf{K}_{n \times n}$ = stiffness matrix.

Example 4.2: Second-Order Matrix Form

Express the equations of motion of the system in Example 4.1 in the second-order matrix form.

Solution

Using the matrix and vector notation, the equations of motion can be written as

$$\begin{bmatrix} m_1 & 0 \\ 0 & m_2 \end{bmatrix} \begin{Bmatrix} \ddot{x}_1 \\ \ddot{x}_2 \end{Bmatrix} + \begin{bmatrix} c & 0 \\ 0 & 0 \end{bmatrix} \begin{Bmatrix} \dot{x}_1 \\ \dot{x}_2 \end{Bmatrix} + \begin{bmatrix} k_1 + k_2 & -k_2 \\ -k_2 & k_2 \end{bmatrix} \begin{Bmatrix} x_1 \\ x_2 \end{Bmatrix} = \begin{Bmatrix} 0 \\ f(t) \end{Bmatrix}.$$

Consequently, the pertinent vectors and matrices can be properly indentified as

$$\mathbf{M} = \begin{bmatrix} m_1 & 0 \\ 0 & m_2 \end{bmatrix}, \quad \mathbf{C} = \begin{bmatrix} c & 0 \\ 0 & 0 \end{bmatrix}, \quad \mathbf{K} = \begin{bmatrix} k_1 + k_2 & -k_2 \\ -k_2 & k_2 \end{bmatrix}, \quad \mathbf{x} = \begin{Bmatrix} x_1 \\ x_2 \end{Bmatrix}, \quad \mathbf{f} = \begin{Bmatrix} 0 \\ f(t) \end{Bmatrix}.$$

PROBLEM SET 4.1

1. The mechanical system in Figure 4.2 is described by its equations of motion

$$\begin{cases} m_1 \ddot{x}_1 + k_1 x_1 - k_2(x_2 - x_1) - c(\dot{x}_2 - \dot{x}_1) = f(t) \\ m_2 \ddot{x}_2 + k_2(x_2 - x_1) + c(\dot{x}_2 - \dot{x}_1) = 0 \end{cases}$$

System Model Representation

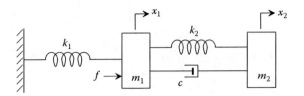

FIGURE 4.2 Mechanical system (Problem 1).

and subjected to initial conditions $x_1(0) = x_{10}$, $x_2(0) = x_{20}$, $\dot{x}_1(0) = \dot{x}_{10}$, and $\dot{x}_2(0) = \dot{x}_{20}$. Obtain the configuration form.

2. The equations of motion for the mechanical system in Figure 4.3 are

$$\begin{cases} m_1\ddot{x}_1 + c_1\dot{x}_1 + k_1 x_1 - k_2(x_2 - x_1) - c_2(\dot{x}_2 - \dot{x}_1) = f_1(t) \\ m_2\ddot{x}_2 + k_2(x_2 - x_1) + c_2(\dot{x}_2 - \dot{x}_1) = f_2(t) \end{cases}$$

subjected to initial conditions $x_1(0) = x_{10}$, $x_2(0) = x_{20}$, $\dot{x}_1(0) = \dot{x}_{10}$, and $\dot{x}_2(0) = \dot{x}_{20}$. Obtain the configuration form.

3. Derive the configuration form for a system described by

$$\begin{cases} \ddot{\theta} + 2\dot{\theta} + \theta + a(\theta - x) = \sin t \\ 2\ddot{x} + \dot{x} - a(\theta - x) = 0 \end{cases} \quad (a = \text{const})$$

subjected to initial conditions $\theta(0) = \theta_0$, $x(0) = x_0$, $\dot{\theta}(0) = \dot{\theta}_0$, and $\dot{x}(0) = \dot{x}_0$.

4. A dynamic system model is derived as

$$\begin{cases} 3\ddot{I}_1 + \dot{I}_1 + 2I_1 + 2(\dot{I}_1 - \dot{I}_2) + \frac{1}{2}(I_1 - I_2) = 0 \\ \ddot{I}_2 + 2(\dot{I}_2 - \dot{I}_1) + \frac{1}{2}(I_2 - I_1) = \sin 2t \end{cases}$$

subjected to $I_1(0) = I_{10}$, $I_2(0) = I_{20}$, $\dot{I}_1(0) = \dot{I}_{10}$, and $\dot{I}_2(0) = \dot{I}_{20}$. Obtain the configuration form.

In each case, express the governing equations in the standard, second-order matrix form.

5. Problem 1
6. Problem 2
7. Problem 3
8. Problem 4

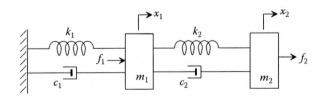

FIGURE 4.3 Mechanical system (Problem 2).

9. $\begin{cases} m\ddot{x}_1 + kx_1 - k(x_2 - x_1) - c(\dot{x}_2 - \dot{x}_1) = 0 \\ m\ddot{x}_2 - k(x_3 - x_2) + k(x_2 - x_1) + c(\dot{x}_2 - \dot{x}_1) = 0 \\ m\ddot{x}_3 + kx_3 + k(x_3 - x_2) + c\dot{x}_3 = f(t) \end{cases}$

10. $\begin{cases} m\ddot{x}_1 + kx_1 - k(x_2 - x_1) - c(\dot{x}_2 - \dot{x}_1) = f_1(t) \\ m\ddot{x}_2 - k(x_3 - x_2) + k(x_2 - x_1) + c(\dot{x}_2 - \dot{x}_1) - c(\dot{x}_3 - \dot{x}_2) = 0 \\ m\ddot{x}_3 + kx_3 + k(x_3 - x_2) + c(\dot{x}_3 - \dot{x}_2) = f_2(t) \end{cases}$

4.2 STATE-SPACE FORM

Once the mathematical model of a dynamic system has been derived (Chapters 5 through 7), it should be expressed in a form that is convenient for analysis and simulation. One of the most common forms to represent a system model is the state-space form. And the first step toward that goal is the selection of a set of state variables.

4.2.1 STATE VARIABLES, STATE-VARIABLE EQUATIONS, AND STATE EQUATION

State variables form the smallest set of independent variables that completely describe the state of a system. More exactly, knowledge of the state variables at some fixed (reference) time t_0 and system inputs at all $t \geq t_0$ translates to knowing the state variables and system outputs at all $t \geq t_0$. A key ingredient here is that state variables are independent, and thus they cannot be expressible as algebraic functions of one another and the system inputs. Moreover, a set of state variables is not unique so that more than one set can be identified for a dynamic system. Given a system model, the state variables are determined as follows:

- The number of state variables is equal to the number of initial conditions needed to completely solve the system's governing equations.
- The state variables are precisely those variables for which initial conditions are needed.

Notation: State variables are denoted by x_i ($i = 1, 2, \ldots, n$).

Example 4.3: State Variables

The mechanical system shown in Figure 4.4 is modeled by its equation of motion

$$\ddot{x} + \dot{x} + 2x = f(t).$$

Identify the state variables.

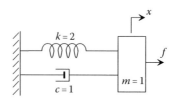

FIGURE 4.4 Mechanical system (Example 4.3).

System Model Representation

SOLUTION

Two initial conditions are required to completely solve the differential equation: $x(0)$ and $\dot{x}(0)$. Therefore, there are two state variables. And based on the notation convention, they are denoted x_1 and x_2. But the state variables are those variables for which initial conditions are needed, that is, x and $\dot{x}$. As a result, we have $x_1 = x$ and $x_2 = \dot{x}$.

4.2.1.1 State-Variable Equations

There are as many state-variable equations as there are state variables. Each one is a first-order differential equation whose left side is the first derivative of a state variable and whose right side is an algebraic function of the state variables, system inputs, and possibly time t. Suppose a dynamic system has n state variables $x_1, x_2, \ldots, x_n$ and m inputs $u_1, u_2, \ldots, u_m$. Then the state-variable equations take the general form

$$\begin{cases} \dot{x}_1 = f_1(x_1, \ldots, x_n; u_1, \ldots, u_m; t) \\ \dot{x}_2 = f_2(x_1, \ldots, x_n; u_1, \ldots, u_m; t) \\ \quad \vdots \\ \dot{x}_n = f_n(x_1, \ldots, x_n; u_1, \ldots, u_m; t) \end{cases}, \quad (4.3)$$

where $f_1, f_2, \ldots, f_n$ are the algebraic functions of the state variables and inputs and are generally nonlinear.

Example 4.4: State-Variable Equations (Example 4.3 Continued)

Referring to Example 4.3, since there are two state variables, there must be two state-variable equations in the form

$$\begin{cases} \dot{x}_1 = \cdots, \\ \dot{x}_2 = \cdots. \end{cases}$$

Recall that only state variables and inputs may appear on the right side of each differential equation. We know $x_1 = x$ so that $\dot{x}_1 = \dot{x}$. But $\dot{x} = x_2$, hence $\dot{x}_1 = \dot{x} = x_2$. This means that the first equation is simply $\dot{x}_1 = x_2$, which is valid because the right side contains only a state variable. Next, we know $x_2 = \dot{x}$ so that $\dot{x}_2 = \ddot{x}$. But $\ddot{x}$ is obtained from the equation of motion as

$$\dot{x}_2 = \ddot{x} \underset{=}{\overset{\text{From the equation of motion}}{=}} -\dot{x} - 2x + f(t) \underset{x_1 = x,\ x_2 = \dot{x}}{\overset{\text{Use state variables}}{=}} -x_2 - 2x_1 + f(t).$$

With this, the state-variable equations can be written as

$$\begin{cases} \dot{x}_1 = x_2, \\ \dot{x}_2 = -x_2 - 2x_1 + f(t). \end{cases}$$

Noting that $f(t)$ is the only system input here, this agrees with Equation 4.3.

4.2.1.2 State Equation

If $f_1, f_2, \ldots, f_n$ in Equation 4.3 are nonlinear algebraic functions, then Equation 4.3 may be expressed in vector form as

$$\dot{\mathbf{x}} = \mathbf{f}(\mathbf{x}, \mathbf{u}, t), \quad (4.4)$$

where

$$\mathbf{x} = \begin{Bmatrix} x_1 \\ \cdot \\ \cdot \\ \cdot \\ x_n \end{Bmatrix}_{n \times 1}, \quad \mathbf{u} = \begin{Bmatrix} u_1 \\ \cdot \\ \cdot \\ \cdot \\ u_m \end{Bmatrix}_{m \times 1}, \quad \mathbf{f} = \begin{Bmatrix} f_1 \\ \cdot \\ \cdot \\ \cdot \\ f_n \end{Bmatrix}_{n \times 1}.$$

However, if all elements of a dynamic system happen to be linear, then $f_1, f_2, \ldots, f_n$ in Equation 4.3 will be linear combinations of $x_1, x_2, \ldots, x_n$ and $u_1, u_2, \ldots, u_m$:

$$\text{Linear} \begin{cases} \dot{x}_1 = a_{11}x_1 + \cdots + a_{1n}x_n + b_{11}u_1 + \cdots + b_{1m}u_m \\ \dot{x}_2 = a_{21}x_1 + \cdots + a_{2n}x_n + b_{21}u_1 + \cdots + b_{2m}u_m \\ \quad \vdots \\ \dot{x}_n = a_{n1}x_1 + \cdots + a_{nn}x_n + b_{n1}u_1 + \cdots + b_{nm}u_m \end{cases} \qquad (4.5)$$

Rewriting Equation 4.5 in matrix form yields

$$\begin{Bmatrix} \dot{x}_1 \\ \dot{x}_2 \\ \cdot \\ \cdot \\ \dot{x}_n \end{Bmatrix} = \begin{bmatrix} a_{11} & a_{12} & \cdots & a_{1n} \\ a_{21} & a_{22} & \cdots & a_{2n} \\ \cdot & & & \cdot \\ \cdot & & & \cdot \\ a_{n1} & a_{n2} & \cdots & a_{nn} \end{bmatrix}_{n \times n} \begin{Bmatrix} x_1 \\ x_2 \\ \cdot \\ \cdot \\ x_n \end{Bmatrix}_{n \times 1} + \begin{bmatrix} b_{11} & b_{12} & \cdots & b_{1m} \\ b_{21} & b_{22} & \cdots & b_{2m} \\ \cdot & & & \cdot \\ \cdot & & & \cdot \\ b_{n1} & b_{n2} & \cdots & b_{nm} \end{bmatrix}_{n \times m} \begin{Bmatrix} u_1 \\ u_2 \\ \cdot \\ \cdot \\ u_m \end{Bmatrix}_{m \times 1}.$$

Finally, this can be conveniently expressed as

$$\dot{\mathbf{x}} = \mathbf{A}\mathbf{x} + \mathbf{B}\mathbf{u} \qquad (4.6)$$

known as the state equation, where

$$\mathbf{x} = \begin{Bmatrix} x_1 \\ x_2 \\ \cdot \\ \cdot \\ x_n \end{Bmatrix}_{n \times 1} = \text{state vector}, \quad \mathbf{u} = \begin{Bmatrix} u_1 \\ u_2 \\ \cdot \\ \cdot \\ u_m \end{Bmatrix}_{m \times 1} = \text{input vector},$$

$$\mathbf{A} = \begin{bmatrix} a_{11} & a_{12} & \cdots & a_{1n} \\ a_{21} & a_{22} & \cdots & a_{2n} \\ \cdot & & & \cdot \\ \cdot & & & \cdot \\ a_{n1} & a_{n2} & \cdots & a_{nn} \end{bmatrix}_{n \times n} = \text{state matrix},$$

$$\mathbf{B} = \begin{bmatrix} b_{11} & b_{12} & \cdots & b_{1m} \\ b_{21} & b_{22} & \cdots & b_{2m} \\ \cdot & & & \cdot \\ \cdot & & & \cdot \\ b_{n1} & b_{n2} & \cdots & b_{nm} \end{bmatrix}_{n \times m} = \text{input matrix}.$$

System Model Representation

Example 4.5: State Equation (Example 4.4 Continued)

The state-variable equations at the conclusion of Example 4.4 are expressed in matrix form, as

$$\begin{Bmatrix} \dot{x}_1 \\ \dot{x}_2 \end{Bmatrix} = \begin{bmatrix} 0 & 1 \\ -2 & -1 \end{bmatrix} \begin{Bmatrix} x_1 \\ x_2 \end{Bmatrix} + \begin{bmatrix} 0 \\ 1 \end{bmatrix} f(t).$$

Therefore, the state equation is

$$\dot{\mathbf{x}} = \mathbf{A}\mathbf{x} + \mathbf{B}u, \quad \mathbf{x} = \begin{Bmatrix} x_1 \\ x_2 \end{Bmatrix}, \quad \mathbf{A} = \begin{bmatrix} 0 & 1 \\ -2 & -1 \end{bmatrix}, \quad \mathbf{B} = \begin{bmatrix} 0 \\ 1 \end{bmatrix}, \quad u = f(t).$$

Since there is only one input $f(t)$, input vector $\mathbf{u}$ is scalar and is denoted by u.

4.2.2 Output Equation, State-Space Form

Consider a dynamic system with state variables $x_1, x_2, \ldots, x_n$ and inputs $u_1, u_2, \ldots, u_m$, as before. Suppose the system has p outputs $y_1, y_2, \ldots, y_p$. Outputs are occasionally called measured outputs, referring to physical quantities that are being measured. Then, the output equations generally appear in the form

$$\begin{cases} y_1 = g_1(x_1, \ldots, x_n; u_1, \ldots, u_m; t) \\ y_2 = g_2(x_1, \ldots, x_n; u_1, \ldots, u_m; t) \\ \quad \vdots \\ y_p = g_p(x_1, \ldots, x_n; u_1, \ldots, u_m; t) \end{cases}, \quad (4.7)$$

where $g_1, g_2, \ldots, g_p$ are the algebraic functions of the state variables and inputs and are generally nonlinear.

4.2.2.1 Output Equation

If $g_1, g_2, \ldots, g_p$ in Equation 4.7 are nonlinear algebraic functions, then Equation 4.7 may be expressed in vector form as

$$\mathbf{y} = \mathbf{g}(\mathbf{x}, \mathbf{u}, t), \quad (4.8)$$

where

$$\mathbf{y} = \begin{Bmatrix} y_1 \\ \vdots \\ y_p \end{Bmatrix}_{p \times 1}, \quad \mathbf{x} = \begin{Bmatrix} x_1 \\ \vdots \\ x_n \end{Bmatrix}_{n \times 1}, \quad \mathbf{u} = \begin{Bmatrix} u_1 \\ \vdots \\ u_m \end{Bmatrix}_{m \times 1}, \quad \mathbf{g} = \begin{Bmatrix} g_1 \\ \vdots \\ g_p \end{Bmatrix}_{p \times 1}.$$

But if all elements are linear, then $g_1, g_2, \ldots, g_p$ in Equation 4.7 are linear combinations of $x_1, x_2, \ldots, x_n$ and $u_1, u_2, \ldots, u_m$:

$$\text{Linear} \begin{cases} y_1 = c_{11}x_1 + \cdots + c_{1n}x_n + d_{11}u_1 + \cdots + d_{1m}u_m \\ y_2 = c_{21}x_1 + \cdots + c_{2n}x_n + d_{21}u_1 + \cdots + d_{2m}u_m \\ \quad \vdots \\ y_p = c_{p1}x_1 + \cdots + c_{pn}x_n + d_{p1}u_1 + \cdots + d_{pm}u_m \end{cases}$$

This is conveniently expressed as

$$\mathbf{y} = \mathbf{C}\mathbf{x} + \mathbf{D}\mathbf{u} \tag{4.9}$$

known as the output equation, where $\mathbf{y}$, $\mathbf{x}$, and $\mathbf{u}$ are defined as before, and

$$\mathbf{C} = \begin{bmatrix} c_{11} & c_{12} & \cdots & c_{1n} \\ c_{21} & c_{22} & \cdots & c_{2n} \\ \vdots & & & \vdots \\ c_{p1} & c_{p2} & \cdots & c_{pn} \end{bmatrix}_{p \times n} = \text{output matrix,}$$

$$\mathbf{D} = \begin{bmatrix} d_{11} & d_{12} & \cdots & d_{1m} \\ d_{21} & d_{22} & \cdots & d_{2m} \\ \vdots & & & \vdots \\ d_{p1} & d_{p2} & \cdots & d_{pm} \end{bmatrix}_{p \times m} = \text{direct transmission matrix.}$$

Example 4.6: Output Equation

For the mechanical system studied in Examples 4.3 through 4.5, suppose the output is the displacement x of the block. Find the output equation.

SOLUTION

Since there is only one output, the output vector $\mathbf{y}$ is 1×1, hence denoted by y. The output is x, therefore $y = x$. But for this system $x = x_1$, thus $y = x_1$. Finally, Equation 4.9 may be written as

$$y = [1 \ 0] \begin{Bmatrix} x_1 \\ x_2 \end{Bmatrix} + 0 \cdot u$$

so that $\mathbf{C} = [1 \ 0]$ and $D = 0$. Note that the direct transmission matrix $\mathbf{D}$ is 1×1, hence denoted by D.

4.2.2.2 State-Space Form

The combination of the state equation and output equation is called the state-space form. For a linear system with state variables $x_1, \ldots, x_n$, inputs $u_1, \ldots, u_m$, and outputs $y_1, \ldots, y_p$, the state-space form is

$$\begin{cases} \dot{\mathbf{x}} = \mathbf{A}_{n \times n} \mathbf{x}_{n \times 1} + \mathbf{B}_{n \times m} \mathbf{u}_{m \times 1} \\ \mathbf{y}_{p \times 1} = \mathbf{C}_{p \times n} \mathbf{x}_{n \times 1} + \mathbf{D}_{p \times m} \mathbf{u}_{m \times 1} \end{cases}. \tag{4.10}$$

Example 4.7: State-Space Form

The equations of motion for the mechanical system in Figure 4.5 are

$$\begin{cases} \ddot{x}_1 + x_1 - (x_2 - x_1) - (\dot{x}_2 - \dot{x}_1) = f_1 \\ 2\ddot{x}_2 + (x_2 - x_1) + (\dot{x}_2 - \dot{x}_1) = f_2 \end{cases}.$$

Assume that the (measured) outputs are x_1 and $\dot{x}_1$. Obtain the state-space form.

System Model Representation

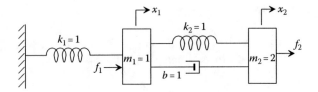

FIGURE 4.5 Mechanical system (Example 4.7).

Solution

The system model comprises two second-order differential equations; hence a total of four initial conditions are needed for complete solution. There are therefore four state variables:

$$x_1 = x_1,$$
$$x_2 = x_2,$$
$$x_3 = \dot{x}_1,$$
$$x_4 = \dot{x}_2.$$

When designating state variables, the above arrangement is the most commonly used. The derivatives of variables follow the variables.

The state-variable equations are then formed as

$$\begin{cases} \dot{x}_1 = x_3, \\ \dot{x}_2 = x_4, \\ \dot{x}_3 = -2x_1 + x_2 + x_4 - x_3 + f_1, \\ \dot{x}_4 = \frac{1}{2}[-x_2 + x_1 - x_4 + x_3 + f_2]. \end{cases}$$

The state equation is subsequently obtained as

$$\dot{\mathbf{x}} = \mathbf{A}\mathbf{x} + \mathbf{B}\mathbf{u},$$

where

$$\mathbf{x} = \begin{Bmatrix} x_1 \\ x_2 \\ x_3 \\ x_4 \end{Bmatrix}_{4\times1}, \quad \mathbf{A} = \begin{bmatrix} 0 & 0 & 1 & 0 \\ 0 & 0 & 0 & 1 \\ -2 & 1 & -1 & 1 \\ \frac{1}{2} & -\frac{1}{2} & \frac{1}{2} & -\frac{1}{2} \end{bmatrix}_{4\times4}, \quad \mathbf{B} = \begin{bmatrix} 0 & 0 \\ 0 & 0 \\ 1 & 0 \\ 0 & \frac{1}{2} \end{bmatrix}_{4\times2}, \quad \mathbf{u} = \begin{Bmatrix} f_1 \\ f_2 \end{Bmatrix}_{2\times1}.$$

Since the outputs are x_1 and $\dot{x}_1$, we have

$$\mathbf{y} = \begin{Bmatrix} x_1 \\ \dot{x}_1 \end{Bmatrix} = \begin{Bmatrix} x_1 \\ x_3 \end{Bmatrix}.$$

As a result, the output equation is

$$\mathbf{y}_{2\times1} = \mathbf{C}_{2\times4}\mathbf{x}_{4\times1} + \mathbf{D}_{2\times2}\mathbf{u}_{2\times1},$$

where

$$\mathbf{C} = \begin{bmatrix} 1 & 0 & 0 & 0 \\ 0 & 0 & 1 & 0 \end{bmatrix}_{2\times4}, \quad \mathbf{D} = \mathbf{0}_{2\times2}.$$

Finally, combining the state equation and the output equation yields the state-space form.

Example 4.8: State-Space Form

The state-space form in Example 4.7 can be stored in MATLAB® under an assigned name, say, sys, as shown below. This stored form can then be used for analysis and simulation.

```
% Input matrices A,B,C and D
>> A=[0 0 1 0;0 0 0 1;-2 1 -1 1;1/2 -1/2 1/2 -1/2];
>> B=[0 0;0 0;1 0;0 1/2]; C=[1 0 0 0;0 0 1 0]; D=[0 0;0 0];
>> sys=ss(A,B,C,D)   % state-space form
a =
           x1    x2    x3    x4
      x1    0     0     1     0
      x2    0     0     0     1
      x3   -2     1    -1     1
      x4   0.5  -0.5   0.5  -0.5

b =
           u1    u2
      x1    0     0
      x2    0     0
      x3    1     0
      x4    0    0.5

c =
           x1   x2   x3   x4
      y1    1    0    0    0
      y2    0    0    1    0

d =
           u1   u2
      y1    0    0
      y2    0    0
Continuous-time model.
```

4.2.3 DECOUPLING THE STATE EQUATION

We mentioned earlier that more than one set of state variables can be selected for a system model. In other words, a state vector $\tilde{\mathbf{x}}$, different from $\mathbf{x}$, still leads to a state-space form that is in the standard form of Equation 4.10. But any such set of state variables is still coupled through the entries of the resulting state matrix. The decoupling of the state equation is possible through the use of the modal matrix (Section 3.2.1) corresponding to the state matrix as follows. Given the state-space form Equation 4.10, assume $\mathbf{A}_{n \times n}$ has eigenvalues $\lambda_1, \ldots, \lambda_n$ and independent eigenvectors $v_1, \ldots, v_n$, so that the modal matrix $\mathbf{V}_{n \times n} = [\mathbf{v}_1 \cdots \mathbf{v}_n]$ diagonalizes matrix $\mathbf{A}$, that is, $\mathbf{V}^{-1}\mathbf{A}\mathbf{V} = \tilde{\mathbf{D}}$ (see Equation 3.9). Note that we have changed the notation of $\mathbf{D}$ to $\tilde{\mathbf{D}}$ to avoid confusion with the direct transmission matrix. Consider the transformation $\mathbf{x} = \mathbf{V}\tilde{\mathbf{x}}$ and substitute into the state-space form to obtain

$$\begin{cases} \mathbf{V}\dot{\tilde{\mathbf{x}}} = \mathbf{A}\mathbf{V}\tilde{\mathbf{x}} + \mathbf{B}\mathbf{u} \\ \mathbf{y} = \mathbf{C}\mathbf{V}\tilde{\mathbf{x}} + \mathbf{D}\mathbf{u} \end{cases} \xrightarrow[\text{by } \mathbf{V}^{-1}]{\text{Premultiply the state equation}} \begin{cases} \mathbf{V}^{-1}\mathbf{V}\dot{\tilde{\mathbf{x}}} = \mathbf{V}^{-1}\mathbf{A}\mathbf{V}\tilde{\mathbf{x}} + \mathbf{V}^{-1}\mathbf{B}\mathbf{u}, \\ \mathbf{y} = \mathbf{C}\mathbf{V}\tilde{\mathbf{x}} + \mathbf{D}\mathbf{u}. \end{cases}$$

Noting $\mathbf{V}^{-1}\mathbf{V} = \mathbf{I}$ and $\mathbf{V}^{-1}\mathbf{A}\mathbf{V} = \tilde{\mathbf{D}}$, and denoting $\tilde{\mathbf{B}} = \mathbf{V}^{-1}\mathbf{B}$ and $\tilde{\mathbf{C}} = \mathbf{C}\mathbf{V}$, the above becomes

$$\begin{cases} \dot{\tilde{\mathbf{x}}} = \tilde{\mathbf{D}}\tilde{\mathbf{x}} + \tilde{\mathbf{B}}\mathbf{u}, \\ \mathbf{y} = \tilde{\mathbf{C}}\tilde{\mathbf{x}} + \mathbf{D}\mathbf{u}. \end{cases} \quad (4.11)$$

System Model Representation

Since $\tilde{D}$ is diagonal—with entries $\lambda_1, \ldots, \lambda_n$—the new state equation in Equation 4.11 is clearly decoupled, each row a first-order differential equation in one state variable, independent of the others. And the new state vector is $\tilde{x}$.

Example 4.9: Decoupled State Equation

The state-space form for a system model is

$$\begin{cases} \dot{x} = Ax + Bu, \\ y = Cx + Du, \end{cases}$$

where

$$x = \begin{Bmatrix} x_1 \\ x_2 \end{Bmatrix}, \quad A = \begin{bmatrix} 0 & 1 \\ -2 & -3 \end{bmatrix}, \quad B = \begin{bmatrix} 0 \\ 1 \end{bmatrix}, \quad u = u, \quad C = [1\ 0], \quad D = 0.$$

a. Find the decoupled state equation together with the corresponding output equation.
b. Confirm in MATLAB.

Solution

a. Proceeding as in Section 3.2, we find $\lambda(A) = -1, -2$ and the modal matrix
$V = \begin{bmatrix} 1 & 1 \\ -1 & -2 \end{bmatrix}$ so that

$$V^{-1} = \begin{bmatrix} 2 & 1 \\ -1 & -1 \end{bmatrix}, \quad V^{-1}AV = \tilde{D} = \begin{bmatrix} -1 & 0 \\ 0 & -2 \end{bmatrix}, \quad \tilde{B} = V^{-1}B = \begin{bmatrix} 1 \\ -1 \end{bmatrix}, \quad \tilde{C} = CV = [1\ 1].$$

It is then readily seen that the decoupled state-space form is

$$\begin{cases} \dot{\tilde{x}} = \begin{bmatrix} -1 & 0 \\ 0 & -2 \end{bmatrix} \tilde{x} + \begin{bmatrix} 1 \\ -1 \end{bmatrix} u, \\ y = [1\ 1]\tilde{x} + 0 \cdot u. \end{cases}$$

b.
```
>> A = [0 1;-2 -3];
>> [V,Dtilda] = eig(A)
V =
      0.7071   -0.4472    % Note that eigenvectors are normalized
     -0.7071    0.8944
Dtilda =
     -1    0
      0   -2
>> B = [0;1];C = [1 0];D = 0;
>> Btilda = V\B;
>> Ctilda = C*V;
>> dec_sys = ss(Dtilda,Btilda,Ctilda,D)    % Define decoupled system
a =
            x1    x2
     x1    -1     0
     x2     0    -2
```

```
b =
            u1
    x1   1.414
    x2   2.236
c =
            x1        x2
    y1   0.7071   -0.4472
d =
            u1
    y1    0
Continuous-time model.
```

Because matrix **V** generated by the ``eig'' command is different from the one obtained in the solution of (a), the eventual numerical results also do not agree with (a), but the new system is decoupled nevertheless. Note that matrix **A** is unique.

PROBLEM SET 4.2

In each case, find a suitable set of state variables, obtain the state-variable equations, and form the state equation.

1. $\ddot{x} + a\dot{x} + 2x = 2f(t)$, a is a constant > 0

2. $3\ddot{x} + 2\dot{x} + x = \sin t$

3. $2\dddot{x} + \ddot{x} + 3\dot{x} + 2x = f(t)$

4. $\begin{cases} \ddot{z}_1 + \dot{z}_1 + 2(z_1 - z_2) = f(t) \\ \dot{z}_2 + z_2 + 2(z_2 - z_1) = 0 \end{cases}$

5. $\begin{cases} \ddot{x} + 2\dot{x} + x = z \\ \dot{z} + 3z + x = f(t) \end{cases}$

6. $\begin{cases} 2\ddot{x}_1 + \dot{x}_1 + 2(x_1 - x_2) = f(t) \\ x_2 - 2(x_1 - x_2) = 0 \end{cases}$

7. $\begin{cases} \ddot{z}_1 + 2\dot{z}_1 + 3(z_1 - z_2) = e^{-t} \\ z_2 = 3(z_1 - z_2) \end{cases}$

8. $\begin{cases} \ddot{x}_1 + \dot{x}_1 + 3x_1 - \dot{x}_2 - 2x_2 = f_1(t) \\ \ddot{x}_2 - \dot{x}_1 - 2x_1 + \dot{x}_2 + 3x_2 = f_2(t) \end{cases}$

A linear dynamic system is stable if the homogeneous solution of its governing equations, subjected to its initial conditions, decays. Equivalently, a system is stable if the eigenvalues of its state matrix all have negative real parts, that is, they all lie in the left half-plane.

9. In Problem 1, determine the range of values of a for which the system is stable.

10. Decide whether the system in Problem 2 is stable.

11. Find the range of values of a such that a system described by $2\ddot{x} - (a+1)\dot{x} + x = f$ is stable.

12. Find the range of values of a for which a system described by $\ddot{z} - (a-1)\dot{z} - z = f$ is stable.

In each case, a system model and the outputs are specified. Find the state-space form.

System Model Representation

13. The system in Example 4.7. The outputs are the displacements x_1 and x_2 of the two blocks.

14. $\begin{cases} 2\ddot{x}_1 + \dot{x}_1 + 2(x_1 - x_2) = f(t), \\ \dot{x}_2 + x_2 - 2(x_1 - x_2) = 0, \end{cases}$ outputs are x_1 and $\dot{x}_1$.

15. $\begin{cases} \ddot{\theta}_1 + \frac{1}{2}\dot{\theta}_1 + 2(\theta_1 - \theta_2) = T(t), \\ \theta_2 = 2(\theta_1 - \theta_2), \end{cases}$ the output is θ_2. (*Note*: θ_1 must be one of the state variables.)

16. $\begin{cases} \ddot{q}_1 + \frac{1}{3}(q_1 - q_2) + \dot{q}_1 + 2q_1 = v(t), \\ \dot{q}_2 + \frac{1}{3}(q_2 - q_1) = 0, \end{cases}$ outputs are q_2 and $\dot{q}_1$.

17. $\begin{cases} \ddot{x}_1 + 2(x_1 - x_3) - 2(\dot{x}_2 - \dot{x}_1) - \frac{1}{2}(x_2 - x_1) = f(t), \\ \ddot{x}_2 + 2(\dot{x}_2 - \dot{x}_1) + \frac{1}{2}(x_2 - x_1) = 0, \\ x_3 - 2(x_1 - x_3) = 0, \end{cases}$ outputs are x_1 and $\dot{x}_1$.

18. $\begin{cases} \ddot{x}_1 + 2(x_1 - x_3) - 2(\dot{x}_2 - \dot{x}_1) - \frac{1}{2}(x_2 - x_1) = 0, \\ \ddot{x}_2 + 2(\dot{x}_2 - \dot{x}_1) + \frac{1}{2}(x_2 - x_1) = f(t), \\ \dot{x}_3 = 2(x_1 - x_3), \end{cases}$ outputs are x_1 and $\dot{x}_1$.

19. A dynamic system model is described by $\ddot{x} + 4\dot{x} + 3x = f(t)$. Assume that $\dot{x}$ is the output.
    a. Find the state-space form.
    b. Decouple the state equation and write the transformed state-space form.

20. The governing equations for a system are given as

    $\begin{cases} \ddot{x}_1 - \dot{x}_1 - x_1 + 3x_2 = f(t), \\ \dot{x}_2 = -x_1 - 3x_2, \end{cases}$ outputs are x_1 and x_2.

    a. Find the state-space form.
    b. ▲ Decouple the state equation obtained in (a) and present the transformed state-space form.

4.3 INPUT–OUTPUT EQUATION, TRANSFER FUNCTION

An input–output (I/O) equation is a differential equation that relates a system input, a system output, and their derivatives (with respect to time). If $u(t)$ is an input and $y(t)$ is an output, then the I/O equation is in the general form

$$y^{(n)} + a_1 y^{(n-1)} + \cdots + a_{n-1}\dot{y} + a_n y = b_0 u^{(m)} + b_1 u^{(m-1)} + \cdots + b_{m-1}\dot{u} + b_m u, \quad m \leq n, \quad (4.12)$$

where $a_1, \ldots, a_n$ and $b_0, b_1, \ldots, b_m$ are constants, and $y^{(n)} = d^n y/dt^n$. A single-input–single-output (SISO) system, therefore, has only one I/O equation. A multi-input–multi-output (MIMO) system, on the other hand, has several I/O equations, one for each pair of input/output. In particular, a system with q inputs and r outputs has a total of qr I/O equations.

4.3.1 INPUT–OUTPUT EQUATIONS FROM THE SYSTEM MODEL

Because the generalized coordinates in a system model are normally coupled through the governing equations, finding one or more I/O equations is usually a very difficult task. However, there is a systematic approach that can be used for this purpose. The idea is to take the Laplace transform of the governing equations—assuming zero initial conditions—and eliminate the unwanted variables in the ensuing algebraic system. The new data are subsequently transformed back to the time domain and interpreted as one or more differential equations, which in turn are the desired I/O equations.

Example 4.10: I/O Equation

The mathematical model for the simple mechanical system in Example 4.3 is

$$\ddot{x} + \dot{x} + 2x = f(t),$$

where $f(t)$ is regarded as the input. Find the I/O equation if x is the output.

SOLUTION

This is a very trivial case and does not require Laplace transformation. In fact, the governing equation is already in the form of Equation 4.12.

Example 4.11: MIMO System

A system is described by its model

$$\begin{cases} \ddot{x}_1 + \dot{x}_1 + 2(x_1 - x_2) = f(t), \\ \dot{x}_2 - 2(x_1 - x_2) = 0, \end{cases}$$

where f is the input and x_1 and x_2 are the outputs. Derive all possible I/O equations.

SOLUTION

To begin with, since there are two outputs and one input, we expect two I/O equations. Assuming zero initial conditions, Laplace transformation of the governing equations yields

$$\begin{cases} (s^2 + s + 2)X_1(s) - 2X_2(s) = F(s), \\ -2X_1(s) + (s + 2)X_2(s) = 0. \end{cases}$$

Since both x_1 and x_2 are outputs, we solve the above once for $X_1(s)$ and a second time for $X_2(s)$ by means of Cramer's rule (Section 3.1.1):

$$X_1(s) = \frac{\begin{vmatrix} F(s) & -2 \\ 0 & s+2 \end{vmatrix}}{\begin{vmatrix} s^2+s+2 & -2 \\ -2 & s+2 \end{vmatrix}} = \frac{(s+2)F(s)}{s^3+3s^2+4s} \overset{\text{Cross multiply}}{\Longrightarrow} (s^3 + 3s^2 + 4s)X_1(s) = (s+2)F(s),$$

$$X_2(s) = \frac{\begin{vmatrix} s^2+s+2 & F(s) \\ -2 & 0 \end{vmatrix}}{\begin{vmatrix} s^2+s+2 & -2 \\ -2 & s+2 \end{vmatrix}} = \frac{2F(s)}{s^3+3s^2+4s} \overset{\text{Cross multiply}}{\Longrightarrow} (s^3 + 3s^2 + 4s)X_2(s) = 2F(s).$$

Interpretation of these two equations in the time domain results in the two desired I/O equations:

$$\dddot{x}_1 + 3\ddot{x}_1 + 4\dot{x}_1 = \dot{f} + 2f, \qquad \dddot{x}_2 + 3\ddot{x}_2 + 4\dot{x}_2 = 2f.$$

System Model Representation

4.3.2 Transfer Functions from the System Model

A transfer function is defined as the ratio of the Laplace transforms of an output and an input, assuming zero initial conditions. That said, if $y(t)$ is an output and $u(t)$ is an input, the corresponding transfer function is

$$G(s) = \frac{\mathcal{L}\{y(t)\}}{\mathcal{L}\{u(t)\}} = \frac{Y(s)}{U(s)}.$$

Taking the Laplace transform of the I/O equation, Equation 4.12, with zero initial conditions, we find

$$(s^n + a_1 s^{n-1} + \cdots + a_{n-1} s + a_n) Y(s) = (b_0 s^m + b_1 s^{m-1} + \cdots + b_{m-1} s + b_m) U(s).$$

As a result, the transfer function is formed as

$$\frac{Y(s)}{U(s)} = \frac{b_0 s^m + b_1 s^{m-1} + \cdots + b_{m-1} s + b_m}{s^n + a_1 s^{n-1} + \cdots + a_{n-1} s + a_n}, \quad m \leq n. \tag{4.13}$$

There is a transfer function for each input/output pair. Therefore, a SISO system has only one transfer function, but a MIMO system has several, one for each input/output pair. If a system has q inputs and r outputs, then there are a total of qr transfer functions, assembled in an $r \times q$ transfer function matrix (or a transfer matrix), denoted by $\mathbf{G}(s) = [G_{ij}(s)]$, where $i = 1, 2, \ldots, r$ and $j = 1, 2, \ldots, q$.

Example 4.12: Transfer Function

The system model in Example 4.3 is given as

$$\ddot{x} + \dot{x} + 2x = f(t).$$

Assuming f is the input and x is the output, find the transfer function.

Solution

Taking the Laplace transform of the equation, yields

$$(s^2 + s + 2)X(s) = F(s).$$

The transfer function is then found as

$$\frac{X(s)}{F(s)} = \frac{1}{s^2 + s + 2}.$$

Example 4.13: MIMO System

The mechanical system in Figure 4.5 (Example 4.7) is described by equations of motion

$$\begin{cases} \ddot{x}_1 + x_1 - (x_2 - x_1) - (\dot{x}_2 - \dot{x}_1) = f_1, \\ 2\ddot{x}_2 + (x_2 - x_1) + (\dot{x}_2 - \dot{x}_1) = f_2. \end{cases}$$

If f_1 and f_2 are the inputs and x_1 and x_2 are the outputs, find the transfer matrix.

Solution

We expect a 2×2 transfer matrix with the specific structure

$$\mathbf{G}(s) = \begin{bmatrix} G_{11}(s) & G_{12}(s) \\ G_{21}(s) & G_{22}(s) \end{bmatrix} = \begin{bmatrix} \left.\frac{X_1(s)}{F_1(s)}\right|_{F_2=0} & \left.\frac{X_1(s)}{F_2(s)}\right|_{F_1=0} \\ \left.\frac{X_2(s)}{F_1(s)}\right|_{F_2=0} & \left.\frac{X_2(s)}{F_2(s)}\right|_{F_1=0} \end{bmatrix}_{2 \times 2}.$$

To find the four transfer functions listed above, take the Laplace transform of the governing equations, with zero initial conditions,

$$\begin{cases} (s^2 + s + 2)X_1(s) - (s+1)X_2(s) = F_1(s), \\ -(s+1)X_1(s) + (2s^2 + s + 1)X_2(s) = F_2(s). \end{cases}$$

Solving for $X_1(s)$, we find

$$X_1(s) = \frac{\begin{vmatrix} F_1(s) & -(s+1) \\ F_2(s) & 2s^2+s+1 \end{vmatrix}}{\Delta(s)} = \frac{2s^2+s+1}{\Delta(s)} F_1(s) + \frac{s+1}{\Delta(s)} F_2(s), \quad \text{(a)}$$

where

$$\Delta(s) = \begin{vmatrix} s^2+s+2 & -(s+1) \\ -(s+1) & 2s^2+s+1 \end{vmatrix} = 2s^4 + 3s^3 + 5s^2 + s + 1.$$

Solving for $X_2(s)$, we find

$$X_2(s) = \frac{\begin{vmatrix} s^2+s+2 & F_1(s) \\ -(s+1) & F_2(s) \end{vmatrix}}{\Delta(s)} = \frac{s^2+s+2}{\Delta(s)} F_2(s) + \frac{s+1}{\Delta(s)} F_1(s). \quad \text{(b)}$$

The four transfer functions are then determined as follows.

Equation (a): $\left.\dfrac{X_1(s)}{F_1(s)}\right|_{F_2=0} = \dfrac{2s^2+s+1}{\Delta(s)}$, $\left.\dfrac{X_1(s)}{F_2(s)}\right|_{F_1=0} = \dfrac{s+1}{\Delta(s)}$,

Equation (b): $\left.\dfrac{X_2(s)}{F_1(s)}\right|_{F_2=0} = \dfrac{s+1}{\Delta(s)}$, $\left.\dfrac{X_2(s)}{F_2(s)}\right|_{F_1=0} = \dfrac{s^2+s+2}{\Delta(s)}$.

Ultimately, the transfer matrix can be formed as

$$\mathbf{G}(s) = \begin{bmatrix} \dfrac{2s^2+s+1}{\Delta(s)} & \dfrac{s+1}{\Delta(s)} \\ \dfrac{s+1}{\Delta(s)} & \dfrac{s^2+s+2}{\Delta(s)} \end{bmatrix}.$$

PROBLEM SET 4.3

I/O Equation

In each case, the governing equations of a system are provided. Find all possible I/O equations.

1. $\begin{cases} \ddot{x}_1 + 2\dot{x}_1 + x_1 - x_2 = 0, \\ \ddot{x}_2 + \dot{x}_2 - x_1 + x_2 = f(t), \end{cases}$ $f(t) = $ input, $x_1 = $ output

2. $\begin{cases} \ddot{x}_1 + 2\dot{x}_1 + x_1 - x_2 = f(t), \\ \ddot{x}_2 + \dot{x}_2 - x_1 + x_2 = 0, \end{cases}$ $f(t) = $ input, $x_1, x_2 = $ outputs

3. $\begin{cases} \ddot{x}_1 + \dot{x}_1 + 2(x_1 - x_2) = f_1(t), \\ \ddot{x}_2 - 2(x_1 - x_2) = f_2(t), \end{cases}$ $f_1(t), f_2(t) = $ inputs, $x_1 = $ output

4. $\begin{cases} \ddot{x}_1 + \dot{x}_1 + 2(x_1 - x_2) = f_1(t), \\ \ddot{x}_2 - 2(x_1 - x_2) = f_2(t), \end{cases}$ $f_1(t), f_2(t) = $ inputs, $x_1, x_2 = $ outputs

System Model Representation

5. $\begin{cases} \ddot{\theta}_1 + \dot{\theta}_1 + \frac{1}{3}(\theta_1 - \theta_2) = h(t), \\ \theta_2 = \frac{1}{3}(\theta_1 - \theta_2), \end{cases}$ $\quad h(t) = \text{input}, \theta_2 = \text{output}$

6. $\begin{cases} \ddot{q}_1 + \dot{q}_1 + \frac{1}{2}(q_1 - q_2) + 2q_1 = v(t), \\ \dot{q}_2 = \frac{1}{2}(q_1 - q_2), \end{cases}$ $\quad v(t) = \text{input}, q_1, q_2 = \text{outputs}$

7. $\begin{cases} \ddot{x}_1 + x_1 - x_3 - (\dot{x}_2 - \dot{x}_1) - \frac{1}{2}(x_2 - x_1) = u(t), \\ \ddot{x}_2 + \dot{x}_2 - \dot{x}_1 + \frac{1}{2}(x_2 - x_1) = 0, \\ \dot{x}_3 = x_1 - x_3, \end{cases}$ $\quad u(t) = \text{input}, x_1 = \text{output}$

8. $\begin{cases} \ddot{x}_1 + x_1 - x_3 - (\dot{x}_2 - \dot{x}_1) - \frac{1}{2}(x_2 - x_1) = 0, \\ \ddot{x}_2 + (\dot{x}_2 - \dot{x}_1) + \frac{1}{2}(x_2 - x_1) = u(t), \\ \dot{x}_3 = x_1 - x_3, \end{cases}$ $\quad u(t) = \text{input}, x_3 = \text{output}$

Transfer Function
In Problems 9 through 12, a system model as well as its inputs and outputs are provided. Find the transfer matrix or transfer function, whichever applicable. Do *not* cancel any terms involving s in the numerator and denominator.

9. Example 4.11

10. $\begin{cases} \ddot{x}_1 + \dot{x}_1 + \frac{1}{3}(x_1 - x_2) = f(t), \\ \ddot{x}_2 + \frac{1}{2}\dot{x}_2 + \frac{1}{3}(x_2 - x_1) = 0, \end{cases}$ $\quad f(t) = \text{input}, x_2 = \text{output}$

11. $\begin{cases} \ddot{q}_1 + \dot{q}_1 + \frac{1}{2}(q_1 - q_2) + q_1 = 0, \\ \dot{q}_2 - \frac{1}{2}(q_1 - q_2) = v(t), \end{cases}$ $\quad v(t) = \text{input}, q_2 = \text{output}$

12. $\begin{cases} \ddot{x}_1 + x_1 - x_3 - 2(\dot{x}_2 - \dot{x}_1) = u_1(t), \\ \ddot{x}_2 + 2(\dot{x}_2 - \dot{x}_1) = u_2(t), \\ \dot{x}_3 = x_1 - x_3, \end{cases}$ $\quad u_1(t), u_2(t) = \text{inputs}, x_1, x_3 = \text{outputs}$

13. In Example 4.11, find the transfer matrix directly from the two I/O equations that were derived. Compare with the result of Problem 9.

14. In Example 4.13, find all possible I/O equations directly from the transfer matrix.

4.4 RELATIONS BETWEEN STATE-SPACE FORM, INPUT–OUTPUT EQUATION, AND TRANSFER FUNCTION

So far we have learned how to derive the state-space form, I/O equation(s), and transfer function(s) directly from the system's governing equations (or model). But these forms can also be obtained from one another. In this section, we develop and implement two systematic techniques for these purposes.

4.4.1 INPUT–OUTPUT EQUATION TO STATE-SPACE FORM

Consider the I/O equation

$$y^{(n)} + a_1 y^{(n-1)} + \cdots + a_{n-1}\dot{y} + a_n y = b_0 u^{(n)} + b_1 u^{(n-1)} + \cdots + b_{n-1}\dot{u} + b_n u, \qquad (4.14)$$

where y is the output, u is the input, $a_1,\ldots,a_n$, and $b_0, b_1,\ldots,b_n$ are all constants, and $y^{(n)} = d^n y/dt^n$. Note that this agrees with Equation 4.12, except that the same highest order of differentiation for y and u is now allowed, that is, $m = n$. The goal here is to derive the state-space form directly from the I/O equation, Equation 4.14. With the assumption of zero initial conditions, the transfer function is readily obtained as

$$\frac{Y(s)}{U(s)} = \frac{b_0 s^n + b_1 s^{n-1} + \cdots + b_{n-1} s + b_n}{s^n + a_1 s^{n-1} + \cdots + a_{n-1} s + a_n}.$$

Rewrite this as

$$\frac{Y(s)}{U(s)} = \frac{Y(s)}{V(s)} \frac{V(s)}{U(s)} = (b_0 s^n + b_1 s^{n-1} + \cdots + b_{n-1} s + b_n) \left(\frac{1}{s^n + a_1 s^{n-1} + \cdots + a_{n-1} s + a_n} \right)$$

so that

$$\frac{Y(s)}{V(s)} = b_0 s^n + b_1 s^{n-1} + \cdots + b_{n-1} s + b_n, \qquad \frac{V(s)}{U(s)} = \frac{1}{s^n + a_1 s^{n-1} + \cdots + a_{n-1} s + a_n}.$$

The time-domain interpretation of the first yields

$$y = b_0 v^{(n)} + b_1 v^{(n-1)} + \cdots + b_{n-1}\dot{v} + b_n v. \qquad (4.15)$$

And the second one yields

$$v^{(n)} + a_1 v^{(n-1)} + \cdots + a_{n-1}\dot{v} + a_n v = u. \qquad (4.16)$$

Equation 4.16 is an nth-order differential equation in v; hence n initial conditions are required for complete solution. Based on the findings in Section 4.2, the n state variables are selected as

$$x_1 = v,$$
$$x_2 = \dot{v},$$
$$\vdots$$
$$x_{n-1} = v^{(n-2)},$$
$$x_n = v^{(n-1)}.$$

The resulting state-variable equations are then formed as

$$\begin{aligned}
\dot{x}_1 &= x_2, \\
\dot{x}_2 &= x_3, \\
&\vdots \\
\dot{x}_{n-1} &= x_n, \\
\dot{x}_n &= -a_n x_1 - a_{n-1} x_2 + \cdots - a_1 x_n + u.
\end{aligned} \qquad (4.17)$$

System Model Representation

The state equation is then written as

$$\dot{\mathbf{x}} = \mathbf{A}\mathbf{x} + \mathbf{B}u,$$

where

$$\mathbf{x} = \begin{Bmatrix} x_1 \\ x_2 \\ \vdots \\ x_{n-1} \\ x_n \end{Bmatrix}, \quad \mathbf{A} = \begin{bmatrix} 0 & 1 & 0 & 0 & \cdots & 0 \\ 0 & 0 & 1 & 0 & \cdots & 0 \\ \vdots & & & & & \vdots \\ 0 & 0 & 0 & 0 & \cdots & 1 \\ -a_n & -a_{n-1} & \cdots & \cdots & -a_2 & -a_1 \end{bmatrix}_{n \times n}, \quad \mathbf{B} = \begin{bmatrix} 0 \\ 0 \\ \vdots \\ 0 \\ 1 \end{bmatrix}_{n \times 1}, \quad u = u.$$

(4.18)

The state matrix in the above form is called the lower companion matrix. The output is given by Equation 4.15. Using the state variables in that equation, we find

$$y = b_0 v^{(n)} + b_1 v^{(n-1)} + \cdots + b_{n-1}\dot{v} + b_n v$$
$$= b_0 \dot{x}_n + b_1 x_n + \cdots + b_{n-1} x_2 + b_n x_1.$$

But the output equation cannot contain $\dot{x}_n$. We substitute for $\dot{x}_n$ using the last equation in Equation 4.17. This yields

$$y = b_0(-a_n x_1 - a_{n-1} x_2 + \cdots - a_1 x_n + u) + b_1 x_n + \cdots + b_{n-1} x_2 + b_n x_1$$

$$\stackrel{\text{Collect like terms}}{=} (-b_0 a_n + b_n)x_1 + (-b_0 a_{n-1} + b_{n-1})x_2 + \cdots + (-b_0 a_1 + b_1)x_n + b_0 u.$$

(4.19)

Finally, the output equation is obtained as

$$y = \mathbf{C}\mathbf{x} + Du,$$

where

$$\mathbf{C} = [-b_0 a_n + b_n \quad -b_0 a_{n-1} + b_{n-1} \quad \cdots \quad \cdots \quad -b_0 a_1 + b_1]_{1 \times n}, \quad D = b_0. \quad (4.20)$$

Equations 4.18 and 4.20 describe all four matrices involved in the state-space form.

4.4.1.1 Controller Canonical Form

The procedure outlined above can be performed in MATLAB using the "tf2ss" command—transfer function to state-space. It calls for the transfer function, which is readily available from the I/O equation, and returns the state-space form in controller canonical form. This form is different from that in Equations 4.18 and 4.20 because it is based on the state variables chosen in the reverse order of what we have become accustomed to. As a result, the state and input matrices appear as

$$\mathbf{A} = \begin{bmatrix} -a_1 & -a_2 & \cdots & 0 & -a_{n-1} & -a_n \\ 1 & 0 & \cdots & 0 & 0 & 0 \\ \vdots & & & & & \\ & & & & & \vdots \\ 0 & 0 & \cdots & 1 & 0 & 0 \\ 0 & 0 & \cdots & 0 & 1 & 0 \end{bmatrix}, \quad \mathbf{B} = \begin{bmatrix} 1 \\ 0 \\ \vdots \\ 0 \\ 0 \end{bmatrix},$$

where the state matrix is known as the upper companion matrix.

Example 4.14: I/O Equation to State-Space Form

A system's I/O equation is provided as

$$\ddot{y} + 2\dot{y} + y = \ddot{u} + 3\dot{u} + 2u.$$

a. Find the state-space representation.
b. Repeat in MATLAB.

Solution

a. A comparison with Equation 4.14 reveals $n = 2$, $a_1 = 2$, $a_2 = 1$, $b_0 = 1$, $b_1 = 3$, and $b_2 = 2$. Therefore, in the state equation, Equation 4.18 yields

$$\mathbf{x} = \begin{Bmatrix} x_1 \\ x_2 \end{Bmatrix}, \quad \mathbf{A} = \begin{bmatrix} 0 & 1 \\ -1 & -2 \end{bmatrix}, \quad \mathbf{B} = \begin{bmatrix} 0 \\ 1 \end{bmatrix}, \quad u = u.$$

In the output equation, Equation 4.20 yields

$$\mathbf{C} = [1 \ 1], \quad D = 1.$$

In summary, the state-space form is

$$\begin{cases} \dot{\mathbf{x}} = \begin{bmatrix} 0 & 1 \\ -1 & -2 \end{bmatrix} \mathbf{x} + \begin{bmatrix} 0 \\ 1 \end{bmatrix} u, \\ y = [1 \ 1] \mathbf{x} + u. \end{cases}$$

b.
The transfer function is directly obtained from the I/O equation,

$$\frac{s^2 + 3s + 2}{s^2 + 2s + 1}.$$

```
>> Num = [1 3 2];    % Define numerator
>> Den = [1 2 1];    % Define denominator
>> [A,B,C,D] = tf2ss(Num,Den)
A =
     -2   -1
      1    0    % Upper companion matrix
B =
      1
      0
C =
      1    1
D =
      1
```

Example 4.15

Find the state-space form, if the I/O equation is given as

$$\dddot{y} + \ddot{y} + \dot{y} + 2y = 2\dot{u} + u.$$

Solution

Comparing with Equation 4.14, we find $n = 3$, $a_1 = 1$, $a_2 = 1$, $a_3 = 2$, $b_0 = 0$, $b_1 = 0$, $b_2 = 2$, and $b_3 = 1$. By Equation 4.18,

$$\mathbf{x} = \begin{Bmatrix} x_1 \\ x_2 \\ x_3 \end{Bmatrix}, \quad \mathbf{A} = \begin{bmatrix} 0 & 1 & 0 \\ 0 & 0 & 1 \\ -2 & -1 & -1 \end{bmatrix}, \quad \mathbf{B} = \begin{bmatrix} 0 \\ 0 \\ 1 \end{bmatrix}, \quad u = u.$$

By Equation 4.20,

$$\mathbf{C} = [1 \ 2 \ 0], \quad D = 0.$$

In summary,

$$\begin{cases} \dot{\mathbf{x}} = \begin{bmatrix} 0 & 1 & 0 \\ 0 & 0 & 1 \\ -2 & -1 & -1 \end{bmatrix} \mathbf{x} + \begin{bmatrix} 0 \\ 0 \\ 1 \end{bmatrix} u, \\ y = [1 \ 2 \ 0] \mathbf{x}. \end{cases}$$

4.4.2 State-Space Form to Transfer Function

The transfer function (for SISO systems) or the transfer matrix (for MIMO systems) can be methodically determined from the state-space form of the model. Consider the state-space form as in Equation 4.10,

$$\begin{cases} \dot{\mathbf{x}} = \mathbf{A}_{n \times n} \mathbf{x}_{n \times 1} + \mathbf{B}_{n \times m} \mathbf{u}_{m \times 1}, \\ \mathbf{y}_{p \times 1} = \mathbf{C}_{p \times n} \mathbf{x}_{n \times 1} + \mathbf{D}_{p \times m} \mathbf{u}_{m \times 1}. \end{cases}$$

The system has m inputs and p outputs so that there are a total of mp transfer functions. The transfer matrix $\mathbf{G}(s)$ is therefore $p \times m$ and derived as follows. But first, we mention that the Laplace transform of a vector such as $\mathbf{x}$ is treated as

$$\mathbf{x} = \begin{Bmatrix} x_1 \\ \cdot \\ \cdot \\ x_n \end{Bmatrix}, \quad \mathbf{X}(s) = \mathcal{L}\{\mathbf{x}\} = \begin{Bmatrix} \mathcal{L}\{x_1\} \\ \cdot \\ \cdot \\ \mathcal{L}\{x_n\} \end{Bmatrix}.$$

Taking the Laplace transform of the state-space form, assuming zero initial state vector $\mathbf{x}(0) = \mathbf{0}_{n \times 1}$, we have

$$\begin{cases} s\mathbf{X}(s) = \mathbf{A}\mathbf{X}(s) + \mathbf{B}\mathbf{U}(s), \\ \mathbf{Y}(s) = \mathbf{C}\mathbf{X}(s) + \mathbf{D}\mathbf{U}(s). \end{cases}$$

The first equation is manipulated as

$$(s\mathbf{I} - \mathbf{A})\mathbf{X}(s) = \mathbf{B}\mathbf{U}(s) \quad \overset{\text{Premultiply by}}{\underset{(s\mathbf{I}-\mathbf{A})^{-1}}{\Longrightarrow}} \quad \mathbf{X}(s) = (s\mathbf{I} - \mathbf{A})^{-1} \mathbf{B}\mathbf{U}(s).$$

Inserting this into the second equation results in

$$\mathbf{Y}(s) = \mathbf{C}(s\mathbf{I} - \mathbf{A})^{-1} \mathbf{B}\mathbf{U}(s) + \mathbf{D}\mathbf{U}(s)$$

$$= \left[\mathbf{C}(s\mathbf{I} - \mathbf{A})^{-1} \mathbf{B} + \mathbf{D} \right] \mathbf{U}(s).$$

Recall from Section 4.3 that for a SISO system with input u and output y, the transfer function is $G(s) = Y(s)/U(s)$ so that $Y(s) = G(s)U(s)$. Extending this idea to MIMO systems, the transfer matrix can be obtained from the above equation as

$$\mathbf{G}(s) = \mathbf{C}(s\mathbf{I} - \mathbf{A})^{-1}\mathbf{B} + \mathbf{D}. \tag{4.21}$$

The fact that the size of $\mathbf{G}(s)$ is $p \times m$ can also be readily verified.

Example 4.16: State-Space Form to Transfer Function

A system's state-space representation is

$$\begin{cases} \dot{\mathbf{x}} = \begin{bmatrix} 0 & 1 \\ -2 & -3 \end{bmatrix} \mathbf{x} + \begin{bmatrix} 0 \\ 2 \end{bmatrix} u, \\ y = [0 \ 1]\mathbf{x}. \end{cases}$$

a. Find the transfer function.
b. Repeat in MATLAB.

Solution

a. This is a SISO system since it has one input, u, and one output because y is 1×1. Therefore, a single transfer function $G(s)$ is expected. Noting that $D = 0$, by Equation 4.21, we have

$$G(s) = \mathbf{C}(s\mathbf{I} - \mathbf{A})^{-1}\mathbf{B}.$$

Using the adjoint matrix (Section 3.1), we find

$$(s\mathbf{I} - \mathbf{A})^{-1} = \frac{1}{(s+1)(s+2)} \begin{bmatrix} s+3 & 1 \\ 2 & s \end{bmatrix}.$$

With this, the transfer function is obtained as

$$G(s) = \mathbf{C}(s\mathbf{I} - \mathbf{A})^{-1}\mathbf{B} = [0 \ 1] \frac{1}{(s+1)(s+2)} \begin{bmatrix} s+3 & 1 \\ 2 & s \end{bmatrix} \begin{bmatrix} 0 \\ 2 \end{bmatrix} = \frac{2s}{(s+1)(s+2)}.$$

b.
The above procedure can be followed step by step in MATLAB:

```
>> syms s
>> A = [0 1;-2 -3]; B = [0;2]; C = [0 1];
>> TF = C*inv(s*eye(2)-A)*B % Transfer function
TF =
2*s/(s^2+3*s+2)
```

The other option is to use the ``ss2tf'' command—state-space to transfer function—which directly finds the transfer function or matrix.

```
>> A = [0 1;-2 -3]; B = [0;2]; C = [0 1]; D = 0;
>> [Num, Den] = ss2tf(A,B,C,D)    % Num = numerator, Den = denominator
Num =
         0    2.0000   -0.0000   % Num = 2s

Den =
         1    3    2   % Den = s^2 + 3s + 2 = (s+1)(s+2)
```

System Model Representation

Example 4.17: State-Space Form to Transfer Matrix

Find the transfer matrix for a system whose state-space form is

$$\begin{cases} \dot{x} = Ax + Bu, \\ y = Cx + Du, \end{cases}$$

where

$$A = \begin{bmatrix} 0 & 0 & 1 \\ 1 & -1 & 0 \\ -1 & 1 & -2 \end{bmatrix}, \quad B = \begin{bmatrix} 0 & 0 \\ 1 & 0 \\ 0 & 1 \end{bmatrix}, \quad C = \begin{bmatrix} 1 & 0 & 0 \\ 0 & 1 & 0 \end{bmatrix}, \quad D = 0_{2\times 2}.$$

Solution

It is observed that **y** and **u** are 2×1 each so that there are two outputs and two inputs. This is therefore classified as a MIMO system and the transfer matrix **G**(s), generated by Equation 4.21, is expected to be 2×2.

```
>> A = [0 0 1;1 -1 0;-1 1 -2]; B = [0 0;1 0;0 1];
>> C = [1 0 0;0 1 0]; D = [0 0;0 0];
```

For systems with multiple inputs, the ``ss2tf'' command requires the input number as the last argument.

```
>> [Num1, Den1] = ss2tf(A,B,C,D,1)    % From input 1
Num1 =
         0    0.0000    0.0000    1.0000
         0    1.0000    2.0000    1.0000
Den1 =
    1.0000    3.0000    3.0000   -0.0000
>> [Num2, Den2] = ss2tf(A,B,C,D,2)    % From input 2
Num2 =
         0   -0.0000    1.0000    1.0000
         0    0.0000    0.0000    1.0000
Den2 =
    1.0000    3.0000    3.0000   -0.0000
```

The first row of Num1 yields 1, and the second one $s^2 + 2s + 1$. The denominator is $s^3 + 3s^2 + 3s$. The first row of Num2 yields $s + 1$, and the second one 1. The denominator is as before. The transfer matrix is therefore found as

$$G(s) = \frac{1}{s^3 + 3s^2 + 3s} \begin{bmatrix} 1 & s^2 + 2s + 1 \\ s+1 & 1 \end{bmatrix}_{2\times 2}.$$

PROBLEM SET 4.4

In each case, find the state-space form directly from the I/O equation provided.

1. $\ddot{y} + \dot{y} + y = 2u$

2. $\ddot{y} + \dot{y} + y = 2\dot{u}$

3. $2\ddot{y} + 3\ddot{y} + \dot{y} + y = 3\ddot{u} + u$

4. $3\ddot{y} + \dot{y} + y = 3\ddot{u} + u$

5. $2\ddddot{y} + \dddot{y} + 2\dot{y} + 3y = \ddot{u} + 2\dot{u}$

6. $\dddot{y} + \ddot{y} + 2\dot{y} + \frac{1}{2}y = 2u$

Given each transfer function $Y(s)/U(s)$,

   a. Find the I/O equation.

   b. Find the state-space form directly from the I/O equation.

7. $\dfrac{s+2}{s^2+s+3}$

8. $\dfrac{s^2+s}{2s^3+s^2+s+2}$

9. $\dfrac{1}{s^2+2s+2}$

10. $\dfrac{2s^2+1}{3s^2+s+2}$

Given each set of state-space matrices **A**, **B**, **C**, and **D**, find the transfer function or matrix

   a. By using Equation 4.21.

   b. In MATLAB using the ``ss2tf'' command.

11. $\mathbf{A} = \begin{bmatrix} 0 & 1 \\ -\frac{2}{3} & -\frac{1}{3} \end{bmatrix}$, $\mathbf{B} = \begin{bmatrix} 0 \\ \frac{1}{3} \end{bmatrix}$, $\mathbf{C} = [0\ 1]$, $D = 0$

12. $\mathbf{A} = \begin{bmatrix} 0 & 1 \\ -\frac{3}{4} & 0 \end{bmatrix}$, $\mathbf{B} = \begin{bmatrix} 0 \\ 1 \end{bmatrix}$, $\mathbf{C} = \begin{bmatrix} \frac{1}{3} & 0 \end{bmatrix}$, $D = 0$

13. $\mathbf{A} = \begin{bmatrix} 0 & 1 & 0 \\ 0 & 0 & 1 \\ -1 & -\frac{3}{2} & -\frac{1}{2} \end{bmatrix}$, $\mathbf{B} = \begin{bmatrix} 0 \\ 0 \\ \frac{1}{2} \end{bmatrix}$, $\mathbf{C} = \begin{bmatrix} 1 & 0 & 0 \\ 0 & 1 & 0 \end{bmatrix}$, $\mathbf{D} = \begin{bmatrix} 0 \\ 0 \end{bmatrix}$

14. $\mathbf{A} = \begin{bmatrix} 0 & 1 & 0 \\ 0 & 0 & 1 \\ -3 & -2 & -1 \end{bmatrix}$, $\mathbf{B} = \begin{bmatrix} 0 & 0 \\ 1 & 0 \\ 0 & 1 \end{bmatrix}$, $\mathbf{C} = \begin{bmatrix} 1 & 0 & 0 \\ 0 & 0 & 1 \end{bmatrix}$, $\mathbf{D} = \mathbf{0}_{2\times 2}$

15. $\mathbf{A} = \begin{bmatrix} 0 & 0 & 1 \\ 1 & -1 & 0 \\ -1 & 1 & -2 \end{bmatrix}$, $\mathbf{B} = \begin{bmatrix} 0 & 0 \\ 1 & 0 \\ 0 & 1 \end{bmatrix}$, $\mathbf{C} = [1\ 0\ 0]$, $\mathbf{D} = [0\ 0]$

16. $\mathbf{A} = \begin{bmatrix} 0 & 1 \\ -\frac{2}{3} & -\frac{1}{3} \end{bmatrix}$, $\mathbf{B} = \begin{bmatrix} 0 \\ 1 \end{bmatrix}$, $\mathbf{C} = \begin{bmatrix} -\frac{1}{9} & -\frac{2}{9} \end{bmatrix}$, $D = \frac{2}{3}$

In each case, an I/O equation is provided. Find

   a. The state-space form.

   b. The transfer function from the state-space form in (a).

   c. The transfer function from the I/O equation and compare with the result in (b).

System Model Representation

17. $\ddot{y} + \dot{y} + 3y = \dot{u} + 3u$
18. $\ddot{y} + \dot{y} + 3y = \ddot{u} + 3\dot{u} + u$
19. $2\ddot{y} + \dot{y} + 2y = \ddot{u} + u$
20. $\dddot{y} + \ddot{y} + \dot{y} + y = \dot{u} + u$

4.5 BLOCK DIAGRAM REPRESENTATION

A block diagram representing a dynamic system is essentially an interconnection of blocks, each block corresponding to an operation carried out by a component, such that the block diagram as a whole agrees with the system's mathematical model. Each block is identified with a transfer function $G(s) = O(s)/I(s)$, also called the gain of the block, as shown in Figure 4.6.

The output of the block is therefore

$$O(s) = G(s)I(s).$$

4.5.1 BLOCK DIAGRAM OPERATIONS

The primary operations in block diagrams include signal amplification, algebraic sum of signals, integration of signals, replacing series and parallel block combinations with equivalent blocks, and treatment of loops.

4.5.1.1 Summing Junction

The output of a summing junction (or a summer) is the algebraic sum of signals that enter the summer. Each signal is accompanied by a positive or negative sign (Figure 4.7). A summing junction may have as many inputs (with the same units) as desired, but only one single output.

4.5.1.2 Series Combinations of Blocks

Figure 4.8 shows two blocks with transfer functions $G_1(s)$ and $G_2(s)$ in a series combination. The block $G_1(s)$ has input $U(s)$ and output $X(s)$, which is the input to the block $G_2(s)$. The output of $G_2(s)$ is $Y(s)$. The objective is to replace this arrangement with a single block that has $U(s)$ as input and $Y(s)$ as output, that is, a single block with the transfer function $Y(s)/U(s)$.

The output of the first block is

$$X(s) = G_1(s)U(s).$$

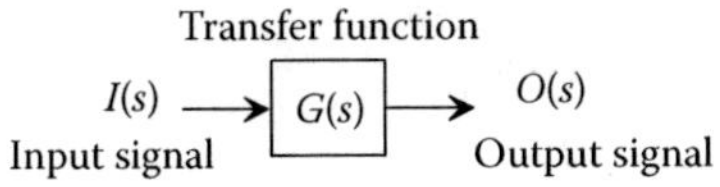

FIGURE 4.6 Schematic of a block.

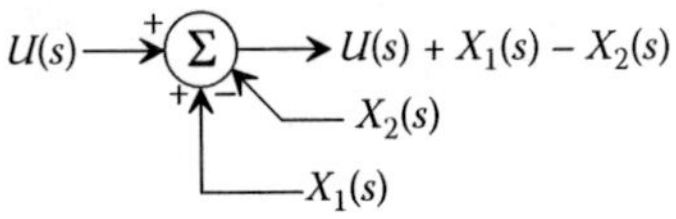

FIGURE 4.7 Summing junction.

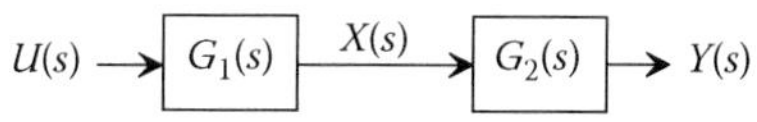

U(s) → [G₁(s)] —X(s)→ [G₂(s)] → Y(s)

FIGURE 4.8 Blocks in series.

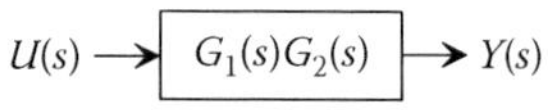

U(s) → [G₁(s)G₂(s)] → Y(s)

FIGURE 4.9 Equivalent block—blocks in series.

The output of the second block is

$$Y(s) = G_2(s)X(s) \stackrel{X(s)=G_1(s)U(s)}{=} G_2(s)[G_1(s)U(s)] = G_1(s)G_2(s)U(s).$$

Therefore,

$$\frac{Y(s)}{U(s)} = G_1(s)G_2(s).$$

Based on this, the series configuration can be replaced with the one in Figure 4.9.

Example 4.18: Blocks in Series

Consider two blocks in a series connection with

$$G_1(s) = \frac{2}{s+3}, \quad G_2(s) = \frac{4}{s^2 + 2s + 2}.$$

The transfer function of the equivalent single block is then determined in MATLAB as follows.

```
>> Num1 = 2; Den1 = [1 3];
>> Num2 = 4; Den2 = [1 2 2];
>> sysG1 = tf(Num1, Den1);      % Define system w/ transfer function G₁(s)
>> sysG2 = tf(Num2, Den2);      % Define system w/ transfer function G₂(s)
>> sysEq = series(sysG1,sysG2)  % Find the equivalent single TF Transfer function:
            8
   --------------------         % Result agrees w/ G₁(s)*G₂(s)
   s^3 + 5 s^2 + 8 s + 6
```

4.5.1.3 Parallel Combinations of Blocks

Consider Figure 4.10 showing two blocks with transfer functions $G_1(s)$ and $G_2(s)$ in a parallel combination. Once again, the goal is to replace the arrangement with a single block that has $U(s)$ as input and $Y(s)$ as output, that is, a single block with the transfer function $Y(s)/U(s)$. The point B in Figure 4.10 is called a branch point.

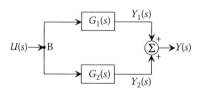

FIGURE 4.10 Blocks in parallel.

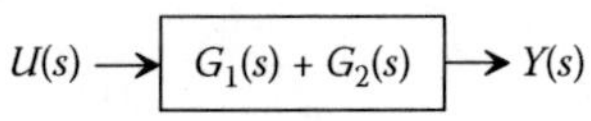

System Model Representation

$$U(s) \rightarrow \boxed{G_1(s) + G_2(s)} \rightarrow Y(s)$$

FIGURE 4.11 Equivalent block—blocks in parallel.

The outputs of the two blocks are simply

$$Y_1(s) = G_1(s)U(s), \quad Y_2(s) = G_2(s)U(s).$$

Consequently,

$$Y(s) = Y_1(s) + Y_2(s) = G_1(s)U(s) + G_2(s)U(s) = [G_1(s) + G_2(s)]U(s),$$

which implies

$$\frac{Y(s)}{U(s)} = G_1(s) + G_2(s).$$

The parallel combination can then be replaced with the single block shown in Figure 4.11.

Example 4.19: ◢ **Blocks in Parallel**

Suppose the two blocks in Example 4.18 are in a parallel connection. The transfer function for the equivalent single block is found in MATLAB as follows.

```
>> Num1 = 2; Den1 = [1 3];
>> Num2 = 4; Den2 = [1 2 2];
>> sysG1 = tf(Num1, Den1);
>> sysG2 = tf(Num2, Den2);
>> sysEq = parallel(sysG1,sysG2)
Transfer function:
   2 s^2 + 8 s + 16
 ---------------------        % Result agrees w/ G_1(s)+G_2(s)
 s^3 + 5 s^2 + 8 s + 6
```

4.5.1.4 Integration

Suppose a signal $u(t)$ is integrated from initial time 0 to the current time t to yield $y(t)$,

$$y(t) = \int_0^t u(t)\,dt.$$

Then, as we learned in Section 2.3,

$$Y(s) = \frac{1}{s}U(s) \quad \overset{\text{Integrator transfer function}}{\Longrightarrow} \quad \frac{Y(s)}{U(s)} = \frac{1}{s}.$$

An integrator is therefore represented as in Figure 4.12.

4.5.1.5 Closed-Loop Systems

A closed-loop (or feedback) system is shown in Figure 4.13. The output $Y(s)$ is fed back through the feedback element $H(s)$ and the outcome $C(s)$ is compared with the input $U(s)$ at the summing junction. The difference, $E(s) = U(s) - C(s)$, is known as the error signal. Because of the negative

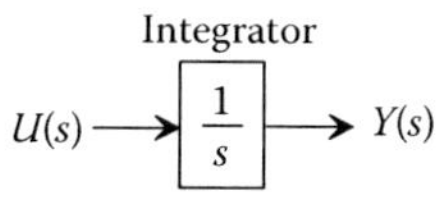

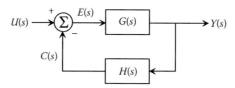

FIGURE 4.12 Integrator.

FIGURE 4.13 A closed-loop system with a feedback element.

sign associated with $C(s)$ at the summing junction, the configuration in Figure 4.13 is called a negative feedback system.

Two important transfer functions here are

$$\text{Feedforward transfer function} = \frac{Y(s)}{E(s)} = G(s)$$

and

$$\text{Open-loop transfer function} = \frac{C(s)}{E(s)} = G(s)H(s).$$

4.5.1.5.1 Closed-Loop Transfer Function

The closed-loop transfer function (CLTF) provides the direct relation between the input $U(s)$ and the output $Y(s)$ and is determined as follows. Referring to Figure 4.13,

$$Y(s) = G(s)E(s) \stackrel{E(s)=U(s)-C(s)}{=} G(s)[U(s) - C(s)] \stackrel{C(s)=H(s)Y(s)}{=} G(s)[U(s) - H(s)Y(s)].$$

Manipulating the above equation, we find

$$[1 + G(s)H(s)]Y(s) = G(s)U(s).$$

Finally, the CLTF is formed as

$$\text{Negative feedback} \quad \frac{Y(s)}{U(s)} = \frac{G(s)}{1 + G(s)H(s)}. \tag{4.22}$$

In the case of a positive feedback, it can easily be shown (see Problem Set 4.5) that

$$\text{Positive feedback} \quad \frac{Y(s)}{U(s)} = \frac{G(s)}{1 - G(s)H(s)}. \tag{4.23}$$

Example 4.20: Negative Feedback

Consider the negative feedback system as in Figure 4.13, where

$$G(s) = \frac{3}{2s+1}, \quad H(s) = s+2.$$

a. Find the CLTF.
b. Repeat in MATLAB.

System Model Representation

SOLUTION

a. By Equation 4.22,

$$\frac{Y(s)}{U(s)} = \frac{G(s)}{1 + G(s)H(s)} = \frac{3/(2s+1)}{1 + (3(s+2))/(2s+1)} = \frac{3}{5s+7}.$$

b.

```
>> NumG = 3;DenG = [2 1];
>> sysG = tf(NumG,DenG);
>> NumH = [1 2];DenH = 1;
>> sysH = tf(NumH,DenH);
>> sysEq = feedback(sysG,sysH)
Transfer function:
     3
  -------
  5 s + 7
```

4.5.2 BLOCK DIAGRAM REDUCTION TECHNIQUES

As dynamic systems get more complex in nature, so do their block diagram representations. The block diagram can then contain several summing junctions, blocks in series or parallel connection, and positive and negative feedback loops. There are a few basic rules that facilitate the process of simplifying a block diagram. These include, among others, moving a branch point and moving a summing junction, as explained below.

4.5.2.1 Moving a Branch Point

A branch point B in Figure 4.14a can be moved to the right side of the block $G(s)$ as shown. The key here is that signals $X(s)$ and $Y(s)$ should be the same before and after B is moved. It is readily seen that in both Figures 4.14a and b, $Y(s) = G(s)U(s)$. Also $X(s) = U(s)$ before B was moved, and $X(s) = [1/G(s)]G(s)U(s) = U(s)$ after the move.

4.5.2.2 Moving a Summing Junction

The summing junction in Figure 4.15a is to be moved to the left side of the block $G_1(s)$ as shown.

The key, once again, is that signal $Y(s)$ goes unaffected by the move. In Figure 4.15a, we have $Y(s) = G_1(s)U_1(s) + G_2(s)U_2(s)$. In Figure 4.15b,

$$Y(s) = G_1(s)\left[U_1(s) + \frac{G_2(s)}{G_1(s)}U_2(s)\right] = G_1(s)U_1(s) + G_2(s)U_2(s).$$

This validates the equivalence of the two arrangements.

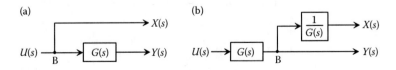

FIGURE 4.14 (a) Branch point and (b) branch point moved.

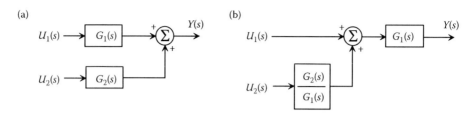

FIGURE 4.15 (a) Summing junction and (b) summing junction moved.

Example 4.21: Block Diagram Reduction

The block diagram in Figure 4.16 can be simplified in several ways, ultimately revealing the overall transfer function $Y(s)/U(s)$. For instance, we can move the summing junction (on the far right) to the left side of the block $1/s$ to generate Figure 4.17. The negative feedback is replaced with a single block with the transfer function

$$\frac{1/(s+1)}{1+1/(s+1)} = \frac{1}{s+2}.$$

The parallel connection is replaced with a single block $2s + 1$. This is shown in Figure 4.18. The series connection in Figure 4.18 is next replaced with a single block whose transfer function is the product of the individual block transfer functions:

$$\frac{2s+1}{s(s+2)}.$$

This yields Figure 4.19, from which the overall transfer function is easily found as

$$\frac{Y(s)}{U(s)} = \frac{2s+1}{s(s+2)}.$$

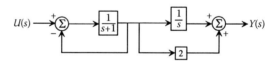

FIGURE 4.16 Block diagram in Example 4.21.

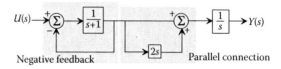

FIGURE 4.17 Summing junction moved.

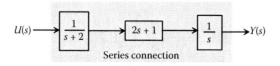

FIGURE 4.18 Feedback loop and parallel connection replaced.

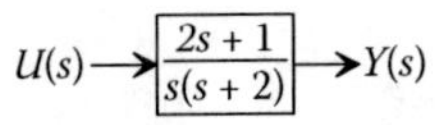

System Model Representation

$$U(s) \rightarrow \boxed{\frac{2s+1}{s(s+2)}} \rightarrow Y(s)$$

FIGURE 4.19 Simplified block diagram in Example 4.21.

4.5.2.3 Mason's Rule

So far we have learned that when a block diagram contains several loops, each loop can be replaced with a single block with a transfer function given by either Equation 4.22 or Equation 4.23 for negative and positive feedbacks, respectively. This, in conjunction with other tactics mentioned earlier, can then help us find the overall transfer function for the block diagram. An alternative approach, however, is to employ Mason's rule outlined below. We first define a forward path as one that originates from the overall input leading to the overall output, never moving in the opposite direction. A loop path (or a loop) is one that originates from a certain variable and returns to the same variable. The gain of a forward path or a loop path is the product of the gains of the individual blocks that constitute the path.

4.5.2.3.1 Mason's Rule: Special Case

Suppose that all forward paths and loops in a block diagram are coupled, that is, they all have a common segment. Then, the overall transfer function is determined as

$$\text{Overall transfer function} = \frac{\Sigma \text{Forward path gains}}{1 - \Sigma \text{Loop gains}}. \qquad (4.24)$$

Example 4.22: Mason's Rule (Special Case)

Find the overall transfer function $Y(s)/U(s)$ for the block diagram shown in Figure 4.20.

SOLUTION

Path segments have been assigned numbers for easier identification. There are two forward paths and two loops:

Forward Path	Gain	Loop	Gain
12356	$G_1 G_2$	2372	$G_1 H_1$
12346	$G_1 G_3$	23582	$-G_1 G_2 H_2$

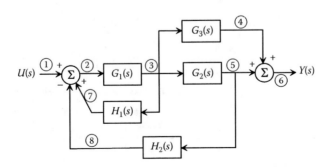

FIGURE 4.20 Block diagram in Example 4.22.

Note that the loop labeled 23582 is a negative feedback, thus its gain is negative. Since all paths have a common segment, labeled 23, the overall transfer function is found via Equation 4.24 as

$$\frac{Y(s)}{U(s)} = \frac{G_1 G_2 + G_1 G_3}{1 - G_1 H_1 + G_1 G_2 H_2}.$$

4.5.2.3.2 Mason's Rule: General Case

In the most general case, when all forward paths and loops are not coupled, the overall transfer function is obtained as

$$\text{Overall transfer function} = \frac{\sum_{k=1}^{m} F_k D_k}{D}, \quad m = \text{number of forward paths,} \qquad (4.25)$$

where F_k is the gain of the kth forward path and

$$D = 1 - \sum \text{single-loop gains} + \sum \text{gain products of all non-touching two-loops}$$
$$- \sum \text{gain products of all non-touching three-loops} + \cdots$$

D_k is the same as D when the block diagram is restricted to the portion not touching the kth forward path.

Example 4.23: Mason's Rule (General Case)

Find the overall transfer function $Y(s)/U(s)$ for the block diagram shown in Figure 4.21.

SOLUTION

Not all forward paths and loops are coupled; in particular, the two single loops do not share a common segment. Therefore, the overall transfer function will be found via the general Mason's rule, Equation 4.25. Two forward paths and two loops are identified:

Forward Path	Gain	Loop	Gain
1234568	$G_1 G_2 G_4$	23(10)2	$-G_1 H_1$
12378	$G_1 G_3$	4594	$-G_2 H_2$

The quantities in Equation 4.25 are determined as follows:

$$F_1 = G_1 G_2 G_4, \quad F_2 = G_1 G_3,$$
$$D = 1 + \underbrace{G_1 H_1 + G_2 H_2}_{\text{Single-loop gains}} + \underbrace{G_1 H_1 G_2 H_2}_{\substack{\text{Gain product of} \\ \text{nontouching two-loops}}}.$$

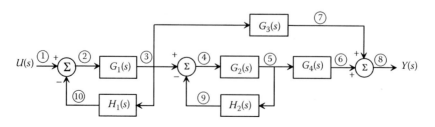

FIGURE 4.21 Block diagram in Example 4.23.

System Model Representation

To find D_1, consider the portion of the block diagram that does not touch the first forward path, 1234568. Since there are no forward paths or loops in the restricted segment, we conclude

$$D_1 = 1.$$

For D_2, imagine the portion of the block diagram that does not touch the second forward path, 12378. The restricted section contains only one single loop, 4594. Therefore,

$$D_2 = 1 + G_2 H_2.$$

Finally, Equation 4.25 yields

$$\frac{Y(s)}{U(s)} = \frac{G_1 G_2 G_4 + G_1 G_3 [1 + G_2 H_2]}{1 + G_1 H_1 + G_2 H_2 + G_1 H_1 G_2 H_2}.$$

4.5.3 Block Diagram Construction from a System Model

Block diagrams reveal many characteristics of dynamic systems that may not be apparent by their mathematical models, such as the interrelation between the different components and variables. In what follows, we will see how to construct partial blocks based on the model of a dynamic system, and then assemble them appropriately to generate the complete block diagram. Based on the block diagram, a model can then be constructed in Simulink® for analysis and simulation purposes (see Section 1.7). This process will be systematically used in the subsequent chapters.

Example 4.24: Block Diagram from a System Model

A first-order system with an input $x(t)$ and an output $y(t)$ is governed by

$$2\dot{y} + 3y = 3x.$$

a. Construct a block diagram that includes a feedback loop.
b. Using the block diagram, find the transfer function $Y(s)/X(s)$.

Solution

a. We start with Figure 4.22a: $X(s)$ is the overall input and appears on the far left of the diagram. The output $Y(s)$ is on the far right of the diagram. There is also a feedback loop, whose sign is to be determined. Rewrite the governing equation as $\dot{y} = \frac{3}{2}(x - y)$ and take the Laplace transform to find

$$sY(s) = \frac{3}{2}[X(s) - Y(s)].$$

Referring to Figure 4.22a, this equation suggests that the signal joining $X(s)$ at the summing junction is $Y(s)$ with a negative sign. The output of the summing junction is then $X(s) - Y(s)$. This output is next multiplied by 3/2 to generate $sY(s)$ (Figure 4.22b). Finally, $sY(s)$ is multiplied by $1/s$ to generate $Y(s)$, which in turn is fed back to complete the diagram (Figure 4.22c).

b. Since there is one forward path and one loop, and they are coupled, Mason's rule (special case) may be applied to find the transfer function as

$$\frac{Y(s)}{X(s)} = \frac{3/2s}{1 + (3/2s)} = \frac{3}{2s + 3}.$$

Note that this agrees with the transfer function directly obtained from the governing equation.

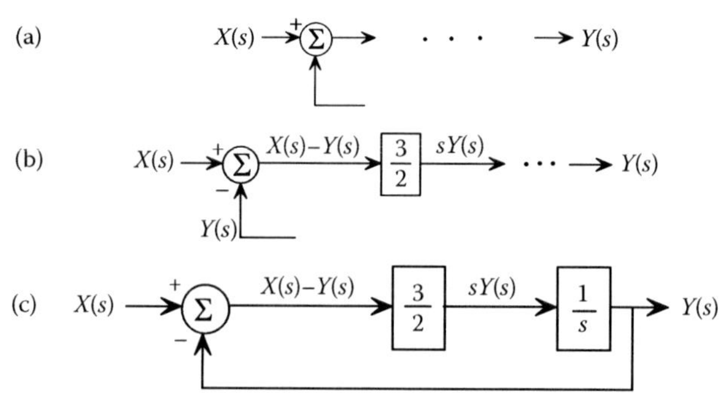

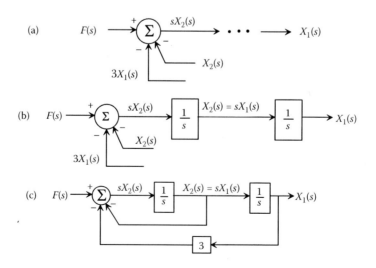

FIGURE 4.22 (a) Initiate block diagram, (b) partial block, and (c) complete diagram.

Example 4.25: Block Diagram from a System Model

A dynamic system is described by

$$\ddot{x} + \dot{x} + 3x = f(t).$$

a. If f is the input and x is the output, find the state-variable equations and the output equation.
b. Construct a block diagram using the information in (a).
c. Build a model in Simulink based on the block diagram of Part (b).

Solution

a. With $x_1 = x$ and $x_2 = \dot{x}$, the state-variable equations and the output equation are

$$\begin{aligned} \dot{x}_1 &= x_2 \\ \dot{x}_2 &= -x_2 - 3x_1 + f \\ y &= x_1 \end{aligned} \quad \overset{\text{Laplace transform}}{\Longrightarrow} \quad \begin{aligned} sX_1 &= X_2, \\ sX_2 &= -X_2 - 3X_1 + F, \\ Y &= X_1. \end{aligned}$$

FIGURE 4.23 (a) Initiate block diagram, (b) partial block, and (c) complete block diagram.

System Model Representation

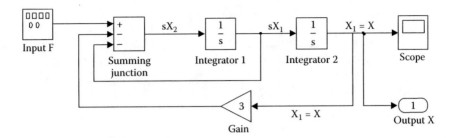

FIGURE 4.24 Simulink model based on Figure 4.23(c).

b. The input $F(s)$ will be on the far left of the diagram, while the output $X_1(s)$ will be on the far right. Starting with the second equation, we build the partial block in Figure 4.23a. The output of the summing junction is $sX_2(s)$, so it needs to be integrated (multiplied by $1/s$) to yield $X_2(s)$. But by the first equation, we have $X_2(s) = sX_1(s)$. Therefore, this signal needs to be integrated to generate $X_1(s)$ (Figure 4.23b). Finally, the feedback loops can be completed and the block diagram is constructed as in Figure 4.23c.

c. Figure 4.24 shows a Simulink model based on the block diagram in Figure 4.23c. The input is represented by a signal generator (Sources library). The output is stored in port 1 (Commonly Used Blocks library) as well as in Scope for simulation purposes. Using this model, the output corresponding to a specified input is easily generated in MATLAB.

PROBLEM SET 4.5

1. Verify the CLTF for a positive feedback, Equation 4.23.
2. Find the CLTF for the feedback system in Figure 4.25.
3. Consider the block diagram in Figure 4.26. Find the transfer function $Y(s)/U(s)$ using
   a. Block diagram reduction techniques.
   b. Mason's rule.
4. In Example 4.21, find the overall transfer function by Mason's rule.
5. Using Mason's rule, find $Y(s)/U(s)$ in Figure 4.27.
6. Reconsider the block diagram in Figure 4.27. Use block diagram reduction techniques as instructed below to find the overall transfer function.
   i. Move the summing junction of the (negative) loop containing $H_2(s)$ outside of the (positive) loop containing $H_1(s)$.

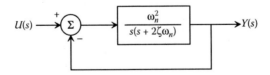

FIGURE 4.25 Problem 2.

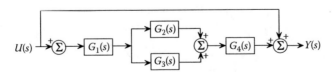

FIGURE 4.26 Problem 3.

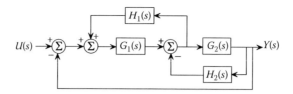

FIGURE 4.27 Problems 5 and 6.

    ii. Replace the loop containing $H_1(s)$ with a single block.
    iii. Treat the two remaining loops in a similar manner in succession.
7. In Figure 4.28, find $Y(s)/U(s)$ using block diagram reduction techniques.
8. Apply Mason's rule to the block diagram in Figure 4.28 to find $Y(s)/U(s)$.
9. In Figure 4.29, find the overall transfer function by using Mason's rule.
10. For the block diagram in Figure 4.30, find $Y(s)/U(s)$ using
    a. Block diagram reduction techniques.
    b. Mason's rule.
11. Consider the block diagram in Figure 4.31 where $V(s)$ represents an external disturbance.
    a. Find the transfer function $Y(s)/U(s)$ by setting $V(s) = 0$.
    b. Find $Y(s)/V(s)$ by setting $U(s) = 0$.
12. Find the overall transfer function for the block diagram in Figure 4.32.
13. A system with input $u(t)$ and output $y(t)$ is described by

$$2\dot{y} + y = 3u.$$

    a. Construct a block diagram containing a feedback loop.
    b. Find the transfer function $Y(s)/U(s)$.

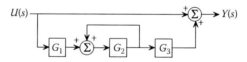

FIGURE 4.28 Problems 7 and 8.

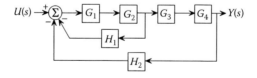

FIGURE 4.29 Problem 9.

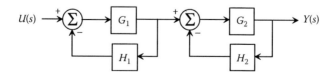

FIGURE 4.30 Problem 10.

System Model Representation

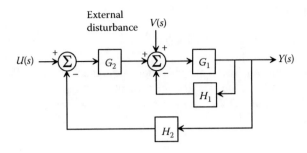

FIGURE 4.31 Problem 11.

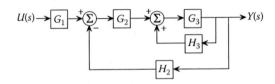

FIGURE 4.32 Problem 12.

14. A system is described by
$$\dot{x}_1 = x_2,$$
$$\dot{x}_2 = -x_1 - 2x_2 + u,$$
$$y = x_1 + x_2,$$
where u is the input, y is the output, and x_1 and x_2 are the state variables. Build the corresponding block diagram.

In problems 15 through 18, an I/O equation is provided.
a. If u is the input and y is the output, find the state-variable equations as well as the output equation.
b. Construct the block diagram.
c. Determine the transfer function $Y(s)/U(s)$ and compare with the one directly obtained from the I/O equation.

15. $2\ddot{y} + \dot{y} + 3y = u$
16. $\ddot{y} + 2\dot{y} + 2y = \dot{u} + 2u$
17. $\dddot{y} + \ddot{y} + \dot{y} + 2y = 2\ddot{u} + u$
18. $\dddot{y} + \ddot{y} + \dot{y} + 2y = \dot{u} + u$
19. A system is described by its transfer function
$$\frac{Y(s)}{U(s)} = \frac{s+2}{2s^2 + s + 3}.$$

a. Find the I/O equation.
b. Find the state-space form directly from the I/O equation.
c. Build a block diagram and find the transfer function, subsequently. Does the result agree with the given transfer function?

20. The state-variable equations and the output equation for a dynamic system are given as
$$\dot{x}_1 = x_2 - 2x_1 + 2u,$$
$$\dot{x}_2 = x_1 - 3x_2,$$
$$y = x_2,$$
where u and y denote the input and the output, respectively. Construct a block diagram and derive the transfer function $Y(s)/U(s)$.

4.6 LINEARIZATION

Up to now, we have routinely studied systems that were assumed to be linear, which of course made their treatment somewhat straightforward. In many practical situations, however, systems contain elements that are inherently nonlinear, which cannot be treated as linear except for a restricted range of operating conditions. In this section, we present a systematic approach to derive a linear approximation of a nonlinear model. In Section 8.5, we will see how the linearized model compares with the nonlinear model.

4.6.1 Linearization of a Nonlinear Element

Consider a nonlinear function f of a single variable x, as shown in Figure 4.33a. The linearization of $f(x)$ will be done with respect to a specific point $P : (\bar{x}, \bar{f})$ known as an operating point. For now we assume that the (constant) values of $\bar{x}$ and $\bar{f}$ are available, but will see shortly how to determine an operating point for a system.

The values $\bar{x}$ and $\bar{f} = f(\bar{x})$ are called the nominal values of x and f, respectively. For any typical point (x, f) on the graph of $f(x)$, we can write (Figure 4.33b)

$$x(t) = \bar{x} + \Delta x(t), \quad f(t) = \bar{f} + \Delta f(t), \tag{4.26}$$

where $\Delta x(t)$ and $\Delta f(t)$ are time varying and called the incremental variables, respectively, for x and f. Graphically, the linear approximation of $f(x)$ is provided by the tangent line to the curve at the operating point P, with a reasonably good accuracy in a small neighborhood of P, that is, as long as $\Delta x(t)$ and $\Delta f(t)$ assume small values. Analytically, this is handled by writing the Taylor series expansion of $f(x)$ about the operating point as

$$f(x) = \underbrace{f(\bar{x}) + \left.\frac{df}{dx}\right|_P (x - \bar{x})}_{\text{Linear terms}} + \frac{1}{2!}\left.\frac{d^2f}{dx^2}\right|_P (x - \bar{x})^2 + \cdots.$$

Assuming $x - \bar{x}$ is small, the linear approximation of $f(x)$ is obtained by retaining the first two terms, neglecting the remaining terms that are of higher order in $x - \bar{x}$, so that

$$f(x) \cong f(\bar{x}) + \left.\frac{df}{dx}\right|_{\bar{x}} \Delta x. \tag{4.27}$$

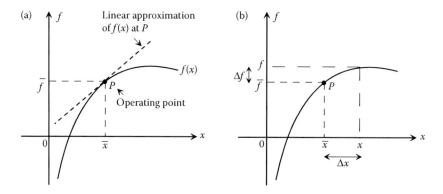

FIGURE 4.33 (a) Linearization about an operating point and (b) incremental variables.

System Model Representation

Example 4.26: Function of a Single Variable

Linearize $f(x) = x|x|$ about the operating point $P : (2, 4)$. Examine the accuracy of the linear approximation for $x = 2.1$ and $x = 1.9$.

Solution

We first note that

$$f(x) = x|x| = \begin{cases} x^2 & \text{if } x \geq 0 \\ -x^2 & \text{if } x < 0 \end{cases} \xRightarrow{\text{Differentiate}} \frac{df}{dx} = \begin{cases} 2x & \text{if } x \geq 0, \\ -2x & \text{if } x < 0. \end{cases}$$

Since $\bar{x} = 2$, Equation 4.27 yields

$$f(x) \cong f(2) + [2x]_{x=2} \Delta x = 4 + 4\Delta x.$$

For $x = 2.1$, we have $\Delta x = x - \bar{x} = 2.1 - 2 = 0.1$ and thus

$$f(2.1) \cong 4 + 4(0.1) = 4.40 \qquad [\text{Exact value} = (2.1)^2 = 4.41].$$

For $x = 1.9$, we have $\Delta x = 1.9 - 2 = -0.1$ and thus

$$f(1.9) \cong 4 + 4(-0.1) = 3.60 \qquad [\text{Exact value} = (1.9)^2 = 3.61].$$

Therefore, as expected, if x does not deviate too much from $\bar{x}$, the accuracy is reasonably good.

4.6.1.1 Functions of Two Variables

Suppose the nonlinear function is $f(x, y)$, a function of two independent variables. In this case, the operating point is represented by $(\bar{x}, \bar{y}, \bar{f})$ and the incremental variables are defined as

$$\Delta x(t) = x(t) - \bar{x}, \quad \Delta y(t) = y(t) - \bar{y}, \quad \Delta f(t) = f(t) - \bar{f}.$$

Taylor series expansion about the operating point yields

$$f(x, y) = \underbrace{f(\bar{x}, \bar{y}) + \left.\frac{\partial f}{\partial x}\right|_{\bar{x}, \bar{y}} (x - \bar{x}) + \left.\frac{\partial f}{\partial y}\right|_{\bar{x}, \bar{y}} (y - \bar{y})}_{\text{Linear terms}}$$

$$+ \frac{1}{2!} \left.\frac{\partial^2 f}{\partial x^2}\right|_{\bar{x}, \bar{y}} (x - \bar{x})^2 + \frac{1}{2!} \left.\frac{\partial^2 f}{\partial y^2}\right|_{\bar{x}, \bar{y}} (y - \bar{y})^2 + \left.\frac{\partial^2 f}{\partial x \partial y}\right|_{\bar{x}, \bar{y}} (x - \bar{x})(y - \bar{y}) + \cdots.$$

Assuming $x - \bar{x}$ and $y - \bar{y}$ are small, the linear approximation of $f(x, y)$ in a small neighborhood of the operating point is given by

$$f(x, y) \cong f(\bar{x}, \bar{y}) + \left.\frac{\partial f}{\partial x}\right|_{\bar{x}, \bar{y}} \Delta x + \left.\frac{\partial f}{\partial y}\right|_{\bar{x}, \bar{y}} \Delta y. \qquad (4.28)$$

4.6.2 Linearization of a Nonlinear Model

Linearization of a nonlinear system model can be performed systematically by following a standard procedure explained shortly. And a key part of this procedure involves the determination of the operating point(s).

4.6.2.1 Operating Point

To find the operating point, we first replace the dependent variables such as $x(t)$ with $\bar{x} = \text{const}$ to be determined. It is also desired that a system operates not far from its equilibrium state. For that, we set the time-varying portion of the input $u(t)$ equal to zero. The result is an algebraic equation that can be solved for variables such as $\bar{x}$. Note that a system may have more than one operating point.

Example 4.27: Operating Point

Find the operating point(s) for a nonlinear system whose model is described by

$$2\ddot{x} + \dot{x} + x|x| = \underbrace{u(t)}_{\text{Unit-step function}}, \quad t \geq 0.$$

SOLUTION

Since $x(t)$ is the only dependent variable involved, the operating point is identified by $\bar{x}$. Also, in the range of $t \geq 0$, the unit-step function is simply 1. Thus, the time-varying part of the input here is already zero. To find the operating point, we replace x with $\bar{x}$ and u with 1. Noting that $\bar{x} = \text{const}$, we have $\ddot{\bar{x}} = 0$ and $\dot{\bar{x}} = 0$. Therefore,

$$\bar{x}|\bar{x}| = 1. \tag{a}$$

We solve this equation as follows.

Case (1) If $\bar{x} > 0$, then $|\bar{x}| = \bar{x}$ and Equation (a) reduces to $\bar{x}^2 = 1$. This has two solutions, $\bar{x} = \pm 1$. But the assumption in this case is $\bar{x} > 0$, so that only the positive solution is acceptable, that is,

$$\bar{x} = 1.$$

Case (2) If $\bar{x} < 0$, then $|\bar{x}| = -\bar{x}$ and Equation (a) becomes $-\bar{x}^2 = 1$, which has no real solution. Therefore, the only valid operating point is given by $\bar{x} = 1$.

4.6.2.2 Linearization Procedure

A nonlinear system model is linearized as follows:

1. Find the operating point as previously explained.
2. Linearize the nonlinear term(s) about the operating point by means of Taylor series expansions; Equation 4.27 for functions of a single variable and Equation 4.28 for two variables.
3. In the original nonlinear model, replace variables such as x with $\bar{x} + \Delta x$, nonlinear terms with their linear approximations of Step 2, and include the time-varying portions of the input that were previously set to zero to calculate the operating point in Step 1. The resulting system is linear in the incremental variables such as Δx.
4. Finally, use the initial conditions of the original model to calculate those for the linearized model. For instance, knowing $x(0)$ for the original system, find $\Delta x(0)$ by noting that $\Delta x(t) = x(t) - \bar{x}$ so that $\Delta x(0) = x(0) - \bar{x}$.

Example 4.28: Linearization Procedure

A nonlinear system model is given as

$$\ddot{x} + \dot{x} + x|x| = 1 + \sin t, \quad x(0) = 0, \quad \dot{x}(0) = 1.$$

Derive a linearized model.

SOLUTION

We will follow the procedure outlined above.

System Model Representation

1. To find the operating point, we replace x with $\bar{x}$ and set $\sin t$ to zero. As a result,

$$\bar{x}|\bar{x}| = 1.$$

    This equation is then solved as in Example 4.27 to give $\bar{x} = 1$.

2. The only nonlinear term here is $f(x) = x|x|$. We will linearize it about the operating point via Equation 4.27:

$$f(x) \cong f(1) + [2x]_{x=1}\Delta x = 1 + 2\Delta x.$$

    (See also Example 4.26.)

3. In the original model, replace x with $1 + \Delta x$ (since $\bar{x} = 1$), the nonlinear term with $1 + 2\Delta x$, and bring back $\sin t$, which had previously been set to zero, to obtain

$$\frac{d^2(1+\Delta x)}{dt^2} + \frac{d(1+\Delta x)}{dt} + (1+2\Delta x) = 1 + \sin t.$$

    This simplifies to

$$\Delta\ddot{x} + \Delta\dot{x} + 2\Delta x = \sin t.$$

4. The initial conditions are adjusted as follows:

$$\Delta x(t) = x(t) - 1 \xRightarrow{\text{Differentiate}} \Delta\dot{x}(t) = \dot{x}(t) \Rightarrow \begin{array}{l} \Delta x(0) = x(0) - 1 = -1, \\ \Delta\dot{x}(0) = \dot{x}(0) = 1. \end{array}$$

    In summary, the linearized model is derived as

$$\Delta\ddot{x} + \Delta\dot{x} + 2\Delta x = \sin t, \quad \Delta x(0) = -1, \quad \Delta\dot{x}(0) = 1.$$

    This can easily be solved to generate the solution $\Delta x(t)$. It is important to note, however, that $\Delta x(t)$ is not compatible with the solution $x(t)$ of the original nonlinear system. To make them compatible, we must recall that $x(t) = 1 + \Delta x(t)$ in this problem. Therefore, the solution $\Delta x(t)$ of the linear model needs to be raised by 1 unit to be compatible with $x(t)$. We will elaborate on this and many other related issues in Section 8.5.

Example 4.29: Linearization Procedure

A system is governed by its nonlinear state-variable equations as

$$\begin{cases} \dot{x}_1 = 2x_2 - 2 \\ \dot{x}_2 = (x_1 - 1)^3 x_2 + 1 \end{cases} \quad \text{subjected to} \quad \begin{array}{l} x_1(0) = 1, \\ x_2(0) = -1. \end{array}$$

Derive a linearized model.

Solution

We will follow the standard procedure outlined earlier.

1. Replace x_1 and x_2 with $\bar{x}_1$ and $\bar{x}_2$, respectively. Since there are no time-varying segments in the input(s), no further modification is needed.

$$\begin{cases} 0 = 2\bar{x}_2 - 2 \\ 0 = (\bar{x}_1 - 1)^3 \bar{x}_2 + 1 \end{cases} \Longrightarrow \begin{cases} \bar{x}_2 = 1 \\ (\bar{x}_1 - 1)^3 = -1 \end{cases} \Longrightarrow \begin{array}{l} \bar{x}_2 = 1, \\ \bar{x}_1 = 0. \end{array}$$

Therefore, the operating point is $(\bar{x}_1, \bar{x}_2) = (0, 1)$.

2. The only nonlinear element is $f(x_1, x_2) = (x_1 - 1)^3 x_2$, which will be linearized about the operating point $(0, 1)$ following Equation 4.28,

$$f(x_1, x_2) = f(0, 1) + \left.\frac{\partial f}{\partial x_1}\right|_{(0,1)} \Delta x_1 + \left.\frac{\partial f}{\partial x_2}\right|_{(0,1)} \Delta x_2$$

$$= -1 + \left[3(x_1 - 1)^2 x_2\right]_{(0,1)} \Delta x_1 + \left[(x_1 - 1)^3\right]_{(0,1)} \Delta x_2$$

$$= -1 + 3\Delta x_1 - \Delta x_2.$$

3. In the original model, replace x_1 with Δx_1 (since $\bar{x}_1 = 0$), x_2 with $1 + \Delta x_2$ (since $\bar{x}_2 = 1$) and the nonlinear term with its linear approximation. No other adjustments need be made as no time-varying portions of input were set to zero.

$$\begin{cases} \Delta \dot{x}_1 = 2(1 + \Delta x_2) - 2 \\ \Delta \dot{x}_2 = -1 + 3\Delta x_1 - \Delta x_2 + 1 \end{cases} \implies \begin{cases} \Delta \dot{x}_1 = 2\Delta x_2, \\ \Delta \dot{x}_2 = 3\Delta x_1 - \Delta x_2. \end{cases}$$

4. The initial conditions are modified as

$$\begin{aligned} \Delta x_1(0) &= x_1(0) - \bar{x}_1 \\ \Delta x_2(0) &= x_2(0) - \bar{x}_2 \end{aligned} \implies \begin{aligned} \Delta x_1(0) &= 1, \\ \Delta x_2(0) &= -2. \end{aligned}$$

The linearized model is therefore

$$\begin{cases} \Delta \dot{x}_1 = 2\Delta x_2, \\ \Delta \dot{x}_2 = 3\Delta x_1 - \Delta x_2, \end{cases} \quad \begin{aligned} \Delta x_1(0) &= 1, \\ \Delta x_2(0) &= -2. \end{aligned}$$

This system is subsequently solved for $\Delta x_1(t)$ and $\Delta x_2(t)$. Once again, it must be noted that $\Delta x_1(t)$ and $\Delta x_2(t)$ are not (generally) compatible with the coordinates $x_1(t)$ and $x_2(t)$ of the original nonlinear system. Compatibility is achieved by taking into account that $x_1 = \Delta x_1$ and $x_2 = 1 + \Delta x_2$.

4.6.2.3 Small-Angle Linearization

We now turn our attention to dynamic systems whose mathematical models contain nonlinear terms that are in terms of trigonometric functions (sine and cosine) of an angle. In practice, such cases arise when dealing with mechanical systems experiencing rotational motion (Section 5.4) through angle θ. These types of models can be approximated as linear as long as the angle is small. In particular, when $\theta \ll 1$ rad, then

$$\begin{aligned} \sin \theta &\cong \theta, \\ \cos \theta &\cong 1, \\ \theta \dot{\theta}^2 &\cong 0. \end{aligned} \tag{4.29}$$

Example 4.30: Small-Angle Linearization

The governing equations for a dynamic system have been derived as

$$\begin{cases} 2\ddot{x} + \ddot{\theta} - \dot{\theta}^2 \sin\theta + \dot{x} + x = f(t), \\ \ddot{\theta} + \ddot{x}\cos\theta + 10\sin\theta = 0. \end{cases}$$

a. Derive the linearized model for $\theta \ll 1$ rad.
b. Obtain the state-variable equations for the linearized model and write the state equation.

System Model Representation

SOLUTION

a. Using the first two approximations in Equation 4.29, we have

$$\begin{cases} 2\ddot{x} + \ddot{\theta} - \dot{\theta}^2\theta + \dot{x} + x = f(t) \\ \ddot{\theta} + \ddot{x} + 10\theta = 0 \end{cases} \xrightarrow{\dot{\theta}^2\theta \cong 0} \begin{array}{l}(a) \\ (b)\end{array} \begin{cases} 2\ddot{x} + \ddot{\theta} + \dot{x} + x = f(t), \\ \ddot{\theta} + \ddot{x} + 10\theta = 0. \end{cases}$$

b. In its present form, the linearized model above cannot be transformed into state-variable equations. This is mainly because both $\ddot{x}$ and $\ddot{\theta}$ appear in the same equation. The remedy, however, is to manipulate the two equations, labeled Equation (a) and Equation (b), to eliminate the unwanted variables, as follows. From Equation (b), find $\ddot{\theta} = -10\theta - \ddot{x}$ and insert into Equation (a) and simplify:

$$\ddot{x} + \dot{x} + x - 10\theta = f(t). \qquad (c)$$

This equation is now in the correct form. Next, from Equation (b), find $\ddot{x} = -10\theta - \ddot{\theta}$ and insert into Equation (a) and simplify:

$$\ddot{\theta} + 20\theta - \dot{x} - x = -f(t). \qquad (d)$$

In summary,

$$\begin{array}{l}(c) \\ (d)\end{array} \begin{cases} \ddot{x} + \dot{x} + x - 10\theta = f(t), \\ \ddot{\theta} + 20\theta - \dot{x} - x = -f(t). \end{cases}$$

We may now proceed as in Section 4.2 by selecting the state variables as $x_1 = x$, $x_2 = \theta$, $x_3 = \dot{x}$, and $x_4 = \dot{\theta}$. The state-variable equations are then formed as

$$\begin{cases} \dot{x}_1 = x_3, \\ \dot{x}_2 = x_4, \\ \dot{x}_3 = -x_3 - x_1 + 10x_2 + f(t), \\ \dot{x}_4 = -20x_2 + x_3 + x_1 - f(t). \end{cases}$$

Finally, the state equation is obtained as

$$\dot{\mathbf{x}} = \mathbf{A}\mathbf{x} + \mathbf{B}u, \quad \mathbf{x} = \begin{Bmatrix} x_1 \\ x_2 \\ x_3 \\ x_4 \end{Bmatrix}, \quad \mathbf{A} = \begin{bmatrix} 0 & 0 & 1 & 0 \\ 0 & 0 & 0 & 1 \\ -1 & 10 & -1 & 0 \\ 1 & -20 & 1 & 0 \end{bmatrix}, \quad \mathbf{B} = \begin{bmatrix} 0 \\ 0 \\ 1 \\ -1 \end{bmatrix}, \quad u = f(t).$$

These types of systems will also be discussed in greater detail in Section 8.5.

PROBLEM SET 4.6

In each case, the mathematical model of a nonlinear dynamic system is given, together with the initial conditions. Follow the procedure outlined in this section to find the operating point(s) and subsequently, derive the linearized model.

1. $\ddot{x} + 2\dot{x} + x^3 = 1$, $x(0) = -1$, $\dot{x}(0) = 0$
2. $\ddot{x} + \dot{x} + 2x|x| = -3 + 2\cos t$, $x(0) = 0$, $\dot{x}(0) = 1$
3. $2\ddot{x} + \dot{x} + x|x| = 2 + \sin 3t$, $x(0) = \sqrt{2}$, $\dot{x}(0) = 1$
4. $\ddot{x} + \dot{x} + x\sqrt{|x|} = 1 + \sin t$, $x(0) = 1$, $\dot{x}(0) = 0$
5. $\ddot{x} + \dot{x} + g(x) = 1 + \cos t$, $x(0) = 1$, $\dot{x}(0) = 0$, $g(x) = \begin{cases} 2\sqrt{x} & \text{if } x \geq 0 \\ -2\sqrt{|x|} & \text{if } x < 0 \end{cases}$

6. $3\ddot{x} + \dot{x} + g(x) = 2 + \cos t, \quad x(0) = 1, \dot{x}(0) = 0, g(x) = \begin{cases} 2(1-e^{-x}) & \text{if } x \geq 0 \\ -2(1-e^{-x}) & \text{if } x < 0 \end{cases}$

7. $\begin{cases} \dot{x}_1 = x_2, & x_1(0) = 0 \\ \dot{x}_2 = -x_1 - x_1|x_1| - x_2^3 - 2 + 0.1\sin t, & x_2(0) = 1 \end{cases}$

8. $\begin{cases} \dot{x}_1 = x_2 - x_1, & x_1(0) = -2 \\ \dot{x}_2 = 2/x_2 + 1 + 0.9\cos t, & x_2(0) = -1 \end{cases}$

9. $\begin{cases} \dot{x}_1 = -x_1 + 2x_2^3, & x_1(0) = 0 \\ \dot{x}_2 = x_1 + 16 + 2\sin t, & x_2(0) = -2 \end{cases}$

10. $\begin{cases} \dot{x}_1 = -x_2 - x_1|x_1| - 3 + \cos 2t, & x_1(0) = -1 \\ \dot{x}_2 = x_1 - x_2 - 3, & x_2(0) = 1 \end{cases}$

4.7 SUMMARY

A set of generalized coordinates completely describes the motion of a system. Suppose a dynamic system has n generalized coordinates $q_1, q_2, \ldots, q_n$ and its model is described by

$$\begin{cases} \ddot{q}_1 = f_1(q_1, q_2, \ldots, q_n, \dot{q}_1, \dot{q}_2, \ldots, \dot{q}_n, t) \\ \ddot{q}_2 = f_2(q_1, q_2, \ldots, q_n, \dot{q}_1, \dot{q}_2, \ldots, \dot{q}_n, t) \\ \quad \vdots \\ \ddot{q}_n = f_n(q_1, q_2, \ldots, q_n, \dot{q}_1, \dot{q}_2, \ldots, \dot{q}_n, t) \end{cases},$$

where $\dot{q}_1, \dot{q}_2, \ldots, \dot{q}_n$ are the generalized velocities and $f_1, f_2, \ldots, f_n$ are the algebraic functions of $q_i, \dot{q}_i, (i = 1, 2, \ldots, n)$ and time t, known as the generalized forces. This model, subjected to initial generalized coordinates and initial generalized velocities, is called the configuration form.

Models of dynamic systems governed by n-dimensional systems of second-order differential equations can be expressed in the standard, second-order matrix form

$$\mathbf{M}\ddot{\mathbf{x}} + \mathbf{C}\dot{\mathbf{x}} + \mathbf{K}\mathbf{x} = \mathbf{f},$$

where

- $\mathbf{x}_{n \times 1}$ = configuration vector, $\mathbf{f}_{n \times 1}$ = vector of external forces,
- $\mathbf{M}_{n \times n}$ = mass matrix, $\mathbf{C}_{n \times n}$ = damping matrix, $\mathbf{K}_{n \times n}$ = stiffness matrix.

The smallest set of independent variables that completely describe the state of a system is the set of state variables, denoted by $x_i (i = 1, 2, \ldots, n)$. The number of state variables is the number of initial conditions needed to completely solve the system's model. The state variables are those variables for which initial conditions are needed. For a linear system with state variables $x_1, \ldots, x_n$, inputs $u_1, \ldots, u_m$, and outputs $y_1, \ldots, y_p$, the state-space form is

$$\begin{cases} \dot{\mathbf{x}} = \mathbf{A}_{n \times n} \mathbf{x}_{n \times 1} + \mathbf{B}_{n \times m} \mathbf{u}_{m \times 1}, \\ \mathbf{y}_{p \times 1} = \mathbf{C}_{p \times n} \mathbf{x}_{n \times 1} + \mathbf{D}_{p \times m} \mathbf{u}_{m \times 1}, \end{cases}$$

System Model Representation

where

- $\mathbf{x}_{n\times 1}$ = state vector, $\mathbf{A}_{n\times n}$ = state matrix, $\mathbf{B}_{n\times m}$ = input matrix, $\mathbf{u}_{m\times 1}$ = input vector
- $\mathbf{C}_{p\times n}$ = output matrix, $\mathbf{D}_{p\times m}$ = direct transmission matrix, $\mathbf{y}_{p\times 1}$ = output vector

If $u(t)$ is an input and $y(t)$ is an output of a system, then the I/O equation is

$$y^{(n)} + a_1 y^{(n-1)} + \cdots + a_{n-1}\dot{y} + a_n y = b_0 u^{(m)} + b_1 u^{(m-1)} + \cdots + b_{m-1}\dot{u} + b_m u, \quad m \le n,$$

where $a_1,\ldots,a_n$ and $b_0, b_1,\ldots,b_m$ are constants, and $y^{(n)} = d^n y/dt^n$.

A transfer function is the ratio of the Laplace transforms of the output and input. If a system has q inputs and r outputs, the qr transfer functions are assembled in an $r \times q$ transfer matrix, denoted by $\mathbf{G}(s) = [G_{ij}(s)]$, where $i = 1, 2, \ldots, r$ and $j = 1, 2, \ldots, q$.

A block diagram is an interconnection of blocks, each block corresponding to an operation carried out by a component, such that the block diagram as a whole agrees with the system's mathematical model.

A nonlinear system model is linearized as follows:

- Find the operating point.
- Linearize the nonlinear term(s) about the operating point.
- In the original nonlinear model, express variables in terms of incremental variables, and replace nonlinear terms with their linear approximations. The resulting system is linear in the incremental variables.
- Find the initial conditions for the incremental variables.

REVIEW PROBLEMS

1. The governing equations for a dynamic system are given as

$$\begin{cases} 3\ddot{x}_1 + \dot{x}_1 + x_1 - 2x_2 = f(t), \\ 2\dot{x}_2 - x_1 - 2x_2 = 0. \end{cases}$$

   a. Assuming that x_2 is the output and $f(t)$ is the input, obtain the state-space form.
   b. ◂ Determine whether the system is stable.
   c. Find the transfer function $X_2(s)/F(s)$ directly from the governing equations.

2. A dynamic system is described by

$$4\ddot{x} + 2\dot{x} + x = f(t).$$

   a. If $f(t)$ and $\dot{x}$ are the system input and output, respectively, find the state-space form.
   b. Find the transfer function $G(s)$ directly from the state-space form.
   c. ◂ Decide if the system is stable by examining the state matrix in (a). Also verify that the eigenvalues of the state matrix are the same as the poles of the transfer function $G(s)$.

3. A system's transfer function is defined as

$$G(s) = \frac{Y(s)}{U(s)} = \frac{4}{3s^2 + s + 2}.$$

   a. Find the system's I/O equation.
   b. Find the state-space form directly from the I/O equation.
   c. Find the transfer function directly from the state-space form, and compare with $G(s)$.

4. A system is described by its governing equations

$$\begin{cases} \ddot{x}_1 + 2x_1 - 2(x_2 - x_1) = f_1(t), \\ \ddot{x}_2 + 2(x_2 - x_1) = 0, \end{cases}$$

where f_1 is the input, and x_1 and x_2 are the outputs.

a. Obtain the state-space form.
b. ◢ Find the transfer matrix directly from the state-space form.
c. Find the system's transfer matrix using the governing equations, and compare with (b).

5. A system's I/O equation is given as

$$\ddot{y} + \frac{1}{2}\dot{y} + y = 2\dot{u} + u.$$

a. Find the state-space form.
b. Find the transfer function directly from the state-space form.

6. Repeat Problem 5 for

$$\frac{1}{2}\ddot{y} + \dot{y} + 2y = \ddot{u} + 3u.$$

7. A system's transfer function is given as

$$G(s) = \frac{Y(s)}{U(s)} = \frac{s+3}{s^3 + s^2 + 2s + 3}.$$

a. Find the system's I/O equation.
b. Find the state-space form directly from the I/O equation.
c. ◢ Find the transfer function from the state-space form and compare with $G(s)$.

8. A SISO dynamic system is modeled through its state-space form as

$$\begin{cases} \dot{\mathbf{x}} = \mathbf{A}\mathbf{x} + \mathbf{B}u, \\ y = \mathbf{C}\mathbf{x} + Du, \end{cases}$$

with

$$\mathbf{A} = \begin{bmatrix} 0 & 1 \\ -1 & -2 \end{bmatrix}, \quad \mathbf{B} = \begin{bmatrix} 0 \\ 1 \end{bmatrix}, \quad u = u, \quad \mathbf{C} = \begin{bmatrix} 1 & 2 \end{bmatrix}, \quad D = 0.$$

Find the system's I/O equation.

9. Repeat Problem 8 for

$$\mathbf{A} = \begin{bmatrix} 0 & 1 \\ -\frac{1}{3} & -1 \end{bmatrix}, \quad \mathbf{B} = \begin{bmatrix} 0 \\ \frac{1}{3} \end{bmatrix}, \quad u = u, \quad \mathbf{C} = \begin{bmatrix} 1 & 1 \end{bmatrix}, \quad D = 0.$$

10. A MIMO system is modeled through its state-space form as

$$\begin{cases} \dot{\mathbf{x}} = \mathbf{A}\mathbf{x} + \mathbf{B}u, \\ \mathbf{y} = \mathbf{C}\mathbf{x} + \mathbf{D}u, \end{cases}$$

with

$$\mathbf{A} = \begin{bmatrix} 0 & 1 \\ -3 & -1 \end{bmatrix}, \quad \mathbf{B} = \begin{bmatrix} 0 \\ 2 \end{bmatrix}, \quad u = u, \quad \mathbf{C} = \begin{bmatrix} 1 & 0 \\ 0 & 2 \end{bmatrix}, \quad \mathbf{D} = \begin{bmatrix} 0 \\ 0 \end{bmatrix}.$$

Labeling the system outputs y_1 and y_2, find all possible I/O equations.

11. Find the transfer function, using Mason's rule, for the system shown in Figure 4.34.
12. Using Mason's rule, find the transfer function $Y(s)/U(s)$ in Figure 4.35 where k is a constant.
13. In Figure 4.36, find $Y(s)/U(s)$ using Mason's rule.
14. For the system in Figure 4.37, find the transfer function using Mason's rule.
15. For the system in Figure 4.38, find $Y(s)/U(s)$ using Mason's rule.
16. In Problem 15, find the transfer function by means of block diagram reduction.

System Model Representation

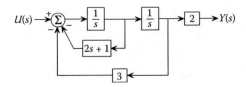

FIGURE 4.34 Problem 11.

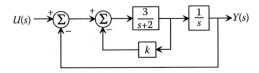

FIGURE 4.35 Problem 12.

17. A system's I/O equation is given as

$$\dddot{x} + 3\dot{x} + 3x = 2\dot{u} + u.$$

   a. Obtain the state-space form.
   b. Construct a block diagram using (a).
   c. Find the transfer function using Mason's rule.
18. Repeat Problem 17 for the I/O equation

$$\ddot{y} + \dot{y} + 2y = \ddot{u} + 3u.$$

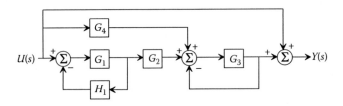

FIGURE 4.36 Problem 13.

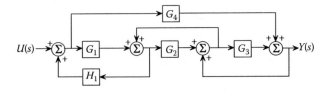

FIGURE 4.37 Problem 14.

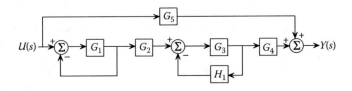

FIGURE 4.38 Problem 15.

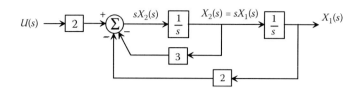

FIGURE 4.39 Problem 19.

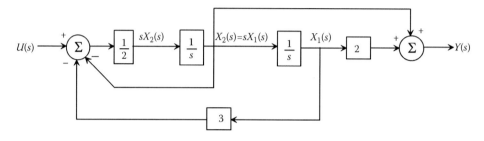

FIGURE 4.40 Problem 20.

19. Consider the block diagram in Figure 4.39.
    a. Find the state-space form directly from the block diagram.
    b. Find the system's transfer function directly from the state-space form.
    c. Find the system's transfer function directly from the block diagram and compare with the result in (b).
20. Repeat Problem 19 for the block diagram in Figure 4.40.
21. A nonlinear system is modeled as

$$\begin{cases} \dot{x}_1 = x_1|x_1| + x_2 - 1 + \cos 2t, & x_1(0) = 3, \\ \dot{x}_2 = -x_1 - x_2 - 1, & x_2(0) = -1. \end{cases}$$

    Derive the linearized model.
22. Repeat Problem 21 for

$$\begin{cases} \dot{x}_1 = x_2 + \sin t, & x_1(0) = 0, \\ \dot{x}_2 = x_1^3 + 3x_2 - 8, & x_2(0) = 1. \end{cases}$$

5 Mechanical Systems

The modeling techniques for mechanical systems are discussed in this chapter. Mechanical systems are in either translational or rotational motion, or both. We begin this chapter by introducing mechanical elements, which include mass elements, spring elements, and damper elements. The concept of equivalence is discussed, which simplifies the modeling of systems in many applications. We then review Newton's second law and apply it to translational systems. For rotational systems, the moment equations are used to obtain dynamic models. Following that we discuss the situation of general plane motion involving both translation and rotation. The chapter concludes with a coverage of gear–train systems.

5.1 MECHANICAL ELEMENTS

The objective of this chapter is to show how one can obtain mathematical models of mechanical systems. Since a real mechanical system is usually complicated, simplifying assumptions must be made to reduce the system to an idealized model, which consists of interconnected elements. The behavior of the mathematical model can then approximate that of the real system.

A mathematical model of a mechanical system can be constructed based on physical laws (such as Newton's laws and the conservation of energy) that the elements and their interconnections must obey. Elements can be broadly divided into three classes according to whether element forces are proportional to accelerations, proportional to displacements, or proportional to velocities. Correspondingly, they can be divided into elements that store and release kinetic energy, store and release potential energy, and dissipate energy. In this section, element equations relating the external forces to the associated element variables are presented.

5.1.1 Mass Elements

Figure 5.1 shows a mass m traveling with a velocity v. The basic variables used to describe the dynamic behavior of a translational mechanical system are the acceleration vector $\mathbf{a}$, the velocity vector $\mathbf{v}$, and the position vector $\mathbf{r}$. They are related by the time derivatives

$$\mathbf{a} = \frac{d\mathbf{v}}{dt} = \frac{d^2\mathbf{r}}{dt^2}, \qquad (5.1)$$

which can also be represented in the simple dot notation

$$\mathbf{a} = \dot{\mathbf{v}} = \ddot{\mathbf{r}}. \qquad (5.2)$$

Assume that the motion of the mass in Figure 5.1 is under the influence of an externally applied force, and the motion of the mass is constrained in only one direction. According to Newtonian mechanics, the resulting force f acting on the mass is equal to the time rate of change of momentum. For a constant mass, Newton's second law is expressed as

$$f = \frac{d}{dt}(mv) = m\frac{dv}{dt} = ma. \qquad (5.3)$$

Note that the acceleration a is absolute and must be measured with respect to an inertial reference frame. For ordinary systems at or near the surface of the earth, the ground can be approximated as a reference for motion.

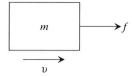

FIGURE 5.1 A mass traveling with a velocity v.

Mass elements store mechanical energy. The energy stored in a mass is kinetic energy if the mass is in motion. The kinetic energy is expressed as

$$T = \frac{1}{2}mv^2, \tag{5.4}$$

which implies that the mass stores kinetic energy as its velocity increases, and releases kinetic energy as its velocity decreases. If a mass has a vertical displacement relative to a reference position, the energy stored in the mass is potential energy given by

$$V_g = mgh, \tag{5.5}$$

where g is the gravitational acceleration (9.81 m/s^2 or 32.2 ft/s^2) and h is the height measured from the reference position or datum to the center of mass. Subscript g is used to denote that the potential energy is associated with gravity.

For rotational mechanical systems, the basic variables used to describe system dynamics are the angular acceleration vector $\boldsymbol{\alpha}$, the angular velocity vector $\boldsymbol{\omega}$, and the angular position vector $\boldsymbol{\theta}$. The direction of an angular vector can be determined using the right-hand rule as shown in Figure 5.2. The sense of rotation follows the curve of the four fingers, and the rotational vector points in the direction of the thumb. In this chapter, we consider the rigid bodies that are constrained to rotate about only one axis. Then in scalar form, we have

$$\alpha = \frac{d\omega}{dt} = \frac{d^2\theta}{dt^2} \tag{5.6}$$

or

$$\alpha = \dot{\omega} = \ddot{\theta}. \tag{5.7}$$

Figure 5.3 shows a disk rotating about an axis through a fixed point O. The relation between the torque τ about the fixed point O and the angular acceleration α of the disk about O is

$$\tau = I_O \alpha, \tag{5.8}$$

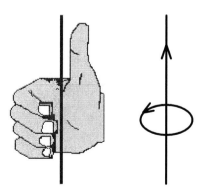

FIGURE 5.2 Right-hand rule.

Mechanical Systems

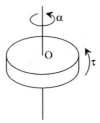

FIGURE 5.3 A disk rotating about an axis through a fixed point O.

where I_O is the mass moment of inertia of the body about the fixed point O, and common units used are kg·m² or slug·ft². Similar to a translational mass, a rotational mass can store kinetic energy and potential energy. The kinetic energy for a rotational mass about a fixed point O is expressed as

$$T = \frac{1}{2}I_O\omega^2. \tag{5.9}$$

The potential energy for a rotational mass has the same form as Equation 5.5.

5.1.2 Spring Elements

Figure 5.4a shows a translational spring element, which is fixed at one end and is subjected to a tensile (or compressive) force f at the other end. The spring has a free length x_0, and the deflection of the spring caused by the force f is denoted by x. Assume that the spring is massless, or of negligible mass. For a linear spring, Hooke's law states that

$$f = kx, \tag{5.10}$$

where k is the spring stiffness in units of N/m or lb/ft. When both ends of a spring are displaced by x_1 and x_2, as shown in Figure 5.4b, the forces at both ends are equal in magnitude but opposite in direction. If $x_2 > x_1 > 0$, then the spring is under elongation, and the force applied to the spring is

$$f = k(x_2 - x_1), \tag{5.11}$$

where $x_{\text{rel}} = x_2 - x_1$, and it is the relative displacement between the two ends of the spring. If the spring is connected to a mass, because of Newton's third law, the force exerted on the mass by the spring has the same magnitude as f, but opposite in direction.

When a spring is stretched or compressed, potential energy is stored in the spring and is given by

$$V_e = \frac{1}{2}kx^2, \tag{5.12}$$

where subscript e denotes that the potential energy is associated with elastic elements.

For a torsional spring as shown in Figure 5.5a, we have

$$\tau = K\theta, \tag{5.13}$$

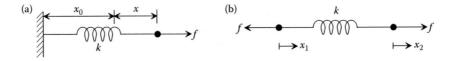

FIGURE 5.4 A translational spring element with (a) one fixed end and (b) two free ends.

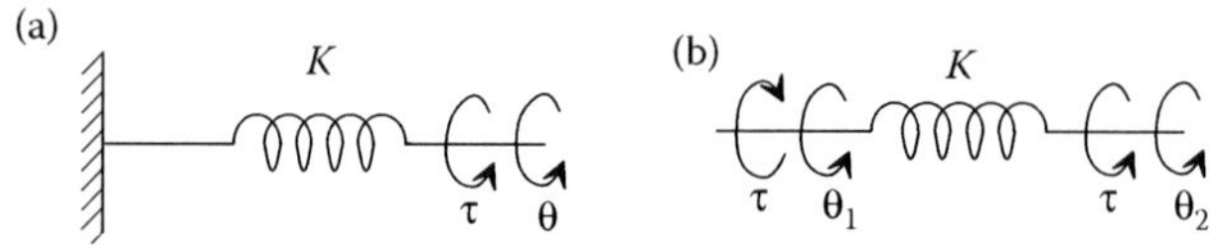

FIGURE 5.5 A torsional spring element with (a) one fixed end and (b) two free ends.

where τ is the applied torque, K is the torsional spring stiffness in units of N·m/rad or ft·lb/rad, and θ is the angular deformation of the spring. Figure 5.5b shows a torsional spring with both ends twisted. Assume that θ_1 and θ_2 are the angular displacements of respective ends corresponding to the applied torque. If $\theta_2 > \theta_1 > 0$, then

$$\tau = K(\theta_2 - \theta_1) \tag{5.14}$$

and the spring is twisted in the counterclockwise direction when viewed from the right-hand side. The potential energy stored in a torsional spring element is expressed as

$$V_e = \frac{1}{2}K\theta^2. \tag{5.15}$$

5.1.3 Damper Elements

A spring element exerts a reaction force that is dependent on the relative displacement between two ends of the spring. In contrast, a force that depends on the relative velocity between two bodies is modeled by a damper element. Figure 5.6 shows a mass sliding on a fixed horizontal surface, where the two surfaces are separated by a film of liquid. The mass is subjected to a friction force generated between the two surfaces, and the friction caused by the liquid is called viscous damping. The direction of the damping force is opposite to the direction of the motion and its magnitude depends on the nature of fluid flow between the two surfaces. The exact viscous damping force is complex; thus for modeling in system dynamics, we use a linear relationship

$$f = bv, \tag{5.16}$$

where the symbol b is used to denote the viscous damping coefficient in units of N·s/m or lb·s/ft. The damping force exerted on the mass in Figure 5.6 is to the left. Note that the symbol c is also often used to denote the viscous damping coefficient. Therefore, both b and c will be used interchangeably in this book.

The viscous friction can be modeled using a viscous damper (or a dashpot). The symbol in Figure 5.7a is the representation of a viscous damper, which is like a piston moving through a liquid-filled cylinder as shown in Figure 5.7b. There are small holes in the piston through which the

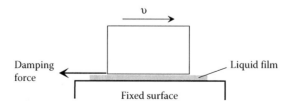

FIGURE 5.6 A mass sliding on a lubricated fixed surface.

Mechanical Systems

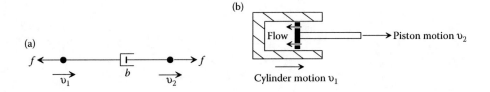

FIGURE 5.7 A viscous damper: (a) symbol and (b) physical system.

liquid flows as the parts move relative to each other. If $v_2 > v_1 > 0$, then the right end of the damper moves to the right with respect to the left end. The force applied to the right end is dependent on the relative velocity $v_{rel} = v_2 - v_1$. The force has a magnitude of

$$f = b(v_2 - v_1) \tag{5.17}$$

and points to the right. Assume that the damper is massless, or of negligible mass. Then the forces at both ends of the damper are equal in magnitude but opposite in direction. If the damper is connected to a mass, because of Newton's third law, the damping force exerted on the mass has the same magnitude but opposite direction.

For a torsional damper as shown in Figure 5.8a, the linear relationship between the externally applied torque and the angular velocity is given by

$$\tau = B\omega, \tag{5.18}$$

where B is the rotational viscous damping coefficient in units of N·m·s/rad or ft·lb·s/rad. The symbol in Figure 5.8b represents a rotational viscous damper, which can be used to model the viscous friction between two rotating surfaces separated by a film of liquid. If $\omega_2 > \omega_1 > 0$, the magnitude of the applied torque is

$$\tau = B(\omega_2 - \omega_1) \tag{5.19}$$

and the direction is as shown.

Note that the damping dissipates the energy of the system. Besides viscous damping, there are two other types of damping in engineering mechanics: Coulomb damping associated with dry friction and structural damping. The former will be discussed in Chapter 9, and the latter is beyond the scope of this text.

5.1.4 Equivalence

In many mechanical systems, multiple springs or dampers are used. In such cases, an equivalent spring stiffness constant or damping coefficient can be obtained to represent the combined elements.

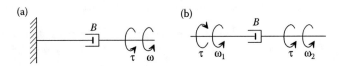

FIGURE 5.8 A rotational viscous damper with (a) one fixed end and (b) two free ends.

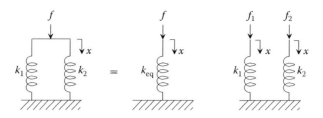

FIGURE 5.9 Equivalence for two springs in parallel.

Example 5.1: Springs in Parallel

Consider a system of two springs, k_1 and k_2, in parallel as shown in Figure 5.9. Prove that the system is equivalent to a single spring whose stiffness is

$$k_{eq} = k_1 + k_2.$$

Proof

Because of parallel interconnection, the bottom ends of the springs are attached to the same fixed body, and their top ends are also attached to a common body. This implies that both springs have the same deflection x. Assume that the forces applied to the two springs are f_1 and f_2, respectively. Since the system is in static equilibrium, the total force is given by

$$f = f_1 + f_2 = k_1 x + k_2 x = (k_1 + k_2)x.$$

Comparing it with the equivalent system,

$$f = k_{eq} x,$$

we obtain the equivalent spring stiffness

$$k_{eq} = k_1 + k_2.$$

The result can be extended to n springs. For a system of n springs in parallel, the equivalent spring stiffness k_{eq} is equal to the sum of all the individual spring stiffnesses k_i:

$$k_{eq} = k_1 + k_2 + \cdots + k_n.$$

Example 5.2: Springs in Series

Consider a system of two springs, k_1 and k_2, in series as shown in Figure 5.10. Prove that the equivalent spring stiffness of the system is

$$k_{eq} = \frac{k_1 k_2}{k_1 + k_2}.$$

Proof

Since both springs are in static equilibrium, they are subjected to the same force f. Assume that the two springs are deformed by x_1 and x_2, respectively. Note that the total deformation of the system is given by

$$x = x_1 + x_2 = \frac{f}{k_1} + \frac{f}{k_2} = f\left(\frac{1}{k_1} + \frac{1}{k_2}\right).$$

Mechanical Systems

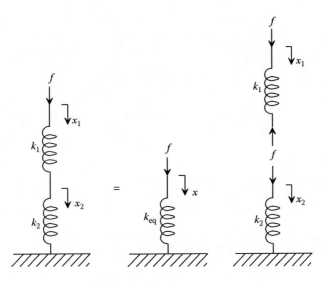

FIGURE 5.10 Equivalence for two springs in series.

For the equivalent system, the deformation is

$$x = \frac{f}{k_{eq}}.$$

Thus,

$$\frac{1}{k_{eq}} = \frac{1}{k_1} + \frac{1}{k_2}$$

or

$$k_{eq} = \frac{k_1 k_2}{k_1 + k_2}.$$

The result can also be extended to n springs. For a system of n springs in series, the reciprocal of the equivalent spring stiffness k_{eq} is equal to the sum of all the reciprocals of the individual spring stiffnesses k_i:

$$\frac{1}{k_{eq}} = \frac{1}{k_1} + \frac{1}{k_2} + \cdots + \frac{1}{k_n}.$$

The above two examples show how one can derive the equivalent spring stiffness for springs connected in parallel or in series. For a system of dampers, the equivalent damping coefficient can be derived using the same logic and similar steps.

Springs are the most familiar elastic elements. However, many engineering applications involving elastic elements do not contain springs but other mechanical elements, such as beams and rods, which can be modeled as springs. The equivalent spring constants can be determined using the results from the study of mechanics of materials [2,11].

Example 5.3: Equivalent Spring Constant of a Cantilever Beam

Consider a uniform cantilever beam of length L, width b, and thickness h in Figure 5.11. Assume that the force f is applied to the free end of the beam, and the corresponding deflection is x. Derive the equivalent spring constant k_{eq}.

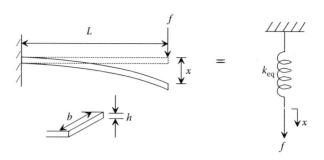

FIGURE 5.11 A beam in bending under a transverse force.

Solution

The force–deflection relation of a cantilever beam can be found in mechanics of materials references. The relation is

$$x = \frac{L^3}{3EI_A} f,$$

where x is the deflection at the free end of the beam, f is the force applied at the free end of the beam, E is the modulus of elasticity of beam material, and I_A is the area moment of inertia about the beam's longitudinal axis. For a beam having a rectangular cross section with a width b and thickness h, the area moment of inertia is

$$I_A = \frac{bh^3}{12}.$$

Thus the force–deflection relation reduces to

$$x = \frac{4L^3}{Ebh^3} f.$$

For the equivalent system, the force–deflection relation is

$$x = \frac{f}{k_{eq}}.$$

Thus, the equivalent spring stiffness is

$$k_{eq} = \frac{Ebh^3}{4L^3}.$$

PROBLEM SET 5.1

1. If the 75-kg block in Figure 5.12 is released from rest at A, determine its kinetic energy and velocity after it slides 10 m down the plane. Assume that the plane is smooth.
2. Repeat Problem 1 if the coefficient of kinetic friction between the block and the plane is $\mu_k = 0.2$.

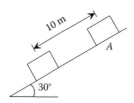

FIGURE 5.12 Problem 1.

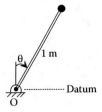

FIGURE 5.13 Problem 3.

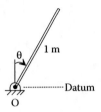

FIGURE 5.14 Problem 4.

3. The ball in Figure 5.13 has a mass of 7.5 kg and is fixed to a rod having a negligible mass. Assume that the ball is released from rest when $\theta = 0°$.
   a. Determine the gravitational potential energy of the ball when $\theta = 30°$. The datum is shown in Figure 5.13.
   b. Determine the kinetic energy and the velocity of the ball when $\theta = 30°$.
4. The 7.5-kg slender rod in Figure 5.14 is released from rest when $\theta = 0°$.
   a. Determine the gravitational potential energy of the rod when $\theta = 30°$. The datum is shown in Figure 5.14.
   b. Determine the kinetic energy and the angular velocity of the rod when $\theta = 30°$. The mass moment of inertia of the slender rod about the fixed point O is $I_O = \frac{1}{3}mL^2$, where L is the length of the rod.
5. Determine the elastic potential energy of the system shown in Figure 5.15 if the 3-kg block is displaced by 0.1 m. Assume that the springs are originally unstretched.
6. If the disk in Figure 5.16 rotates in the clockwise direction by 5°, determine the elastic potential energy of the system. Assume that the springs are originally unstretched.

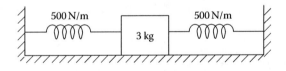

FIGURE 5.15 Problem 5.

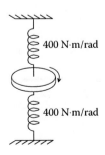

FIGURE 5.16 Problem 6.

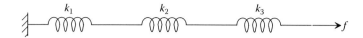

FIGURE 5.17 Problem 7.

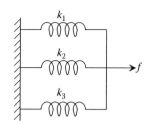

FIGURE 5.18 Problem 8.

7. Determine the equivalent spring constant for the system shown in Figure 5.17.
8. Determine the equivalent spring constant for the system shown in Figure 5.18.
9. Determine the equivalent spring constant for the system shown in Figure 5.19.
10. Determine the equivalent spring constant for the system shown in Figure 5.20.
11. Derive the spring constant expression for the cylindrical rod shown in Figure 5.21. Assume an axial force f is applied to the free end of the rod, and the corresponding deflection is x. The modulus of elasticity of the rod material is E.
12. The uniform circular shaft in Figure 5.22 acts as a torsional spring. Derive the equivalent spring constant corresponding to a torque applied at the free end. Assume that the shear modulus is G.

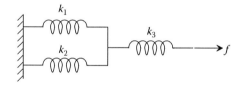

FIGURE 5.19 Problem 9.

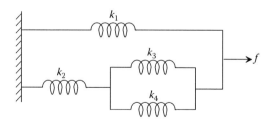

FIGURE 5.20 Problem 10.

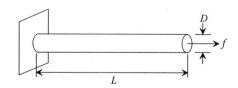

FIGURE 5.21 Problem 11.

Mechanical Systems

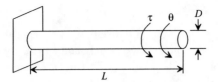

FIGURE 5.22 Problem 12.

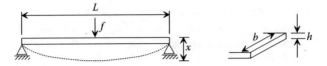

FIGURE 5.23 Problem 13.

13. Derive the spring constant expression of the simply supported beam in Figure 5.23. Assume that the force f and the deflection x are at the center of the beam.
14. A rod is made of two uniform sections, as shown in Figure 5.24. The two sections are made of the same material, and the modulus of elasticity of the rod material is E. The areas for the two sections are A_1 and A_2, respectively. Derive the equivalent spring constant corresponding to a tensile force applied at the free end.
15. Determine the equivalent damping coefficient for the system shown in Figure 5.25.
16. Determine the equivalent damping coefficient for the system shown in Figure 5.26.

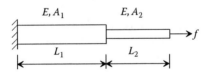

FIGURE 5.24 Problem 14.

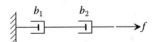

FIGURE 5.25 Problem 15.

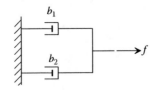

FIGURE 5.26 Problem 16.

5.2 TRANSLATIONAL SYSTEMS

With appropriate simplifying assumptions, a translational mechanical system can be modeled as a system of interconnected mechanical elements. The dynamic behavior of the system must obey the physical laws, and the dynamic equations of motion can be obtained by applying these physical laws, such as Newton's second law or D'Alembert's principle. The number of equations of motion is determined by the number of degrees of freedom of the system.

The number of degrees of freedom of a dynamic system is defined as the number of independent generalized coordinates that specify the configuration of the system. Generalized coordinates need not be restricted only to the actual position coordinates, which are physical coordinates. They could be anything, for example, position coordinate, translational displacement, rotational displacement, pressure, voltage, or current. The generalized coordinates of a system need not be of the same type.

Figure 5.27a shows a translational mechanical system, in which the mass m moves in the horizontal direction, and x is the displacement measured from the static equilibrium position of the mass. The displacement x is the generalized coordinate. If the origin of the coordinate system is at the static equilibrium position, then x is also the position coordinate. When a pendulum consisting of a massless rod of length L and a point mass of M is attached to the block of mass m, one displacement coordinate x is not enough to describe the motion of the system. On the one hand, the pendulum moves together with the block in the horizontal direction. On the other hand, the pendulum rotates, and the rotational motion can be described using an angular displacement θ. Thus, for the system in Figure 5.27b, x and θ are the two independent generalized coordinates, and two equations of motion are obtained if Newton's second law or D'Alembert's principle is applied.

5.2.1 Newton's Second Law

Newton's second law states that the acceleration of a mass is proportional to the resultant force vector acting on it and is in the direction of this force. Assume that the translational motion of a particle or a rigid body is restricted in a plane. For a particle, which is a mass of negligible dimensions, Newton's second law can be expressed in vector form

$$\sum \mathbf{F} = m\mathbf{a} \tag{5.20}$$

or in scalar form

$$\sum F_x = ma_x, \quad \sum F_y = ma_y, \tag{5.21}$$

where $\sum F_x$ and $\sum F_y$ are summations of the applied forces decomposed along the x and y directions, respectively, and a_x and a_y are the x and y components of the acceleration of the particle, respectively.

For a rigid body, Newton's second law is given by

$$\sum \mathbf{F} = m\mathbf{a}_C \tag{5.22}$$

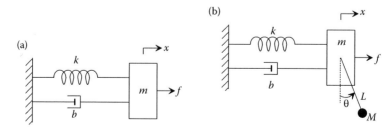

FIGURE 5.27 A mechanical system with (a) displacement as the generalized coordinate and (b) mixed types of generalized coordinates.

Mechanical Systems

or

$$\sum F_x = ma_{Cx}, \quad \sum F_y = ma_{Cy}, \tag{5.23}$$

where subscript C denotes the center of mass. In many engineering applications, the gravity field is considered to be uniform, and the center of mass coincides with the center of gravity.

5.2.2 Free-Body Diagram

To apply Newton's second law to a mechanical system, it is useful to draw a free-body diagram for each mass in the system, showing all external forces. The non-input forces can be described in terms of displacements or velocities using the expressions associated with the basic spring or damper elements. Drawing correct free-body diagrams is the most important step in analyzing mechanical systems by the force/moment approach (as opposed to the energy approach).

Let us consider a simple system consisting of a block of mass m, a spring of stiffness k, and a viscous damper of viscous damping coefficient b. Figure 5.28 shows the physical mass–spring–damper system and the free-body diagram drawn for the mass. Note that the motion of the system can be described using the displacement variable x, which is chosen as the generalized coordinate. The positive direction is the direction shown by the arrow next to the displacement. Assume that the positive direction is to the right as shown. This sign convention implies that the displacement x, velocity $\dot{x}$, and acceleration $\ddot{x}$ are positive to the right.

Three forces included in the free-body diagram are the applied force f, the force exerted by the spring f_k, and the force exerted by the damper f_b. The magnitudes of the forces are shown in the free-body diagram, and their physical directions are indicated by the arrows. The force f is externally applied to the mass–spring–damper system, and the positive direction is given to the right. In order to determine the forces f_k and f_b, we can imagine the mass to be displaced along the positive direction, $x > 0$. Thus, the spring is in tension, and there must be a tensile force $f_k = kx$ applied to the right end of the spring. Because of Newton's third law, the mass is subjected to a reaction force with the same magnitude but in the opposite direction (i.e., to the left). Similarly, the assumption of $\dot{x} > 0$ indicates that the right end of the damper moves to the right with a velocity $\dot{x}$. There must be a force $f_b = b\dot{x}$ applied to the right end of the damper and directed to the right. A reaction force f_b on the mass is to the left. Remember that the damping force for a moving mass is always opposite to the direction of motion.

The above analysis shows that an assumption about the motion of all masses in a mechanical system must be made in order to draw the free-body diagrams. It is customary to assume that all displacements are in the assumed positive directions when determining the proper magnitudes and directions for the forces. Applying Newton's second law to the correct free-body diagrams leads to differential equations of motion, which can be converted to other system representations, such as the transfer function form and the state-space form.

Example 5.4: A Single-Degree-of-Freedom Mass–Spring–Damper System

Consider the simple mass–spring–damper system subjected to an input force f, as shown in Figure 5.28a.

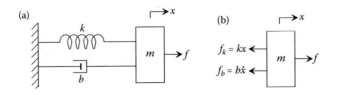

FIGURE 5.28 A mass–spring–damper system: (a) physical system and (b) free-body diagram.

a. Apply Newton's second law to derive the differential equation of motion.
b. Determine the transfer function $X(s)/F(s)$. Assume that the system output is the displacement x and the initial conditions are $x(0) = 0$ and $\dot{x}(0) = 0$.
c. Determine the state-space representation. Assume that the system output is the displacement x and the state variables are $x_1 = x$ and $x_2 = \dot{x}$.
d. Assume that $m = 2$ kg, $b = 1.6$ N·s/m, and $k = 8$ N/m. Use MATLAB® commands to define the system in the state-space form, and then convert it to the transfer function form.

Solution

a. Let us choose the displacement of the mass as the coordinate x. The free-body diagram of the mass is shown in Figure 5.28b. Applying Newton's second law in the x direction gives

$$+\rightarrow x : \sum F = ma_x,$$
$$f(t) - kx - b\dot{x} = m\ddot{x},$$

which can be rearranged into the standard input–output differential equation form

$$m\ddot{x} + b\dot{x} + kx = f(t).$$

b. Taking the Laplace transform of both sides of the preceding equation with zero initial conditions results in

$$(ms^2 + bs + k)X(s) = F(s).$$

Thus the transfer function relating the input $f(t)$ to the output $x(t)$ is

$$\frac{X(s)}{F(s)} = \frac{1}{ms^2 + bs + k}.$$

c. As specified, the state, the input, and the output are

$$\mathbf{x} = \begin{Bmatrix} x_1 \\ x_2 \end{Bmatrix} = \begin{Bmatrix} x \\ \dot{x} \end{Bmatrix}, \quad u = f, \quad y = x.$$

Taking the time derivative of each state variable and expressing them in terms of the state variables and the input variable gives

$$\dot{x}_1 = \dot{x} = x_2,$$
$$\dot{x}_2 = \ddot{x} = -\frac{k}{m}x - \frac{b}{m}\dot{x} + \frac{1}{m}f$$
$$= -\frac{k}{m}x_1 - \frac{b}{m}x_2 + \frac{1}{m}u.$$

The output equation is

$$y = x = x_1.$$

Writing the state-variable equations and the output equation in matrix form yields

$$\begin{Bmatrix} \dot{x}_1 \\ \dot{x}_2 \end{Bmatrix} = \begin{bmatrix} 0 & 1 \\ -\frac{k}{m} & -\frac{b}{m} \end{bmatrix} \begin{Bmatrix} x_1 \\ x_2 \end{Bmatrix} + \begin{bmatrix} 0 \\ \frac{1}{m} \end{bmatrix} u,$$

$$y = \begin{bmatrix} 1 & 0 \end{bmatrix} \begin{Bmatrix} x_1 \\ x_2 \end{Bmatrix} + 0 \cdot u.$$

d. The following is the MATLAB session.

Mechanical Systems

```
>> m = 2;
>> b = 1.6;
>> k = 8;
>> A = [0 1; -k/m -b/m];
>> B = [0; 1/m];
>> C = [1 0];
>> D = 0;
>> sys_ss = ss(A,B,C,D);
>> sys_tf = tf(sys_ss);
```

The command `tf` returns the transfer function from the input f to the output x.

$$\frac{X(s)}{F(s)} = \frac{0.5}{s^2 + 0.8s + 4}.$$

Note that the command `tf` can be used either to create a transfer function or to convert a defined system to the transfer function form. To convert the state-space form to the transfer function form, we can also use the command `ss2tf` to obtain the coefficients of the numerator and the denominator, and then use the command `tf` to create the transfer function.

```
>> [num,den] = ss2tf(A,B,C,D);
>> sys_tf = tf(num,den);
```

For this simple mass–spring–damper system, one coordinate, x, is enough to describe the system dynamics. Such a system is called a single-degree-of-freedom system.

5.2.3 Static Equilibrium Position and Coordinate Reference

In Example 5.4, we specified the static equilibrium position as the coordinate origin. For this mass–spring–damper system moving only in the horizontal direction, the mass is in equilibrium when the spring is at its free length. Note that it is advantageous to choose the static equilibrium position as the coordinate origin, as this choice can simplify the equation of motion by eliminating static forces. The advantage is obvious when the motion along the vertical direction is involved. The following example shows that the gravity term does not enter into the governing differential equation if the displacement is measured from the static equilibrium.

Consider the mass–spring system shown in Figure 5.29, where the mass is assumed to move only in the vertical direction. The free length of the spring is y_0. Due to gravity, the spring is stretched by δ_{st} when the mass is in static equilibrium and $mg = k\delta_{st}$. Imagine the mass to be displaced downward by a distance of x. If we choose the undeformed position in Figure 5.29a as the origin of the coordinate y, applying Newton's second law to the free-body diagram in Figure 5.29c gives

$$+\downarrow y : \sum F_y = ma_y,$$
$$mg - ky = m\ddot{y},$$
$$m\ddot{y} + ky = mg.$$

Note that the gravity term mg appears in the equation of motion. Now let us choose the static equilibrium position in Figure 5.29b as the origin of the coordinate x. The equation of motion is

$$+\downarrow x : \sum F_x = ma_x,$$
$$mg - k(x + \delta_{st}) = m\ddot{x},$$
$$m\ddot{x} + kx = 0,$$

where the gravity term mg and the static spring force $k\delta_{st}$ cancel each other out, resulting in a simpler equation.

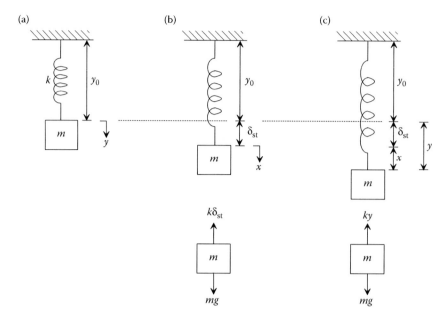

FIGURE 5.29 Choice of coordinate origins for a mass–spring system: (a) undeformed position, (b) static equilibrium position, and (c) dynamic position.

Example 5.5: A Two-Degree-of-Freedom Quarter-Car Model

Consider a quarter-car model shown in Figure 5.30a, where m_1 is the mass of one-fourth of the car body and m_2 is the mass of the wheel–tire–axle assembly. The spring k_1 represents the elasticity of the suspension and the spring k_2 represents the elasticity of the tire. $z(t)$ is the displacement input due to the surface of the road.

a. Draw the necessary free-body diagrams and derive the differential equations of motion.
b. Determine the state-space representation. Assume that the displacements of the two masses, x_1 and x_2, are the outputs and the state variables are $x_1 = x_1$, $x_2 = x_2$, $x_3 = \dot{x}_1$, and $x_4 = \dot{x}_2$.

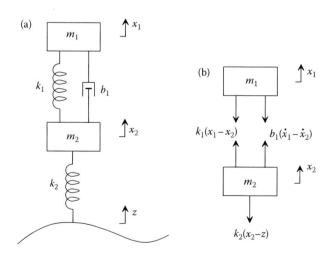

FIGURE 5.30 A quarter-car model: (a) physical system and (b) free-body diagram.

Mechanical Systems

c. The parameter values are $m_1 = 290$ kg, $m_2 = 59$ kg, $b_1 = 1000$ N·s/m, $k_1 = 16{,}182$ N/m, and $k_2 = 19{,}000$ N/m. Use MATLAB commands to define the system in the state-space form and then convert it to the transfer function form. Note that there are two transfer functions $X_1(s)/Z(s)$ and $X_2(s)/Z(s)$. Assume that all the initial conditions are zero.

Solution

a. We choose the displacements of the two masses x_1 and x_2 as the generalized coordinates. The static equilibrium positions of m_1 and m_2 are set as the coordinate origins. Assume

$$x_1 > x_2 > z > 0,$$

which implies that the springs are in tension and

$$\dot{x}_1 > \dot{x}_2 > \dot{z} > 0.$$

The free-body diagrams of m_1 and m_2 are shown in Figure 5.30b. Note that the gravitational forces, $m_1 g$ and $m_2 g$, are not included in the free-body diagrams.

Applying Newton's second law to the masses m_1 and m_2, respectively, gives

$$+\uparrow x : \sum F_x = ma_x,$$
$$-k_1(x_1 - x_2) - b_1(\dot{x}_1 - \dot{x}_2) = m_1 \ddot{x}_1,$$
$$k_1(x_1 - x_2) + b_1(\dot{x}_1 - \dot{x}_2) - k_2(x_2 - z) = m_2 \ddot{x}_2.$$

Rearranging the equations into the standard input–output form,

$$m_1 \ddot{x}_1 + b_1 \dot{x}_1 - b_1 \dot{x}_2 + k_1 x_1 - k_1 x_2 = 0,$$
$$m_2 \ddot{x}_2 - b_1 \dot{x}_1 + b_1 \dot{x}_2 - k_1 x_1 + (k_1 + k_2) x_2 = k_2 z,$$

which can be expressed in second-order matrix form (Section 4.1) as

$$\begin{bmatrix} m_1 & 0 \\ 0 & m_2 \end{bmatrix} \begin{Bmatrix} \ddot{x}_1 \\ \ddot{x}_2 \end{Bmatrix} + \begin{bmatrix} b_1 & -b_1 \\ -b_1 & b_1 \end{bmatrix} \begin{Bmatrix} \dot{x}_1 \\ \dot{x}_2 \end{Bmatrix} + \begin{bmatrix} k_1 & -k_1 \\ -k_1 & k_1 + k_2 \end{bmatrix} \begin{Bmatrix} x_1 \\ x_2 \end{Bmatrix} = \begin{bmatrix} 0 \\ k_2 \end{bmatrix} z.$$

b. Note that the input to the system is the displacement $z(t)$. The state, the input, and the output are

$$\mathbf{x} = \begin{Bmatrix} x_1 \\ x_2 \\ x_3 \\ x_4 \end{Bmatrix} = \begin{Bmatrix} x_1 \\ x_2 \\ \dot{x}_1 \\ \dot{x}_2 \end{Bmatrix}, \quad u = z, \quad \mathbf{y} = \begin{Bmatrix} x_1 \\ x_2 \end{Bmatrix}.$$

Taking the time derivative of each state variable, we obtain a set of first-order differential equations

$$\dot{x}_1 = x_3,$$
$$\dot{x}_2 = x_4,$$
$$\dot{x}_3 = \ddot{x}_1 = -\frac{k_1}{m_1} x_1 + \frac{k_1}{m_1} x_2 - \frac{b_1}{m_1} \dot{x}_1 + \frac{b_1}{m_1} \dot{x}_2$$
$$= -\frac{k_1}{m_1} x_1 + \frac{k_1}{m_1} x_2 - \frac{b_1}{m_1} x_3 + \frac{b_1}{m_1} x_4,$$
$$\dot{x}_4 = \ddot{x}_2 = \frac{k_1}{m_2} x_1 - \frac{k_1 + k_2}{m_2} x_2 + \frac{b_1}{m_2} \dot{x}_1 - \frac{b_1}{m_2} \dot{x}_2 + \frac{k_2}{m_2} z$$
$$= \frac{k_1}{m_2} x_1 - \frac{k_1 + k_2}{m_2} x_2 + \frac{b_1}{m_2} x_3 - \frac{b_1}{m_2} x_4 + \frac{k_2}{m_2} u.$$

The output equation is

$$\mathbf{y} = \begin{Bmatrix} x_1 \\ x_2 \end{Bmatrix}.$$

Thus, the state-space representation is

$$\begin{Bmatrix} \dot{x}_1 \\ \dot{x}_2 \\ \dot{x}_3 \\ \dot{x}_4 \end{Bmatrix} = \begin{bmatrix} 0 & 0 & 1 & 0 \\ 0 & 0 & 0 & 1 \\ -\dfrac{k_1}{m_1} & \dfrac{k_1}{m_1} & -\dfrac{b_1}{m_1} & \dfrac{b_1}{m_1} \\ \dfrac{k_1}{m_2} & -\dfrac{k_1+k_2}{m_2} & \dfrac{b_1}{m_2} & -\dfrac{b_1}{m_2} \end{bmatrix} \begin{Bmatrix} x_1 \\ x_2 \\ x_3 \\ x_4 \end{Bmatrix} + \begin{bmatrix} 0 \\ 0 \\ 0 \\ \dfrac{k_2}{m_2} \end{bmatrix} u,$$

$$\mathbf{y} = \begin{bmatrix} 1 & 0 & 0 & 0 \\ 0 & 1 & 0 & 0 \end{bmatrix} \begin{Bmatrix} x_1 \\ x_2 \\ x_3 \\ x_4 \end{Bmatrix} + \begin{bmatrix} 0 \\ 0 \end{bmatrix} u.$$

c. The following is the MATLAB session.
```
>> m1 = 290;
>> m2 = 59;
>> b1 = 1000;
>> k1 = 16182;
>> k2 = 19000;
>> A = [0 0 1 0;
   0 0 0 1;
   -k1/m1 k1/m1 -b1/m1 b1/m1;
   k1/m2 -(k1+k2)/m2 b1/m2 -b1/m2];
>> B = [0 0 0 k2/m2]';
>> C = [1 0 0 0; 0 1 0 0];
>> D = [0 0]';
>> sys_ss = ss(A,B,C,D);
>> sys_tf = tf(sys_ss);
```

The command **tf** returns two transfer functions from the input z to the two outputs x_1 and x_2:

$$\frac{X_1(s)}{Z(s)} = \frac{1110s + 17{,}970}{s^4 + 20.4s^3 + 652.1s^2 + 1110s + 17{,}970},$$

$$\frac{X_2(s)}{Z(s)} = \frac{322s^2 + 1110s + 17{,}970}{s^4 + 20.4s^3 + 652.1s^2 + 1110s + 17{,}970}.$$

If the MATLAB command **ss2tf** is used,

```
>> [num,den] = ss2tf(A,B,C,D);
```

the numerator coefficients are returned in matrix **num** with as many rows as the outputs **y**. The transfer function from the input to the first or the second output can be created using the command **tf** with the first or the second row of **num**.

```
>> sys1_tf = tf(num(1,:),den);
>> sys2_tf = tf(num(2,:),den);
```

Note that the gravity terms in Example 5.5 do not appear in the equations of motion since the static equilibrium positions are chosen as the coordinate origins. Two independent coordinates, x_1 and x_2, are required to specify the system dynamics. Such a system is called a two-degree-of-freedom system, which is a special case of multiple-degree-of-freedom systems.

Mechanical Systems

5.2.4 D'Alembert's Principle

Newton's second law can be reformulated as

$$\sum \mathbf{F} - m\mathbf{a}_C = 0, \quad (5.24)$$

which is known as D'Alembert's principle. The equation has $\sum \mathbf{F}$ as the sum of all the physical forces and $-m\mathbf{a}_C$ as the inertial force, which is a fictitious force. The minus sign associated with the inertial force indicates that the force acts in the negative direction when $\mathbf{a}_C > 0$. If the inertial force is included with the external forces, the mass can be considered to be in equilibrium. D'Alembert's principle is completely equivalent to the formulation of Newton's second law, although it looks like a classical static force balance. To use D'Alembert's principle correctly, the inertial force must be shown in the free-body diagrams correctly. The following is a simple example that shows the derivation of the differential equation of motion using two different methods: Newton's second law and D'Alembert's principle.

Example 5.6: A Simple Pendulum

Consider the simple pendulum shown in Figure 5.31. Assume that the dimension of the mass is negligible and that the string is massless and inextensible. Neglecting the friction, draw a free-body diagram and derive the differential equation of motion using

a. Newton's second law
b. D'Alembert's principle.

Solution

The angular displacement θ is chosen as the generalized coordinate. The origin is set at the static equilibrium position, which is shown in Figure 5.31 using a dashed line. Assume that the mass is displaced by an angle $\theta > 0$. From the results of kinematics, the tangential acceleration of the mass is $a = L\ddot{\theta}$.

a. For Newton's second law, the free-body diagram of the mass is shown in Figure 5.32a, where T is the tension. Applying Newton's second law in the x direction, which is perpendicular to the length of the string, we obtain

$$+\nearrow x: \quad \sum F_x = ma_x,$$
$$-mg\sin\theta = mL\ddot{\theta},$$

which can be rearranged as

$$mL\ddot{\theta} + mg\sin\theta = 0$$

or

$$\ddot{\theta} + \frac{g}{L}\sin\theta = 0.$$

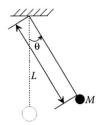

FIGURE 5.31 A simple pendulum system.

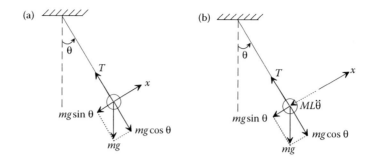

FIGURE 5.32 Free-body diagram for (a) Newton's second law and (b) D'Alembert's principle.

b. For D'Alembert's principle, the free-body diagram of the mass is shown in Figure 5.32b, where the inertial force is shown by a dashed line. Note that we assumed $\theta > 0$, which implies $\ddot{\theta} > 0$. The inertial force acts in the negative direction with a magnitude of $mL\ddot{\theta}$. Applying D'Alembert's principle in the x direction results in

$$+\nearrow x: \quad \sum F_x - ma_x = 0,$$
$$-mg\sin\theta - mL\ddot{\theta} = 0,$$

or

$$\ddot{\theta} + \frac{g}{L}\sin\theta = 0,$$

which is the same as the one obtained previously.

Note that the gravity term is not canceled out in this example, where the effect of gravity is like a torque causing the mass to rotate about the point O. Because the resulting torque is not constant, $mgL\sin\theta$, the gravity term remains in the equation of motion.

5.2.5 Massless Junctions

A system of massless junctions is a system of springs and dampers without any masses. The differential equations of motion for such a system can be derived using either Newton's second law or D'Alembert's principle and simply letting the masses be zero at the junctions.

Example 5.7: A Two-Degree-of-Freedom System with Massless Junctions

Consider the system of massless junctions shown in Figure 5.33a. An external force f is applied to the junction A. Draw the free-body diagrams and derive the differential equations of motion.

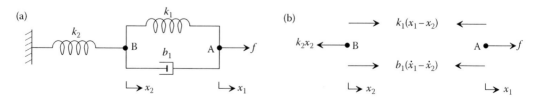

FIGURE 5.33 A two-degree-of-freedom system with massless junctions: (a) physical system and (b) free-body diagram.

Mechanical Systems

Solution

Two massless junctions, A and B, are included in this system. We choose the displacements of the two junctions as the generalized coordinates, which are denoted by x_1 and x_2. Assume that $x_1 > x_2 > 0$. This implies that the two springs are in extension. The free-body diagrams at the two massless junctions are shown in Figure 5.33b. Applying Newton's second law to each massless junction gives

$$+\rightarrow x: \sum F_x = ma_x = 0,$$
$$\text{A: } -k_1(x_1 - x_2) - b_1(\dot{x}_1 - \dot{x}_2) + f = 0,$$
$$\text{B: } k_1(x_1 - x_2) - k_2 x_2 + b_1(\dot{x}_1 - \dot{x}_2) = 0.$$

The equations can be rearranged as

$$b_1 \dot{x}_1 - b_1 \dot{x}_2 + k_1 x_1 - k_1 x_2 = f,$$
$$-b_1 \dot{x}_1 + b_1 \dot{x}_2 - k_1 x_1 + (k_1 + k_2) x_2 = 0,$$

or in matrix form

$$\begin{bmatrix} b_1 & -b_1 \\ -b_1 & b_1 \end{bmatrix} \begin{Bmatrix} \dot{x}_1 \\ \dot{x}_2 \end{Bmatrix} + \begin{bmatrix} k_1 & -k_1 \\ -k_1 & k_1 + k_2 \end{bmatrix} \begin{Bmatrix} x_1 \\ x_2 \end{Bmatrix} = \begin{Bmatrix} f \\ 0 \end{Bmatrix}.$$

Note that the system in Example 5.7 is a two-degree-of-freedom system. The dynamic behavior of the system is described by two first-order differential equations of motion, and thus it is a second-order system. If the two massless junctions are replaced by two masses as shown in Figure 5.34, the resulting system is still a two-degree-of-freedom system, but a fourth-order system. The reader can derive the differential equations of motion for the new system in Figure 5.34 as an exercise.

In Examples 5.5 and 5.7, the differential equations of motion are also given in second-order matrix form as $\mathbf{M\ddot{x}} + \mathbf{C\dot{x}} + \mathbf{Kx} = \mathbf{f}$. By observation, we find the following:

1. All the mass, damping, and stiffness matrices are symmetric with respect to the main diagonal.
2. All the elements on the main diagonals are nonnegative.
3. The off-diagonal elements of both the damping and the stiffness matrices are nonpositive.
4. The off-diagonal elements of the mass matrix are nonnegative.

These results are true for stable mechanical systems with purely translational or rotational motion. The reader can use them as necessary conditions to check the correctness of the differential equations of motion.

PROBLEM SET 5.2

1. For the system shown in Figure 5.35, the input is the force f and the output is the displacement x of the mass.

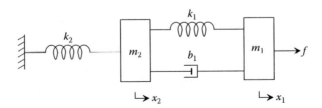

FIGURE 5.34 A system obtained by replacing the massless junctions in Figure 5.33 by the masses.

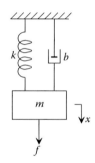

FIGURE 5.35 Problem 1.

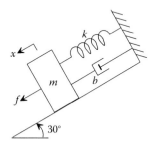

FIGURE 5.36 Problem 2.

  a. Draw the necessary free-body diagram and derive the differential equation of motion.
  b. Using the differential equation obtained in Part (a), determine the transfer function. Assume initial conditions $x(0) = 0$ and $\dot{x}(0) = 0$.
  c. Using the differential equation obtained in Part (a), determine the state-space representation.
2. Repeat Problem 1 for the system shown in Figure 5.36.
3. Repeat Problem 1 for the system shown in Figure 5.37.
4. Repeat Problem 1 for the system shown in Figure 5.38.
5. For the system shown in Figure 5.39, the input is the force f and the output is the displacement x of the mass.
  a. Draw the necessary free-body diagram and derive the differential equation of motion.
  b. Using the differential equation obtained in Part (a), determine the state-space representation.
  c. ◆ Use MATLAB commands to convert the state-space representation to the transfer function form. Assume initial conditions $x(0) = 0$ and $\dot{x}(0) = 0$.

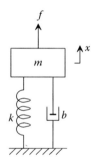

FIGURE 5.37 Problem 3.

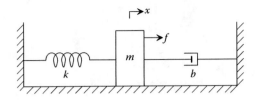

FIGURE 5.38 Problem 4.

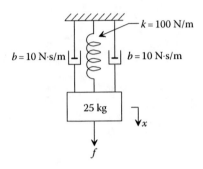

FIGURE 5.39 Problem 5.

6. Repeat Problem 5 for the system shown in Figure 5.40.
7. The system shown in Figure 5.41 simulates a vehicle traveling on a rough road. The input is the displacement z.
   a. Draw the necessary free-body diagram and derive the differential equation of motion.
   b. Assuming zero initial conditions, determine the transfer function for two different cases of output: (1) displacement x and (2) velocity $\dot{x}$.

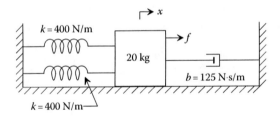

FIGURE 5.40 Problem 6.

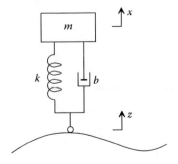

FIGURE 5.41 Problem 7.

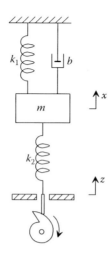

FIGURE 5.42 Problem 8.

    c. Determine the state-space representation for two different cases of output: (1) displacement x and (2) velocity $\dot{x}$.

8. The cam and follower shown in Figure 5.42 impart a displacement z to the lower end of the system. The input is the displacement z.
   a. Draw the necessary free-body diagram and derive the differential equation of motion.
   b. Assuming zero initial conditions, determine the transfer function for two different cases of output: (1) displacement x and (2) velocity $\dot{x}$.
   c. Determine the state-space representation for two different cases of output: (1) displacement x and (2) velocity $\dot{x}$.
9. For the system shown in Figure 5.43, the input is the force f and the outputs are the displacements x_1 and x_2 of the masses.
   a. Draw the necessary free-body diagrams and derive the differential equations of motion.
   b. Write the differential equations of motion in the second-order matrix form.
   c. Using the differential equations obtained in Part (a), determine the state-space representation.
10. Repeat Problem 9 for the system shown in Figure 5.44.

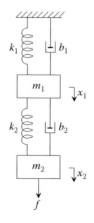

FIGURE 5.43 Problem 9.

Mechanical Systems

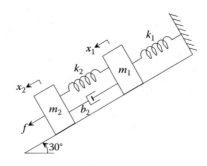

FIGURE 5.44 Problem 10.

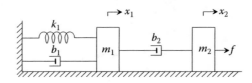

FIGURE 5.45 Problem 11.

11. Repeat Problem 9 for the system shown in Figure 5.45.
12. Repeat Problem 9 for the system shown in Figure 5.46.
13. ▲ For Problems 9 through 12, use MATLAB commands to define the systems in the state-space form and then convert to the transfer function form. All initial conditions are assumed to be zero. The masses are $m_1 = 10$ kg and $m_2 = 30$ kg. The spring constants are $k_1 = 15$ kN/m and $k_2 = 30$ kN/m. The viscous damping coefficients are $b_1 = 560$ N·s/m and $b_2 = 180$ N·s/m. Note that each system has two transfer functions, $X_1(s)/F(s)$ and $X_2(s)/F(s)$.
14. For the system in Figure 5.47, the inputs are the forces f_1 and f_2 applied to the masses and the outputs are the displacements x_1 and x_2 of the masses.
    a. Draw the necessary free-body diagrams and derive the differential equations of motion.
    b. Write the differential equations of motion in the second-order matrix form.
    c. Using the differential equations obtained in Part (a), determine the state-space representation.
15. Repeat Problem 14 for the system shown in Figure 5.48.

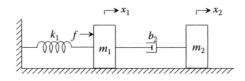

FIGURE 5.46 Problem 12.

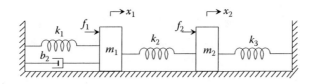

FIGURE 5.47 Problem 14.

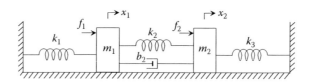

FIGURE 5.48 Problem 15.

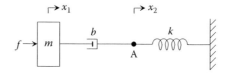

FIGURE 5.49 Problem 17.

16. ◢ For Problems 14 and 15, use MATLAB commands to define the systems in the state-space form and then convert to the transfer function form. All initial conditions are assumed to be zero. The masses are $m_1 = 10$ kg and $m_2 = 30$ kg. The spring constants are $k_1 = 15$ kN/m, $k_2 = 30$ kN/m, and $k_3 = 60$ kN/m. The viscous damping coefficients are $b_1 = 560$ N·s/m and $b_2 = 180$ N·s/m. Note that each system has four transfer functions, $X_1(s)/F_1(s)$, $X_1(s)/F_2(s)$, $X_2(s)/F_1(s)$, and $X_2(s)/F_2(s)$.

17. For the system in Figure 5.49, the input is the force f and the outputs are the displacement x_1 of the mass and the displacement x_2 of the massless junction A.
    a. Draw the necessary free-body diagrams and derive the differential equations of motion. Determine the number of degrees of freedom and the order of the system.
    b. Write the differential equations of motion in the second-order matrix form.
    c. Using the differential equation obtained in Part (a), determine the state-space representation.

18. Repeat Problem 17 for the system shown in Figure 5.50. The input is the force f and the outputs are the displacement x_1 of the mass and the displacement x_2 of the massless junction A.

19. For the system in Figure 5.51, the input is the displacement z and the outputs are the displacements x_1 and x_2.
    a. Draw the necessary free-body diagrams and derive the differential equations of motion.
    b. Write the differential equations of motion in the second-order matrix form.
    c. Using the differential equations obtained in Part (a), determine the state-space representation.

20. Repeat Problem 19 for the system shown in Figure 5.52.

5.3 ROTATIONAL SYSTEMS

In this section, we consider the derivation of a mathematical model for a rotational mechanical system. When a rigid body moves arbitrarily in three-dimensional space, the axis of rotation keeps

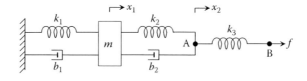

FIGURE 5.50 Problem 18.

Mechanical Systems

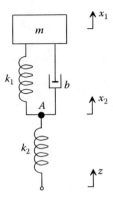

FIGURE 5.51 Problem 19.

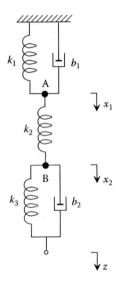

FIGURE 5.52 Problem 20.

changing. This makes modeling rather complex. Thus, the discussion of rigid bodies in three dimensions is not covered in this text, and we are mainly concerned with two-dimensional or plane motion. Fundamentals of rigid bodies in three dimensions are included only to help the interested reader in further studies.

5.3.1 General Moment Equation

The moment equation is applicable to systems of particles and rigid bodies in three dimensions and two dimensions. For a system of particles connected rigidly or a rigid body in arbitrary motion, using Newton's second law leads to the general moment equation given by

$$\sum \mathbf{M}_P = \dot{\mathbf{H}}_P + m \mathbf{r}_{C/P} \times \mathbf{a}_P, \qquad (5.25)$$

where subscript P denotes an arbitrary accelerating point, subscript C denotes the mass center of the system of particles or the rigid body, $\sum \mathbf{M}_P$ is the sum of all externally applied moments about point P, $\mathbf{H}_P$ is the angular momentum vector about point P, $\dot{\mathbf{H}}_P$ is the time rate of change of $\mathbf{H}_P$, m is the total mass of the system of particles or the mass of the rigid body, $\mathbf{r}_{C/P}$ is the position vector of the mass center C with respect to point P, and $\mathbf{a}_P$ is the acceleration vector of point P.

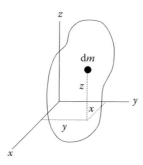

FIGURE 5.53 A differential element of a rigid body.

It is usually difficult to obtain $\dot{\mathbf{H}}_P$ for complex systems. For a rigid body, the angular momentum is related to mass moments of inertia and the angular velocity,

$$\mathbf{H}_P = \mathbf{I}_P \boldsymbol{\omega} \tag{5.26}$$

or

$$\begin{Bmatrix} H_x \\ H_y \\ H_z \end{Bmatrix}_P = \begin{bmatrix} I_{xx} & I_{xy} & I_{xz} \\ I_{yx} & I_{yy} & I_{yz} \\ I_{zx} & I_{zy} & I_{zz} \end{bmatrix}_P \begin{Bmatrix} \omega_x \\ \omega_y \\ \omega_z \end{Bmatrix}, \tag{5.27}$$

where $\mathbf{H}_P$, the angular momentum about point P, and the angular velocity $\boldsymbol{\omega}$ are 3×1 vectors. The 3×3 matrix $\mathbf{I}_P$ is known as the mass moment of inertia tensor about point P, where

$$I_{xx} = \int (y^2 + z^2)\,dm, \quad I_{yy} = \int (x^2 + z^2)\,dm, \quad I_{zz} = \int (x^2 + y^2)\,dm, \tag{5.28}$$

$$I_{xy} = I_{yx} = -\int xy\,dm, \quad I_{yz} = I_{zy} = -\int yz\,dm, \quad I_{xz} = I_{zx} = -\int xz\,dm. \tag{5.29}$$

As shown in Figure 5.53, the integrals containing squares represent the mass moments of inertia of the body about the x, y, and z axis, respectively. The integrals containing products of coordinates represent the mass products of inertia of the body.

In general, the time derivatives of the nine elements in the matrix $\mathbf{I}_P$ are nonzero for a non-symmetric rigid body, and thus it is difficult to study the dynamics of a rigid body in three dimensions. However, if the motion of the rigid body is restricted to a plane, the complexity is reduced significantly.

5.3.2 Modeling of Rigid Bodies in Plane Motion

In many engineering applications, the motion of rigid bodies is primarily in two dimensions. A rigid body is in plane motion if it can translate in two dimensions and can rotate only about an axis that is perpendicular to the plane. For a rigid body in plane motion, the mass moment of inertia is a scalar quantity and the time rate of change of the angular momentum reduces to

$$\dot{\mathbf{H}}_P = I_P \alpha, \tag{5.30}$$

where I_P is the mass moment of inertia of the body about an axis through a point P and α is the angular acceleration. Then Equation 5.25 becomes

$$\sum \mathbf{M}_P = I_P \alpha + m\mathbf{r}_{C/P} \times \mathbf{a}_P. \tag{5.31}$$

Note that the rigid body in plane motion is constrained to rotate about only one axis. Therefore, the net moment $\sum \mathbf{M}_P$ and the angular acceleration α are essentially scalars, whose signs signify

Mechanical Systems

the directions (e.g., clockwise or counterclockwise). Since the cross product term $m\mathbf{r}_{C/P} \times \mathbf{a}_P = \mathbf{r}_{C/P} \times (m\mathbf{a}_P)$, it can be considered as the effective moment caused by a fictitious force $m\mathbf{a}_P$, and the direction of the effective moment can be denoted by its sign.

If the rigid body rotates about a fixed axis through a point O, Equation 5.31 can be simplified as

$$\sum M_O = I_O \alpha \qquad (5.32)$$

with P = O and $a_P = a_O = 0$. I_O is the mass moment of inertia of the body about the point O. If the axis of rotation is not fixed, the model of the rigid body can be derived using Equation 5.31 or

$$\sum M_C = I_C \alpha \qquad (5.33)$$

with P = C and $r_{C/P} = 0$. I_C is the mass moment of inertia of the body about the mass center C. Equation 5.33 is applicable regardless of whether the axis of rotation is fixed or not.

Example 5.8: A Single-Degree-of-Freedom Rotational Mass–Spring–Damper System

Consider a simple disk–shaft system shown in Figure 5.54a, where the disk rotates about a fixed axis through the point O. A single-degree-of-freedom torsional mass–spring–damper system in Figure 5.54b can be used to approximate the dynamic behavior of the disk–shaft system. I_O is the mass moment of inertia of the disk about the point O, K represents the elasticity of the shaft, and B represents torsional viscous damping. Derive the differential equation of motion.

Solution

The free-body diagram of the disk is shown in Figure 5.54c. Because the disk rotates about a fixed axis, we can apply Equation 5.32 about the fixed point O. Assuming that counterclockwise is the positive direction, we have

$$+\curvearrowleft: \sum M_O = I_O \alpha,$$
$$\tau - K\theta - B\dot{\theta} = I_O \ddot{\theta}.$$

Thus

$$I_O \ddot{\theta} + B\dot{\theta} + K\theta = \tau.$$

Example 5.9: A Two-Degree-of-Freedom Rotational Mass–Spring System

Consider the disk–shaft system shown in Figure 5.55a. The mass moments of inertia of disks are I_1 and I_2, respectively. The shafts can be modeled as massless torsional springs. The torsional mass–spring model is shown in Figure 5.55b.

a. Draw the necessary free-body diagrams and derive the differential equations of motion.
b. Determine the transfer functions $\Theta_1(s)/T(s)$ and $\Theta_2(s)/T(s)$. All initial conditions are assumed to be zero.

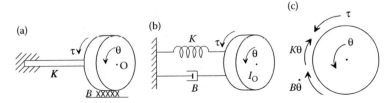

FIGURE 5.54 A rotational mass–spring–damper system: (a) physical system, (b) system model, and (c) free-body diagram.

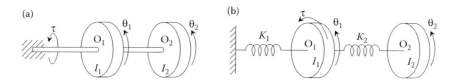

FIGURE 5.55 A disk–shaft system: (a) physical system and (b) mass–spring model.

SOLUTION

a. We choose the angular displacements θ_1 and θ_2 as the generalized coordinates. Assume that $\theta_2 > \theta_1 > 0$. The free-body diagrams are shown in Figure 5.56. Applying Equation 5.32 about the fixed points O_1 and O_2, respectively, gives

$$+\curvearrowleft: \quad \sum M_O = I_O \alpha,$$
$$\tau - K_1 \theta_1 + K_2(\theta_2 - \theta_1) = I_1 \ddot{\theta}_1,$$
$$-K_2(\theta_2 - \theta_1) = I_2 \ddot{\theta}_2.$$

Rearranging the equations results in

$$I_1 \ddot{\theta}_1 + (K_1 + K_2)\theta_1 - K_2 \theta_2 = \tau,$$
$$I_2 \ddot{\theta}_2 - K_2 \theta_1 + K_2 \theta_2 = 0.$$

b. Taking Laplace transform gives

$$I_1 s^2 \Theta_1(s) + (K_1 + K_2)\Theta_1(s) - K_2 \Theta_2(s) = T(s),$$
$$I_2 s^2 \Theta_2(s) - K_2 \Theta_1(s) + K_2 \Theta_2(s) = 0,$$

or in matrix form

$$\begin{bmatrix} I_1 s^2 + K_1 + K_2 & -K_2 \\ -K_2 & I_2 s^2 + K_2 \end{bmatrix} \begin{Bmatrix} \Theta_1(s) \\ \Theta_2(s) \end{Bmatrix} = \begin{Bmatrix} T(s) \\ 0 \end{Bmatrix}.$$

Using Cramer's rule, we can solve for $\Theta_1(s)/T(s)$ and $\Theta_2(s)/T(s)$ as

$$\frac{\Theta_1(s)}{T(s)} = \frac{I_2 s^2 + K_2}{(I_1 s^2 + K_1 + K_2)(I_2 s^2 + K_2) - K_2^2} = \frac{I_2 s^2 + K_2}{I_1 I_2 s^4 + (I_1 K_2 + I_2 K_1 + I_2 K_2)s^2 + K_1 K_2},$$

$$\frac{\Theta_2(s)}{T(s)} = \frac{K_2}{(I_1 s^2 + K_1 + K_2)(I_2 s^2 + K_2) - K_2^2} = \frac{K_2}{I_1 I_2 s^4 + (I_1 K_2 + I_2 K_1 + I_2 K_2)s^2 + K_1 K_2}.$$

5.3.3 Mass Moment of Inertia

As shown in Equation 5.28, the mass moment of inertia of a rigid body about a specified axis of rotation is defined as

$$I = \int r^2 \, dm, \tag{5.34}$$

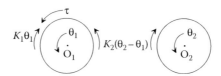

FIGURE 5.56 Free-body diagrams for the disk–shaft system in Figure 5.55.

Mechanical Systems

where r is the distance between the axis of rotation and the mass element dm. The mass moments of inertia for some rigid bodies with common shapes are given in Table 5.1, where all masses are assumed to be uniformly distributed and the axes of rotation all pass through the mass centers. If

TABLE 5.1
Mass Moments of Inertia of Common Geometric Shapes

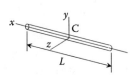

Slender rod

$$I_y = I_z = \frac{1}{12}mL^2$$
$$I_x = 0$$

Thin disk

$$I_x = \frac{1}{2}mr^2$$
$$I_y = I_z = \frac{1}{4}mr^2$$

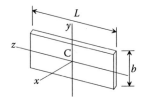

Thin rectangular plate

$$I_x = \frac{1}{12}m(L^2 + b^2)$$
$$I_y = \frac{1}{12}mL^2$$
$$I_z = \frac{1}{12}mb^2$$

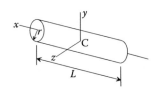

Circular cylinder

$$I_x = \frac{1}{2}mr^2$$
$$I_y = I_z = \frac{1}{12}m(3r^2 + L^2)$$

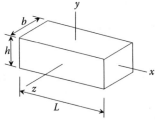

Rectangular prism

$$I_x = \frac{1}{12}m(b^2 + h^2)$$
$$I_y = \frac{1}{12}m(L^2 + b^2)$$
$$I_z = \frac{1}{12}m(L^2 + h^2)$$

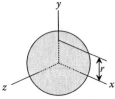

Sphere

$$I_x = I_y = I_z = \frac{2}{5}mr^2$$

Source: Beer, F.P., Johnston, E.R. Jr., and Cornwell, P.J., *Vector Mechanics for Engineers: Dynamics.* 9th ed., McGraw-Hill, 2009; Hibbeler, R.C., *Engineering Mechanics: Dynamics.* 12th ed., Prentice Hall, Upper Saddle River, NJ, 2010.

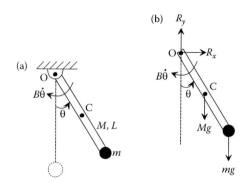

FIGURE 5.57 A pendulum–bob system: (a) physical system and (b) free-body diagram.

the axis of rotation does not coincide with the axis through the mass center, but is parallel to it, the parallel-axis theorem can be applied to obtain the corresponding moment of inertia,

$$I = I_C + md^2, \qquad (5.35)$$

where d is the distance between the two parallel axes.

Example 5.10: A Pendulum–Bob System

Consider the pendulum system shown in Figure 5.57a, where a point mass m is attached at the tip of a uniform slender rod of mass M and length L. The mass center of the rod is located at point C. The pendulum system rotates about an axis through the joint O. The friction at the joint is modeled as a torsional viscous damper of coefficient B.

  a. Determine the mass moment of inertia of the system about point O.
  b. Draw the free-body diagram for the pendulum system and obtain the nonlinear equation of motion.
  c. Linearize the equation of motion for small angles θ.
  d. Construct a Simulink® block diagram to find the output $\theta(t)$ of the linearized system with initial displacement $\theta(0) = 0.1$ rad and initial velocity $\dot\theta(0) = 0.1$ rad/s. The parameter values are $m = 0.1$ kg, $M = 1.2$ kg, $L = 0.6$ m, $B = 0.5$ N·s/m, and $g = 9.81$ m/s².

Solution

  a. The system consists of a point mass and a slender rod. The total mass moment of inertia about point O is

$$I_O = I_{O\_mass} + I_{O\_rod},$$

where

$$I_{O\_mass} = mL^2$$

and I_{O_rod} can be obtained using the parallel-axis theorem,

$$I_{O\_rod} = I_{C\_rod} + Md^2 = \frac{1}{12}ML^2 + M\left(\frac{L}{2}\right)^2 = \frac{1}{3}ML^2.$$

Then

$$I_O = \left(m + \frac{M}{3}\right)L^2.$$

Mechanical Systems

b. For the pendulum system, the free-body diagram is shown in Figure 5.57b, where R_x and R_y are the x and y components of the reaction force at the joint O, respectively. Note that the system rotates about a fixed axis through the point O. Applying Equation 5.32 about the fixed point O gives

$$+\curvearrowleft: \quad \sum M_O = I_O \alpha,$$

$$-\frac{1}{2} L \sin\theta \cdot Mg - L \sin\theta \cdot mg - B\dot{\theta} = I_O \ddot{\theta}.$$

Substituting I_O obtained in Part (a) into the equation, and rearranging it in the input–output differential equation form, we obtain

$$\left(m + \frac{M}{3}\right) L^2 \ddot{\theta} + B\dot{\theta} + \left(m + \frac{M}{2}\right) gL \sin\theta = 0.$$

c. For small angles θ, $\sin\theta \approx \theta$. The equation of motion becomes

$$\left(m + \frac{M}{3}\right) L^2 \ddot{\theta} + B\dot{\theta} + \left(m + \frac{M}{2}\right) gL\theta = 0,$$

which is a linear equation in terms of θ.

d. Substituting the parameter values into the linearized equation obtained in Part (c) yields

$$0.18 \ddot{\theta} + 0.5 \dot{\theta} + 4.12 \theta = 0.$$

Solving for the highest derivative of the output θ gives

$$\ddot{\theta} = \frac{1}{0.18}(-0.5\dot{\theta} - 4.12\theta),$$

which can be represented using the block diagram shown in Figure 5.58. Two `Integrator` blocks are used to form the angular velocity $\dot{\theta}$ and the angular displacement θ, both of which are fed back to form the angular acceleration $\ddot{\theta}$. Note that the nonzero initial displacement causes the motion of the system. Double-click on the block with the name `Integrator 2` and type `0.1` for the `Initial condition` to define the initial angular displacement $\theta(0) = 0.1$ rad. Run the simulation. Figure 5.59 shows the resulting angular displacement output $\theta(t)$.

Note that the output $\theta(t)$ can also be found analytically. Because of the nonzero initial conditions, taking the Laplace transform for the differential equation gives

$$0.18 \left[s^2 \Theta(s) - s\theta(0) - \dot{\theta}(0)\right] + 0.5 \left[s\Theta(s) - \theta(0)\right] + 4.12 \Theta(s) = 0,$$

$$\Theta(s) = \frac{0.18 s\theta(0) + 0.18\dot{\theta}(0) + 0.5\theta(0)}{0.18s^2 + 0.5s + 4.12} = \frac{0.018s + 0.05}{0.18s^2 + 0.5s + 4.12},$$

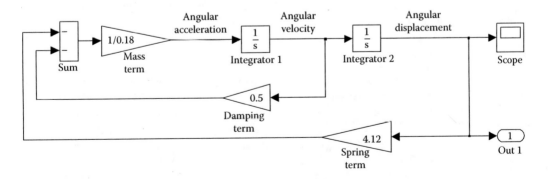

FIGURE 5.58 A Simulink block diagram to represent the linearized system in Figure 5.57.

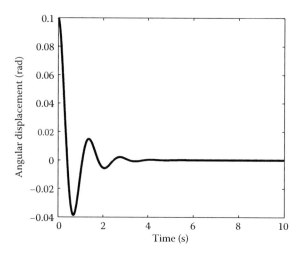

FIGURE 5.59 Angular displacement output $\theta(t)$ for $\theta(0) = 0.1$ rad.

which can be rewritten as
$$\Theta(s) = \frac{0.1(s+1.39) + 0.03(4.58)}{(s+1.39)^2 + (4.58)^2}.$$

Taking the inverse Laplace transform yields
$$\theta(t) = 0.1e^{-1.39t}\cos(4.58t) + 0.03e^{-1.39t}\sin(4.58t).$$

The reader can use MATLAB to plot the analytical solution $\theta(t)$, and the same curve as that shown in Figure 5.59 can be obtained.

Let us examine the nonlinear differential equation of motion in Part (b). Note that the unknown reaction forces R_x and R_y at the joint O do not appear in the moment equation $\sum M_O = I_O \alpha$. If an arbitrary nonfixed point P is used, we have to apply Equation 5.31. The moments caused by the unknown forces R_x and R_y appear in the equation, and auxiliary equations are required to eliminate the terms related to R_x and R_y. This is the advantage of choosing the fixed point to apply the moment equation if a rigid body rotates about a fixed axis.

If the mass of the rod can be neglected compared with the concentrated mass m, and the damping is negligible, the system reduces to the one in Example 5.6. The equation of motion in Example 5.6 is derived using Newton's second law, and it is exactly the same as that obtained using the moment equation $\sum M_O = I_O \alpha$. In general, for systems involving rotational motion, it may be more efficient to use the moment equation.

5.3.4 Pure Rolling Motion

Wheels are common mechanical systems involving general plane motion. Figure 5.60a shows a uniform disk rolling on a horizontal surface. If there is no slipping between the disk and the surface, the disk undergoes pure rolling motion. The contact point is the instantaneous center (IC), where the velocity is zero. From the results of kinematics, the acceleration of the IC is $r\dot\theta^2$, and its direction points from the IC to the mass center C (or for a uniform disk, the geometric center). The problem of finding a_{IC} is left to the reader as an exercise.

The free-body diagram of the disk is shown in Figure 5.60b, where reaction forces at the contact point include the normal force N and the friction force f. Note that a fixed point does not exist for the rolling motion; thus we cannot use the moment equation $\sum M_O = I_O \alpha$. Also, it is inconvenient

Mechanical Systems

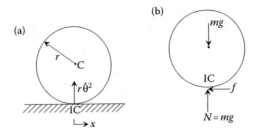

FIGURE 5.60 A pure rolling disk: (a) physical system and (b) free-body diagram.

to sum the moments about the mass center using the equation $\sum M_C = I_C \alpha$ because the unknown reaction force f inevitably appears in the equation and needs to be eliminated with the help of auxiliary equations. However, if the point IC is used, the moment equation

$$\sum \mathbf{M}_{IC} = I_{IC}\alpha + m\mathbf{r}_{C/IC} \times \mathbf{a}_{IC} \tag{5.36}$$

reduces to

$$\sum M_{IC} = I_{IC}\alpha. \tag{5.37}$$

Because the position vector $\mathbf{r}_{C/IC}$ and the acceleration vector $\mathbf{a}_{IC}$ are parallel, their cross product is zero, $m\mathbf{r}_{C/IC} \times \mathbf{a}_{IC} = 0$. Therefore, we do not need to deal with the moment caused by the friction force at the contact point. This is a special way of using Equation 5.31.

In summary, if there is a point S such that the position vector $\mathbf{r}_{C/S}$ is parallel to the acceleration vector $\mathbf{a}_S$, we can apply

$$\sum M_S = I_S \alpha \tag{5.38}$$

to model the dynamics of the rigid body. I_S is the mass moment of inertia of the body about the point S. For the pure rolling disk in Figure 5.60, the IC at the contact point has the property of $\mathbf{r}_{C/IC} \| \mathbf{a}_{IC}$.

Example 5.11: A Pure Rolling Sphere

Consider the system shown in Figure 5.61a, where a uniform sphere of mass m and radius r rolls along an inclined plane of 30°. A translational spring of stiffness k is attached to the sphere. Assuming that there is no slipping between the sphere and the surface, derive the differential equation of motion.

SOLUTION

The free-body diagram of the system is shown in Figure 5.61b, where the normal force N and the friction force f are reaction forces at the contact point. Assuming that the sphere rolls down

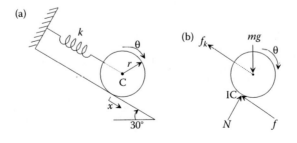

FIGURE 5.61 A pure rolling sphere: (a) physical system and (b) free-body diagram.

the incline, the spring is in tension and f_k is the spring force. When the sphere is at the static equilibrium position, we have $f_k = k\delta_{st}$, where δ_{st} is the static deformation of the spring. Then,

$$+\curvearrowleft: \quad \sum M_{IC} = 0,$$
$$r\sin 30° \cdot mg - r \cdot k\delta_{st} = 0,$$

or

$$0.5mg = k\delta_{st}.$$

We choose the static equilibrium position as the origin. Because of no slipping, the contact point is the IC, and $\mathbf{r}_{C/IC} \| \mathbf{a}_{IC}$. Applying Equation 5.37 gives

$$+\curvearrowleft: \quad \sum M_{IC} = I_{IC}\alpha,$$
$$r\sin 30° \cdot mg - r \cdot k(x + \delta_{st}) = I_{IC}\ddot{\theta},$$

where

$$I_{IC} = I_C + md^2 = \frac{2}{5}mr^2 + mr^2 = \frac{7}{5}mr^2.$$

Introducing the static equilibrium condition, $0.5mg = k\delta_{st}$, and the assumption of no slipping, $x = r\theta$, we obtain the differential equation of motion

$$\frac{7}{5}mr^2\ddot{\theta} + kr^2\theta = 0.$$

Strictly speaking, the disk in Figure 5.60 or the sphere in Example 5.11 is not a purely rotational system, and the rolling motion involves both translation and rotation. Such a system can also be modeled using both Newton's second law and the moment equation. We will present the corresponding discussion in the next section.

PROBLEM SET 5.3

1. Consider the rotational system shown in Figure 5.62. The system consists of a massless shaft and a uniform thin disk of mass m and radius r. The disk is constrained to rotate about a fixed longitudinal axis along the shaft. The shaft is equivalent to a torsional spring of stiffness K. Draw the necessary free-body diagram and derive the differential equation of motion.
2. Repeat Problem 1 for the system shown in Figure 5.63.
3. The system shown in Figure 5.64 consists of a uniform rod of mass m and length L and a translational spring of stiffness k at the rod's tip. The friction at the joint O is modeled as a damper with coefficient of torsional viscous damping B. The input is the force f and the output is the angle θ. The position $\theta = 0$ corresponds to the equilibrium position when $f = 0$.
   a. Determine the mass moment of inertia of the rod about the point O.

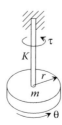

FIGURE 5.62 Problem 1.

Mechanical Systems

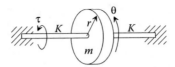

FIGURE 5.63 Problem 2.

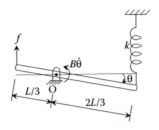

FIGURE 5.64 Problem 3.

b. Draw the necessary free-body diagram and derive the differential equation of motion for small angles θ.
c. Using the linearized differential equation obtained in Part (b), determine the transfer function $\Theta(s)/F(s)$. Assume that the initial conditions are $\theta(0) = 0$ and $\dot{\theta}(0) = 0$.
d. Using the differential equation obtained in Part (b), determine the state-space representation.

4. Repeat Problem 3 for the system shown in Figure 5.65.
5. ◢ Example 5.10 Part (d) shows how one can represent a linear system in Simulink based on the differential equation of the system. A linear system can also be represented in transfer function or state-space form. The corresponding blocks in Simulink are `Transfer Fcn` and `State-Space`, respectively. Refer to Problem 3. The parameter values are $m = 1.2$ kg, $L = 0.6$ m, $k = 100$ N/m, $B = 0.5$ N·s/m, and $g = 9.81$ m/s^2. Construct a Simulink block diagram to find the output $\theta(t)$ of the system if the input force is $f = \sin t$ N, where the system is represented using
   a. The linearized differential equation of motion obtained in Problem 3, Part (b)
   b. The transfer function obtained in Problem 3, Part (c)
   c. The state-space representation obtained in Problem 3, Part (d).
6. ◢ Repeat Problem 5 using the linearized differential equation of motion, the transfer function, and the state-space representation obtained in Problem 4.
7. Consider the uniform slender rod pivoted at point O as shown in Figure 5.66. The friction at the joint is modeled as a damper with coefficient of torsional viscous damping B. The position $\theta = 0$ corresponds to the equilibrium position.

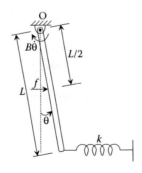

FIGURE 5.65 Problem 4.

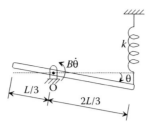

FIGURE 5.66 Problem 7.

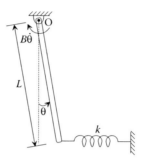

FIGURE 5.67 Problem 8.

   a. Draw the necessary free-body diagram and derive the differential equation of motion for small angles θ.
   b. Construct a Simulink block diagram to find the output $\theta(t)$ of the linearized system for the first 2 s if the initial displacement is $\theta(0) = 0.1$ rad. The parameter values are $m = 1.2$ kg, $L = 0.6$ m, $k = 100$ N/m, $B = 0.5$ N·s/m, and $g = 9.81$ m/s^2.
   c. Use Laplace and inverse Laplace transforms to find the analytical solution of $\theta(t)$. Plot the solution $\theta(t)$ and compare it with the one obtained using Simulink.
8. Repeat Problem 7 for the system shown in Figure 5.67.
9. Consider the inverted pendulum system shown in Figure 5.68. The system consists of a bob of mass m and a uniform rod of mass M and length L. The inverted pendulum is pivoted at the joint O. Draw the necessary free-body diagram and derive the differential equation of motion. Assume small angles for θ.
10. Consider the pendulum system shown in Figure 5.69. The system consists of a bob of mass m and a uniform rod of mass M and length L. The input is the force applied to the bob at the tip of the rod. The output is the angular displacement θ. A translational spring of stiffness k is connected to the middle of the rod. When $\theta = 0$ and $f = 0$, the spring is at its free length. Draw the necessary free-body diagram and derive the differential equation of motion. Assume small angles for θ.

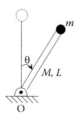

FIGURE 5.68 Problem 9.

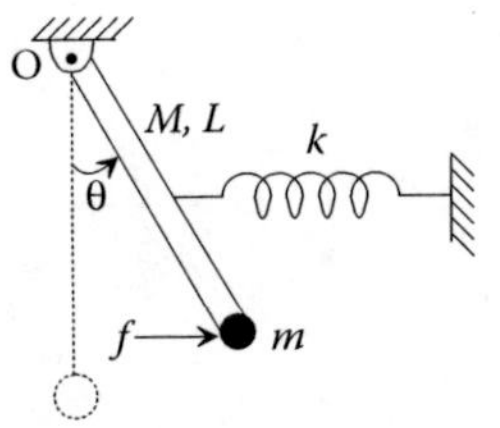

Mechanical Systems

FIGURE 5.69 Problem 10.

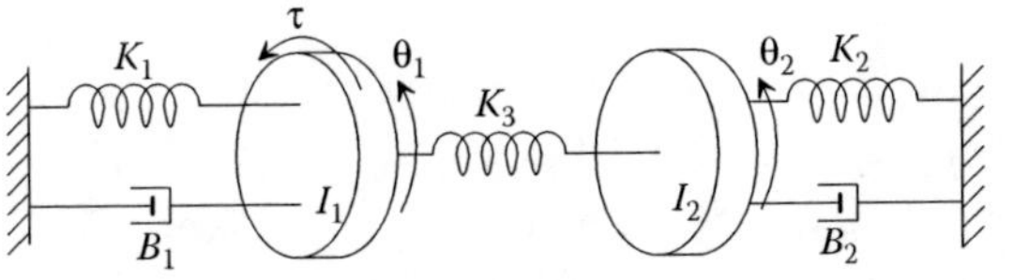

FIGURE 5.70 Problem 11.

11. Consider the torsional mass–spring–damper system in Figure 5.70. The mass moments of inertia of the two disks about their longitudinal axes are I_1 and I_2, respectively. The massless torsional springs represent the elasticity of the shafts and the torsional viscous dampers represent the fluid coupling.
    a. Draw the necessary free-body diagrams and derive the differential equations of motion. Provide the equations in the second-order matrix form.
    b. Determine the transfer functions $\Theta_1(s)/T(s)$ and $\Theta_2(s)/T(s)$. All the initial conditions are assumed to be zero.
    c. Determine the state-space representation with the angular displacements θ_1 and θ_2 as the outputs.
12. Repeat Problem 11 for the system shown in Figure 5.71. The input is the angular displacement ϕ at the end of the shaft.
13. Consider the two-degree-of-freedom system shown in Figure 5.72, where two simple pendulums are connected by a translational spring of stiffness k. Each pendulum

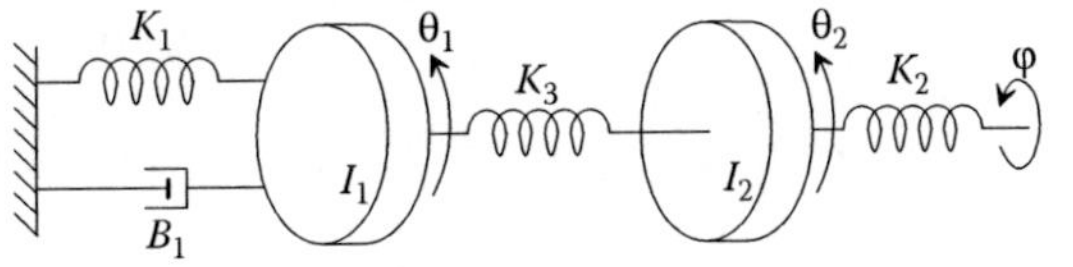

FIGURE 5.71 Problem 12.

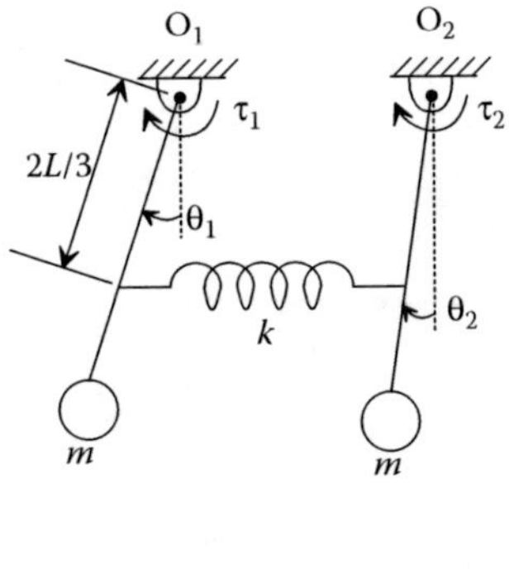

FIGURE 5.72 Problem 13.

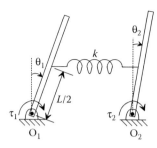

FIGURE 5.73 Problem 14.

consists of a point mass m concentrated at the tip of a massless rod of length L. The inputs are the torques τ_1 and τ_2 applied to the pivot points O_1 and O_2, respectively. The outputs are the angular displacements θ_1 and θ_2 of the pendulums. When $\theta_1 = 0$, $\theta_2 = 0$, $\tau_1 = 0$, and $\tau_2 = 0$, the spring is at its free length.
 a. Draw the necessary free-body diagrams and derive the differential equations of motion. Assume small angles for θ_1 and θ_2. Provide the equations in the second-order matrix form.
 b. Using the differential equations obtained in Part (a), determine the state-space representation with the angular velocities $\dot\theta_1$ and $\dot\theta_2$ as the outputs.
14. Repeat Problem 13 for the system shown in Figure 5.73, where each pendulum is a uniform slender rod of mass m and length L.
15. Consider the system shown in Figure 5.74. Assume that the thin disk rolls without slipping. Draw the necessary free-body diagram and derive the differential equation of motion.
16. Consider the system shown in Figure 5.75. Assume that the uniform circular cylinder of length L rolls down an inclined plane without slipping. Draw the necessary free-body diagram and derive the differential equation of motion.

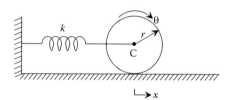

FIGURE 5.74 Problem 15.

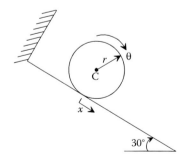

FIGURE 5.75 Problem 16.

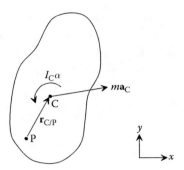

FIGURE 5.76 A rigid body in plane motion.

5.4 MIXED SYSTEMS: TRANSLATIONAL AND ROTATIONAL

For a system involving both translational and rotational motions, Newton's second law and the moment equation can be used to model the system. This section provides only one method to obtain the differential equation of motion using the force and moment equations, which may be applied in different ways.

5.4.1 Force and Moment Equations

Consider a mechanical system in plane motion, which involves translations along the x and y directions and rotation about one axis perpendicular to the x–y plane. For a system of a single mass, applying Newton's second law to the translational directions gives the force equations

$$\sum F_x = ma_{Cx}, \quad \sum F_y = ma_{Cy}. \tag{5.39}$$

The moment equation is in the form of

$$\sum M_C = I_C \alpha \tag{5.40}$$

or

$$\sum M_P = I_C \alpha + M_{\text{eff\_}ma_C}, \tag{5.41}$$

where the symbol $M_{\text{eff\_}ma_C}$ is used to represent the effective moment caused by the fictitious force ma_C. Although Equation 5.41 looks different from the general moment equation 5.31 given in Section 5.3, they are essentially equivalent. Figure 5.76 shows a rigid body in plane motion. Applying Equation 5.31 gives the net moment about point P as $\sum M_P = I_P \alpha + m\mathbf{r}_{C/P} \times \mathbf{a}_P$. According to the parallel-axis theorem, we have $I_P = I_C + mr_{C/P}^2$. Then, $I_P \alpha = I_C \alpha + mr_{C/P}^2 \alpha$ and it follows the direction of α, that is, the counterclockwise direction. Also, from the kinematics of rigid bodies, the acceleration of point P is $\mathbf{a}_P = \mathbf{a}_C + \mathbf{a}_{P/C,t} + \mathbf{a}_{P/C,n}$, as shown in Figure 5.77. Note that $m\mathbf{r}_{C/P} \times \mathbf{a}_C = \mathbf{r}_{C/P} \times (m\mathbf{a}_C)$, which can be considered as the effective moment caused by a fictitious force ma_C and is denoted as $M_{\text{eff\_}ma_C}$. The tangential component of the relative acceleration

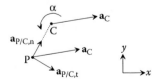

FIGURE 5.77 A kinematic diagram of $\mathbf{a}_P$ and $\mathbf{a}_C$.

$\mathbf{a}_{P/C,t}$ is perpendicular to the position vector $\mathbf{r}_{C/P}$ with a magnitude of $\alpha r_{C/P}$. Thus, the magnitude of the cross product $m\mathbf{r}_{C/P} \times \mathbf{a}_{P/C,t}$ is $mr_{C/P}^2 \alpha$ and its direction is clockwise. The normal component $\mathbf{a}_{P/C,n}$ is parallel to the position vector $\mathbf{r}_{C/P}$, and thus $m\mathbf{r}_{C/P} \times \mathbf{a}_{P/C,n} = 0$. Assuming that the positive direction is counterclockwise, we have $\Sigma M_P = I_C \alpha + mr_{C/P}^2 \alpha + M_{\text{eff}\_m\mathbf{a}_C} - mr_{C/P}^2 \alpha$, which is the same as Equation 5.41.

For a system of multiple masses, the force equations become

$$\Sigma F_x = \sum_{i=1}^{n} m_i (a_{Ci})_x, \quad \Sigma F_y = \sum_{i=1}^{n} m_i (a_{Ci})_y, \tag{5.42}$$

where n is the number of masses and a_{Ci} is the acceleration of the mass center of the ith mass. Because it may be a challenge to find the mass center for the entire system of multiple masses, we use an arbitrary point P to sum the moments,

$$\Sigma M_P = \sum_{i=1}^{n} I_{Ci} \alpha_i + \sum_{i=1}^{n} M_{\text{eff}\_m_i \mathbf{a}_{Ci}}, \tag{5.43}$$

where I_{Ci} is the moment of inertia of the ith mass about its center of mass and α_i is the angular acceleration of the ith mass.

In order to derive the differential equation of motion correctly, we recommend drawing two diagrams before applying the force and moment equations. One is the free-body diagram that shows all external forces and moments applied to the system, and the other is the kinematic diagram that indicates the acceleration at the mass center of each mass. The left-hand sides of the force and moment equations 5.42 and 5.43 are written based on the free-body diagram, and the right-hand sides are written based on the kinematic diagram.

Example 5.12: A Pulley System

Consider the pulley system shown in Figure 5.78a. A block of mass m is connected to a translational spring of stiffness k through a cable, which passes by a pulley with mass m_p and radius r. The pulley rotates about the fixed mass center O. The moment of inertia of the pulley about its mass center is I_O. Draw the free-body diagram and kinematic diagram, and derive the equation of motion.

SOLUTION

The block undergoes translational motion along the vertical direction, and the pulley rotates about its mass center that is fixed. We choose the displacement of the block x and the angular displacement of the pulley θ as the generalized coordinates. Note that x is measured from the static

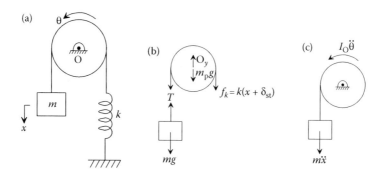

FIGURE 5.78 A pulley system: (a) physical system, (b) free-body diagram, and (c) kinematic diagram.

equilibrium position, which is set as the origin. The free-body diagram of the system is shown in Figure 5.78b. At static equilibrium, we have

$$+\curvearrowleft: \quad \sum M_O = 0,$$
$$r \cdot mg - r \cdot k\delta_{st} = 0,$$
$$mg = k\delta_{st},$$

where δ_{st} is the static deformation of the spring. Assume that the block and the pulley are displaced in their positive directions. The spring force is $f_k = k(x + \delta_{st})$. The kinematic diagram is shown in Figure 5.78c. For the block (translation only), applying the force equation in the x direction gives

$$+\downarrow x: \sum F_x = ma_{Cx},$$
$$mg - T = m\ddot{x}.$$

For the pulley (rotation only), applying the moment equation about the fixed point O gives

$$+\curvearrowleft: \quad \sum M_O = I_O \alpha,$$
$$r \cdot T - r \cdot k(x + \delta_{st}) = I_O \ddot{\theta}.$$

Combining the two equations and eliminating the unknown T results in

$$r(mg - m\ddot{x}) - rk(x + \delta_{st}) = I_O \ddot{\theta}.$$

Introducing the geometric constraint, $x = r\theta$, and the static equilibrium condition, $mg = k\delta_{st}$, the equation becomes

$$(I_O + mr^2)\ddot{\theta} + kr^2\theta = 0.$$

Example 5.13: A Cart–Inverted-Pendulum System

Consider the mechanical system shown in Figure 5.79, where a uniform rod of mass m and length L is pivoted on a cart of mass M. An external force f is applied to the cart. Assume that the pendulum is constrained to move in a vertical plane, and the cart moves without slipping along a horizontal line. Denote the displacement of the cart as x and the angular displacement of the pendulum as θ.

  a. Draw the necessary free-body diagram and kinematic diagram, and derive the nonlinear equations of motion.
  b. Linearize the equations of motion for small angular motions, and determine the state-space form with x and θ as the outputs.

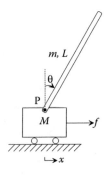

FIGURE 5.79 A cart–inverted-pendulum system.

Solution

a. This is a mixed system with two masses. The motion of the cart is purely translational, and the inverted pendulum undergoes both translation and rotation. The free-body diagram and the kinematic diagram are given in Figure 5.80. Note that the acceleration of the mass center of the rod consists of three components. On the one hand, the pendulum moves together with the cart along the horizontal line at an acceleration of $\ddot{x}$. On the other hand, it rotates about the pivot P on the cart. The relative rotational accelerations include the tangential component $\frac{L}{2}\ddot{\theta}$ and the normal component $\frac{L}{2}\dot{\theta}^2$.

Applying the force equation to the whole system along the x direction gives

$$+\rightarrow x: \sum F_x = \sum_{i=1}^{2} m_i(a_{Ci})_x,$$

$$f = M\ddot{x} + m\ddot{x} - m\frac{L}{2}\dot{\theta}^2 \cdot \sin\theta + m\frac{L}{2}\ddot{\theta} \cdot \cos\theta.$$

The forces at the pivot are canceled out because they are internal forces between the cart and the pendulum. Applying the moment equation to the pendulum about the point P results in

$$+\curvearrowleft: \quad \sum M_P = I_C\alpha + M_{\text{eff\_}ma_C},$$

$$\frac{L}{2}\sin\theta \cdot mg = \frac{1}{12}mL^2\ddot{\theta} + \frac{L}{2}\cos\theta \cdot m\ddot{x} + \frac{L}{2} \cdot m\frac{L}{2}\ddot{\theta}.$$

Rearranging the two equations into the standard input–output form

$$(M+m)\ddot{x} + \frac{1}{2}mL\cos\theta \cdot \ddot{\theta} - \frac{1}{2}mL\dot{\theta}^2 \cdot \sin\theta = f,$$

$$\frac{1}{3}mL^2\ddot{\theta} + \frac{1}{2}mL\cos\theta \cdot \ddot{x} - \frac{1}{2}mgL\sin\theta = 0.$$

b. For small angular motions, $\cos\theta \approx 1$, $\sin\theta \approx \theta$, $\dot{\theta}^2\theta \approx 0$ (see Section 4.6). The linearized equations are

$$(M+m)\ddot{x} + \frac{1}{2}mL\ddot{\theta} = f,$$

$$\frac{1}{3}mL^2\ddot{\theta} + \frac{1}{2}mL\ddot{x} - \frac{1}{2}mgL\theta = 0.$$

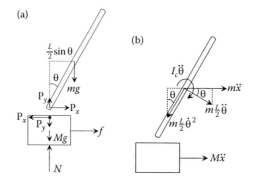

FIGURE 5.80 A cart–inverted-pendulum system: (a) free-body diagram and (b) kinematic diagram.

Mechanical Systems

The state, the input, and the output are specified as

$$\mathbf{x} = \begin{Bmatrix} x_1 \\ x_2 \\ x_3 \\ x_4 \end{Bmatrix} = \begin{Bmatrix} x \\ \theta \\ \dot{x} \\ \dot{\theta} \end{Bmatrix}, \quad u = f, \quad \mathbf{y} = \begin{Bmatrix} x \\ \theta \end{Bmatrix}.$$

We then take the time derivative of each state variable. For the first two,

$$\dot{x}_1 = \dot{x} = x_3,$$
$$\dot{x}_2 = \dot{\theta} = x_4.$$

Note that the two linearized equations are coupled with each other. The derivatives $\ddot{x}$ and $\ddot{\theta}$ cannot be solved using only one of the equations. To find $\dot{x}_3$ and $\dot{x}_4$, we rewrite the two linearized differential equations as

$$\begin{bmatrix} M+m & \frac{1}{2}mL \\ \frac{1}{2}mL & \frac{1}{3}mL^2 \end{bmatrix} \begin{Bmatrix} \ddot{x} \\ \ddot{\theta} \end{Bmatrix} = \begin{Bmatrix} f \\ \frac{1}{2}mgL\theta \end{Bmatrix},$$

from which $\ddot{x}$ and $\ddot{\theta}$ can be solved using Cramer's rule:

$$\ddot{x} = \frac{(1/3)mL^2 f - (1/4)m^2L^2 g\theta}{(1/12)mL^2(4M+m)},$$

$$\ddot{\theta} = \frac{-(1/2)mL f + (M+m)(1/2)mgL\theta}{(1/12)mL^2(4M+m)}.$$

Simplifying the equations gives

$$\dot{x}_3 = \frac{4}{4M+m}u - \frac{3mg}{4M+m}x_2,$$

$$\dot{x}_4 = \frac{-6}{L(4M+m)}u + \frac{6(M+m)g}{L(4M+m)}x_2.$$

Thus, the state-space equation in matrix form is

$$\begin{Bmatrix} \dot{x}_1 \\ \dot{x}_2 \\ \dot{x}_3 \\ \dot{x}_4 \end{Bmatrix} = \begin{bmatrix} 0 & 0 & 1 & 0 \\ 0 & 0 & 0 & 1 \\ 0 & -\frac{3mg}{4M+m} & 0 & 0 \\ 0 & \frac{6(M+m)g}{L(4M+m)} & 0 & 0 \end{bmatrix} \begin{Bmatrix} x_1 \\ x_2 \\ x_3 \\ x_4 \end{Bmatrix} + \begin{bmatrix} 0 \\ 0 \\ \frac{4}{4M+m} \\ -\frac{6}{L(4M+m)} \end{bmatrix} u$$

$$\mathbf{y} = \begin{bmatrix} 1 & 0 & 0 & 0 \\ 0 & 1 & 0 & 0 \end{bmatrix} \begin{Bmatrix} x_1 \\ x_2 \\ x_3 \\ x_4 \end{Bmatrix} + \begin{bmatrix} 0 \\ 0 \end{bmatrix} u.$$

5.4.2 Energy Method

The differential equation of motion of a mechanical system can be obtained using the force/moment approach, which is based on Newtonian mechanics. Free-body diagrams are necessary in order to apply the force and moment equations correctly. An alternative way of obtaining the differential equation of motion is to use the energy method based on analytical mechanics.

For a mass–spring system with negligible friction and damping, the principle of conservation of energy states that

$$T + V = \text{constant} \tag{5.44}$$

or

$$\frac{d}{dt}(T + V) = 0, \tag{5.45}$$

where T is the kinetic energy and V is the potential energy, which includes the gravitational potential energy and the elastic potential energy.

The expression for the kinetic energy of a translational or rotational mass element was given in Section 5.1. In general, the kinetic energy of a rigid body in plane motion is

$$T = \frac{1}{2}mv_C^2 + \frac{1}{2}I_C\omega^2, \tag{5.46}$$

where v_C is the velocity of the mass center C of the body and ω is the angular velocity of the body. Note that the kinetic energy of a rigid body in plane motion can be separated into two parts: (1) the kinetic energy associated with the translational motion of the mass center C of the body, and (2) the kinetic energy associated with the rotation of the body about the mass center C. If a rigid body rotates about a fixed point O with an angular velocity ω, the kinetic energy reduces to Equation 5.5, $T = (1/2)I_O\omega^2$.

Example 5.14: A Pulley System—Energy Method

For the system in Example 5.12, use the energy method to derive the equation of motion.

Solution

The system has two mass elements, one translational block of mass and one pulley rotating about its fixed mass center. Their motions are related to each other by the geometric constraint, $x = r\theta$. The kinetic energy of the system is

$$T = T_{\text{block}} + T_{\text{pulley}} = \frac{1}{2}m\dot{x}^2 + \frac{1}{2}I_O\dot{\theta}^2.$$

The undeformed position, static equilibrium position, and dynamic position of the mass–spring system are shown in Figure 5.81. Choosing the static equilibrium position as the datum for the gravitational potential energy, we have

$$V_g = -mgx.$$

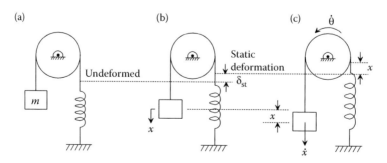

FIGURE 5.81 A pulley system: (a) undeformed position, (b) static equilibrium position, and (c) dynamic position.

Mechanical Systems

The elastic potential energy is

$$V_e = \frac{1}{2}k(x+\delta_{st})^2.$$

The total potential energy is

$$V = V_g + V_e = -mgx + \frac{1}{2}kx^2 + \frac{1}{2}k\delta_{st}^2 + kx\delta_{st}.$$

Because of the static equilibrium condition, $mg = k\delta_{st}$, the expression of potential energy becomes

$$V = \frac{1}{2}kx^2 + \frac{1}{2}k\delta_{st}^2.$$

The total energy of the system is

$$T + V = \frac{1}{2}m\dot{x}^2 + \frac{1}{2}I_O\dot{\theta}^2 + \frac{1}{2}kx^2 + \frac{1}{2}k\delta_{st}^2,$$

for which the time derivative is

$$\frac{d}{dt}(T+V) = m\dot{x}\ddot{x} + I_O\dot{\theta}\ddot{\theta} + kx\dot{x}.$$

Using the geometric constraint, $x = r\theta$, which implies that $\dot{x} = r\dot{\theta}$ and $\ddot{x} = r\ddot{\theta}$, we can rewrite the equation as

$$\frac{d}{dt}(T+V) = mr^2\dot{\theta}\ddot{\theta} + I_O\dot{\theta}\ddot{\theta} + kr^2\theta\dot{\theta}.$$

Applying the principle of conservation of energy gives

$$(I_O + mr^2)\ddot{\theta} + kr^2\theta = 0,$$

which is the same as that obtained in Example 5.12 using the force/moment approach.

Note that the mechanical system in Example 5.14 is a single-degree-of-freedom system, which requires only one generalized coordinate (e.g., x or θ) to describe the system dynamics. The example shows that if we can obtain the expression of $T + V$ in terms of the generalized coordinate, the equation of motion can be derived by taking the time derivative and making it zero.

For an n-degree-of-freedom system, n independent equations of motion can be derived using Lagrange's formulation, which is applicable to both conservative and nonconservative systems. The discussion of applying Lagrange's equations to nonconservative systems is beyond the scope of this text, and we are mainly concerned with conservative systems. One of the expressions of Lagrange's equations for a conservative system is

$$\frac{d}{dt}\left(\frac{\partial T}{\partial \dot{q}_i}\right) - \frac{\partial T}{\partial q_i} + \frac{\partial V}{\partial q_i} = 0, \quad i = 1, 2, \ldots, n, \quad (5.47)$$

where q_i is the ith generalized coordinate and n is the total number of independent generalized coordinates. In general, the kinetic energy is a function of the generalized displacements and the generalized velocities, and

$$T = T(q_1, q_2, \ldots, q_n, \dot{q}_1, \dot{q}_2, \ldots, \dot{q}_n). \quad (5.48)$$

The potential energy is a function of the generalized displacements, and

$$V = V(q_1, q_2, \ldots, q_n). \quad (5.49)$$

Example 5.15: A Double Pendulum

Consider the double pendulum in Figure 5.82, where two point masses of equal mass m are attached to two rigid links of equal length L. The links are assumed to be massless. The motion of the system is constrained in the vertical plane. Neglecting friction, derive the equations of motion using Lagrange's equations. Assume small angles for θ_1 and θ_2.

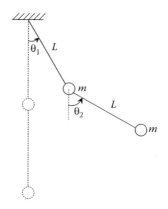

FIGURE 5.82 A double-pendulum system.

SOLUTION

The system is only subjected to gravitational forces, and it is a conservative system. The dynamics of the system can be described using two independent angular displacement coordinates, θ_1 and θ_2. The kinetic energy of the system is

$$T = \frac{1}{2}mv_1^2 + \frac{1}{2}mv_2^2.$$

From kinematics,

$$v_1 = L\dot{\theta}_1$$

and, as shown in Figure 5.83a,

$$v_2^2 = (L\dot{\theta}_1)^2 + (L\dot{\theta}_2)^2 + 2L^2\dot{\theta}_1\dot{\theta}_2\cos(\theta_2 - \theta_1),$$

which is obtained by applying the law of cosines. Thus,

$$T = \frac{1}{2}m(L\dot{\theta}_1)^2 + \frac{1}{2}m\left[(L\dot{\theta}_1)^2 + (L\dot{\theta}_2)^2 + 2L^2\dot{\theta}_1\dot{\theta}_2\cos(\theta_2 - \theta_1)\right].$$

Note that no spring elements are involved in the system. This implies that $V_e = 0$ and $V = V_g$. Using the datum defined in Figure 5.83b, we can obtain the gravitational potential energy

$$V_g = -mgh_1 - mgh_2,$$

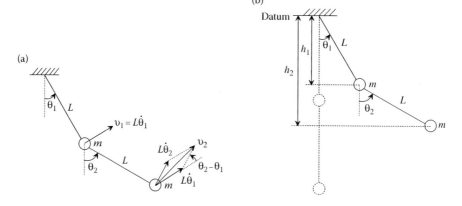

FIGURE 5.83 A double pendulum: (a) kinematic diagram and (b) positions of the mass centers.

where
$$h_1 = L\cos\theta_1,$$
$$h_2 = L\cos\theta_1 + L\cos\theta_2.$$

Thus,
$$V = -2mgL\cos\theta_1 - mgL\cos\theta_2.$$

We then apply Lagrange's equations
$$\frac{d}{dt}\left(\frac{\partial T}{\partial \dot{q}_i}\right) - \frac{\partial T}{\partial q_i} + \frac{\partial V}{\partial q_i} = 0, \quad i = 1, 2.$$

For $i = 1$, $q_1 = \theta_1$, $\dot{q}_1 = \dot{\theta}_1$,
$$\frac{\partial T}{\partial \dot{\theta}_1} = 2mL^2\dot{\theta}_1 + mL^2\dot{\theta}_2\cos(\theta_2 - \theta_1),$$
$$\frac{d}{dt}\left(\frac{\partial T}{\partial \dot{\theta}_1}\right) = 2mL^2\ddot{\theta}_1 + mL^2\ddot{\theta}_2\cos(\theta_2 - \theta_1) - mL^2\dot{\theta}_2\sin(\theta_2 - \theta_1)(\dot{\theta}_2 - \dot{\theta}_1),$$
$$\frac{\partial T}{\partial \theta_1} = -mL^2\dot{\theta}_1\dot{\theta}_2\sin(\theta_2 - \theta_1)(-1),$$
$$\frac{\partial V}{\partial \theta_1} = 2mgL\sin\theta_1.$$

Substituting into Lagrange's equation results in
$$\frac{d}{dt}\left(\frac{\partial T}{\partial \dot{\theta}_1}\right) - \frac{\partial T}{\partial \theta_1} + \frac{\partial V}{\partial \theta_1} = 2mL^2\ddot{\theta}_1 + mL^2\ddot{\theta}_2\cos(\theta_2 - \theta_1) - mL^2\dot{\theta}_2^2\sin(\theta_2 - \theta_1) + 2mgL\sin\theta_1 = 0.$$

Similarly, for $i = 2$, $q_2 = \theta_2$, $\dot{q}_2 = \dot{\theta}_2$,
$$\frac{\partial T}{\partial \dot{\theta}_2} = mL^2\dot{\theta}_2 + mL^2\dot{\theta}_1\cos(\theta_2 - \theta_1),$$
$$\frac{d}{dt}\left(\frac{\partial T}{\partial \dot{\theta}_2}\right) = mL^2\ddot{\theta}_2 + mL^2\ddot{\theta}_1\cos(\theta_2 - \theta_1) - mL^2\dot{\theta}_1\sin(\theta_2 - \theta_1)(\dot{\theta}_2 - \dot{\theta}_1),$$
$$\frac{\partial T}{\partial \theta_2} = -mL^2\dot{\theta}_1\dot{\theta}_2\sin(\theta_2 - \theta_1),$$
$$\frac{\partial V}{\partial \theta_2} = mgL\sin\theta_2.$$

Substituting into Lagrange's equation gives
$$\frac{d}{dt}\left(\frac{\partial T}{\partial \dot{\theta}_2}\right) - \frac{\partial T}{\partial \theta_2} + \frac{\partial V}{\partial \theta_2} = mL^2\ddot{\theta}_2 + mL^2\ddot{\theta}_1\cos(\theta_2 - \theta_1) + mL^2\dot{\theta}_1^2\sin(\theta_2 - \theta_1) + mgL\sin\theta_2 = 0.$$

For small motions ($\theta_1 \approx 0$ and $\theta_2 \approx 0$), the two differential equations of motion after linearization are
$$2mL^2\ddot{\theta}_1 + mL^2\ddot{\theta}_2 + 2mgL\theta_1 = 0,$$
$$mL^2\ddot{\theta}_1 + mL^2\ddot{\theta}_2 + mgL\theta_2 = 0,$$

or in second-order matrix form
$$\begin{bmatrix} 2mL^2 & mL^2 \\ mL^2 & mL^2 \end{bmatrix}\begin{Bmatrix} \ddot{\theta}_1 \\ \ddot{\theta}_2 \end{Bmatrix} + \begin{bmatrix} 2mgL & 0 \\ 0 & mgL \end{bmatrix}\begin{Bmatrix} \theta_1 \\ \theta_2 \end{Bmatrix} = \begin{Bmatrix} 0 \\ 0 \end{Bmatrix}.$$

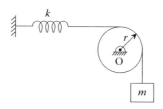

FIGURE 5.84 Problem 1.

PROBLEM SET 5.4

1. Consider the pulley system shown in Figure 5.84. A block of mass m is connected to a translational spring of stiffness k through a cable, which passes by a pulley. The pulley rotates about the fixed mass center O. The moment of inertia of the pulley about its mass center is I_O. Draw the free-body diagram and kinematic diagram, and derive the equation of motion.
2. The double pulley shown in Figure 5.85 has an inner radius of r_1 and an outer radius of r_2. The mass moment of inertia of the pulley about the point O is I_O. A translational spring of stiffness k and a block of mass m are suspended by cables wrapped around the pulley as shown. Draw the free-body diagram and kinematic diagram, and derive the equation of motion.
3. Consider the mechanical system shown in Figure 5.86, where a uniform rod of mass m and length L is pivoted on a cart of mass M. Assume that the pendulum is constrained to move in a vertical plane, and the cart moves on a smooth horizontal surface. Denote the displacement of the cart as x and the angular displacement of the pendulum as θ. Draw the necessary free-body diagram and kinematic diagram, and derive the equations of motion for small angles.
4. Consider the mechanical system shown in Figure 5.87, where a simple pendulum is pivoted on a cart of mass M. The pendulum consists of a point mass m concentrated

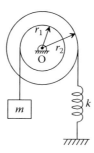

FIGURE 5.85 Problem 2.

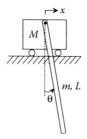

FIGURE 5.86 Problem 3.

Mechanical Systems

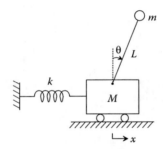

FIGURE 5.87 Problem 4.

at the tip of a massless rod of length L. The cart is connected to a translational spring of stiffness k. Denote the displacement of the cart as x and the angular displacement of the pendulum as θ. Draw the necessary free-body diagram and kinematic diagram, and derive the equations of motion for small angles.

5. Consider the mechanical system shown in Figure 5.88. Draw the necessary free-body diagram and kinematic diagram, and derive the equations of motion for small angles.
6. Consider the mechanical system shown in Figure 5.89. The inputs are the force f_2 applied at the tip of the rod and the force f_1 applied to the cart. The outputs are the displacement x of the cart and the angular displacement θ of the pendulum.
   a. Draw the necessary free-body diagram and kinematic diagram, and derive the equations of motion for small angles.
   b. Using the differential equation obtained in Part (a), determine the state-space representation.
7. Repeat Problem 1 using the energy method.
8. Repeat Problem 2 using the energy method.

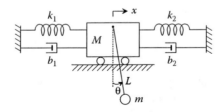

FIGURE 5.88 Problem 5.

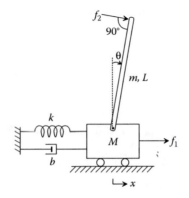

FIGURE 5.89 Problem 6.

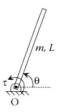

FIGURE 5.90 Problem 13.

9. Repeat Problem 7(a) in Problem Set 5.3 using the energy method. Ignore the torsional viscous damping at the joint.
10. Repeat Problem 8(a) in Problem Set 5.3 using the energy method. Ignore the torsional viscous damping at the joint.
11. Repeat Problem 3 using Lagrange's equations.
12. Repeat Problem 4 using Lagrange's equations.
13. A robot arm consists of rigid links connected by joints allowing the relative motion of neighboring links. The dynamic model for a robot arm can be derived using Lagrange's equations

$$\frac{d}{dt}\left(\frac{\partial T}{\partial \dot{\theta}_i}\right) - \frac{\partial T}{\partial \theta_i} + \frac{\partial V}{\partial \theta_i} = \tau_i, \quad i = 1, 2, \ldots, n,$$

where θ_i is the angular displacement of the ith joint, τ_i is the torque applied to the ith joint, and n is the total number of joints. Consider a single-link planar robot arm as shown in Figure 5.90. Use Lagrange's equations to derive the dynamic model of the robot arm. Assume that the motion of the robot arm is constrained in a vertical plane, and the joint angle varies between $0°$ and $360°$.

14. Repeat Problem 13 for a two-link planar robot arm as shown in Figure 5.91. Assume that the motion of the robot arm is constrained in a horizontal plane, and the joint angles vary between $0°$ and $360°$.

5.5 GEAR–TRAIN SYSTEMS

Gear–train systems are important in many engineering applications. Figure 5.92a shows a pair of ideal gears, which are assumed to be rigid and meshed without backlash. A torque τ_1 produced by a motor is applied to gear 1, which rotates and causes gear 2 to rotate in the opposite direction. The radii of gear 1 and gear 2 are r_1 and r_2, respectively. The relative sizes of the two gears result in a proportionality constant between the angular velocities of the respective shafts.

For the purpose of analysis, a free-body diagram for the rotational gear–train system must be drawn with care. It is convenient to visualize the gears as circles. Figure 5.92b shows the free-body diagram seen from the side of the input shaft. The two gears are tangent at the contact point and rotate without slipping. The torque applied to gear 1 causes an action force F onto gear 2 at the

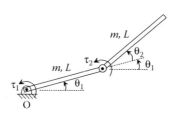

FIGURE 5.91 Problem 14.

Mechanical Systems

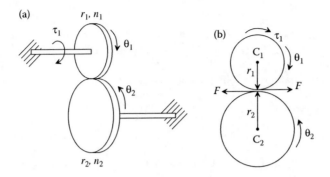

FIGURE 5.92 A gear–train system: (a) physical system and (b) free-body diagram.

contact point. Because of Newton's third law, gear 1 is subjected to a reaction force at the contact point. Use θ_1 and θ_2 to denote the respective angular displacements. The geometric constraint is

$$r_1\theta_1 = r_2\theta_2, \tag{5.50}$$

which can be rewritten as

$$\frac{\theta_2}{\theta_1} = \frac{r_1}{r_2} = N, \tag{5.51}$$

where N is called the gear ratio, which is also the relationship between the numbers of teeth on the two gears:

$$\frac{n_1}{n_2} = N. \tag{5.52}$$

Note that the gear ratio N may be defined differently among authors. Differentiating Equation 5.51 gives

$$\frac{\omega_2}{\omega_1} = N. \tag{5.53}$$

Thus, the gear pair is a speed reducer for $N < 1$. If the gears have negligible inertia or zero angular acceleration, and if the energy loss due to friction between the gear teeth can be neglected, the input work must be equal to the output work. Under these conditions, the output torque is greater than the input torque for a speed reducer.

To obtain the mathematical model of a gear–train system, the fundamental laws are still applied. We need to draw a free-body diagram and apply the moment equation for each gear. The derivation also requires a consideration of the geometric constraint.

Example 5.16: A Single-Degree-of-Freedom Gear–Train System

For the gear–train system shown in Figure 5.92a, derive the differential equation of motion. The mass moments of inertia of the two gears about their respective fixed center are I_{C1} and I_{C2}.

Solution

The free-body diagram is shown in Figure 5.92b. The geometric constraint is

$$\frac{\theta_2}{\theta_1} = \frac{r_1}{r_2}.$$

Applying the moment equation to each gear gives

$$+\curvearrowleft: \quad \sum M_{C1} = I_{C1}\alpha,$$
$$\tau_1 - r_1 F = I_{C1}\ddot{\theta}_1,$$

and

$$+\curvearrowleft: \quad \sum M_{C2} = I_{C2}\alpha,$$
$$r_2 F = I_{C2}\ddot{\theta}_2.$$

Combining the two equations and eliminating F yields

$$\tau_1 - \frac{r_1}{r_2}I_{C2}\ddot{\theta}_2 = I_{C1}\ddot{\theta}_1.$$

Note that the angular displacements θ_1 and θ_2 are dependent on each other through the geometric constraint. Thus, the gear-train in Figure 5.92 is a single-degree-of-freedom system, which requires only one equation of motion in terms of only one coordinate. Assume that the equation is expressed in terms of θ_1. Introducing the geometric constraint results in the equation of motion

$$\tau_1 - \frac{r_1}{r_2}I_{C2}\left(\frac{r_1}{r_2}\ddot{\theta}_1\right) = I_{C1}\ddot{\theta}_1,$$
$$\left(I_{C1} + \frac{r_1^2}{r_2^2}I_{C2}\right)\ddot{\theta}_1 = \tau_1,$$

which is the dynamics seen from the input side.

Example 5.17: A Single-Link Robot Arm

The mechanical model of a single-link robot arm driven by a motor can be represented as a gear-train system, as shown in Figure 5.93, where two rotational subsystems are coupled with a pair of gears with negligible inertia. The mass moments of inertia of the motor and the load are I_m and I, respectively. The coefficients of torsional viscous damping on the motor and the load are B_m and B, respectively. τ_m is the torque generated by the motor. Assume that the gear ratio is $N = r_1/r_2$. Derive the differential equation of motion in terms of the motor variable θ_m.

Solution

The free-body diagrams for the motor, the load, and the gear-train are shown in Figure 5.94, where F represents the contact force between the two gears. The moments caused by the contact force on the motor and on the load are $r_1 F$ and $r_2 F$, respectively. Applying the moment equation to the motor and the load gives

$$+\curvearrowleft: \quad \sum M_O = I_O\alpha,$$
$$\tau_m - B_m\dot{\theta}_m - r_1 F = I_m\ddot{\theta}_m,$$

and

$$+\curvearrowleft: \quad \sum M_O = I_O\alpha,$$
$$-B\dot{\theta} + r_2 F = I\ddot{\theta}.$$

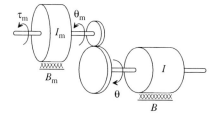

FIGURE 5.93 The mechanical model of a single-link robot arm driven by a motor.

Mechanical Systems

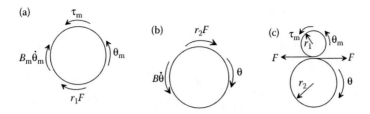

FIGURE 5.94 Free-body diagrams: (a) motor, (b) load, and (c) gear–train.

Solving for F from the equation for the load and substituting it into the equation for the motor results in

$$\tau_m - B_m \dot{\theta}_m - \frac{r_1}{r_2}(I\ddot{\theta} + B\dot{\theta}) = I_m \ddot{\theta}_m.$$

By the geometry of the gears,

$$\dot{\theta} = \frac{r_1}{r_2}\dot{\theta}_m = N\dot{\theta}_m,$$

$$\ddot{\theta} = \frac{r_1}{r_2}\ddot{\theta}_m = N\ddot{\theta}_m,$$

where N is the gear ratio. Substituting them into the previous differential equation gives

$$\tau_m - B_m \dot{\theta}_m - N^2(I\ddot{\theta}_m + B\dot{\theta}_m) = I_m \ddot{\theta}_m.$$

The equation can be rearranged as

$$(I_m + N^2 I)\ddot{\theta}_m + (B_m + N^2 B)\dot{\theta}_m = \tau_m,$$

which is written in terms of the angular displacement of the motor.

PROBLEM SET 5.5

1. Repeat Example 5.16, and determine a mathematical model for the simple one-degree-of-freedom system shown in Figure 5.92a in the form of a differential equation of motion in θ_2.
2. Repeat Example 5.17, and determine a mathematical model for the single-link robot arm shown in Figure 5.93 in the form of a differential equation of motion in the load variable θ.
3. Consider the one-degree-of-freedom system shown in Figure 5.95. The system consists of two gears of mass moments of inertia I_1 and I_2 and radii r_1 and r_2, respectively. The applied torque on gear 1 is τ_1. Assume that the gears are connected with flexible shafts, which can be approximated as two torsional springs of stiffnesses K_1 and K_2, respectively.

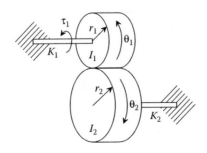

FIGURE 5.95 Problem 3.

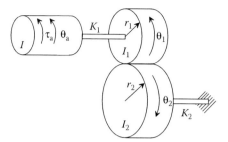

FIGURE 5.96 Problem 4.

   a. Draw the necessary free-body diagrams, and derive the differential equation of motion in θ_1.
   b. Using the differential equation obtained in Part (a), determine the transfer function $\Theta_2(s)/T_1(s)$.
   c. Using the differential equation obtained in Part (a), determine the state-space representation with θ_2 as the output.
4. Consider the gear–train system shown in Figure 5.96. The system consists of a rotational cylinder and a pair of gears. The gear ratio is $N = r_1/r_2$. The applied torque on the cylinder is τ_a. Assume that the gears are connected with flexible shafts, which can be approximated as two torsional springs of stiffnesses K_1 and K_2, respectively.
   a. Draw the necessary free-body diagrams, and derive the differential equations of motion.
   b. Using the differential equation obtained in Part (a), determine the state-space representation. Use θ_a, θ_1, ω_a, and ω_1 as the state variables, and use θ_2 and ω_2 as the output variables.
5. A three-degree-of-freedom gear–train system is shown in Figure 5.97, which consists of four gears of moments of inertia I_1, I_2, I_3, and I_4. Gears 2 and 3 are meshed and their radii are r_2 and r_3, respectively. Gears 1 and 2 are connected by a relatively long shaft, and gears 3 and 4 are connected in the same way. The shafts are assumed to be flexible, and can be approximated by torsional springs. The applied torque and load torque are τ_a and τ_l on gear 1 and gear 4, respectively. The gears are assumed to be rigid and have no backlash. Derive the differential equations of motion.
6. Repeat Problem 5. Assume that the shaft connecting gears 1 and 2 is relatively short and rigid.

5.6 SUMMARY

This chapter was devoted to modeling mechanical systems. Since real systems are usually quite complicated, simplifying assumptions must be made to reduce the system to an idealized model

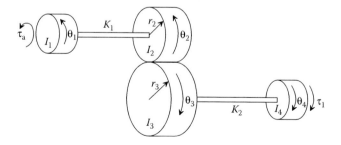

FIGURE 5.97 Problem 5.

Mechanical Systems

consisting of interconnected mass, damper, and spring elements. The relations between the external forces or moments applied to the elements and the associated element variables are given as

- Mass translation: $f = m\ddot{x}$ rotation about fixed O: $\tau = I_O \ddot{\theta}$
- Damper translation: $f = b\dot{x}_{rel}$ rotation: $\tau = B\dot{\theta}_{rel}$
- Spring translation: $f = kx_{rel}$ rotation: $\tau = K\theta_{rel}$

Here the spring force is dependent on the relative displacement between the two ends of the spring, and the damping force depends on the relative velocity between the two ends of the damper.

For a system of interconnected mechanical elements, the dynamic equations of motion can be obtained by applying Newton's second law and/or the moment equation. The number of equations of motion is determined by the number of degrees of freedom of the system. The number of degrees of freedom of a dynamic system is defined as the number of independent generalized coordinates that specify the configuration of the system. The static equilibrium position of a mechanical system is usually chosen as the coordinate origin. This choice can simplify the equation of motion by eliminating static forces.

To apply Newton's second law and/or the moment equation to a mechanical system, it is useful to draw a free-body diagram for each mass in the system, showing all external forces and/or moments. The non-input forces can be described in terms of displacements or velocities using the expressions associated with the basic spring or damper elements. Drawing correct free-body diagrams is the most important step in analyzing mechanical systems from the force/moment approach.

In this chapter, we were mainly concerned with the modeling of mechanical systems in plane motion, which involves translations along the x and y directions and rotation about one axis perpendicular to the x–y plane. Newton's second law is used for modeling translational mechanical systems, whereas the moment equation is used for rotational mechanical systems. Both Newton's second law and the moment equation are used together for mixed translational and rotational systems.

The general form of Newton's second law for a rigid body (or a particle) is given by

$$\sum \mathbf{F} = m\mathbf{a}_C$$

or

$$\sum F_x = ma_{Cx}, \quad \sum F_y = ma_{Cy}.$$

For a rigid body rotating about a fixed axis through a point O, the moment equation is given by

$$\sum M_O = I_O \alpha.$$

If the axis of rotation is not fixed, the moment equation can be set about the mass center C as

$$\sum M_C = I_C \alpha$$

or an arbitrary point P as

$$\sum \mathbf{M}_P = I_P \alpha + m\mathbf{r}_{C/P} \times \mathbf{a}_P,$$

which is equivalent to

$$\sum M_P = I_C \alpha + M_{eff\_ma_C}.$$

The symbol $M_{eff\_ma_C}$ represents the effective moment caused by the fictitious force $m\mathbf{a}_C$.

The mass moments of inertia for some rigid bodies with common shapes were given in Table 5.1, where all masses are assumed to be uniformly distributed and the axes of rotation all pass through the mass centers. If the axis of rotation does not coincide with the axis through the mass center, but is parallel to it, the parallel-axis theorem can be applied to obtain the corresponding moment of inertia,

$$I = I_C + md^2,$$

where d is the distance between the two parallel axes.

For a system of multiple masses, the force equations become

$$\sum F_x = \sum_{i=1}^{n} m_i (a_{Ci})_x, \quad \sum F_y = \sum_{i=1}^{n} m_i (a_{Ci})_y,$$

and the moment equation becomes

$$\sum M_P = \sum_{i=1}^{n} I_{Ci} \alpha_i + \sum_{i=1}^{n} M_{\text{eff}\_m_i a_{Ci}}.$$

The force/moment approach is based on Newtonian mechanics. An alternative way of obtaining the system's equations of motion is to use the energy method based on analytical mechanics. For a single-degree-of-freedom mass–spring system with negligible friction and damping, the principle of conservation of energy states that

$$\frac{d}{dt}(T + V) = 0.$$

The kinetic energy of a rigid body in plane motion can be separated into two parts: (1) the kinetic energy associated with the translational motion of the mass center C of the body, and (2) the kinetic energy associated with the rotation of the body about C. Expressions for the kinetic energy of a rigid body in plane motion are given as

- Translation only: $T = \frac{1}{2} m v^2$
- Rotation about a fixed point O: $T = \frac{1}{2} I_O \omega^2$
- Mixed translation and rotation: $T = \frac{1}{2} m v_C^2 + \frac{1}{2} I_C \omega^2$

The potential energy includes

- Gravitational potential energy: $V_g = mgh$
- Elastic potential energy: $V_e = \frac{1}{2} k x^2$ or $V_e = \frac{1}{2} K \theta^2$

For an n-degree-of-freedom system, n independent equations of motion can be derived using Lagrange's formulation. One of the expressions of Lagrange's equations for a conservative system is

$$\frac{d}{dt}\left(\frac{\partial T}{\partial \dot{q}_i}\right) - \frac{\partial T}{\partial q_i} + \frac{\partial V}{\partial q_i} = 0, \quad i = 1, 2, \ldots, n,$$

where q_i is the ith generalized coordinate and n is the total number of independent generalized coordinates. In contrast to the force/moment approach with Newton's second law and the moment equation, Lagrange's equations do not require a free-body diagram. However, velocity analysis is essential to Lagrange's equations.

REVIEW PROBLEMS

1. For the system shown in Figure 5.98, the input is the force f and the output is the displacement x of the mass.
   a. Draw the necessary free-body diagram and derive the differential equation of motion.
   b. Using the differential equation obtained in Part (a), determine the transfer function $X(s)/F(s)$. Assume that the initial conditions are $x(0) = 0$ and $\dot{x}(0) = 0$.
   c. Using the differential equation obtained in Part (a), determine the state-space representation.
2. For the system shown in Figure 5.99, the input is the force f and the outputs are the displacements x_1 and x_2 of the masses.

Mechanical Systems

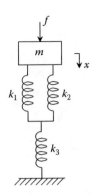

FIGURE 5.98 Problem 1.

  a. Draw the necessary free-body diagrams and derive the differential equations of motion. Write the differential equations of motion in the second-order matrix form.
  b. Using the differential equation obtained in Part (a), determine the transfer functions $X_1(s)/F(s)$ and $X_2(s)/F(s)$. Assume that all initial conditions are zero.
  c. Using the differential equation obtained in Part (a), determine the state-space representation.
3. Consider a quarter-car model shown in Figure 5.100, where m_1 is the mass of the seats including passengers, m_2 is the mass of one-fourth of the car body, and m_3 is the mass of the wheel–tire–axle assembly. The spring k_1 represents the elasticity of the seat supports, k_2 represents the elasticity of the suspension, and k_3 represents the elasticity of the tire. $z(t)$ is the displacement input due to the surface of the road.
  a. Draw the necessary free-body diagrams and derive the differential equations of motion.
  b. Determine the state-space representation, assuming the displacements of the three masses, x_1, x_2, and x_3 are the outputs.
4. For the system in Figure 5.101, the input is the force f and the outputs are the displacements x_1 and x_2 of the massless junctions A and B.
  a. Draw the necessary free-body diagrams and derive the differential equations of motion. Write the differential equations of motion in the second-order matrix form.
  b. Using the differential equation obtained in Part (a), determine the state-space representation.
5. The system shown in Figure 5.102 consists of a uniform rod of mass m and length L and two translational springs of stiffnesses k_1 and k_2. The friction at the joint O is

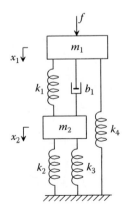

FIGURE 5.99 Problem 2.

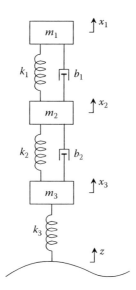

FIGURE 5.100 Problem 3.

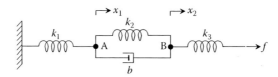

FIGURE 5.101 Problem 4.

modeled as a damper with coefficient of torsional viscous damping B. The input is the force f and the output is the angle θ. The position $\theta = 0$ corresponds to the equilibrium position when $f = 0$.

a. Draw the necessary free-body diagram and derive the differential equation of motion. Assume that θ is small.
b. Using the linearized differential equation obtained in Part (a), determine the transfer function $\Theta(s)/F(s)$. Assume that the initial conditions are $\theta(0) = 0$ and $\dot{\theta}(0) = 0$.
c. Using the linearized differential equation obtained in Part (a), determine the state-space representation.

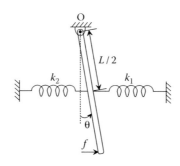

FIGURE 5.102 Problem 5.

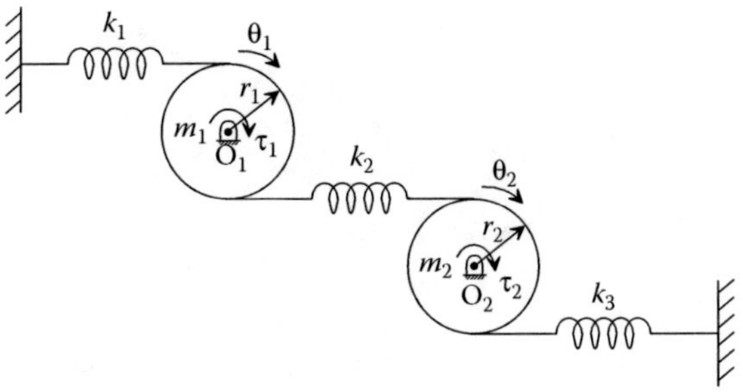

Mechanical Systems

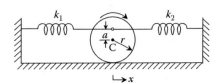

FIGURE 5.103 Problem 6.

6. Consider the two-degree-of-freedom system shown in Figure 5.103, where each disk is pivoted at its mass center. The inputs are the torques τ_1 and τ_2 applied to the pivot points O_1 and O_2. The outputs are the angular displacements θ_1 and θ_2 of the disk. When $\theta_1 = 0$, $\theta_2 = 0$, $\tau_1 = 0$, and $\tau_2 = 0$, each spring is at its free length.
   a. Draw the necessary free-body diagrams and derive the differential equations of motion. Assume small angles for θ_1 and θ_2. Provide the equations in the second-order matrix form.
   b. Using the differential equations obtained in Part (a), determine the state-space representation with the angular velocities $\dot{\theta}_1$ and $\dot{\theta}_2$ as the outputs.
7. Consider the system shown in Figure 5.104. Assume that the cylinder rolls without slipping. Draw the necessary free-body diagram and derive the differential equation of motion for small angles θ.
8. The pulley of mass M shown in Figure 5.105 has a radius of r. The mass moment of inertia of the pulley about the point O is I_O. A translational spring of stiffness k and a block of mass m are connected to the pulley as shown. Assume that the pulley rolls without slipping. Draw the free-body diagram and kinematic diagram, and derive the equation of motion.
9. Repeat Problem 8 using the energy method.

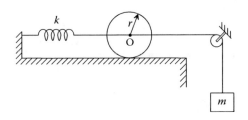

FIGURE 5.104 Problem 7.

FIGURE 5.105 Problem 8.

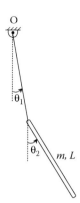

FIGURE 5.106 Problem 10.

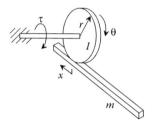

FIGURE 5.107 Problem 12.

10. Consider the mechanical system shown in Figure 5.106, where a uniform rod of mass m and length L is attached to a massless rigid link of equal length. Assume that the system is constrained to move in a vertical plane. Denote the angular displacement of the link as θ_1, and the angular displacement of the rod as θ_2. Draw the necessary free-body diagram and kinematic diagram, and derive the equations of motion for small angles.
11. Repeat Problem 10 using Lagrange's equations.
12. A rack and pinion is a pair of gears that convert rotational motion into translation. As shown in Figure 5.107, a torque τ is applied to the shaft. The pinion rotates and causes the rack to translate. The mass moment of inertia of the pinion is I and the mass of the rack is m. Draw the free-body diagram and derive the differential equation of motion.

6 Electrical, Electronic, and Electromechanical Systems

Many engineering systems have electrical, electronic, or electromechanical subsystems as important components, such as power supplies, motors, sensors, or controllers. In this chapter, we discuss the modeling techniques for these systems. We first introduce the fundamentals of electrical elements, which include resistors, inductors, and capacitors. The two main physical laws, Kirchhoff's voltage law and Kirchhoff's current law, are then reviewed and used to develop mathematical models of electrical systems or electrical circuits. For electronic systems, we take a look at simple operational amplifiers (op-amps), and the op-amp equation is presented, which is useful for obtaining models of amplifiers. Following that, we discuss the modeling of electromechanical systems. The coupling between electrical and mechanical subsystems is established and applied to motor modeling. The chapter concludes with a discussion of the concept of impedance, which is a generalization of the concept of electrical resistance and makes electrical circuit modeling easier.

6.1 ELECTRICAL ELEMENTS

Electrical systems, or electrical circuits, can usually be considered as interconnections of lumped elements, such as sources, resistors, inductors, and capacitors. Sources are active electrical elements, which can provide energy to the circuit and serve as the inputs. Resistors, inductors, and capacitors can store or dissipate energy available in the circuit; however, they cannot produce energy. They are referred to as passive electrical elements.

Two primary variables used to describe the dynamic behavior of an electrical circuit are current and voltage. Current is the time rate of change of charge,

$$i = \frac{dq}{dt}, \tag{6.1}$$

where q is charge in coulomb (C) and i is current in ampere (A). For a two-terminal electrical element, the current entering one end of the element must be equal to the current leaving the other end. As shown in Figure 6.1, an arrow is used to denote the direction in which the positive current (or charge) flows.

The voltage at a point in a circuit is a measure of the electrical potential difference between that point and a reference point called the ground. The unit of voltage is volt (V). If a point has the same electrical potential as the ground, it has a voltage of zero. For a two-terminal electrical element, the voltages at both ends are different. As shown in Figure 6.1, v_1 and v_2 denote the terminal voltages with respect to the ground, and

$$v = v_1 - v_2, \tag{6.2}$$

where v is the voltage across the element. The sense of the voltage v is indicated by plus and minus signs. The terminal with the plus sign has a higher voltage than that with the minus sign. The positive sense of the current associated with an electrical element is defined such that within the element, the positive current is assumed to flow from the high-voltage terminal to the low-voltage terminal. If the current flow has the same direction as the voltage drop, then the energy is supplied to the element. The supplied energy is either stored in the element or dissipated by the element. This type of electrical element is passive. For an active electrical element, current flows in the direction

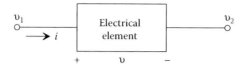

FIGURE 6.1 A two-terminal electrical element.

opposite to the voltage drop, and it supplies energy to the rest of the circuit. The power supplied to a passive element or generated by an active element is

$$P = vi. \qquad (6.3)$$

An active electrical element can be modeled as an ideal current source or an ideal voltage source, as shown in Figure 6.2. An ideal current source provides specified current no matter how much voltage is required by the circuit. An ideal voltage source provides specified voltage no matter how much current flows in the circuit.

To derive the dynamic model of an electrical circuit, it is important to understand the voltage–current relations for all passive electrical elements.

6.1.1 Resistors

The voltage–current relation for a linear resistor as shown in Figure 6.3 is an algebraic relationship,

$$v = Ri, \qquad (6.4)$$

where R is the resistance in units of ohm (Ω). Equation 6.4 is known as Ohm's law, which states that the voltage and current of a linear resistor are directly proportional to each other. It is an empirical formula and can be obtained by a series of measurements.

A resistor dissipates energy by converting it into heat. The power dissipated by a linear resistor is given by

$$P = Ri^2 = \frac{v^2}{R}. \qquad (6.5)$$

In many electrical circuits, multiple resistors are used. They are arranged in different ways, such as series connections, parallel connections, or both. The equivalent resistance for several resistors arranged in any of these ways can be obtained to simplify the modeling procedure.

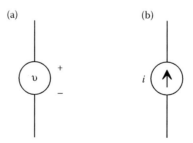

FIGURE 6.2 Active electrical elements: (a) ideal voltage source and (b) ideal current source.

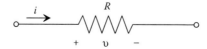

FIGURE 6.3 A resistor and its variables.

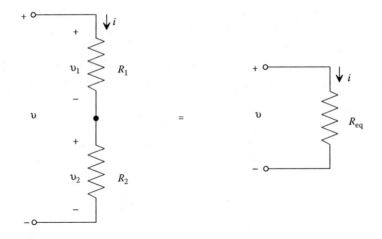

FIGURE 6.4 Equivalence for two resistors in series.

Figure 6.4 shows a circuit with two resistors in series. It is known that the current (or charge) remains unchanged when crossing an electrical element. Thus, for the series connection, the current is the same through each resistor. Ohm's law gives $v_1 = R_1 i$ and $v_2 = R_2 i$, where the voltages v_1 and v_2 also represent a measure of the energy required to move a charge through resistor R_1 or resistor R_2, respectively. The total energy required to move a charge across the two resistors is $v = v_1 + v_2 = (R_1 + R_2)i$. Comparing this result with the equivalent circuit, where $v = R_{eq} i$, we have

$$R_{eq} = R_1 + R_2. \tag{6.6}$$

A circuit with parallel resistors is shown in Figure 6.5. Note that the parallel resistors share the same terminals. Thus, the voltage across each resistor must be the same, $v_1 = v_2$. Ohm's law gives $v = R_1 i_1$ and $v = R_2 i_2$, where i_1 and i_2 are currents through resistors R_1 and R_2, respectively. Because of the conservation of charge, $i = i_1 + i_2 = v(1/R_1 + 1/R_2)$. Comparing this result with the equivalent circuit, where $i = v/R_{eq}$, we have

$$\frac{1}{R_{eq}} = \frac{1}{R_1} + \frac{1}{R_2} \tag{6.7}$$

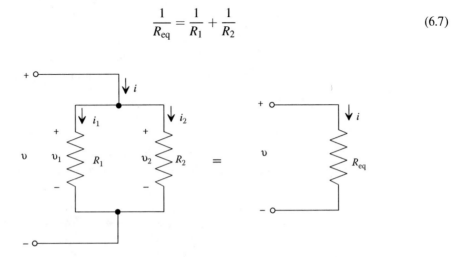

FIGURE 6.5 Equivalence for two resistors in parallel.

or

$$R_{eq} = \frac{R_1 R_2}{R_1 + R_2}. \quad (6.8)$$

The results for series or parallel resistance can be extended to n resistors. For a circuit of n resistors in series, the equivalent resistance is equal to the sum of all the individual resistances R_i:

$$R_{eq} = R_1 + R_2 + \cdots + R_n. \quad (6.9)$$

For a system of n resistors in parallel, the reciprocal of the equivalent resistance R_{eq} is equal to the sum of all the reciprocals of the individual resistances R_i:

$$\frac{1}{R_{eq}} = \frac{1}{R_1} + \frac{1}{R_2} + \cdots + \frac{1}{R_n}. \quad (6.10)$$

6.1.2 Inductors

Figure 6.6 shows the symbol for an inductor. The voltage–current relation for a linear inductor is

$$v = L\frac{di}{dt} \quad (6.11)$$

or

$$i = \frac{1}{L}\int v\, dt, \quad (6.12)$$

where L is the inductance, and the unit is henry (H).

The energy supplied to an inductor is stored in its magnetic field. The stored energy can be derived by performing integration of the power

$$P = vi = \left(L\frac{di}{dt}\right)i \quad (6.13)$$

as follows:

$$E(t) = \int_0^t P(t)\,dt = \int_0^t Li(t)\,di = \frac{1}{2}Li^2\bigg|_0^t = \frac{1}{2}Li^2(t) - \frac{1}{2}Li^2(0) = \frac{1}{2}Li^2(t). \quad (6.14)$$

In Equation 6.14, it is assumed that the current through the inductor is zero at $t = 0$. Note that the energy stored in an inductor is dependent on the square of the current through the inductor and is independent of how the current is established.

6.1.3 Capacitors

Figure 6.7 shows the symbol for a capacitor, where C is the capacitance in units of farad (F). The capacitance is a measure of how much charge can be stored for a given voltage difference across the

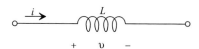

FIGURE 6.6 An inductor and its variables.

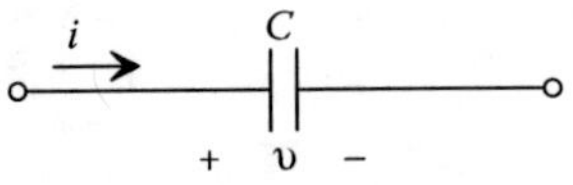

Electrical, Electronic, and Electromechanical Systems

FIGURE 6.7 A capacitor and its variables.

capacitor, and the mathematical description is $q = Cv$ or $v = q/C$. Note that the charge q is related to the current, $i = dq/dt$ or $q = \int i \, dt$. Thus the voltage–current relation for a capacitor is expressed as

$$v = \frac{1}{C} \int i \, dt \tag{6.15}$$

or

$$i = C \frac{dv}{dt}. \tag{6.16}$$

A capacitor is also designed to store energy. The energy supplied to the capacitor is stored in its electrical field and can be derived by performing integration of the power

$$P = vi = v \left(C \frac{dv}{dt} \right) \tag{6.17}$$

as follows:

$$E(t) = \int_0^t P(t) \, dt = \int_0^t Cv(t) \, dv = \frac{1}{2} Cv^2 \Big|_0^t = \frac{1}{2} Cv^2(t) - \frac{1}{2} Cv^2(0) = \frac{1}{2} Cv^2(t). \tag{6.18}$$

In Equation 6.18, it is assumed that the voltage across the capacitor is zero at $t = 0$. Note that the energy stored in a capacitor is dependent on the square of the voltage across the capacitor and is independent of how the voltage is acquired.

PROBLEM SET 6.1

1. Determine the equivalent resistance R_{eq} for the circuit shown in Figure 6.8.
2. Determine the equivalent resistance R_{eq} for the circuit shown in Figure 6.9.
3. Determine the equivalent resistance R_{eq} for the circuit shown in Figure 6.10.
4. Determine the equivalent resistance R_{eq} for the circuit shown in Figure 6.11.

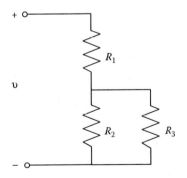

FIGURE 6.8 Problem 1.

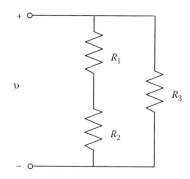

FIGURE 6.9 Problem 2.

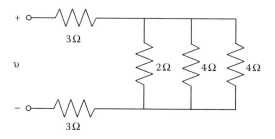

FIGURE 6.10 Problem 3.

5. A potentiometer is a variable resistor with three terminals. Figure 6.12a shows a potentiometer connected to a voltage source. The two end terminals are labeled as 1 and 2, and the adjustable terminal is labeled as 3. The potentiometer acts as a voltage divider, and the total resistance is separated into two parts as shown in Figure 6.12b. R_{13} is the resistance between terminal 1 and terminal 3, and R_{32} is the resistance between terminal 3 and terminal 2. Determine the relationship between the input voltage v_i and the output voltage v_o.
6. Find the output voltage of the voltage divider in Figure 6.13 for two different values of load resistance: (1) $R_L = 10\,\text{k}\Omega$ and (2) $R_L = 100\,\text{k}\Omega$.
7. Consider a circuit of two inductors, L_1 and L_2, in series. Prove that the equivalent inductance of the circuit is $L_{eq} = L_1 + L_2$.
8. Consider a circuit of two inductors, L_1 and L_2, in parallel. Prove that the equivalent inductance of the circuit is $\dfrac{1}{L_{eq}} = \dfrac{1}{L_1} + \dfrac{1}{L_2}$.

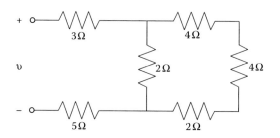

FIGURE 6.11 Problem 4.

Electrical, Electronic, and Electromechanical Systems

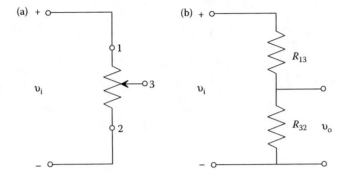

FIGURE 6.12 Problem 5. (a) Potentiometer and (b) voltage divider.

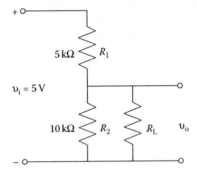

FIGURE 6.13 Problem 6.

9. Consider a circuit of two capacitors, C_1 and C_2, in series. Prove that the equivalent capacitance of the circuit is $\dfrac{1}{C_{eq}} = \dfrac{1}{C_1} + \dfrac{1}{C_2}$.
10. Consider a circuit of two capacitors, C_1 and C_2, in parallel. Prove that the equivalent capacitance of the circuit is $C_{eq} = C_1 + C_2$.
11. The current through an inductor of 10 mH is shown in Figure 6.14. Find the voltage across the inductor. What is the energy stored in the inductor when (1) $t = 1\,\text{s}$ and (2) $t = 2\,\text{s}$?
12. The voltage across a capacitor of 100 µF is shown in Figure 6.15. Find the current through the capacitor. What is the energy stored in the capacitor when (1) $t = 1\,\text{s}$ and (2) $t = 2\,\text{s}$?

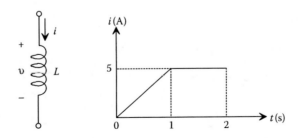

FIGURE 6.14 Problem 11.

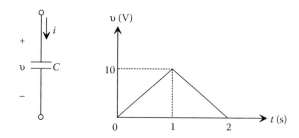

FIGURE 6.15 Problem 12.

6.2 ELECTRIC CIRCUITS

When electrical elements are interconnected to form an electrical circuit, the dynamics model of the circuit can be developed using the voltage–current relations for electrical elements along with two main physical laws. The two laws are known as Kirchhoff's voltage law and Kirchhoff's current law.

6.2.1 Kirchhoff's Voltage Law

For a closed path, or a loop, in a circuit, Kirchhoff's voltage law states that the algebraic sum of voltages around the loop must be zero,

$$\sum_j v_j = 0, \tag{6.19}$$

where v_j is the voltage across the jth element in the loop.

As an example, let us consider a series RLC circuit shown in Figure 6.16, where a resistor, an inductor, and a capacitor are connected in series. An ideal voltage source provides the desired voltage to the circuit. The current flows from the high-voltage terminal of the source, crosses three passive elements, and enters the low-voltage terminal of the source. If we sum the voltages around the loop in a clockwise direction, a voltage drop occurs for each of the three passive elements, and a voltage gain occurs for the source. Assign a positive sign to a voltage drop, and a negative sign to a voltage gain. Kirchhoff's voltage law gives $v_R + v_L + v_C - v_a = 0$ or $v_a = v_R + v_L + v_C$, which implies that the voltage of the source must equal the sum of the voltages across the resistor, the inductor, and the capacitor.

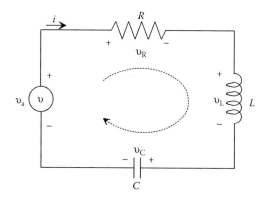

FIGURE 6.16 A series RLC circuit.

Electrical, Electronic, and Electromechanical Systems

Example 6.1: A Series RLC Circuit

Consider the series RLC circuit shown in Figure 6.16.

a. Derive the differential equation for charge q.
b. Determine the transfer function $V_C(s)/V_a(s)$, which relates the source voltage $v_a(t)$ to the capacitor voltage $v_C(t)$. Assume that all the initial conditions are zero.
c. Choose the source voltage $v_a(t)$ as the input, the loop current $i(t)$ as the output, and $x_1 = q(t)$, $x_2 = i(t)$ as the state variables. Determine the state-space representation.

Solution

a. Applying Kirchhoff's voltage law to the single loop along the clockwise direction gives

$$v_R + v_L + v_C - v_a = 0.$$

For the series loop, the same current flows through each element. The expressions for v_R, v_L, and v_C are

$$v_R = Ri,$$

$$v_L = L\frac{di}{dt},$$

$$v_C = \frac{1}{C}\int i\,dt.$$

We then have

$$Ri + L\frac{di}{dt} + \frac{1}{C}\int i\,dt = v_a.$$

Note that the above equation is an integral–differential equation, not a differential equation. Recall that current is the time rate of change of charge, $i = dq/dt$ or $q = \int i\,dt$. Rewriting the current-related terms in terms of q gives

$$L\frac{d^2q}{dt^2} + R\frac{dq}{dt} + \frac{1}{C}q = v_a,$$

which is a second-order differential equation for the charge $q(t)$ with the applied voltage as the forcing function.

b. Note that the capacitor voltage v_C does not appear explicitly in the above integral–differential equation or the differential equation. To determine the transfer function $V_C(s)/V_a(s)$, we first find the transfer function $I(s)/V_a(s)$ and then apply the voltage–current relation for a capacitor.

Taking the Laplace transform of the above integral–differential equation yields

$$RI(s) + LsI(s) + \frac{1}{Cs}I(s) = V_a(s).$$

Thus, the transfer function relating the input voltage $v_a(t)$ and the output current $i(t)$ is

$$\frac{I(s)}{V_a(s)} = \frac{s}{Ls^2 + Rs + (1/C)}.$$

The voltage–current relation for a capacitor is

$$i = C\frac{dv_C}{dt},$$

which gives

$$I(s) = CsV_C(s).$$

Thus, the transfer function relating the input voltage $v_a(t)$ and the capacitor voltage $v_C(t)$ is

$$\frac{V_C(s)}{V_a(s)} = \frac{1}{Cs}\frac{I(s)}{V_a(s)} = \frac{1}{LCs^2 + RCs + 1}.$$

c. As specified, the state variables, the input, and the output are

$$\mathbf{x} = \begin{Bmatrix} x_1 \\ x_2 \end{Bmatrix} = \begin{Bmatrix} q \\ i \end{Bmatrix}, \quad u = v_a, \quad y = i.$$

Calculating the time derivative of each state and expressing it in terms of the states and the input only, we obtain

$$\dot{x}_1 = \frac{dq}{dt} = i = x_2,$$

$$\dot{x}_2 = \frac{di}{dt} = \frac{d^2q}{dt^2} = -\frac{R}{L}\frac{dq}{dt} - \frac{1}{LC}q + \frac{1}{L}v_a$$

$$= -\frac{R}{L}x_2 - \frac{1}{LC}x_1 + \frac{1}{L}u,$$

where $\dot{x}_2$ is solved from the differential equation for the charge q. The output equation is

$$y = i = x_2.$$

Writing the state equation and the output equation in matrix form gives

$$\begin{Bmatrix} \dot{x}_1 \\ \dot{x}_2 \end{Bmatrix} = \begin{bmatrix} 0 & 1 \\ -\frac{1}{LC} & -\frac{R}{L} \end{bmatrix} \begin{Bmatrix} x_1 \\ x_2 \end{Bmatrix} + \begin{bmatrix} 0 \\ \frac{1}{L} \end{bmatrix} u,$$

$$y = \begin{bmatrix} 0 & 1 \end{bmatrix} \begin{Bmatrix} x_1 \\ x_2 \end{Bmatrix} + 0 \cdot u.$$

The circuit in Figure 6.16 can also be modeled using a differential equation for current i by taking the time derivative of both sides of the equation $Ri + L(di/dt) + (1/C)\int i\, dt = v_a$. Rearranging the terms gives

$$L\frac{d^2 i}{dt^2} + R\frac{di}{dt} + \frac{1}{C}i = \frac{dv_a}{dt},$$

which is a second-order differential equation for current i with the time derivative of the applied voltage as the forcing function.

6.2.2 Kirchhoff's Current Law

When the terminals of two or more circuit elements are connected together, the common junction is referred to as a node. For a node in a circuit, Kirchhoff's current law states that the sum of the currents entering the node must be equal to the sum of the currents leaving that node. If we assign a positive sign to the current entering the node and a negative sign to the current leaving the node, the algebraic sum of the currents at the node must be zero,

$$\sum_j i_j = 0, \tag{6.20}$$

where i_j is the current of the jth element at the node.

Example 6.2: A Parallel RLC Circuit

Consider the parallel RLC circuit shown in Figure 6.17, where an ideal current source supplies the desired current to the circuit.

- a. Derive the differential equation relating the input current i_a to the output voltage v_o.
- b. Determine the transfer function $I_L(s)/I_a(s)$, which relates the input current $i_a(t)$ to the current through the inductor $i_L(t)$. Assume that all the initial conditions are zero.
- c. Choose the source current $i_a(t)$ as the input, the voltage $v_o(t)$ as the output, and $x_1 = i_L(t)$, $x_2 = v_C(t)$ as the state variables. Determine the state-space representation.

Solution

a. The currents through the three passive elements i_R, i_L, and i_C are defined in Figure 6.17. Each passive element has one terminal connected to the ground and the other terminal connected to a common node. We can apply Kirchhoff's law to either the ground or node 1. Applying Kirchhoff's current law to node 1 gives

$$i_a - i_R - i_L - i_C = 0.$$

For the parallel connection, the voltages across all three elements are the same. The expressions for i_R, i_L, and i_C are

$$i_R = \frac{v_o}{R},$$

$$i_L = \frac{1}{L}\int v_o \, dt,$$

$$i_C = C\frac{dv_o}{dt}.$$

We then have

$$\frac{v_o}{R} + \frac{1}{L}\int v_o \, dt + C\frac{dv_o}{dt} = i_a,$$

which is an integral–differential equation. To eliminate the integral term, we take the time derivative of both sides of the equation. Rearranging the terms results in

$$C\frac{d^2 v_o}{dt^2} + \frac{1}{R}\frac{dv_o}{dt} + \frac{1}{L}v_o = \frac{di_a}{dt},$$

which is a second-order differential equation for the output voltage $v_o(t)$.

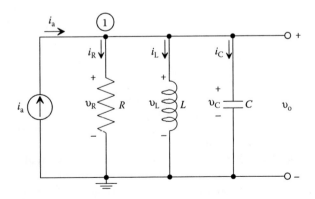

FIGURE 6.17 A parallel RLC circuit.

b. Taking the Laplace transform of the above differential equation yields

$$Cs^2 V_o(s) + \frac{1}{R} s V_o(s) + \frac{1}{L} V_o(s) = s I_a(s).$$

The transfer function relating the input current $i_a(t)$ and the output voltage $v_o(t)$ is

$$\frac{V_o(s)}{I_a(s)} = \frac{s}{Cs^2 + (1/R)s + (1/L)}.$$

To find the transfer function $I_L(s)/I_a(s)$, note that

$$v_o = L \frac{di_L}{dt},$$

which gives

$$V_o(s) = L s I_L(s).$$

Thus,

$$\frac{I_L(s)}{I_a(s)} = \frac{1}{Ls} \frac{V_o(s)}{I_a(s)} = \frac{1}{LCs^2 + (L/R)s + 1}.$$

c. As specified, the state variables, the input, and the output are

$$\mathbf{x} = \begin{Bmatrix} x_1 \\ x_2 \end{Bmatrix} = \begin{Bmatrix} i_L \\ v_C \end{Bmatrix}, \quad u = i_a, \quad y = v_o.$$

Note that the output voltage v_o is the same as the voltage across the resistor, the inductor, or the capacitor. Solving for $\dot{x}_1$ using the voltage–current relation for the inductor gives

$$\dot{x}_1 = \frac{di_L}{dt} = \frac{1}{L} v_C = \frac{1}{L} x_2.$$

From Kirchhoff's current law applied in Part (a), we have

$$\frac{v_C}{R} + i_L + C \frac{dv_C}{dt} = i_a.$$

Solving for $\dot{x}_2$ results in

$$\dot{x}_2 = \frac{dv_C}{dt} = -\frac{1}{C} i_L - \frac{1}{RC} v_C + \frac{1}{C} i_a$$

$$= -\frac{1}{C} x_1 - \frac{1}{RC} x_2 + \frac{1}{C} u.$$

The output equation is

$$y = v_o = v_C = x_2.$$

The state equation and the output equation can be written in matrix form as follows:

$$\begin{Bmatrix} \dot{x}_1 \\ \dot{x}_2 \end{Bmatrix} = \begin{bmatrix} 0 & \frac{1}{L} \\ -\frac{1}{C} & -\frac{1}{RC} \end{bmatrix} \begin{Bmatrix} x_1 \\ x_2 \end{Bmatrix} + \begin{bmatrix} 0 \\ \frac{1}{C} \end{bmatrix} u,$$

$$y = \begin{bmatrix} 0 & 1 \end{bmatrix} \begin{Bmatrix} x_1 \\ x_2 \end{Bmatrix} + 0 \cdot u.$$

The above two simple examples illustrate how one can develop a differential equation for a series or parallel RLC circuit using Kirchhoff's voltage law or Kirchhoff's current law. However, it is not so easy to obtain a set of equations for more complicated circuits by using voltage–current relations and Kirchhoff's laws. A formal method is needed that produces a small, easy set of equations leading directly to the input–output relation. The two commonly used methods are the node method (which relies on Kirchhoff's current law) and the loop method (which relies on Kirchhoff's voltage law).

Electrical, Electronic, and Electromechanical Systems

6.2.3 Node Method

If a node in a circuit is chosen as the reference, any other node can be assigned a voltage, which is defined between this node and the reference. This common reference node is usually referred to as the ground. To apply the node method to a circuit, we start by labeling all currents at each node whose voltage is unknown. The current through each passive circuit element is expressed in terms of the node voltages using the voltage–current relations given in Section 6.1. We then apply Kirchhoff's current law to each node, and the resulting set of equations can be combined to obtain the complete model of the circuit. The following example shows details of the node-voltage method.

Example 6.3: Circuit Modeling Using the Node Method

Consider the circuit shown in Figure 6.18.

a. Derive the differential equations of the system using the node method.
b. Determine the transfer function $V_o(s)/V_a(s)$. Assume that all the initial conditions are zero, and the parameter values are $R_1 = R_2 = 5\,\Omega$, $L_1 = L_2 = 2\,H$, and $C = 0.5\,F$.

Solution

a. Note that the voltages at node 1 and node 2 are unknown. Denote the voltage at node 1 as v_1 and the voltage at node 2 as v_2. All currents entering or leaving node 1 and node 2 are labeled as shown in Figure 6.18.
At node 1, applying Kirchhoff's current law gives

$$i_{L_1} - i_{L_2} - i_{R_1} = 0.$$

Expressing the current through each element in terms of the node voltages, we have

$$\frac{1}{L_1}\int (v_a - v_1)\,dt - \frac{1}{L_2}\int (v_1 - v_2)\,dt - \frac{v_1 - 0}{R_1} = 0.$$

Differentiating the above equation with respect to time results in

$$\frac{v_a - v_1}{L_1} - \frac{v_1 - v_2}{L_2} - \frac{\dot{v}_1}{R_1} = 0,$$

which can be rearranged and gives the first differential equation

$$\frac{1}{R_1}\dot{v}_1 + \left(\frac{1}{L_1} + \frac{1}{L_2}\right)v_1 - \frac{1}{L_2}v_2 = \frac{1}{L_1}v_a.$$

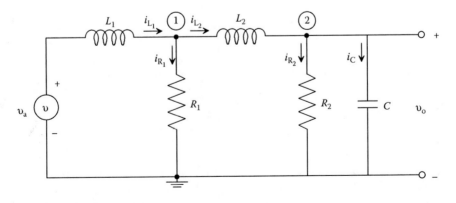

FIGURE 6.18 A circuit containing two nodes.

Similarly, applying Kirchhoff's current law to node 2 gives

$$i_{L_2} - i_{R_2} - i_C = 0.$$

Expressing the current through each element in terms of the node voltages, we have

$$\frac{1}{L_2}\int(v_1 - v_2)dt - \frac{v_2 - 0}{R_2} - C\frac{d}{dt}(v_2 - 0) = 0.$$

Differentiating the above equation with respect to time results in

$$\frac{v_1 - v_2}{L_2} - \frac{\dot{v}_2}{R_2} - C\ddot{v}_2 = 0,$$

which can be rearranged and gives the second differential equation

$$C\ddot{v}_2 + \frac{1}{R_2}\dot{v}_2 - \frac{1}{L_2}v_1 + \frac{1}{L_2}v_2 = 0.$$

Note that the differential equation obtained at node 1 is first-order, and that obtained at node 2 is second-order. Thus, the circuit shown in Figure 6.18 is a third-order system. In second-order matrix form,

$$\begin{bmatrix} 0 & 0 \\ 0 & C \end{bmatrix}\begin{Bmatrix} \ddot{v}_1 \\ \ddot{v}_2 \end{Bmatrix} + \begin{bmatrix} \frac{1}{R_1} & 0 \\ 0 & \frac{1}{R_2} \end{bmatrix}\begin{Bmatrix} \dot{v}_1 \\ \dot{v}_2 \end{Bmatrix} + \begin{bmatrix} \frac{1}{L_1} + \frac{1}{L_2} & -\frac{1}{L_2} \\ -\frac{1}{L_2} & \frac{1}{L_2} \end{bmatrix}\begin{Bmatrix} v_1 \\ v_2 \end{Bmatrix} = \begin{Bmatrix} \frac{1}{L_1}v_a \\ 0 \end{Bmatrix}.$$

b. To simplify the calculations, we substitute the numerical values into the preceding differential equations and then take the Laplace transform.

$$\begin{bmatrix} 0 & 0 \\ 0 & 0.5 \end{bmatrix}\begin{Bmatrix} \ddot{v}_1 \\ \ddot{v}_2 \end{Bmatrix} + \begin{bmatrix} 0.2 & 0 \\ 0 & 0.2 \end{bmatrix}\begin{Bmatrix} \dot{v}_1 \\ \dot{v}_2 \end{Bmatrix} + \begin{bmatrix} 1 & -0.5 \\ -0.5 & 0.5 \end{bmatrix}\begin{Bmatrix} v_1 \\ v_2 \end{Bmatrix} = \begin{Bmatrix} 0.5v_a \\ 0 \end{Bmatrix},$$

$$\begin{bmatrix} 0.2s + 1 & -0.5 \\ -0.5 & 0.5s^2 + 0.2s + 0.5 \end{bmatrix}\begin{Bmatrix} V_1(s) \\ V_2(s) \end{Bmatrix} = \begin{Bmatrix} 0.5V_a(s) \\ 0 \end{Bmatrix}.$$

Applying Cramer's rule gives

$$\frac{V_1(s)}{V_a(s)} = \frac{0.25s^2 + 0.1s + 0.25}{0.1s^3 + 0.54s^2 + 0.3s + 0.25},$$

$$\frac{V_2(s)}{V_a(s)} = \frac{0.25}{0.1s^3 + 0.54s^2 + 0.3s + 0.25}.$$

Note that the output voltage is the same as the voltage at node 2, that is, $v_o = v_2$. Thus, the transfer function relating the output voltage v_o and the applied voltage v_a is

$$\frac{V_o(s)}{V_a(s)} = \frac{2.5}{s^3 + 5.4s^2 + 3s + 2.5}.$$

6.2.4 Loop Method

In a circuit carrying some current, there is at least one loop. Starting from any loop, a current circulating that loop can be assigned. For any additional loop containing at least one new element that is not in any previous loops, a new loop current can be assigned. To apply the loop method to a circuit, we start by assigning each loop current. Generally, assume that all unknown currents flow

Electrical, Electronic, and Electromechanical Systems

in the clockwise direction, and all known currents follow the directions of the current sources. The voltage across each passive element is expressed in terms of loop currents using the voltage–current relations given in Section 6.1. We then apply Kirchhoff's voltage law to each loop with unknown current, and the resulting set of equations can be combined to obtain the complete model of the circuit. The following example shows the details of the loop-current method.

Example 6.4: Circuit Modeling Using the Loop Method

Reconsider the circuit in Figure 6.18 and solve Example 6.3 using the loop method.

Solution

a. Assign loop currents as shown in Figure 6.19. Note that there are three loops with unknown currents.
   For loop 1, applying Kirchhoff's voltage law gives

$$v_{L_1} + v_{R_1} - v_a = 0.$$

Expressing the voltage across each element in terms of the loop currents, we have

$$L_1 \frac{di_1}{dt} + R_1(i_1 - i_2) - v_a = 0.$$

Similarly, applying Kirchhoff's voltage law to loop 2 and loop 3 gives

$$v_{L_2} + v_{R_2} + v_{R_1} = 0,$$

$$L_2 \frac{di_2}{dt} + R_2(i_2 - i_3) + R_1(i_2 - i_1) = 0,$$

and

$$v_C + v_{R_2} = 0,$$

$$\frac{1}{C} \int i_3 \, dt + R_2(i_3 - i_2) = 0.$$

Differentiating the above equation with respect to time results in

$$\frac{1}{C} i_3 + R_2 \left(\frac{di_3}{dt} - \frac{di_2}{dt} \right) = 0.$$

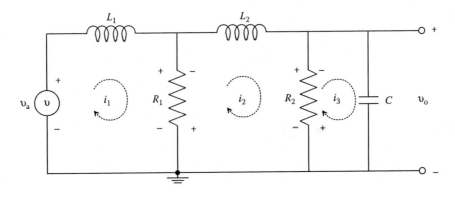

FIGURE 6.19 A circuit containing three loops.

Note that the differential equation obtained in each loop is first-order, and thus the circuit is a third-order system. The three first-order differential equations can be written in matrix form as follows:

$$\begin{bmatrix} L_1 & 0 & 0 \\ 0 & L_2 & 0 \\ 0 & -R_2 & R_2 \end{bmatrix} \begin{Bmatrix} di_1/dt \\ di_2/dt \\ di_3/dt \end{Bmatrix} + \begin{bmatrix} R_1 & -R_1 & 0 \\ -R_1 & R_1+R_2 & -R_2 \\ 0 & 0 & 1/C \end{bmatrix} \begin{Bmatrix} i_1 \\ i_2 \\ i_3 \end{Bmatrix} = \begin{Bmatrix} v_a \\ 0 \\ 0 \end{Bmatrix}.$$

b. Substituting the numerical values into the preceding differential equations and then taking the Laplace transform gives

$$\begin{bmatrix} 2s+5 & -5 & 0 \\ -5 & 2s+10 & -5 \\ 0 & -5s & 5s+2 \end{bmatrix} \begin{Bmatrix} I_1(s) \\ I_2(s) \\ I_3(s) \end{Bmatrix} = \begin{Bmatrix} V_a(s) \\ 0 \\ 0 \end{Bmatrix}.$$

Note that the output voltage v_o is related to the current through the capacitor, that is, $v_o = \frac{1}{C}\int i_3\,dt$.

Using Cramer's rule to solve for $I_3(s)$, we have

$$I_3(s) = \frac{1.25s}{s^3 + 5.4s^2 + 3s + 2.5} V_a(s).$$

Thus, the transfer function relating the output voltage v_o and the applied voltage v_a is

$$\frac{V_o(s)}{V_a(s)} = \frac{1}{Cs}\frac{I_3(s)}{V_a(s)} = \frac{2.5}{s^3 + 5.4s^2 + 3s + 2.5},$$

which is the same as that obtained previously in Example 6.3.

Example 6.4 shows that the loop method is similar to the node method. The choice between the two methods is often made based on the circuit at hand. For the circuit in Examples 6.3 and 6.4, there are two independent nodes, but three independent loops. Therefore, the node method is expected to be easier to apply. We will emphasize the node method to obtain mathematical models of circuits in this book.

6.2.5 State Variables of Circuits

In order to represent a circuit model in state-space form, we need to choose an appropriate set of state variables, whose time derivatives are expressed in terms of the state variables and inputs. The choice of state variables is not unique. For instance, we selected the charge q and the current i as state variables in Example 6.1, whereas in Example 6.2 the inductor current i_L and the capacitor voltage v_C were selected as state variables. We can even select the node voltages (in Example 6.3) or the loop currents (in Example 6.4) as state variables.

It is difficult to identify the appropriate states for expressing a circuit in state-space form. We here introduce a normal choice of state variables by identifying the energy storage elements. As stated in Section 6.1, both inductors and capacitors can store energy. In a given circuit, a knowledge of the voltage signals across capacitors and of the current signals through inductors is sufficient enough to calculate other circuit variables using only algebraic equations. Generally, inductor currents and capacitor voltages are continuous in nature and are often chosen as the state variables. In order to determine the state-space form of an electric circuit, we need to find the expression of di_L/dt or dv_C/dt for each inductor or capacitor. Based on the voltage–current relations, we have

$$\frac{di_L}{dt} = \frac{v_L}{L} \tag{6.21}$$

Electrical, Electronic, and Electromechanical Systems

and
$$\frac{dv_C}{dt} = \frac{i_C}{C}, \tag{6.22}$$

where v_L is the inductor voltage and i_C is the capacitor current. Thus, the problem is converted to expressing v_L and i_C in terms of state variables and inputs using Kirchhoff's laws and voltage–current relations for electrical elements.

Example 6.5: State-Variable Model of the Circuit in Example 6.3

Reconsider the circuit shown in Figure 6.18.

a. Derive the state-variable model with inductor currents and capacitor voltages as states. The input is the applied voltage v_a, and the output is the voltage across the capacitor C.
b. Determine the transfer function $V_o(s)/V_a(s)$. Assume that all the initial conditions are zero.

SOLUTION

a. We first label the nodes and currents as we did in Example 6.3. Note that the circuit has three independent energy storage elements, L_1, L_2, and C. This implies that three states are needed, and they are

$$x_1 = i_{L_1}, \quad x_2 = i_{L_2}, \quad x_3 = v_C.$$

Their time derivatives are

$$\dot{x}_1 = \frac{d}{dt} i_{L_1} = \frac{1}{L_1} v_{L_1},$$

$$\dot{x}_2 = \frac{d}{dt} i_{L_2} = \frac{1}{L_2} v_{L_2},$$

$$\dot{x}_3 = \frac{d}{dt} v_C = \frac{1}{C} i_C.$$

We need to express the voltage across each inductor, v_{L_1} and v_{L_2}, and the current through the capacitor, i_C, in terms of the state variables and the input. Note that

$$v_{L_1} = v_a - v_1 = v_a - R_1 i_{R_1},$$
$$v_{L_2} = v_1 - v_2 = R_1 i_{R_1} - v_C.$$

Applying Kirchhoff's current law to node 1 and node 2 gives

$$i_{R_1} = i_{L_1} - i_{L_2},$$
$$i_C = i_{L_2} - i_{R_2} = i_{L_2} - \frac{v_C}{R_2}.$$

Thus, the complete set of three state-variable equations is

$$\dot{x}_1 = \frac{1}{L_1} \left[v_a - R_1 \left(i_{L_1} - i_{L_2} \right) \right] = -\frac{R_1}{L_1} x_1 + \frac{R_1}{L_1} x_2 + \frac{1}{L_1} u,$$

$$\dot{x}_2 = \frac{1}{L_2} \left[R_1 \left(i_{L_1} - i_{L_2} \right) - v_C \right] = \frac{R_1}{L_2} x_1 - \frac{R_1}{L_2} x_2 - \frac{1}{L_2} x_3,$$

$$\dot{x}_3 = \frac{1}{C} \left(i_{L_2} - \frac{1}{R_2} v_C \right) = \frac{1}{C} x_2 - \frac{1}{R_2 C} x_3.$$

The output equation is

$$y = v_o = v_C = x_3.$$

The state equation and the output equation can be written in matrix form as follows:

$$\begin{Bmatrix} \dot{x}_1 \\ \dot{x}_2 \\ \dot{x}_3 \end{Bmatrix} = \begin{bmatrix} -\dfrac{R_1}{L_1} & \dfrac{R_1}{L_1} & 0 \\ \dfrac{R_1}{L_2} & -\dfrac{R_1}{L_2} & -\dfrac{1}{L_2} \\ 0 & \dfrac{1}{C} & -\dfrac{1}{R_2 C} \end{bmatrix} \begin{Bmatrix} x_1 \\ x_2 \\ x_3 \end{Bmatrix} + \begin{bmatrix} 1/L_1 \\ 0 \\ 0 \end{bmatrix} u,$$

$$y = \begin{bmatrix} 0 & 0 & 1 \end{bmatrix} \begin{Bmatrix} x_1 \\ x_2 \\ x_3 \end{Bmatrix} + 0 \cdot u.$$

b. Note that $V_o(s) = Y(s)$ and $V_a(s) = U(s)$. As presented in Section 4.4, state-space equations can be converted to a transfer function using

$$\frac{Y(s)}{U(s)} = \mathbf{C}(s\mathbf{I} - \mathbf{A})^{-1}\mathbf{B} + D.$$

Substituting **A**, **B**, **C**, and *D* matrices obtained in Part (a) with numbers inserted gives

$$\frac{V_o(s)}{V_a(s)} = \begin{bmatrix} 0 & 0 & 1 \end{bmatrix} \begin{bmatrix} s+2.5 & -2.5 & 0 \\ -2.5 & s+2.5 & 0.5 \\ 0 & -2 & s+0.4 \end{bmatrix}^{-1} \begin{bmatrix} 0.5 \\ 0 \\ 0 \end{bmatrix},$$

which returns the same transfer function as the one obtained in Examples 6.3 and 6.4.

PROBLEM SET 6.2

1. For the first-order RC circuit shown in Figure 6.20, derive the input–output differential equation relating v_C and v_a.
2. Repeat Problem 1 for the first-order RC circuit shown in Figure 6.21.
3. For the first-order RL circuit shown in Figure 6.22, derive the input–output differential equation relating i_L and i_a.
4. Repeat Problem 3 for the first-order RL circuit shown in Figure 6.23.
5. Consider the circuit shown in Figure 6.24.
    a. Use the node method to derive the input–output differential equation relating v_o and v_a.
    b. Determine the transfer function $V_o(s)/V_a(s)$. Assume that all the initial conditions are zero.
6. Repeat Problem 5 using the loop method.
7. Consider the circuit shown in Figure 6.25.

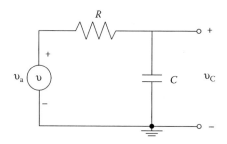

FIGURE 6.20 Problem 1.

Electrical, Electronic, and Electromechanical Systems

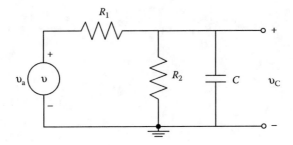

FIGURE 6.21 Problem 2.

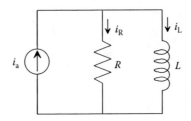

FIGURE 6.22 Problem 3.

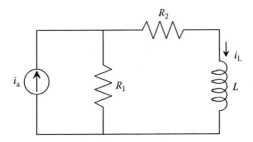

FIGURE 6.23 Problem 4.

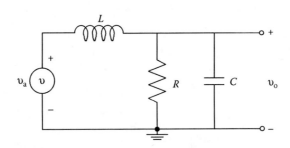

FIGURE 6.24 Problem 5.

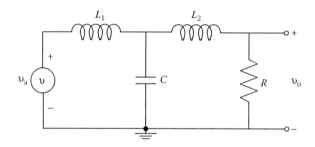

FIGURE 6.25 Problem 7.

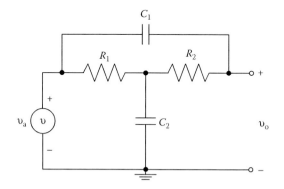

FIGURE 6.26 Problem 11.

    a. Use the node method to derive the differential equations.
    b. Determine the transfer function $V_o(s)/V_a(s)$. Assume that all initial conditions are zero.
8. Repeat Problem 7 using the loop method.
9. Consider the circuit shown in Figure 6.24. Determine a suitable set of state variables and obtain the state-space representation with v_o as the output.
10. Repeat Problem 9 for the circuit shown in Figure 6.25.
11. Repeat Problem 9 for the circuit shown in Figure 6.26.
12. Repeat Problem 9 for the circuit shown in Figure 6.27.
13. Repeat Problem 9 for the circuit shown in Figure 6.28.
14. Repeat Problem 9 for the circuit shown in Figure 6.29.

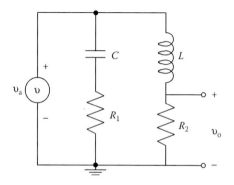

FIGURE 6.27 Problem 12.

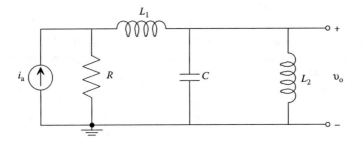

FIGURE 6.28 Problem 13.

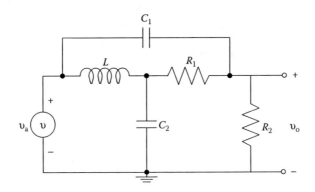

FIGURE 6.29 Problem 14.

6.3 OPERATIONAL AMPLIFIERS

An op-amp is an electronic element that is used to amplify electrical signals and drive physical devices. Figure 6.30 shows the schematic diagram of an op-amp, which is a voltage amplifier with a high gain K. Unlike the electrical elements discussed in earlier sections, op-amps have more than two terminals. The diagram in Figure 6.30 does not show all of the terminals connected to the physical devices. It only shows a pair of input terminals and one output terminal. The output voltage is

$$v_o = K(v_+ - v_-), \tag{6.23}$$

where K is a very large positive number, typically 10^5–10^6. Because the output voltage v_o must be a finite number and K is very large, the voltage difference between the input terminals must approach zero. Thus,

$$v_+ \approx v_-, \tag{6.24}$$

which is considered to be the op-amp equation.

Note that the diagram in Figure 6.30 is a simple symbol for an op-amp, which typically contains many resistors, inductors, and capacitors built on an integrated chip.

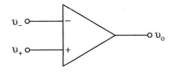

FIGURE 6.30 An op-amp.

Example 6.6: An Op-amp Multiplier

Consider the op-amp circuit shown in Figure 6.31, where one resistor R_2 is in parallel connection with an op-amp, and the resulting parallel circuit is in series connection with another resistor R_1. Determine the relation between the input voltage v_i and the output voltage v_o. Assume that the current drawn by the op-amp is very small.

Solution

Label the currents at the nodes with unknown voltages. The system has only one significant node: node 1. Applying Kirchhoff's current law to node 1 gives

$$i_1 - i_2 - i_3 = 0.$$

Because the current drawn by the op-amp is very small, that is, $i_3 \approx 0$, we have

$$i_1 \approx i_2.$$

Using the voltage–current relation for each resistor yields

$$\frac{v_i - v_1}{R_1} = \frac{v_1 - v_o}{R_2}.$$

Note that the input terminal marked with the plus sign is connected to the ground. From the op-amp equation $v_+ \approx v_-$, we have

$$v_1 = v_- \approx v_+ = 0.$$

Thus, the relation between the input voltage v_i and the output voltage v_o is

$$\frac{v_i}{R_1} = -\frac{v_o}{R_2}$$

or

$$v_o = -\frac{R_2}{R_1} v_i.$$

This circuit is known as an op-amp multiplier and is widely used in control systems. Op-amps can also be used for integrating and differentiating signals.

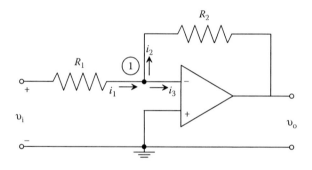

FIGURE 6.31 An op-amp multiplier circuit.

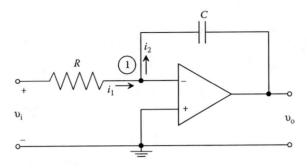

FIGURE 6.32 An op-amp integrator circuit.

Example 6.7: An Op-Amp Integrator

Consider the op-amp circuit shown in Figure 6.32.

a. Derive the differential equation relating the input voltage v_i and the output voltage v_o.
b. Determine the transfer function $V_o(s)/V_i(s)$. Assume that all initial conditions are zero.
c. ▲ Using the transfer function obtained in Part (b), construct a Simulink® block diagram to find the output $v_o(t)$ of the circuit if the input voltage is $v_i = -0.1$ V. The parameter values are $R = 1\,\text{M}\Omega$ and $C = 1\,\mu\text{F}$.

Solution

a. Note that the current drawn by the op-amp is very small. Applying Kirchhoff's current law to node 1 gives

$$i_1 = i_2,$$

$$\frac{v_i - v_1}{R} = C\frac{d}{dt}(v_1 - v_o).$$

Because the input terminal marked with the plus sign is connected to the ground, the op-amp equation yields $v_1 = v_- \approx v_+ = 0$. Thus, the differential equation for the op-amp integrator is

$$\frac{v_i}{R} = -C\dot{v}_o$$

or

$$\dot{v}_o = -\frac{1}{RC}v_i.$$

b. Taking the Laplace transform of both sides of the above differential equation gives

$$\frac{V_o(s)}{V_i(s)} = -\frac{1}{RCs}.$$

c. ▲ The block diagram is shown in Figure 6.33. The constant input voltage is represented using a Constant block. A Transfer Fcn block is used to represent the op-amp circuit. Run the simulation and the resulting voltage output $v_o(t)$ is shown in Figure 6.34. Note that the output voltage is a linear function of time with a slope of 0.1, that is, $v_o(t) = 0.1t$. The output $v_o(t)$ can also be found analytically. Note that the differential equation in Part (a) can be written as an integral equation

$$v_o = -\frac{1}{RC}\int_0^t v_i\,dt = -\int_0^t -0.1\,dt = 0.1t.$$

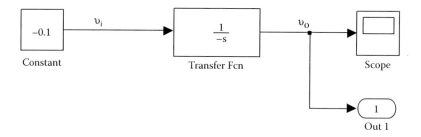

FIGURE 6.33 A Simulink block diagram to represent the integrator circuit in Figure 6.31.

This implies that the output voltage v_o is proportional to the integral of the input voltage v_i. The circuit in Figure 6.32 is therefore called an integrator. Switching the resistor and the capacitor results in an op-amp differentiator as shown in Figure 6.35. The relation between the output voltage and the input voltage is $v_o = -RC\dot{v}_i$. We leave the derivation as an exercise for the reader.

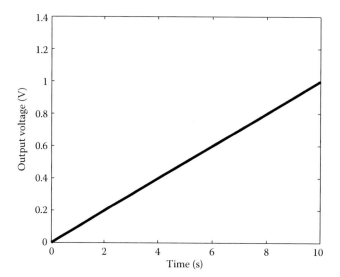

FIGURE 6.34 Voltage output $v_o(t)$ for $v_i(t) = -0.1$ V.

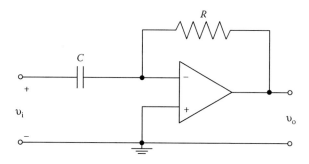

FIGURE 6.35 An op-amp differentiator.

Electrical, Electronic, and Electromechanical Systems

Example 6.8: An Op-amp Circuit

Consider the op-amp circuit shown in Figure 6.36. Derive the differential equation relating the output voltage $v_o(t)$ and the input voltage $v_i(t)$.

SOLUTION

Note that the current flowing into the input terminals of the op-amp is very small. Applying Kirchhoff's current law to node 1 gives

$$i_{R_1} + i_{C_1} - i_{R_2} - i_{C_2} = 0.$$

Using the voltage–current relations for electrical elements to express each term in the equation, we obtain

$$\frac{v_i - v_1}{R_1} + C_1 \frac{d}{dt}(v_i - v_1) - \frac{v_1 - v_o}{R_2} - C_2 \frac{d}{dt}(v_1 - v_o) = 0.$$

Because $v_+ = 0$, the op-amp equation yields $v_1 = v_- \approx v_+ = 0$. Substituting this into the previous equation results in

$$\frac{v_i}{R_1} + C_1 \dot{v}_i - \frac{-v_o}{R_2} - C_2(-\dot{v}_o) = 0,$$

which can be rearranged into

$$C_2 \dot{v}_o + \frac{1}{R_2} v_o = -C_1 \dot{v}_i - \frac{1}{R_1} v_i.$$

PROBLEM SET 6.3

1. The op-amp circuit shown in Figure 6.37 is a noninverting amplifier. Determine the relation between the input voltage v_i and the output voltage v_o.
2. Consider the op-amp circuit shown in Figure 6.38. Prove that the relation between the input voltages v_1, v_2, v_3, and the output voltage v_o is

$$v_o = -R_4 \left(\frac{v_1}{R_1} + \frac{v_2}{R_2} + \frac{v_3}{R_3} \right).$$

3. Consider the op-amp circuit shown in Figure 6.35.
   a. Derive the differential equation relating the input voltage v_i and the output voltage v_o.

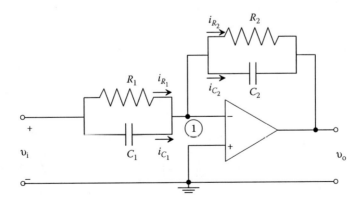

FIGURE 6.36 An op-amp circuit.

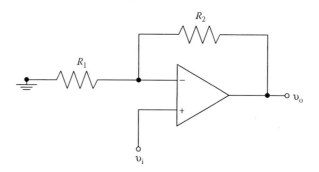

FIGURE 6.37 Problem 1.

  b. Determine the transfer function $V_o(s)/V_i(s)$. Assume that all initial conditions are zero.
  c. ▲ Construct a Simulink block diagram to find the output $v_o(t)$ of the circuit if the input voltage is $v_i = -0.1t$ V. What can you expect at the output? The parameter values are $R = 1\,\text{M}\Omega$ and $C = 1\,\mu\text{F}$.
 4. Repeat Problem 3 for the op-amp circuit shown in Figure 6.39. The input voltage is $v_i = -0.1$ V, and the parameter values are $R = 1\,\text{k}\Omega$ and $C = 1\,\mu\text{F}$.
 5. Consider the op-amp circuit shown in Figure 6.40. Derive the differential equation relating the output voltage $v_o(t)$ and the input voltage $v_i(t)$.
 6. Repeat Problem 5 for the op-amp circuit shown in Figure 6.41.

6.4 ELECTROMECHANICAL SYSTEMS

Many useful devices are constructed by combining electrical elements and mechanical elements, such as motors, generators, speakers, microphones, and accelerometers. For such an electromechanical system, we must apply electrical principles (e.g., Kirchhoff's laws) and mechanical principles (e.g., Newton's second law) to develop the dynamics model of the system. In this section, we discuss the modeling of direct-current (DC) motors, which can generate forces or torques using electrical subsystems, and are essential actuators in control systems.

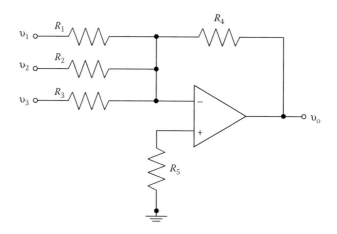

FIGURE 6.38 Problem 2.

Electrical, Electronic, and Electromechanical Systems

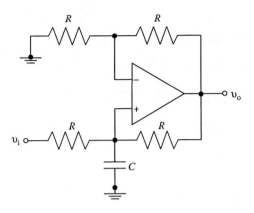

FIGURE 6.39 Problem 4.

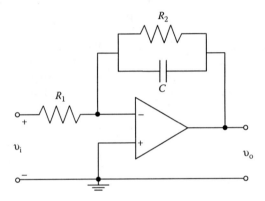

FIGURE 6.40 Problem 5.

6.4.1 Elemental Relations of Electromechanical Systems

For a variety of electromechanical systems, their electrical subsystems and mechanical subsystems are coupled by a magnetic field. Figure 6.42 shows a DC motor, which consists of basic elements (including the stator, the rotor, the armature, and the commutator). The stator provides a magnetic field across the rotor. The current is conducted to coils attached to the rotor via brushes, and the rotor is free to rotate. The combined unit of coils attached to the rotor is called the armature. The brushes are in contact with the rotating commutator, which causes the current to always be in the proper conductor windings so as to produce a torque and keep it in the proper direction. The magnetic coupling relations between the electrical and mechanical subsystems in a DC motor can be derived using fundamental electromagnetic laws in introductory physics textbooks [4].

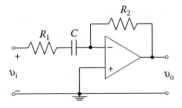

FIGURE 6.41 Problem 6.

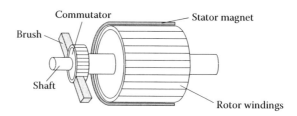

FIGURE 6.42 A DC brush motor.

For simplicity, let us first consider a wire carrying a current within a magnetic field. Assume that the wire is either a straight conductor perpendicular to a uniform magnitude field or a circular conductor in a radial magnetic field. These are two common situations in many applications. Then a force will be exerted on the wire, and the relation between the force f and the current i is

$$f = BLi, \tag{6.25}$$

where B is the magnetic flux density in tesla (1 T = 1 Wb/m^2) and L is the length of the conductor in the magnetic field. The direction of the force can be determined using the right-hand rule as shown in Figure 6.43a. Curl four fingers from the positive current direction to the positive direction of the magnetic field, and the thumb will point to the positive direction of the force.

If the conductor moves relative to the magnetic field, then a voltage will be induced in the conductor. Figure 6.43b shows a straight conductor moving upward in a magnetic field. Assume that the direction of the motion is perpendicular to the direction of the magnetic field. Then the scalar relation between the induced voltage e_b and the velocity v of the conductor is

$$e_b = BLv. \tag{6.26}$$

To avoid confusion, we use e to denote the voltage instead of v as we did in the previous sections. Again, using the right-hand rule, curl four fingers from the positive direction of the velocity to the positive field direction, and the thumb will point to the positive direction of the induced voltage e_b. Note that the induced voltage e_b opposes the current. It is known as back electromotive force (emf, an old term for voltage).

For the DC motor shown in Figure 6.42, assume that the armature current is i and the number of armature coils is n. From Equation 6.25, the force generated on the armature due to the magnetic field is $f = nBLi$. If the radius of the armature is r, then the torque produced by the motor is

$$\tau_m = fr = nBLir = K_t i, \tag{6.27}$$

where $K_t = nBLr$ is the torque constant of the motor. Note that the linear velocity of the coils is proportional to the angular velocity, $v = \omega r$. Then, from Equation 6.26, the back emf generated in

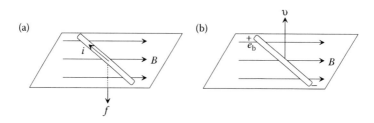

FIGURE 6.43 The direction of (a) the force on a conductor and (b) the voltage induced in a moving conductor.

Electrical, Electronic, and Electromechanical Systems

the armature due to the rotating motion is

$$e_b = nBLv = nBL\omega r = K_e\omega, \tag{6.28}$$

where $K_e = nBLr$ is the back emf constant of the motor. The two constants K_t and K_e have the same expression, and they will have the same numerical value if expressed in the same system of units. Equations 6.27 and 6.28 are used to model the coupling between the electrical and mechanical subsystems in a DC motor. Two primary types of DC motors, armature-controlled DC motors and field-controlled DC motors, are discussed next.

6.4.2 Armature-Controlled Motors

Figure 6.44 shows an electromechanical system with an armature-controlled DC motor. The electrical system is represented by an armature circuit, where v_a is applied armature voltage, R_a is armature resistance, L_a is armature inductance, and e_b is back emf generated in the armature. The mechanical part is represented by a rotational system, where I is the mass moment of inertia due to the rotor and the load, B is the viscous rotational damping associated with the load, τ_m is the torque produced by the motor, and τ_L is an additional torque applied to the load. In general, the load torque acts in the direction opposite to the motor torque. The differential equations of the system can be derived by using Kirchhoff's voltage law, the moment equation, and the electromechanical coupling relations.

For the electrical circuit, applying Kirchhoff's voltage law gives

$$\sum_j v_j = 0,$$

$$R_a i_a + L_a \frac{di_a}{dt} + e_b - v_a = 0. \tag{6.29}$$

For the mechanical part, applying the moment equation gives

$$+\curvearrowleft: \quad \Sigma M_C = I_C \alpha,$$

$$\tau_m - \tau_L - B\dot{\theta} = I\ddot{\theta}. \tag{6.30}$$

Substituting the coupling relations between the electrical and mechanical subsystems, $e_b = K_e\omega = K_e\dot{\theta}$ and $\tau_m = K_t i_a$, into Equations 6.29 and 6.30, we obtain

$$L_a \frac{di_a}{dt} + R_a i_a + K_e\dot{\theta} = v_a, \tag{6.31}$$

$$I\ddot{\theta} + B\dot{\theta} - K_t i_a = -\tau_L. \tag{6.32}$$

Note that the stiffness terms associated with the variable θ do not appear in Equations 6.31 and 6.32. Thus, the system dynamics can be expressed in terms of ω instead of θ. The differential equations of the system are

$$L_a \frac{di_a}{dt} + R_a i_a + K_e\omega = v_a, \tag{6.33}$$

$$I\dot{\omega} + B\omega - K_t i_a = -\tau_L. \tag{6.34}$$

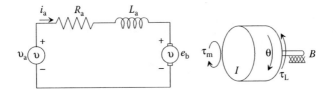

FIGURE 6.44 An electromechanical system with an armature-controlled DC motor.

Assume that all the initial conditions are set to zero. Taking the Laplace transform of Equations 6.33 and 6.34 results in

$$L_a s I_a(s) + R_a I_a(s) = V_a(s) - K_e \Omega(s), \tag{6.35}$$

$$I s \Omega(s) + B \Omega(s) = -T_L(s) + K_t I_a(s). \tag{6.36}$$

Figure 6.45 shows a block diagram of the above system. The transfer function relating the armature voltage $v_a(t)$ and $\omega(t)$, with $\tau_L(t) = 0$, is

$$\frac{\Omega(s)}{V_a(s)} = \frac{(1/(L_a s + R_a)) \cdot K_t \cdot (1/(I s + B))}{1 + (1/(L_a s + R_a)) \cdot K_t \cdot (1/(I s + B)) \cdot K_e} = \frac{K_t}{L_a I s^2 + (L_a B + R_a I) s + R_a B + K_t K_e}. \tag{6.37}$$

The transfer function relating the load torque $\tau_L(t)$ and $\omega(t)$, with $v_a(t) = 0$, is

$$\frac{\Omega(s)}{T_L(s)} = \frac{-(1/(I s + B))}{1 - (1/(I s + B)) \cdot (-K_e) \cdot (1/(L_a s + R_a)) \cdot K_t} = -\frac{L_a s + R_a}{L_a I s^2 + (L_a B + R_a I) s + R_a B + K_t K_e}. \tag{6.38}$$

Note that the two transfer functions have the same denominator, which is the characteristic of the system. The order of the characteristic polynomial implies that the system is second-order. The transfer functions $\Omega(s)/V_a(s)$ and $\Omega(s)/T_L(s)$ can also be solved from Equations 6.35 and 6.36 using Cramer's rule, and we leave this as an exercise for the reader.

The motor model can be represented in state-space form by choosing appropriate states. We choose the armature current i_a and the angular velocity ω as the states. As shown in Figure 6.44, the armature voltage and the load torque are two inputs. Let $u_1 = v_a$ and $u_2 = \tau_L$. Solving for the time derivatives di_a/dt and $\dot{\omega}$ from Equations 6.33 and 6.34 yields

$$\dot{x}_1 = \frac{di_a}{dt} = -\frac{R_a}{L_a} i_a - \frac{K_e}{L_a} \omega + \frac{1}{L_a} v_a, \tag{6.39}$$

$$\dot{x}_2 = \dot{\omega} = \frac{K_t}{I} i_a - \frac{B}{I} \omega - \frac{1}{I} \tau_L. \tag{6.40}$$

Thus, the state equation is

$$\begin{Bmatrix} \dot{x}_1 \\ \dot{x}_2 \end{Bmatrix} = \begin{bmatrix} -\dfrac{R_a}{L_a} & -\dfrac{K_e}{L_a} \\ \dfrac{K_t}{I} & -\dfrac{B}{I} \end{bmatrix} \begin{Bmatrix} x_1 \\ x_2 \end{Bmatrix} + \begin{bmatrix} \dfrac{1}{L_a} & 0 \\ 0 & -\dfrac{1}{I} \end{bmatrix} \begin{Bmatrix} u_1 \\ u_2 \end{Bmatrix}. \tag{6.41}$$

If we select ω as the output, then the output equation is

$$y = \begin{bmatrix} 0 & 1 \end{bmatrix} \begin{Bmatrix} x_1 \\ x_2 \end{Bmatrix} + \begin{bmatrix} 0 & 0 \end{bmatrix} \begin{Bmatrix} u_1 \\ u_2 \end{Bmatrix}. \tag{6.42}$$

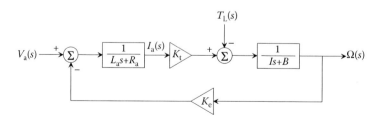

FIGURE 6.45 Block diagram of an armature-controlled DC motor.

Electrical, Electronic, and Electromechanical Systems

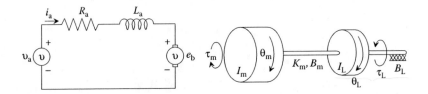

FIGURE 6.46 A more complicated mechanical model of the DC motor.

A more complicated mechanical model of the DC motor is shown in Figure 6.46, where the rotor is connected to an inertial load through a flexible and damped shaft. K_m and B_m represent the torsional stiffness and the torsional viscous damping of the shaft, respectively. The mass moments of inertia of the motor and the load are I_m and I_L, respectively. If the shaft is massless, rigid, and undamped, then the rotor, the shaft, and the load can be lumped together as a single rigid body with combined inertia $I_m + I_L$ rotating at the same speed. This reduces to the simple model shown in Figure 6.44.

To derive differential equations for the more complicated model in Figure 6.46, we specify three independent variables, i_a, θ_m, and θ_L, where θ_m and θ_L are the angular displacements of the rotor and the load, respectively. For the electrical part, the circuit remains unchanged except for the change in the angular displacement,

$$R_a i_a + L_a \frac{di_a}{dt} + K_e \dot{\theta}_m = v_a. \tag{6.43}$$

The mechanical part becomes a two-degree-of-freedom system. Assuming that $\theta_m > \theta_L$ and applying the moment equation to the rotor and the load gives

$$K_t i_a - K_m(\theta_m - \theta_L) - B_m(\dot{\theta}_m - \dot{\theta}_L) = I_m \ddot{\theta}_m, \tag{6.44}$$

$$-\tau_L + K_m(\theta_m - \theta_L) + B_m(\dot{\theta}_m - \dot{\theta}_L) - B_L \dot{\theta}_L = I_L \ddot{\theta}_L, \tag{6.45}$$

or

$$\begin{bmatrix} I_m & 0 \\ 0 & I_L \end{bmatrix} \begin{Bmatrix} \ddot{\theta}_m \\ \ddot{\theta}_L \end{Bmatrix} + \begin{bmatrix} B_m & -B_m \\ -B_m & B_m + B_L \end{bmatrix} \begin{Bmatrix} \dot{\theta}_m \\ \dot{\theta}_L \end{Bmatrix} + \begin{bmatrix} K_m & -K_m \\ -K_m & K_m \end{bmatrix} \begin{Bmatrix} \theta_m \\ \theta_L \end{Bmatrix} = \begin{Bmatrix} K_t i_a \\ -\tau_L \end{Bmatrix}. \tag{6.46}$$

The dynamics model of the system is obtained by combining Equations 6.43 through 6.45, and its second-order matrix form is

$$\begin{bmatrix} 0 & 0 & 0 \\ 0 & I_m & 0 \\ 0 & 0 & I_L \end{bmatrix} \begin{Bmatrix} d^2 i_a/dt^2 \\ \ddot{\theta}_m \\ \ddot{\theta}_L \end{Bmatrix} + \begin{bmatrix} L_a & K_e & 0 \\ 0 & B_m & -B_m \\ 0 & -B_m & B_m + B_L \end{bmatrix} \begin{Bmatrix} di_a/dt \\ \dot{\theta}_m \\ \dot{\theta}_L \end{Bmatrix}$$

$$+ \begin{bmatrix} R_a & 0 & 0 \\ -K_t & K_m & -K_m \\ 0 & -K_m & K_m \end{bmatrix} \begin{Bmatrix} i_a \\ \theta_m \\ \theta_L \end{Bmatrix} = \begin{Bmatrix} v_a \\ 0 \\ -\tau_L \end{Bmatrix}. \tag{6.47}$$

Example 6.9: A Single-Link Robot Arm Driven by a DC Motor

Consider the dynamic system shown in Figure 6.47, where a single-link robot arm is driven by a DC motor. The differential equation of the robot arm in terms of the motor variable θ_m was determined in Example 5.17 to be

$$(I_m + N^2 I)\ddot{\theta}_m + (B_m + N^2 B)\dot{\theta}_m = \tau_m,$$

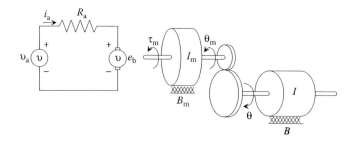

FIGURE 6.47 The model of a single-link robot arm driven by an armature-controlled DC motor.

where I_m and I are the mass moments of inertia of the motor and the load, respectively, B_m and B are the coefficients of the torsional viscous damping of the motor and the load, respectively, τ_m is the torque generated by the motor, and N is the gear ratio. Assume that the armature inductance is negligibly small, that is, $L_a \approx 0$. The torque and the back emf constants of the motor are K_t and K_e, respectively.

a. Derive the differential equation relating the applied voltage v_a and the link angular displacement θ.
b. Determine the transfer function $\Theta(s)/V_a(s)$ using the differential equation obtained in Part (a). Assume that all initial conditions are zero.
c. ◢ Following Figure 6.45, construct a Simulink block diagram to represent this electromechanical system. Assume that the parameter values are $R_a = 0.5\,\Omega$, $K_e = 0.05$ V·s/rad, $K_t = 0.05$ N·m/A, $I_m = 0.00025$ N·m·s²/rad, $B_m = 0.0001$ N·m·s/rad, $I = 0.0015$ N·m·s²/rad, $B = 0.0005$ N·m·s/rad, and $N = 0.5$. Find the output $\dot{\theta}(t)$ of the robot arm if the applied voltage is $v_a = 10$ V.

Solution

a. For the electrical circuit, applying Kirchhoff's voltage law gives

$$R_a i_a + e_b - v_a = 0,$$

where $e_b = K_e \dot{\theta}_m$. With the given gear ratio,

$$\theta = N\theta_m,$$

we have

$$R_a i_a = v_a - K_e \frac{1}{N}\dot{\theta}.$$

Thus, the current i_a can expressed as

$$i_a = \frac{1}{R_a}v_a - \frac{K_e}{R_a}\frac{1}{N}\dot{\theta}.$$

The model of the mechanical part in terms of θ is given by

$$(I_m + N^2 I)\frac{1}{N}\ddot{\theta} + (B_m + N^2 B)\frac{1}{N}\dot{\theta} = \tau_m,$$

where $\tau_m = K_t i_a$. Substituting the expression of the current i_a, we obtain

$$(I_m + N^2 I)\frac{1}{N}\ddot{\theta} + (B_m + N^2 B)\frac{1}{N}\dot{\theta} = K_t\left(\frac{1}{R_a}v_a - \frac{K_e}{R_a}\frac{1}{N}\dot{\theta}\right).$$

Rearranging the equation yields

$$(I_m + N^2 I)\ddot{\theta} + \left(B_m + N^2 B + \frac{K_t K_e}{R_a}\right)\dot{\theta} = \frac{NK_t}{R_a}v_a.$$

Electrical, Electronic, and Electromechanical Systems

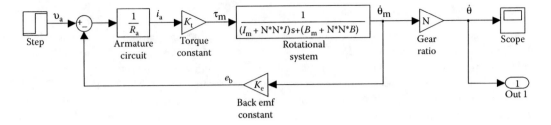

FIGURE 6.48 A Simulink block diagram to represent the robot arm in Figure 6.47.

b. Taking the Laplace transform gives

$$\frac{\Theta(s)}{V_a(s)} = \frac{NK_t/R_a}{(I_m + N^2 I)s^2 + (B_m + N^2 B + K_t K_e/R_a)s}.$$

c. ▲ This electromechanical system includes an armature circuit

$$R_a i_a + e_b - v_a = 0,$$

a rotational system

$$(I_m + N^2 I)\ddot{\theta}_m + (B_m + N^2 B)\dot{\theta}_m = \tau_m,$$

a gear–train system

$$\dot{\theta} = N\dot{\theta}_m,$$

and couplings between the electrical and mechanical subsystems

$$e_b = K_e \dot{\theta}_m, \quad \tau_m = K_t i_a.$$

Following Figure 6.45, we can construct a Simulink block diagram (see Figure 6.48), which shows the major components of the robot arm and their interconnections. The dynamics of the electrical armature circuit and mechanical rotational system are represented using `Transfer Fcn` blocks. The torque constant, back emf constant, and gear ratio are represented using `Gain` blocks. All the parameter values are defined in MATLAB® Workspace. Run the Simulink model. The result can be plotted as shown in Figure 6.49.

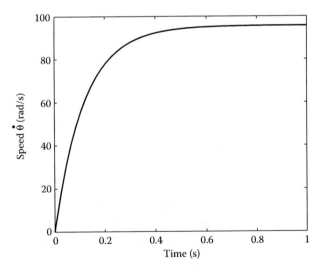

FIGURE 6.49 Speed output $\dot{\theta}(t)$ of the robot arm for $v_a = 10$ V.

6.4.3 FIELD-CONTROLLED MOTORS

In all but the smallest motors, the magnetic field is established by a current in separate field windings on the stator. For an armature-controlled DC motor, a constant current source is supplied to the field windings and the applied armature voltage, v_a, varies. Another way of controlling a DC motor is to keep the armature current i_a constant while varying the voltage applied to the field windings. A simple model of a field-controlled DC motor is shown in Figure 6.50, where the shaft in the mechanical subsystem is assumed to be massless, rigid, and undamped. The electrical part is represented by a field circuit, where R_f is field resistance, L_f is field inductance, v_f is field voltage, and i_f is field current. Note that there is no back emf created in the field circuit. The torque generated by the motor is proportional to the field current,

$$\tau_m = K_t i_f. \tag{6.48}$$

The system under consideration has two inputs, v_f and τ_L. Two independent variables, i_f and θ, can be used to describe the system dynamics. For the electrical part, we apply Kirchhoff's voltage law to the field circuit,

$$L_f \frac{di_f}{dt} + R_f i_f = v_f. \tag{6.49}$$

For the mechanical part, introducing the motor Equation 6.48 and applying the moment equation gives

$$I\ddot{\theta} + B\dot{\theta} - K_t i_f = -\tau_L \tag{6.50}$$

or

$$I\dot{\omega} + B\omega - K_t i_f = -\tau_L. \tag{6.51}$$

The two Equations, 6.49 and 6.50, or 6.49 and 6.51, are the system differential equations of the field-controlled DC motor.

Assuming all the initial conditions to be zero and taking the Laplace transform of Equations 6.49 and 6.51, we have

$$L_f s I_f(s) + R_f I_f(s) = V_f(s), \tag{6.52}$$
$$I s \Omega(s) + B \Omega(s) = K_t I_f(s) - T_L(s). \tag{6.53}$$

A block diagram of the system is shown in Figure 6.51.

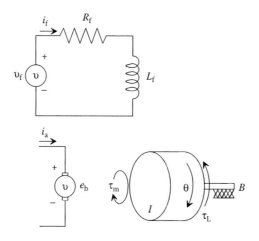

FIGURE 6.50 An electromechanical system with a field-controlled DC motor.

Electrical, Electronic, and Electromechanical Systems

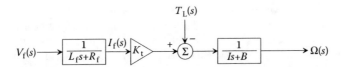

FIGURE 6.51 Block diagram of a field-controlled DC motor.

The transfer functions $\Omega(s)/V_f(s)$ and $\Omega(s)/T_L(s)$ can be easily derived from the above diagram,

$$\frac{\Omega(s)}{V_f(s)} = \frac{K_t}{(L_f s + R_f)(Is + B)} = \frac{K_t}{L_f I s^2 + (L_f B + R_f I) s + R_f B}, \quad (6.54)$$

$$\frac{\Omega(s)}{T_L(s)} = -\frac{1}{Is + B}. \quad (6.55)$$

If we choose the field current and the angular velocity as the state variables ($x_1 = i_f$ and $x_2 = \omega$), the field voltage and the load torque as the inputs ($u_1 = v_f$ and $u_2 = T_L$), and the angular velocity as the output ($y = \omega$), then the state-space form is

$$\begin{Bmatrix} \dot{x}_1 \\ \dot{x}_2 \end{Bmatrix} = \begin{bmatrix} -\dfrac{R_f}{L_f} & 0 \\ \dfrac{K_t}{I} & -\dfrac{B}{I} \end{bmatrix} \begin{Bmatrix} x_1 \\ x_2 \end{Bmatrix} + \begin{bmatrix} \dfrac{1}{L_f} & 0 \\ 0 & -\dfrac{1}{I} \end{bmatrix} \begin{Bmatrix} u_1 \\ u_2 \end{Bmatrix}, \quad y = \begin{bmatrix} 0 & 1 \end{bmatrix} \begin{Bmatrix} x_1 \\ x_2 \end{Bmatrix} + \begin{bmatrix} 0 & 0 \end{bmatrix} \begin{Bmatrix} u_1 \\ u_2 \end{Bmatrix}. \quad (6.56)$$

PROBLEM SET 6.4

1. Reconsider the armature-controlled motor in Figure 6.44.
   a. Derive the transfer functions $\Theta(s)/V_a(s)$ and $\Theta(s)/T_L(s)$. All of the initial conditions are assumed to be zero.
   b. Assuming the angle θ to be the output, draw a block diagram to represent the dynamics of the armature-controlled motor.
   c. Determine the state-space form assuming the angle θ to be the output.
2. Reconsider the field-controlled motor in Figure 6.50.
   a. Derive the transfer functions $\Theta(s)/V_f(s)$ and $\Theta(s)/T_L(s)$. All of the initial conditions are assumed to be zero.
   b. Assuming the angle θ to be the output, draw a block diagram to represent the dynamics of the field-controlled motor.
   c. Determine the state-space form assuming the angle θ to be the output.
3. Consider the electromechanical system shown in Figure 6.52a. It consists of a cart of mass m moving without slipping on a ground track. The cart is equipped with an armature-controlled DC motor, which is coupled to a rack and pinion mechanism to

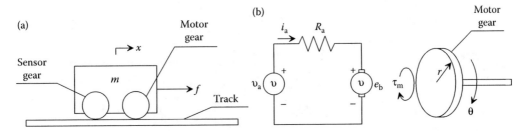

FIGURE 6.52 Problem 3. (a) Physical system and (b) equivalent electrical circuit and mechanical model of DC motor.

convert the rotational motion to translation and to create the driving force for the system. Figure 6.52b shows the equivalent electric circuit and the mechanical model of the DC motor, where r is the radius of the motor gear. The torque and the back emf constants of the motor are K_t and K_e, respectively. Neglect the armature inductance and the mass moment of inertia of the motor. The driving force f is related to the torque by $f = \tau_m/r$.

   a. Derive the differential equation of the system relating the cart position x and the applied voltage v_a.
   b. Determine the transfer function $X(s)/V_a(s)$ using the differential equation obtained in Part (a). Assume that all initial conditions are zero.

4. Consider the single-link robot arm as shown in Figure 6.53. It is driven by a DC motor through spur gears with a total gear ratio $N = 1/70$. The differential equation of the robot arm relating the link angle θ and the motor torque τ_m is given by

$$\left(\frac{I_m}{N^2} + I\right)\ddot{\theta} + \left(\frac{B_m}{N^2} + B\right)\dot{\theta} = \frac{\tau_m}{N}.$$

Assume that the effective inertia I_{eq} as seen at the output of the gearing system is

$$I_{eq} = \frac{I_m}{N^2} + I = 0.0035\,\text{kg·m}^2$$

and the effective damping coefficient B_{eq} as seen at the output of the gearing system is

$$B_{eq} = \frac{B_m}{N^2} + B = 0.004\,\text{N·m·s/rad}.$$

Neglect armature inductance. The motor parameters are $R_a = 2.5\,\Omega$, $K_t = 0.0075\,\text{N·m/A}$, and $K_e = 0.0075\,\text{V·s/rad}$.

   a. Derive the differential equation relating the applied voltage v_a and the link angular displacement θ.
   b. Determine the transfer function $\Theta(s)/V_a(s)$ using the differential equations obtained in Part (a). Assume that all initial conditions are zero.
   c. Construct a Simulink block diagram to find the output $\dot{\theta}(t)$ of the robot arm if the applied voltage is $v_a = 1$ V.

5. A more complicated model of the field-controlled motor is shown in Figure 6.54, where the rotor is connected to an inertial load through a flexible and damped shaft. K_m and B_m represent the torsional stiffness and the torsional viscous damping of the shaft, respectively. The mass moments of inertia of the motor and the load are I_m and I_L, respectively.

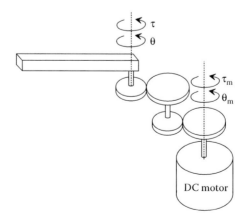

FIGURE 6.53 Problem 4.

Electrical, Electronic, and Electromechanical Systems

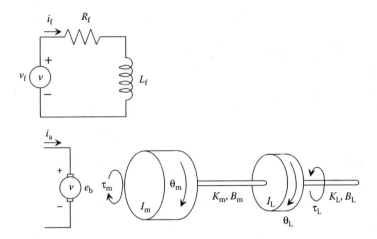

FIGURE 6.54 Problem 5.

a. Derive the transfer functions $\Omega_L(s)/V_f(s)$ and $\Omega_L(s)/T_L(s)$. All of the initial conditions are assumed to be zero.
b. Assuming the angular velocity ω_L to be the output, draw a block diagram to represent the dynamics of the field-controlled motor.
c. Determine the state-space form assuming the angular velocity ω_L to be the output.

6.5 IMPEDANCE METHODS

The concept of impedance is very useful in electrical systems, as it provides an alternative way of obtaining system mathematical models (transfer functions and differential equations).

6.5.1 Impedances of Electric Elements

Impedance is a generalization of the concept of resistance. Mathematically, electrical impedance is defined as the ratio of the voltage to the current in the Laplace domain,

$$Z(s) = \frac{V(s)}{I(s)}. \tag{6.57}$$

For a resistor, $v = Ri$, the impedance is its resistance R,

$$Z(s) = R. \tag{6.58}$$

Assuming that all the initial conditions are zero, for an inductor, $v = L\,di/dt$ or $V(s) = LsI(s)$. Thus, the impedance of an inductor is

$$Z(s) = Ls. \tag{6.59}$$

Similarly, for a capacitor, $i = C\,dv/dt$ or $I(s) = CsV(s)$. Thus, the impedance of a capacitor is

$$Z(s) = \frac{1}{Cs}. \tag{6.60}$$

6.5.2 Series and Parallel Impedances

Because impedance can be viewed as a generalized resistance, it is easy to find the equivalent impedance for series- or parallel-connected electrical elements. Figure 6.55 shows n impedances

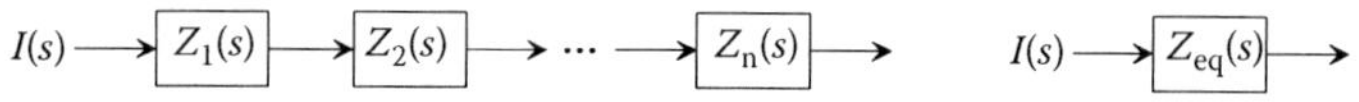

Z₁(s) → Z₂(s) → ⋯ → Zₙ(s) diagram and equivalent Z_eq(s) diagram

FIGURE 6.55 Equivalence for impedances in series.

in series. Note that the same current flows through n impedances and the total voltage drop across them is

$$V(s) = I(s)Z_1(s) + I(s)Z_2(s) + \cdots + I(s)Z_n(s). \tag{6.61}$$

In the equivalent diagram, the relation between the current and the voltage is

$$V(s) = I(s)Z_{eq}(s). \tag{6.62}$$

Thus, we have

$$Z_{eq}(s) = Z_1(s) + Z_2(s) + \cdots + Z_n(s). \tag{6.63}$$

That is, the equivalent impedance Z_{eq} is equal to the sum of all the individual impedances Z_i.

If there are n impedances in parallel as shown in Figure 6.56, then all the impedances have the same voltage drop. The total current through all the elements is

$$I(s) = \frac{V(s)}{Z_1(s)} + \frac{V(s)}{Z_2(s)} + \cdots + \frac{V(s)}{Z_n(s)}. \tag{6.64}$$

For the equivalent impedance,

$$I(s) = \frac{V(s)}{Z_{eq}(s)}. \tag{6.65}$$

Thus, we have

$$\frac{1}{Z_{eq}(s)} = \frac{1}{Z_1(s)} + \frac{1}{Z_2(s)} + \cdots + \frac{1}{Z_n(s)}. \tag{6.66}$$

That is, the reciprocal of the equivalent impedance Z_{eq} is equal to the sum of all the reciprocals of the individual impedances Z_i.

Note that impedance is essentially a transfer function, which has no integral or derivative signs. If we redraw an electrical system in the s domain by replacing passive elements with their corresponding impedances, we can determine the transfer function of the system using Kirchhoff's laws along with series and parallel laws. The differential equation of the system can be obtained by converting the transfer function back from the s domain to the time domain. Thus, the concept of impedance provides another way of modeling electrical systems without writing any time-domain equations.

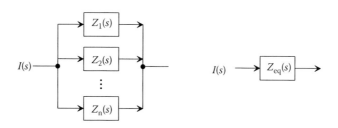

FIGURE 6.56 Equivalence for impedances in parallel.

Electrical, Electronic, and Electromechanical Systems

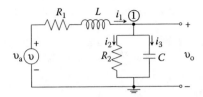

FIGURE 6.57 An RLC circuit.

Example 6.10: Impedance Method

Consider the electrical circuit in Figure 6.57. Assume that all the initial conditions are zero.

a. Using Kirchhoff's laws and the impedance concept, determine the differential equation of the system.
b. Verify the result in Part (a) by directly writing the time-domain equations.

Solution

a. We can replace the passive elements with their impedance representations. The circuit in the s domain is redrawn as shown in Figure 6.58a. Note that the resistor R_1 is in series connection with the inductor L, and the resistor R_2 is in parallel connection with the capacitor C. Therefore, the corresponding equivalent impedances are

$$Z_1(s) = R_1 + Ls,$$
$$\frac{1}{Z_2(s)} = \frac{1}{R_2} + \frac{1}{1/Cs},$$

or

$$Z_2(s) = \frac{R_2}{R_2 Cs + 1}.$$

For the equivalent impedance circuit in Figure 6.58b, we apply Kirchhoff's voltage law,

$$Z_1(s) + Z_2(s) - V_a(s) = 0,$$

where the current is related to the output voltage by

$$V_o(s) = Z_2(s)I(s).$$

Thus, we have

$$Z_1(s)\frac{V_o(s)}{Z_2(s)} + V_o(s) = V_a(s),$$

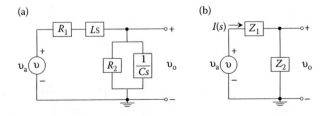

FIGURE 6.58 Equivalent circuit of Figure 6.57 drawn (a) in s domain and (b) using impedances.

which gives the transfer function relating the input voltage v_a and the output voltage v_o,

$$\frac{V_o(s)}{V_a(s)} = \frac{Z_2(s)}{Z_1(s) + Z_2(s)} = \frac{R_2}{R_2LCs^2 + (R_1R_2C + L)s + R_1 + R_2}.$$

By transforming $V_o(s)/V_a(s)$ from the s domain to the time domain with the assumption of zero initial conditions, we obtain the differential equation of the system

$$R_2LC\ddot{v}_o + (R_1R_2C + L)\dot{v}_o + (R_1 + R_2)v_o = R_2v_a.$$

b. Consider the circuit in Figure 6.56. Applying Kirchhoff's voltage law to the left loop yields

$$R_1i_1 + L\frac{di_1}{dt} + v_o = v_a.$$

Applying Kirchhoff's current law to node 1 yields

$$i_1 = i_2 + i_3 = \frac{v_o}{R_2} + C\frac{dv_o}{dt}.$$

Eliminating i_1 from these two relations results in

$$R_1\left(\frac{v_o}{R_2} + C\dot{v}_o\right) + L\left(\frac{\dot{v}_o}{R_2} + C\ddot{v}_o\right) + v_o = v_a.$$

Multiplying both sides by R_2 and rearranging the terms, yields the same differential equation as the one obtained in Part (a).

6.5.3 Mechanical Impedances

Analogous to electrical impedance, mechanical impedance is defined as

$$Z(s) = \frac{V(s)}{F(s)}, \tag{6.67}$$

where $V(s)$ and $F(s)$ are the Laplace transforms of velocity $v(t)$ and force $f(t)$, respectively. The impedance concept can also be used to obtain models of mechanical systems along with the linear graph, which is a topic beyond the scope of this text. Here, we only give the definitions of impedances for fundamental mechanical elements.

For a viscous damper, the damping force is related to the velocity by $f = bv$ or $F(s) = bV(s)$. Thus, the impedance of a damper is

$$Z(s) = \frac{1}{b}. \tag{6.68}$$

For a spring element, the spring force is proportional to the displacement, $f = kx = k\int v\,dt$ or $F(s) = kV(s)/s$. Thus,

$$Z(s) = \frac{s}{k}. \tag{6.69}$$

For a mass element, from Newton's second law, $f = ma = m\dot{v}$ or $F(s) = msV(s)$. Thus, the impedance of a mass element is

$$Z(s) = \frac{1}{ms}. \tag{6.70}$$

When comparing the two sets of equations, Equations 6.58 through 6.60 and 6.68 through 6.70, we note that the corresponding electrical and mechanical elements are not equivalent, although they have similar physical effects. For example, both the resistor and the damper dissipate energy. But the mathematical expressions for their impedances are different.

Electrical, Electronic, and Electromechanical Systems

PROBLEM SET 6.5

1. Reconsider the circuit shown in Figure 6.20.
   a. Use the impedance method to determine the transfer function $V_C(s)/V_a(s)$. All initial conditions are assumed to be zero.
   b. Determine the input–output differential equation relating v_C and v_a.
2. Repeat Problem 1 for the circuit shown in Figure 6.21.
3. Reconsider the circuit shown in Figure 6.22.
   a. Use the impedance method to determine the transfer function $I_L(s)/I_a(s)$. All initial conditions are assumed to be zero.
   b. Derive the input–output differential equation relating i_L and i_a.
4. Repeat Problem 3 for the circuit shown in Figure 6.23.
5. Reconsider the circuit shown in Figure 6.24.
   a. Use the impedance method to determine the transfer function $V_o(s)/V_a(s)$. All initial conditions are assumed to be zero.
   b. Derive the input–output differential equation relating v_o and v_a.
6. Repeat Problem 5 for the circuit shown in Figure 6.25.
7. Reconsider the op-amp circuit shown in Figure 6.40.
   a. Use the impedance method to determine the transfer function $V_o(s)/V_i(s)$. All initial conditions are assumed to be zero.
   b. Derive the input–output differential equation relating the output voltage $v_o(t)$ and the input voltage $v_i(t)$.
8. Repeat Problem 7 for the circuit shown in Figure 6.41.

6.6 SUMMARY

This chapter was devoted to the modeling of electrical, electronic, and electromechanical systems.

An electrical system or electrical circuit can be considered to be an interconnection of active and passive electrical elements. Active electrical elements include ideal current sources and ideal voltage sources, both of which can provide energy to the circuit and serve as the inputs. Passive electrical elements, including resistors, inductors, and capacitors, can either store or dissipate energy available in the circuit, but they cannot produce energy. The voltage–current relations for passive electrical elements are given as follows:

- Resistor: $v = Ri, \quad i = \dfrac{v}{R}$
- Inductor: $v = L\dfrac{di}{dt}, \quad i = \dfrac{1}{L}\int v\,dt$
- Capacitor: $v = \dfrac{1}{C}\int i\,dt, \quad i = C\dfrac{dv}{dt}$

For the modeling of electrical, electronic, and electromechanical systems, Kirchhoff's voltage law and Kirchhoff's current law are the two main physical laws to derive the governing differential equations.

For a closed path, or a loop, in a circuit, Kirchhoff's voltage law states that the algebraic sum of the voltages around the loop must be zero,

$$\sum_j v_j = 0,$$

where v_j is the voltage across the jth element in the loop.

For a node in a circuit, Kirchhoff's current law states that the sum of the currents entering the node must be equal to the sum of the currents leaving that node. If we assign a positive sign to the current entering the node and a negative sign to the current leaving the node, then the algebraic sum of the currents at the node must be zero,

$$\sum_j i_j = 0,$$

where i_j is the current of the jth element at the node.

It is usually not so easy to obtain a set of differential equations for complicated circuits. The two systematic methods, the node method that relies on Kirchhoff's current law and the loop method that relies on Kirchhoff's voltage law, were introduced in Section 6.2. Using these two methods can produce a small, simple set of equations leading directly to the input–output relation.

In order to represent a circuit model in state-space form, an appropriate set of state variables can normally be chosen by identifying the energy storage elements. Both inductors and capacitors can store energy, and expressions for the stored electrical energy are given as follows:

- Inductor $E = \frac{1}{2}Li^2$
- Capacitor $E = \frac{1}{2}Cv^2$

Generally, inductor currents and capacitor voltages are chosen as the state variables. To determine the state-space form of an electric circuit, the expression of di_L/dt or dv_C/dt for each inductor or capacitor is needed. Based on the voltage–current relations, $di_L/dt = \frac{1}{L}v_L$ and $dv_C/dt = \frac{1}{C}i_C$. The problem is thus converted to expressing the inductor voltage v_L and the capacitor current i_C in terms of state variables and inputs using Kirchhoff's laws and voltage–current relations for electrical elements.

For an op-amp, which is an electronic element used to amplify electrical signals and drive physical devices, the differential equation relating the output voltage and the input voltage can be derived by applying Kirchhoff's laws and the op-amp equation

$$v_+ \approx v_-,$$

where v_+ and v_- are the voltages at the two input terminals of the op-amp.

For an electromechanical system, the dynamic model can be derived by applying electrical principles (e.g., Kirchhoff's laws) and mechanical principles (e.g., Newton's second law). The modeling of DC motors was discussed in Section 6.4. For an armature-controlled motor, the torque produced by the motor is

$$\tau_m = K_t i_a$$

and the back emf generated in the armature due to the rotating motion is

$$e_b = K_e \omega = K_e \dot{\theta}.$$

Armature-controlled motors are commonly used, for which a constant current source i_f is supplied to the field windings and the applied armature voltage v_a varies. Field-controlled motors are used in another way, which keeps the armature current i_a constant but varies the voltage v_f applied to the field windings. The torque generated by a field-controlled motor is proportional to the field current,

$$\tau_m = K_t i_f.$$

The impedance concept provides an alternative technique to obtain mathematical models of systems (e.g., transfer functions and differential equations). The electrical impedance is defined as the ratio of the voltage to the current in the Laplace domain. The expressions of impedances for passive electrical elements are given as follows:

- Resistor $Z(s) = R$
- Inductor $Z(s) = Ls$
- Capacitor $Z(s) = \dfrac{1}{Cs}$

Because the impedance can be viewed as a generalized resistance, it is easy to find the equivalent impedance for electrical elements connected in series or parallel and determine mathematical models of systems.

REVIEW PROBLEMS

1. Determine the equivalent resistance R_{eq} for the circuit shown in Figure 6.59.
2. Find R_{13} and R_{32} for the voltage divider shown in Figure 6.60 so that the current is limited to 0.5 A when $v_i = 110$ V and $v_o = 100$ V.
3. Consider the first-order RL circuit shown in Figure 6.61.
   a. Derive the input–output differential equation relating i_L and v_a.
   b. Determine the transfer function $V_L(s)/V_a(s)$. Assume that all the initial conditions are zero.

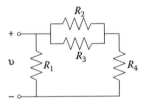

FIGURE 6.59 Problem 1.

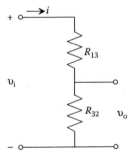

FIGURE 6.60 Problem 2.

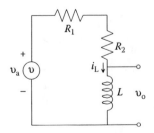

FIGURE 6.61 Problem 3.

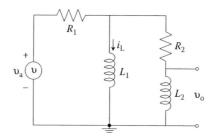

FIGURE 6.62 Problem 4.

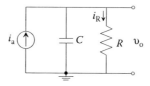

FIGURE 6.63 Problem 5.

4. Consider the circuit shown in Figure 6.62.
   a. Use the loop method to derive the differential equations.
   b. Determine the transfer function $I_L(s)/V_a(s)$. Assume that all the initial conditions are zero.
5. Consider the first-order RC circuit shown in Figure 6.63.
   a. Derive the input–output differential equation relating i_R and i_a.
   b. Determine the transfer function $V_C(s)/I_a(s)$. Assume that all the initial conditions are zero.
6. Consider the circuit shown in Figure 6.64.
   a. Use the node method to derive the differential equations.
   b. Determine the transfer function $I_L(s)/V_a(s)$. Assume that all the initial conditions are zero.
7. Consider the circuit shown in the Figure 6.61. Determine a suitable set of state variables and obtain the state-space representation with v_o as the output.
8. Repeat Problem 7 for the circuit shown in Figure 6.62.
9. Repeat Problem 7 for the circuit shown in Figure 6.63.
10. Repeat Problem 7 for the circuit shown in Figure 6.64.
11. Consider the op-amp circuit shown in Figure 6.65. Derive the differential equation relating the output voltage $v_o(t)$ and the input voltage $v_i(t)$.
12. Repeat Problem 11 for the circuit shown in Figure 6.66.

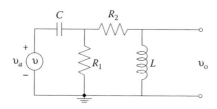

FIGURE 6.64 Problem 6.

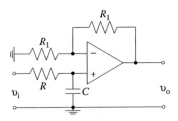

FIGURE 6.65 Problem 11.

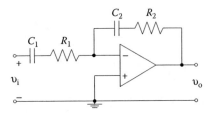

FIGURE 6.66 Problem 12.

13. Consider the armature-controlled motor system as shown in Figure 6.67, where the rotor is connected to an inertial load through a flexible and damped shaft. K and B represent the torsional stiffness and the torsional viscous damping of the shaft, respectively. The mass moments of inertia of the motor and the load are I_m and I_L, respectively. The torque and the back emf constants of the motor are K_t and K_e, respectively.
    a. Derive the differential equations.
    b. Determine the transfer function $\Theta_L(s)/V_a(s)$. All of the initial conditions are assumed to be zero.
    c. Assuming the angle θ_L to be the output, draw a block diagram to represent the dynamics of the electromechanical system.
    d. Determine the state-space form assuming the angle θ_L to be the output.
14. Consider the field-controlled motor system as shown in Figure 6.68, where the rotor is connected to an inertial load through a flexible and damped shaft. K and B represent the torsional stiffness and the torsional viscous damping of the shaft, respectively. The mass moments of inertia of the motor and the load are I_m and I_L, respectively. The torque and the back emf constants of the motor are K_t and K_e, respectively.
    a. Derive the differential equations.
    b. Determine the transfer function $\Theta_L(s)/V_f(s)$. All of the initial conditions are assumed to be zero.
    c. Assuming the angle θ_L to be the output, draw a block diagram to represent the dynamics of the electromechanical system.
    d. Determine the state-space form assuming the angle θ_L to be the output.

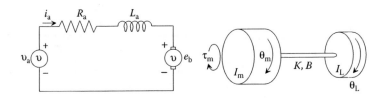

FIGURE 6.67 Problem 13.

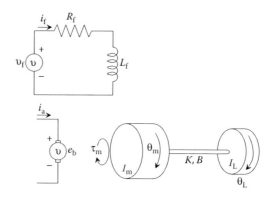

FIGURE 6.68 Problem 14.

15. Reconsider the circuit shown in Figure 6.62.
    a. Use the impedance method to determine the transfer function $I_L(s)/V_a(s)$. All initial conditions are assumed to be zero.
    b. Determine the input–output differential equation relating i_L and v_a.
16. Repeat Problem 15 for the circuit shown in Figure 6.64.

7 Fluid and Thermal Systems

Fluid is a general term used to represent a gas or a liquid. A fluid is said to be incompressible if the density of the fluid does not change with pressure. All gases are considered compressible, while liquids can be considered incompressible. Although real liquids are actually compressible, the changes in their densities are insignificant when pressure is varied. Fluid systems can be divided into pneumatics and hydraulics. A pneumatic system is one in which the fluid is compressible. A hydraulic system is one in which the fluid is incompressible. A general type of incompressible liquid systems is liquid-level systems, which are operated by adjusting the heights or levels of liquids in storage tanks.

A thermal system is one in which thermal energy is stored or transferred. The mathematical model of a thermal system is often complicated because of the complex temperature distribution throughout the system. Partial rather than ordinary differential equations are required for precisely analyzing such a distributed-parameter system. This topic is beyond the scope of this text. To simplify analysis, a lumped-parameter model, rather than a distributed-parameter model, governed by ordinary differential equations may be used to approximate the dynamics of the system.

The modeling of fluid and thermal systems is presented in this chapter. Three major systems that include pneumatic systems, liquid-level systems, and heat-transfer systems are discussed. For each of them, we first introduce the concepts of capacitance and resistance. It is useful to think of fluid and thermal systems as electrical circuits. Along with the basic elements, two main laws, the conservation of mass and the conservation of energy, are then used to develop mathematical models of fluid and thermal systems, respectively.

7.1 PNEUMATIC SYSTEMS

Pneumatic systems are often used in industry, particularly for pneumatic switches, pneumatic actuators, compressed-air engines, air brakes on buses and trucks, and so on. The working medium in a pneumatic system is compressible gas, typically air. To derive the mathematical model of a pneumatic system, it is important to understand the thermodynamic properties of gases.

7.1.1 Ideal Gases

The state of an amount of gas is determined by its pressure, volume, and temperature. In other words, pressure, volume, temperature, and mass are functionally related for gases. The ideal gas law is the model most often used to describe this relation. An ideal gas is a hypothetical gas whose quantity pV/T is constant, where p is the absolute pressure of the gas, V is the volume, and T is the absolute temperature (K or °R). All real gases behave as ideal gases if the pressure is low enough and the temperature is high enough. At low pressure and moderate temperature, real gases may be approximated as ideal gases to simplify calculations.

The ideal gas law states the relationship

$$pV = nR_\mathrm{u}T, \qquad (7.1)$$

where n is the number of moles of the gas and R_u is the universal gas constant. The mole is the unit of amount of substance. A mole of an element or a compound contains 6.023×10^{23} atoms or molecules. The numerical value of the universal gas constant R_u is 8314.3 N·m/(kg-mol·K) or 1545.3 ft·lb/(lb-mol°·R).

The amount of substance can also be given in mass instead of moles. The number of moles n is equal to m/M, where m is the mass and M is the molar mass. Therefore, an alternative form of the ideal gas law is

$$pV = mR_g T, \tag{7.2}$$

where $R_g = R_u/M$ is the specific gas constant that depends on the particular type of gas. For dry air, $R_g = 287.06\,\text{N·m/(kg·K)}$ or $1716.6\,\text{ft·lb/(slug·°R)}$. The ideal gas law can be used to solve for one of the four variables (p, V, T, or m) if the other three are known.

For a particular thermodynamic process from state 1 to state 2, the ideal gas equation can be simplified. Assume that the mass is constant. The following lists five possible processes, where the state number is denoted by the subscript.

1. *Isobaric (or constant-pressure) process* ($p_1 = p_2$): The ideal gas law implies $T_1/V_1 = T_2/V_2$ or $T_1/T_2 = V_1/V_2$. When heat is added to the gas, some of it raises the temperature and some expands the volume to exert external work.
2. *Isochoric (or constant-volume) process* ($V_1 = V_2$): The ideal gas law implies $T_1/p_1 = T_2/p_2$ or $T_1/T_2 = p_1/p_2$. Because the volume is constant, there is no external work done. The heat added to the gas only raises the temperature.
3. *Isothermal (or constant-temperature) process* ($T_1 = T_2$): The ideal gas law implies $p_1 V_1 = p_2 V_2$ or $p_1/p_2 = V_2/V_1$. The heat added to the gas only does external work.
4. *Isentropic process (reversible adiabatic process)*: An adiabatic process is a process in which no heat is transferred to or from the gas. A reversible process is a process that, after it has taken place, can be reversed and causes no change in the thermodynamic conditions of either the system or its surroundings. Any reversible adiabatic process is an isentropic process. This process is described by $p_1 V_1^\gamma = p_2 V_2^\gamma$, where γ is defined as the heat capacity ratio.
5. *Polytropic process:* It is the most general thermodynamic process. The process is described by

$$p\left(\frac{V}{m}\right)^n = \frac{p}{\rho^n} = \text{constant}, \tag{7.3}$$

where ρ is the density of the gas. For an ideal gas with a constant mass, this process reduces to the previous four processes if n is chosen as $0, \infty, 1$, or γ, respectively.

7.1.2 Pneumatic Capacitance

Fluid capacitance is the relation between the stored fluid mass and the resulting pressure caused by the stored mass. Specifically, the pneumatic capacitance C is defined as the ratio of the change in stored gas mass to the change in gas pressure:

$$C = \frac{dm}{dp}. \tag{7.4}$$

For a container of constant volume V with a gas of density ρ, Equation 7.4 can be rewritten as

$$C = \frac{d(\rho V)}{dp} = V\frac{d\rho}{dp}. \tag{7.5}$$

For a polytropic process, we have

$$\frac{dp}{d\rho} = n\rho^{n-1}\left(\frac{p}{\rho^n}\right) = \frac{np}{\rho}. \tag{7.6}$$

Fluid and Thermal Systems

Introducing the ideal gas law presented in Equation 7.2 gives

$$\frac{p}{\rho} = \frac{pV}{m} = R_g T. \tag{7.7}$$

Substituting it into Equation 7.6 gives

$$\frac{d\rho}{dp} = \frac{1}{nR_g T}. \tag{7.8}$$

Thus, the capacitance of the container is

$$C = \frac{V}{nR_g T}. \tag{7.9}$$

Example 7.1: Pneumatic Capacitance

Dry air passes through a valve into a rigid 27 m³ container at a constant temperature of 25°C (298 K). The process is assumed to be an isothermal process. Determine the capacitance of the air container.

Solution

The filling of the container is modeled as an isothermal process. In Equation 7.9, let $n = 1$, and we have

$$C = \frac{V}{R_g T} = \frac{27}{287.06 \times 298} = 3.16 \times 10^{-4}\, \text{kg·m}^2/\text{N}.$$

Note that the same container can have a different pneumatic capacitance. The value of C depends on the type of gas, the temperature of gas, and the type of thermodynamic process.

7.1.3 Modeling of Pneumatic Systems

It is rather difficult to precisely model pneumatic systems due to their highly nonlinear dynamics. To simplify the modeling, each mass storage element in a pneumatic system can be represented using a capacitance element, and the resistance due to a valve, an orifice, or pipe wall friction, can be represented using a resistance element. Such a simple model may be adequate to describe the dynamic behavior of the real system.

Consider a pneumatic system shown in Figure 7.1, where p_i is the inlet pressure, q_i is the volume flow rate (m³/s), and p and ρ are the pressure and the density of the gas in a container of constant volume V. The gas passes through a valve and flows into the rigid container by the pressure difference, $\Delta p = p_i - p$. Note that the gas meets resistance when flowing through the valve. The valve resistance depends on the pressure p and the mass flow rate q_m (kg/s). Generally, the p

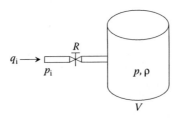

FIGURE 7.1 A pneumatic system with gas flowing into a container of constant volume.

versus q_m curve is nonlinear. Thus, the value of the valve resistance R, which is defined as the slope of the curve, varies with operating conditions. The definition of the resistance R is valid for both pneumatic and hydraulic systems, and more details will be given in Section 7.2.

Although the valve resistance R is nonlinear, it may be linearized about an operating condition. For a constant pressure p_i at the inlet of the valve and a constant volume flow rate q_i through the valve, the resistance R is given as

$$R = \frac{\Delta p}{\Delta q_m} = \frac{p_i - p}{\rho_i q_i}. \tag{7.10}$$

Combining with the capacitance of the container given in Equation 7.9, the pneumatic system can be represented using a resistance–capacitance model.

The differential equation of the system can be derived by applying the law of conservation of mass:

$$\frac{dm}{dt} = q_{mi} - q_{mo}, \tag{7.11}$$

that is, the rate of mass increase in the container equals the mass flow rate into the container minus the mass flow rate out of the container. Note that

$$\frac{dm}{dt} = \frac{dm}{dp}\frac{dp}{dt} = C\frac{dp}{dt} \tag{7.12}$$

and

$$q_{mi} = \rho_i q_i = \frac{p_i - p}{R}. \tag{7.13}$$

For the pneumatic system in Figure 7.1, the mass flow rate out of the container is $q_{mo} = 0$. Thus, Equation 7.11 can be rewritten as

$$C\frac{dp}{dt} = \frac{p_i - p}{R} \tag{7.14}$$

or

$$RC\frac{dp}{dt} + p = p_i, \tag{7.15}$$

which is a first-order ordinary differential equation of the pressure p. Introducing the expression of the capacitance C given by Equation 7.9 gives

$$\frac{RV}{nR_g T}\frac{dp}{dt} + p = p_i. \tag{7.16}$$

Equation 7.15 or 7.16 is the mathematical model of the pneumatic system undergoing a polytropic process with an ideal gas. It is also valid for an isobaric, isochoric, isothermal, or isentropic process if the value of n in the capacitance C is chosen as $0, \infty, 1$, or γ, respectively.

Example 7.2: A Pneumatic System

Dry air at a constant temperature of 20°C (293 K) passes through a valve into a rigid cubic container of 1 m on each side (see Figure 7.2). The pressure p_i at the inlet of the valve is constant, and

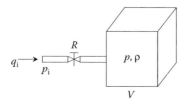

FIGURE 7.2 A pneumatic system with a rigid cubic container.

Fluid and Thermal Systems

it is greater than p. The valve resistance is approximately linear, and $R = 1000$ Pa·s/kg. Assume the filling process is isothermal.

a. Develop a mathematical model of the pressure p in the container.
b. Construct a Simulink® block diagram to find the output $p(t)$ of the pneumatic system if the pressure inside the container initially is 0 Pa and the pressure at the inlet is 101.325 kPa.

Solution

a. Applying the law of conservation of mass gives

$$\frac{dm}{dt} = \rho_i q_i.$$

Note that

$$\frac{dm}{dt} = \frac{dm}{dp}\frac{dp}{dt} = C\frac{dp}{dt}.$$

Air at room temperature and low pressure can be approximated as an ideal gas. For an isothermal process, the pneumatic capacitance of the container is

$$C = \frac{V}{R_{air}T}.$$

The linear valve resistance R is defined as

$$R = \frac{p_i - p}{\rho_i q_i}.$$

Thus, the differential equation of the system is

$$\frac{V}{R_{air}T}\frac{dp}{dt} = \frac{p_i - p}{R}$$

or

$$\frac{RV}{R_{air}T}\frac{dp}{dt} + p = p_i$$

where $RV/(R_{air}T) = 1000 \times 1^3/(287.06 \times 293) = 1.19 \times 10^{-2}$ s.

b. Substituting the parameter values into the differential equation obtained in Part (a), we obtain

$$1.19 \times 10^{-2}\dot{p} + p = 101.325 \times 10^3.$$

Solving for the highest derivative of the output p gives

$$\dot{p} = 84.03(101{,}325 - p),$$

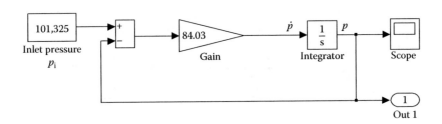

FIGURE 7.3 A Simulink block diagram representing the pneumatic system in Figure 7.2.

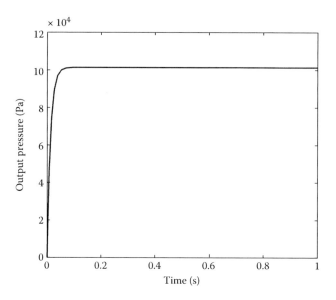

FIGURE 7.4 Pressure output $p(t)$ for constant inlet pressure $p_i = 101.325$ kPa.

which can be represented using the block diagram shown in Figure 7.3. One `Integrator` block is used to form the container pressure p, which is fed back to form the variation rate $\dot{p}$. Note that the system input is the inlet pressure p_i, which is constant and is represented using a `Constant` block. Run the simulation. Double-click the `Scope` block and the resulting output of the pneumatic system $p(t)$ is shown in Figure 7.4.

PROBLEM SET 7.1

1. A car has an internal volume of 2.5 m³. If the sun heats the car from a temperature of 20°C to a temperature of 55°C, what will the pressure inside the car be? Assume the pressure is initially 1 atm.
2. Find the pneumatic capacitance of dry air in a rigid container with volume 13.5 ft³ for an isothermal process. Assume the air is initially at ambient temperature of 68°F.
3. Dry air at a constant temperature of 20°C passes through a valve out of a rigid cubic container of 1 m on each side (see Figure 7.5). The pressure p_o at the outlet of the valve is constant, and it is less than p. The valve resistance is approximately linear, and $R = 1000$ Pa·s/kg. Assume the process is isothermal.
   a. Develop a mathematical model of the pressure p in the container.
   b. Construct a Simulink block diagram to find the output $p(t)$ of the pneumatic system if the pressure inside the container initially is 2 atm and the pressure at the outlet is 1 atm.

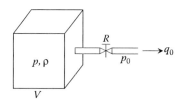

FIGURE 7.5 Problem 3.

Fluid and Thermal Systems

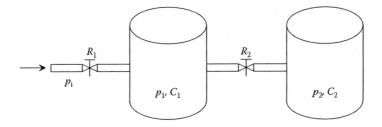

FIGURE 7.6 Problem 4.

4. Figure 7.6 shows a pneumatic system, where the pneumatic capacitances of the two rigid containers are C_1 and C_2, respectively. Dry air at a constant temperature passes through a valve of linear resistance R_1 into container 1. The pressure p_i at the inlet of the valve is constant, and it is greater than p_1. The air flows from container 1 to container 2 through a valve of linear resistance R_2. Derive the differential equations in terms of the pressures p_1 and p_2. Write the equations in second-order matrix form.

7.2 LIQUID-LEVEL SYSTEMS

Unlike gases, most liquids are generally considered incompressible, and this approximation greatly simplifies the modeling of hydraulic systems. A general category of hydraulic systems is liquid-level systems, which often appear in water treatment, water supply, and other chemical processing applications. Such a system usually consists of storage tanks interconnected to other systems through valves, pumps, or pistons.

The dynamic behavior of a liquid-level system can be described using volume flow rate q, pressure p, and liquid height h. Note that the hydrostatic pressure, rather than the dynamic pressure, will be used in the modeling of liquid-level systems. The hydrostatic pressure is defined as the pressure that exists in a fluid at rest. It is caused by the weight of the fluid. For a liquid of density ρ, the absolute pressure p and the liquid height h are related by

$$p = p_a + \rho g h, \qquad (7.17)$$

where p_a is the atmospheric pressure.

7.2.1 Hydraulic Capacitance

As introduced in Section 7.1, fluid capacitance is the ratio of the change in stored mass to the change in pressure. Because the density for an incompressible liquid is constant, the change in mass is equivalent to the change in volume. Some books define the capacitance for hydraulic systems in terms of volume instead of mass,

$$C_V = \frac{dV}{dp}, \qquad (7.18)$$

which is related to Equation 7.4 as $C = \rho C_V$. The definition given in Equation 7.4 is used for both pneumatic and hydraulic systems in this text.

To find the expression of capacitance for a hydraulic system, let us consider a container, whose cross-sectional area varies with the liquid height. The mass stored in the container can be determined by integrating $\rho A(h)$ from the base of the container to the top of the liquid,

$$m = \int_0^h \rho A(y)\,dy. \tag{7.19}$$

From Equation 7.4, we have

$$C = \frac{dm}{dp} = \frac{dm}{dh}\frac{dh}{dp}. \tag{7.20}$$

Note that Equation 7.19 implies that

$$\frac{dm}{dh} = \rho A(h). \tag{7.21}$$

Also, Equation 7.17 gives

$$\frac{dp}{dh} = \frac{d}{dh}(p_a + \rho g h) = \rho g. \tag{7.22}$$

Thus, the hydraulic capacitance of the container is

$$C = \rho A(h)\frac{1}{\rho g} = \frac{A(h)}{g}. \tag{7.23}$$

If the cross-sectional area of the container is constant, then we have $C = A/g$. Different from the pneumatic capacitance (see Equation 7.9), which depends on the type of gas and its temperature, the hydraulic capacitance does not depend on any liquid properties.

Example 7.3: Hydraulic Capacitance of a Conical Tank

Derive the capacitance of the conical tank shown in Figure 7.7a using
a. $C = dm/dp$.
b. $C = A(h)/g$.

Solution

a. From Figure 7.7b, the radius r of the cross section A at an arbitrary height is

$$r = h\tan\alpha = h\frac{R}{H}.$$

Thus, the volume of the liquid is

$$V(h) = \frac{1}{3}\pi r^2 h = \frac{1}{3}\frac{\pi R^2}{H^2}h^3.$$

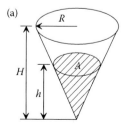 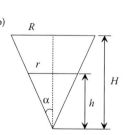

FIGURE 7.7 A conical tank. (a) Three-dimensional view and (b) cross-sectional view.

Fluid and Thermal Systems

and the stored mass is

$$m = \rho V(h) = \frac{1}{3} \frac{\rho \pi R^2}{H^2} h^3.$$

Note that the pressure caused by the height of the liquid is

$$p = p_a + \rho g h,$$

which gives

$$\frac{dp}{dh} = \rho g.$$

Thus, the capacitance of the conical tank is

$$C = \frac{dm}{dp} = \frac{dm}{dh}\frac{dh}{dp} = \left(\frac{\rho \pi R^2}{H^2} h^2\right)\left(\frac{1}{\rho g}\right) = \frac{\pi R^2 h^2}{H^2 g}.$$

b. The hydraulic capacitance can also be derived directly using $C = A(h)/g$, which gives

$$C = \frac{\pi r^2}{g} = \left(\frac{\pi}{g}\right)\left(\frac{R^2 h^2}{H^2}\right) = \frac{\pi R^2 h^2}{H^2 g}.$$

7.2.2 Hydraulic Resistance

When liquid flows through a pipe, a valve, or an orifice, the liquid meets resistance that creates a drop in the pressure of the liquid. The pressure difference is associated with the mass flow rate q_m in a nonlinear relationship, $p = f(q_m)$, as illustrated in Figure 7.8. The slope of the curve is defined as the hydraulic resistance R, which depends on the reference mass flow rate q_{mr} and reference pressure p_r. The expression of the hydraulic resistance R is given by

$$R = \left.\frac{dp}{dq_m}\right|_{(q_{mr}, p_r)}, \qquad (7.24)$$

which is also used for pneumatic systems. Near a reference operating point (q_{mr}, p_r), we can perform linearization and obtain the linearized resistance, which is

$$R = \frac{\Delta p}{\Delta q_m} = \frac{p - p_r}{q_m - q_{mr}}. \qquad (7.25)$$

The resistance due to a valve, an orifice, or pipe wall friction can be represented using the valve-like symbol as shown in Figure 7.9. In many fluid systems, multiple valves, orifices, or pipes are

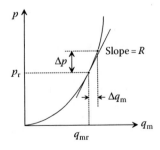

FIGURE 7.8 Linearized resistance near a reference point.

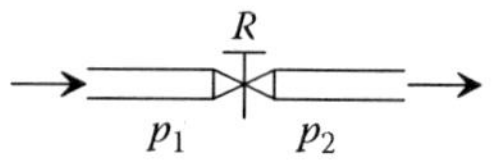

FIGURE 7.9 A symbol for hydraulic resistance.

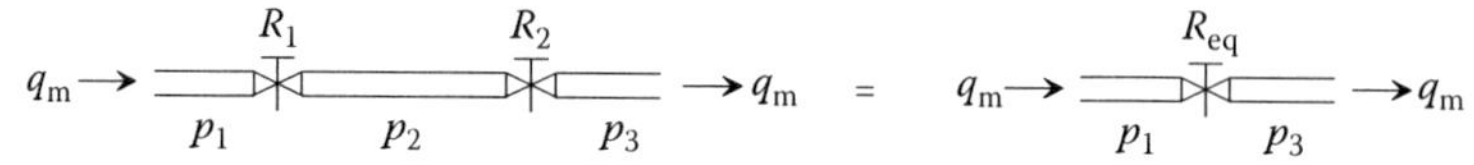

FIGURE 7.10 Equivalence for series hydraulic resistances.

used. They are arranged in different ways, such as in series connections or parallel connections. The equivalent linear hydraulic resistances can be obtained similar to electrical resistances.

Figure 7.10 shows resistances in series. Note that the mass flow rate remains the same through each resistance. The pressure drops across the resistances R_1 and R_2 are $p_1 - p_2 = R_1 q_m$ and $p_2 - p_3 = R_2 q_m$, respectively. Consequently, the total pressure drop across the two resistances in series is $p_1 - p_3 = (R_1 + R_2) q_m$. Comparing this result with the equivalent fluid system, $p_1 - p_2 = R_{eq} q_m$, we have

$$R_{eq} = R_1 + R_2. \tag{7.26}$$

A fluid system with resistances in parallel is shown in Figure 7.11. Note that the pressure drop across each resistance must be the same. The mass flow rates through the resistances R_1 and R_2 are $q_{m1} = (p_1 - p_2)/R_1$ and $q_{m2} = (p_1 - p_2)/R_2$, respectively. Therefore, the total mass flow rate through the two resistances in parallel is $q_{m1} + q_{m2} = (p_1 - p_2)(1/R_1 + 1/R_2)$. Comparing this result with the equivalent fluid system, $q_m = (p_1 - p_2)/R_{eq}$, we have

$$\frac{1}{R_{eq}} = \frac{1}{R_1} + \frac{1}{R_2}. \tag{7.27}$$

7.2.3 Modeling of Liquid-Level Systems

To obtain a simple model of a liquid-level system, we will use a capacitance element to represent each storage tank and a resistance element to represent each valve in the system. The resulting mathematical model may adequately describe the dynamics of the real system. Similar to the modeling of pneumatic systems, the basic law used to derive the differential equation of a liquid-level system is the law of conservation of mass presented in Equation 7.11: $dm/dt = q_{mi} - q_{mo}$. That is, the time rate of change of fluid mass in a tank equals the mass flow rate into the tank minus the mass flow rate out of the tank.

Consider a single tank with a valve as shown in Figure 7.12, where p_a is the atmospheric pressure, and q_i and q_o are the volume flow rates into and out of the tank, respectively. The cross-sectional

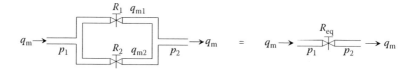

FIGURE 7.11 Equivalence for parallel hydraulic resistances.

Fluid and Thermal Systems

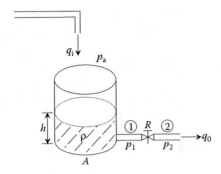

FIGURE 7.12 A single-tank liquid-level system with a valve.

area of the tank is A and the liquid height is h. The liquid leaves the tank through the valve, for which the hydraulic resistance is R. The density of the liquid is constant ρ.

Next, we will show how to derive the differential equation of the system by applying the law of conservation of mass presented in Equation 7.11. Note that the total fluid mass in the tank is $\rho A h$. For constant cross-sectional area and constant density, the left-hand side of Equation 7.11 can be rewritten as

$$\frac{dm}{dt} = \frac{d}{dt}(\rho A h) = \rho A \frac{dh}{dt}. \tag{7.28}$$

The right-hand side of Equation 7.11 can be rewritten as

$$q_{mi} - q_{mo} = \rho q_i - \rho q_o. \tag{7.29}$$

Labeling point 1 at the upstream side of the valve and point 2 at the downstream side of the valve, the hydraulic resistance of the valve can be expressed as

$$R = \frac{\Delta p}{\Delta q_m} = \frac{p_1 - p_2}{\rho q_o}. \tag{7.30}$$

Assume that the pressure p_1 can be approximated as the hydrostatic pressure, $p_1 = p_a + \rho g h$, and the pressure p_2 can be approximated as the atmospheric pressure, $p_2 = p_a$. Substituting p_1 and p_2 into Equation 7.30 gives

$$R = \frac{\rho g h}{\rho q_o}. \tag{7.31}$$

We then have

$$\rho q_i - \rho q_o = \rho q_i - \frac{\rho g h}{R} \tag{7.32}$$

or

$$q_{mi} - q_{mo} = \rho q_i - \frac{\rho g h}{R}. \tag{7.33}$$

Substituting Equations 7.28 and 7.33 into Equation 7.11 results in

$$\rho A \frac{dh}{dt} = \rho q_i - \frac{\rho g h}{R}. \tag{7.34}$$

Rearranging the equation gives

$$\frac{RA}{g}\frac{dh}{dt} + h = \frac{R}{g} q_i, \tag{7.35}$$

which is a first-order ordinary differential equation relating the liquid height h and the inlet pressure q_i. Introducing the expression of the hydraulic capacitance C in Equation 7.23, Equation 7.35 can be written as

$$RC\frac{dh}{dt} + h = \frac{R}{g}q_i. \tag{7.36}$$

Equation 7.36 describes the dynamic behavior of a single-tank liquid-level system with capacitance C and resistance R as shown schematically in Figure 7.12.

Hydraulic systems are usually connected with pumps, which can be considered as pressure sources. The following example shows how to derive the governing differential equation for a single-tank liquid-level system with a pump.

Example 7.4: A Single-Tank Liquid-Level System with a Pump

Consider the single-tank liquid-level system shown in Figure 7.13, where a pump is connected to the bottom of the tank through a valve of linear resistance R. The inlet to the pump is open to the atmosphere, and the pressure of the fluid increases by Δp when crossing the pump. Derive the differential equation relating the liquid height h and the volume flow rate q_o at the outlet. The tank's cross-sectional area A is constant. The density ρ of the liquid is constant.

Solution

We begin by applying the law of conservation of mass to the tank,

$$\frac{dm}{dt} = q_{mi} - q_{mo}.$$

The fluid mass inside the tank is $\rho A h$. For constant fluid density and constant cross-sectional area,

$$\frac{dm}{dt} = \rho A \frac{dh}{dt}.$$

The mass flow rate into the tank is

$$q_{mi} = \frac{p_1 - p_2}{R},$$

where $p_1 = p_a + \Delta p$ and $p_2 = p_a + \rho g h$, which is equal to the hydrostatic pressure at the bottom of the tank. Thus,

$$q_{mi} = \frac{\Delta p - \rho g h}{R}.$$

The mass flow rate out of the tank can be expressed in terms of the volume flow rate q_o as

$$q_{mo} = \rho q_o.$$

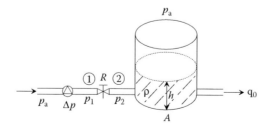

FIGURE 7.13 A single-tank liquid-level system with a pump.

Fluid and Thermal Systems

Substituting these expressions into the law of conservation of mass gives

$$\rho A \frac{dh}{dt} = \frac{\Delta p - \rho g h}{R} - \rho q_o.$$

Rearranging the equation gives

$$\rho A \frac{dh}{dt} + \frac{\rho g}{R} h - \frac{\Delta p}{R} = -\rho q_o.$$

For a liquid-level system with two or more tanks, we apply the law of conservation of mass to each tank.

Example 7.5: A Two-Tank Liquid-Level System

Figure 7.14 shows a liquid-level system, in which two tanks have cross-sectional areas A_1 and A_2, respectively. A pump is connected to the bottom of tank 1 through a valve of linear resistance R_1. The liquid flows from tank 1 to tank 2 through a valve of linear resistance R_2, and leaves tank 2 through a valve of linear resistance R_3. The density ρ of the liquid is constant.

a. Derive the differential equations in terms of the liquid heights h_1 and h_2. Write the equations in second-order matrix form.
b. Assume the pump pressure Δp as the input and the liquid heights h_1 and h_2 as the outputs. Determine the state-space form of the system.
c. Construct a Simulink block diagram to find the outputs $h_1(t)$ and $h_2(t)$ of the liquid-level system. Assume $\rho = 1000$ kg/m^3, $g = 9.81$ m/s^2, $A_1 = 2$ m^2, $A_2 = 3$ m^2, $R_1 = R_2 = R_3 = 400$ N·s/(kg·m^2), and initial liquid heights $h_1(0) = 1$ m and $h_2(0) = 0$ m. The pump pressure Δp is a step function with a magnitude of 0 before $t = 0$ s and a magnitude of 130 kPa after $t = 0$ s.

Solution

a. Applying the law of conservation of mass to tank 1 gives

$$\frac{dm}{dt} = q_{mi} - q_{mo},$$

where

$$\frac{dm}{dt} = \rho A_1 \frac{dh_1}{dt}.$$

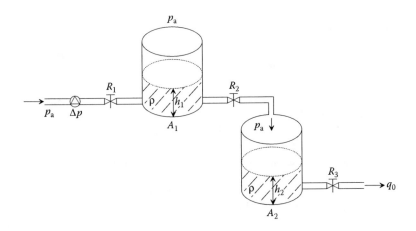

FIGURE 7.14 A two-tank liquid-level system.

The mass flow rate entering and leaving tank 2 can be written as

$$q_{mi} = \frac{(p_a + \Delta p) - (p_a + \rho g h_1)}{R_1} = \frac{\Delta p - \rho g h_1}{R_1}$$

and

$$q_{mo} = \frac{(p_a + \rho g h_1) - p_a}{R_2} = \frac{\rho g h_1}{R_2}.$$

Substituting these expressions results in

$$\rho A_1 \frac{dh_1}{dt} = \frac{\Delta p - \rho g h_1}{R_1} - \frac{\rho g h_1}{R_2},$$

which can be rearranged as

$$\rho A_1 \frac{dh_1}{dt} + \rho g h_1 \left(\frac{1}{R_1} + \frac{1}{R_2}\right) = \frac{\Delta p}{R_1}.$$

Applying the law of conservation of mass to tank 2 gives

$$\frac{dm}{dt} = q_{mi} - q_{mo},$$

where

$$\frac{dm}{dt} = \rho A_2 \frac{dh_2}{dt}.$$

The mass flow rate entering and leaving tank 2 can be written as

$$q_{mi} = \frac{(p_a + \rho g h_1) - p_a}{R_2} = \frac{\rho g h_1}{R_2}$$

and

$$q_{mo} = \frac{(p_a + \rho g h_2) - p_a}{R_3} = \frac{\rho g h_2}{R_3}.$$

Substituting these expressions results in

$$\rho A_2 \frac{dh_2}{dt} = \frac{\rho g h_1}{R_2} - \frac{\rho g h_2}{R_3},$$

which can be rearranged as

$$\rho A_2 \frac{dh_2}{dt} - \frac{\rho g h_1}{R_2} + \frac{\rho g h_2}{R_3} = 0.$$

The system of differential equations in second-order matrix form is found to be

$$\begin{bmatrix} \rho A_1 & 0 \\ 0 & \rho A_2 \end{bmatrix} \begin{Bmatrix} \dfrac{dh_1}{dt} \\ \dfrac{dh_2}{dt} \end{Bmatrix} + \begin{bmatrix} \dfrac{\rho g}{R_1} + \dfrac{\rho g}{R_2} & 0 \\ -\dfrac{\rho g}{R_2} & \dfrac{\rho g}{R_3} \end{bmatrix} \begin{Bmatrix} h_1 \\ h_2 \end{Bmatrix} = \begin{Bmatrix} \dfrac{\Delta p}{R_1} \\ 0 \end{Bmatrix}.$$

b. As specified, the state, the input, and the output are

$$\mathbf{x} = \begin{Bmatrix} x_1 \\ x_2 \end{Bmatrix} = \begin{Bmatrix} h_1 \\ h_2 \end{Bmatrix}, \quad u = \Delta p, \quad \mathbf{y} = \begin{Bmatrix} h_1 \\ h_2 \end{Bmatrix}.$$

Fluid and Thermal Systems

The state-variable equations are

$$\dot{x}_1 = \frac{dh_1}{dt} = -\frac{g}{A_1}\left(\frac{1}{R_1} + \frac{1}{R_2}\right)h_1 + \frac{1}{\rho A_1 R_1}\Delta p,$$

$$\dot{x}_2 = \frac{dh_2}{dt} = \frac{g}{A_2 R_2}h_1 - \frac{g}{A_2 R_3}h_2,$$

or in matrix form

$$\begin{Bmatrix}\dot{x}_1 \\ \dot{x}_2\end{Bmatrix} = \begin{bmatrix} -\frac{g}{A_1}\left(\frac{1}{R_1} + \frac{1}{R_2}\right) & 0 \\ \frac{g}{A_2 R_2} & -\frac{g}{A_2 R_3} \end{bmatrix}\begin{Bmatrix}x_1 \\ x_2\end{Bmatrix} + \begin{bmatrix}\frac{1}{\rho A_1 R_1} \\ 0\end{bmatrix}u.$$

The output equation is

$$\begin{Bmatrix}y_1 \\ y_2\end{Bmatrix} = \begin{bmatrix}1 & 0 \\ 0 & 1\end{bmatrix}\begin{Bmatrix}x_1 \\ x_2\end{Bmatrix} + \begin{bmatrix}0 \\ 0\end{bmatrix}u.$$

c. The Simulink block diagram can be constructed based on either the differential equations obtained in Part (a) or the state-space form obtained in Part (b). Substituting the values of the parameters into the differential equations gives

$$\frac{dh_1}{dt} = -0.0245 h_1 + 1.25 \times 10^{-6}\Delta p,$$

$$\frac{dh_2}{dt} = 0.0082 h_1 - 0.0082 h_2.$$

Figure 7.15 shows the resulting Simulink block diagram, where two Integrator blocks are used to form the time derivatives dh_1/dt and dh_2/dt. Double-clicking on each Integrator block, we can enter the initial liquid level for each tank.

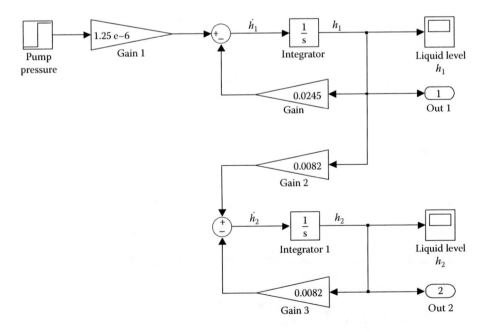

FIGURE 7.15 The Simulink block diagram constructed based on the differential equations.

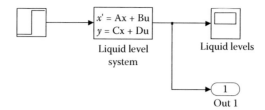

FIGURE 7.16 The Simulink block diagram constructed based on state-space equations.

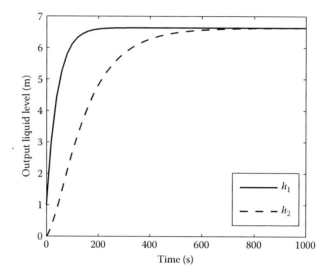

FIGURE 7.17 Liquid level outputs $h_1(t)$ and $h_2(t)$.

The Simulink block diagram based on the state-space form is shown in Figure 7.16, where a **State-Space** block is used to represent the liquid-level system. Double-clicking on the **State-Space** block with the name `Liquid-level system`, we can define the matrices **A**, **B**, **C**, and **D**, which are [−0.0245 0; 0.0082 −0.0082], [1.25e−6; 0], [1 0; 0 1], and [0; 0], respectively. The initial liquid level is a vector, which is [1; 0]. Running either of the two simulations, we can obtain the same results as plotted in Figure 7.17.

PROBLEM SET 7.2

1. Derive the capacitance of the spherical tank as shown in Figure 7.18. There is an opening at the top of the tank at height H, where $R < H < 2R$.
2. Derive the capacitance of the tank shown in Figure 7.19a. The cross section of the tank is shown in Figure 7.19b.
3. Determine the equivalent hydraulic resistance for the system shown in Figure 7.20.
4. Determine the equivalent hydraulic resistance for the system shown in Figure 7.21.
5. Consider the single-tank liquid-level system shown in Figure 7.22, where the volume flow rate into the tank through a pipe is q_i. A pump is connected to the bottom of the tank through a valve of linear resistance R_1. The pressure of the fluid increases by Δp when crossing the pump. The liquid leaves the tank through a valve of linear resistance R_2. Derive the differential equation relating the liquid height h and the volume flow rate q_i at the inlet. The tank's cross-sectional area is constant. The density ρ of the liquid is constant.

Fluid and Thermal Systems

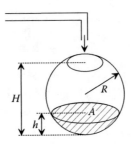

FIGURE 7.18 Problem 1.

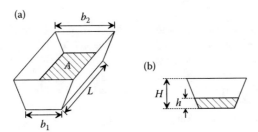

FIGURE 7.19 Problem 2. (a) Three-dimensional view and (b) cross-sectional view.

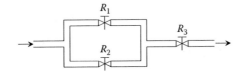

FIGURE 7.20 Problem 3.

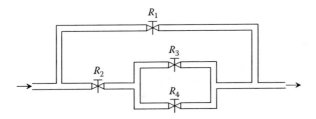

FIGURE 7.21 Problem 4.

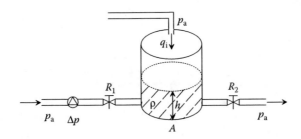

FIGURE 7.22 Problem 5.

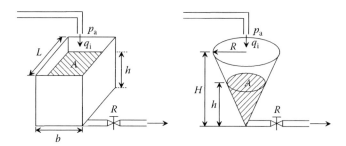

FIGURE 7.23 Problem 6.

6. Consider the rectangular tank in Figure 7.23a and the conical tank in Figure 7.23b. The volume flow rate into each tank through a pipe is q_i. The liquid leaves each tank through a valve of linear resistance R. The density ρ of the liquid is constant.
   a. Derive the dynamic model of the liquid height h for each tank.
   b. For each tank, write the differential equation in terms of hydraulic capacitance and hydraulic resistance. Compare the models for the two single-tank systems.
7. Figure 7.24 shows a liquid-level system in which two tanks have hydraulic capacitances C_1 and C_2, respectively. The volume flow rate into tank 1 is q_i. The liquid flows from tank 1 to tank 2 through a valve of linear resistance R_1 and leaves tank 2 through a valve of linear resistance R_2. The density ρ of the liquid is constant.
   a. Derive the differential equations in terms of the liquid heights h_1 and h_2. Write the equations in second-order matrix form.
   b. Assume the volume flow rate q_i as the input and the liquid heights h_1 and h_2 as the outputs. Determine the state-space form of the system.
   c. Construct a Simulink block diagram to find the outputs $h_1(t)$ and $h_2(t)$ of the liquid-level system. Assume $\rho = 1000 \text{ kg/m}^3$, $g = 9.81 \text{ m/s}^2$, $C_1 = 0.2 \text{ kg·m}^2/\text{N}$,

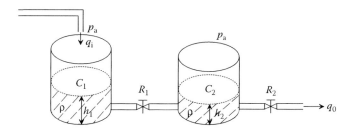

FIGURE 7.24 Problem 7.

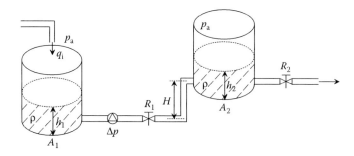

FIGURE 7.25 Problem 8.

Fluid and Thermal Systems

$C_2 = 0.3 \text{ kg·m}^2/\text{N}$, $R_1 = R_2 = 400 \text{ N·s}/(\text{kg·m}^2)$, and initial liquid heights $h_1(0) = 1$ m and $h_2(0) = 0$ m. The volume flow rate q_i is a step function with a magnitude of 0 before $t = 0$ s and a magnitude of $0.5 \text{ m}^3/\text{s}$ after $t = 0$ s.

8. Figure 7.25 shows a liquid-level system, in which two tanks have cross-sectional areas A_1 and A_2, respectively. The volume flow rate into tank 1 is q_i. A pump is connected to the bottom of tank 1 and the pressure of the fluid increases by Δp when crossing the pump. Tank 2 is located higher than tank 1 and the vertical distance between the two tanks is H. The liquid is pumped from tank 1 to tank 2 through a valve of linear resistance R_1 and leaves tank 2 through a valve of linear resistance R_2. The density ρ of the liquid is constant. Derive the differential equations in terms of the liquid heights h_1 and h_2. Write the equations in second-order matrix form.

7.3 THERMAL SYSTEMS

Thermal systems are those that involve the transfer of heat from one object to another. When an object is at a different temperature than its surroundings or another object, heat will transfer from the higher-temperature object to the lower-temperature one, obeying the law of conservation of energy. Examples of thermal systems include heaters, air conditioners, refrigerators, and so on. Just like with fluid systems (either pneumatic or hydraulic), for which fluid capacitance, fluid resistance, and the conservation of mass are the basis for system modeling, we will introduce thermal capacitance, thermal resistance, and the conservation of energy, which together form the basis of modeling for thermal systems.

7.3.1 First Law of Thermodynamics

The first law of thermodynamics is an expression of the more general physical law of conservation of energy. For a system with well-defined boundaries, the law of conservation of energy states

$$\Delta E = Q - W, \tag{7.37}$$

where ΔE is the change in energy of the system, Q is the heat flow into or out of the system, and W is the work done by or on the system. In Equation 7.37, Q is positive if heat is supplied to the system and negative if heat is given off by the system. W is positive if work is done by the system and negative if work is done to the system. Based on this sign convention, we have

$$\Delta E = (Q_{in} - Q_{out}) - (W_{out} - W_{in}), \tag{7.38}$$

where Q_{in}, Q_{out}, W_{in}, and W_{out} are all positive quantities.

In actuality, the net amount of energy added to the system is equal to the net increase in the energy stored internally in the system and any change in the mechanical energy of the system's center of mass,

$$\Delta E = \Delta U + \Delta ME_C. \tag{7.39}$$

U is the internal energy (or internal thermal energy), which is the energy stored at the molecular level. It includes the kinetic energy due to the motion of molecules and the potential energy that holds the atoms together. ME_C stands for the mechanical energy, which includes the kinetic energy and the potential energy associated with the system's mass center. For systems where the change in mechanical energy is negligible,

$$\Delta U = Q - W = (Q_{in} - Q_{out}) - (W_{out} - W_{in}), \tag{7.40}$$

which is the mathematical expression of the first law of thermodynamics. It basically states that the change in internal energy is equal to the amount of energy gained by heating minus the amount lost by doing work on the environment.

Heat Q is the energy transfer at the molecular level. Work W is the energy transfer that is capable of producing macroscopic mechanical motion of the system's mass center. For thermal systems with pure heat transfer and no work involved, that is, $W_{\text{in}} = W_{\text{out}} = 0$, the law of conservation of energy presented in Equation 7.40 can be rewritten as

$$\Delta U = Q = Q_{\text{in}} - Q_{\text{out}} \tag{7.41}$$

or

$$\frac{dU}{dt} = q_{\text{hi}} - q_{\text{ho}}, \tag{7.42}$$

where $q_h = dQ/dt$ is the heat flow rate having units of J/s, which is a watt or ft·lb/s.

7.3.2 Thermal Capacitance

For an object, the thermal capacitance C is defined as the ratio of the change in heat flow to the change in the object's temperature,

$$C = \frac{dQ}{dT}, \tag{7.43}$$

where C has units of J/K, J/°C, or ft·lb/°F. The thermal capacitance is a measure of the heat required to increase the temperature of an object by a certain temperature interval.

Strictly speaking, the value of the thermal capacitance of a substance depends on thermodynamic processes. For a constant-volume process, no work is involved and all the heat goes into the internal energy of the substance,

$$Q = \Delta U = mc_v \Delta T, \tag{7.44}$$

where m is the mass of the substance, c_v is the constant-volume specific heat capacity of the substance in units of J·K/kg, J°C/kg, or ft·lb°F/slug, and ΔT is the change in temperature of the substance. For a constant-pressure process,

$$Q = \Delta H = mc_p \Delta T, \tag{7.45}$$

where H is the enthalpy and c_p is the constant-pressure specific heat capacity. Combining Equation 7.43 with Equation 7.44 or 7.45 gives

$$C = mc_v \tag{7.46}$$

or

$$C = mc_p. \tag{7.47}$$

For incompressible liquids and solids, since the volume cannot expand, the heat flow in a constant-pressure process is equal to the internal energy: $Q = \Delta U$ and $c_p = c_v$. Assuming that the density and the volume of the mass are ρ and V, respectively, we have

$$C = mc = \rho V c. \tag{7.48}$$

where c is the specific heat capacity and $c = c_p = c_v$. The subscripts p and v will be dropped in the rest of the chapter for simplicity.

Note that the value of the specific heat capacity c depends on the substance of the object, while the thermal capacitance C is an extensive property because its value is proportional to the mass of the object. For instance, the specific heat capacity of water at room temperature (25°C) is a constant value of 4186 J/(kg°C). However, the thermal capacitance for a bathtub of water is greater than that for a cup of water.

Fluid and Thermal Systems

7.3.3 Thermal Resistance

There are three mechanisms by which heat is transported: conduction, convection, and radiation. Conduction is the transfer of heat between substances that are in direct contact with each other. Convection is the transfer of heat due to a flowing fluid, which can be a gas or a liquid. Radiation is the transfer of heat through empty space. Here, we only consider conduction and convection.

The thermal resistance R for heat transfer is defined as the ratio of the change in temperature difference to the change in heat flow rate,

$$R = \frac{dT}{dq_h}, \tag{7.49}$$

The thermal resistance R has units of K·s/J, °C·s/J, or °F·s/(ft·lb).

For simple one-dimensional conduction as shown in Figure 7.26, Fourier's law, also known as the law of heat conduction, gives

$$q_h = kA\frac{\Delta T}{L} = kA\frac{T_1 - T_2}{L}, \tag{7.50}$$

where L is the length of the body in the direction of heat flow, A is the cross-sectional area normal to the heat flow direction, ΔT is the temperature difference along its length, and k is the thermal conductivity of the material in $W/(m \cdot K)$ or $W/(m \cdot °C)$. Note that the heat flow is in the direction of decreasing temperature. Fourier's equation is often used only for solids, although it is valid for both solids and fluids. Combining Equations 7.49 and 7.50 gives the thermal resistance for conduction

$$R = \frac{L}{kA}. \tag{7.51}$$

For convective heat transfer, Newton's law of cooling states that the rate of heat flow of a body is proportional to the difference in temperatures between the body and its surroundings or environment. The mathematical expression is

$$q_h = hA\Delta T = hA(T_s - T_{env}), \tag{7.52}$$

where A is the surface area, from which the heat is being transferred, T_s is the temperature of the body's surface, T_{env} is the temperature of the environment, and h is the heat transfer coefficient in $W/(m^2 \cdot K)$ or $W/(m^2 \cdot °C)$. Combining Equations 7.49 and 7.52 gives the thermal resistance for convection

$$R = \frac{1}{hA}. \tag{7.53}$$

It is very useful to utilize the concept of thermal resistance and represent heat transfer by thermal circuits. The heat flow rate q_h is analogous to the current, the temperature difference ΔT is analogous to the voltage, and the thermal resistance is analogous to the electric resistance.

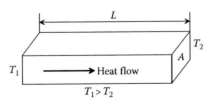

FIGURE 7.26 One-dimensional conduction: heat flow from higher to lower temperature.

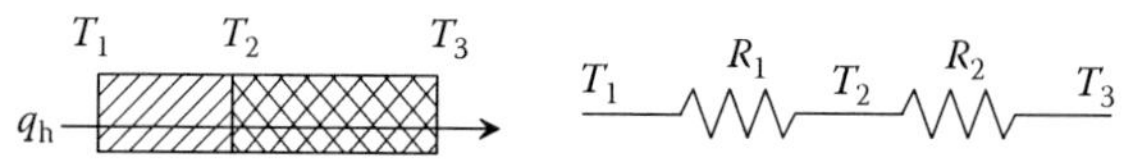

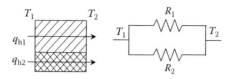

FIGURE 7.27 Heat transfer across series thermal resistances.

FIGURE 7.28 Heat transfer across parallel thermal resistances.

Figure 7.27 shows the heat transfer across a composite slab, which can be represented using a thermal resistance network with series interconnection. The heat flow rate remains the same through each component. Assume that the thermal resistances are R_1 and R_2. As a result, the temperature differences across the resistances R_1 and R_2 are $T_1 - T_2 = R_1 q_h$ and $T_2 - T_3 = R_2 q_h$, respectively. The total temperature difference across the composite slab is $T_1 - T_3 = (R_1 + R_2) q_h$. Thus, the equivalent thermal resistance for a series interconnection is

$$R_{eq} = R_1 + R_2. \tag{7.54}$$

Figure 7.28 shows heat transfer across a wall with different materials and how it can be represented by using a thermal resistance network with parallel interconnection. Note that the temperature drop across each material must be the same. Consequently, the heat flow rates through the resistances R_1 and R_2 are $q_{h1} = (T_1 - T_2)/R_1$ and $q_{h2} = (T_1 - T_2)/R_2$, respectively. The total heat flow rate through the wall is $q_{h1} + q_{h2} = (T_1 - T_2)(1/R_1 + 1/R_2)$. Thus, the equivalent thermal resistance for a parallel interconnection is

$$\frac{1}{R_{eq}} = \frac{1}{R_1} + \frac{1}{R_2}. \tag{7.55}$$

If there are different heat transfer modes in a system, a thermal circuit with thermal resistances representing the different modes of heat transfer can be used to analyze the system.

Example 7.6: Thermal Resistance

Consider heat transfer through an insulated wall as shown in Figure 7.29. The wall is made of a layer of brick with thermal conductivity k_1 and two layers of foam with thermal conductivity k_2

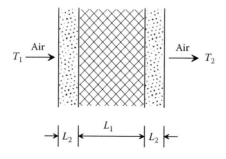

FIGURE 7.29 Heat transfer through an insulated wall.

Fluid and Thermal Systems

for insulation. The left surface of the wall is at temperature T_1 and exposed to air with heat transfer coefficient h_1. The right surface of the wall is at temperature T_2 and exposed to air with heat transfer coefficient h_2. Assume that $k_1 = 0.5 \, \text{W/(m·°C)}$, $k_2 = 0.17 \, \text{W/(m·°C)}$, $h_1 = h_2 = 10 \, \text{W/(m}^2\text{·°C)}$, $T_1 = 38°C$, and $T_2 = 20°C$. The thickness of the brick layer is 0.1 m, the thickness of each foam layer is 0.03 m, and the cross-sectional area of the wall is 16 m².

a. Determine the heat flow rate through the wall.
b. Determine the temperature distribution through the wall.

Solution

a. The heat transfer through the insulated wall can be represented using a thermal circuit with five thermal resistances connected in series as shown in Figure 7.30. Two modes of heat transfer, conduction and convection, are involved. The corresponding thermal resistances are

$$R_1 = \frac{1}{h_1 A} = \frac{1}{10 \times 16} = 6.25 \times 10^{-3} \, °\text{C·s/J},$$

$$R_2 = R_4 = \frac{L_2}{k_2 A} = \frac{0.03}{0.17 \times 16} = 1.10 \times 10^{-2} \, °\text{C·s/J},$$

$$R_3 = \frac{L_1}{k_1 A} = \frac{0.1}{0.5 \times 16} = 1.25 \times 10^{-2} \, °\text{C·s/J},$$

$$R_5 = \frac{1}{h_2 A} = \frac{1}{10 \times 16} = 6.25 \times 10^{-3} \, °\text{C·s/J}.$$

The total thermal resistance is

$$R_{eq} = \sum_{i=1}^{5} R_i = 4.70 \times 10^{-2} \, °\text{C·s/J}$$

Thus, the heat flow rate through the insulated wall is

$$q_h = \frac{\Delta T}{R_{eq}} = \frac{38 - 20}{4.70 \times 10^{-2}} = 382.98 \, \text{W}$$

b. Note that the heat flow rate stays the same through the insulated wall. Thus, from left to right, the heat flow rate through each layer is

$$\text{Air:} \quad q_h = h_1 A(T_1 - T_3)$$

$$\text{Foam:} \quad q_h = \frac{k_2 A}{L_2}(T_3 - T_4)$$

$$\text{Brick:} \quad q_h = \frac{k_1 A}{L_1}(T_4 - T_5)$$

$$\text{Foam:} \quad q_h = \frac{k_2 A}{L_2}(T_5 - T_6)$$

$$\text{Air:} \quad q_h = h_2 A(T_6 - T_2)$$

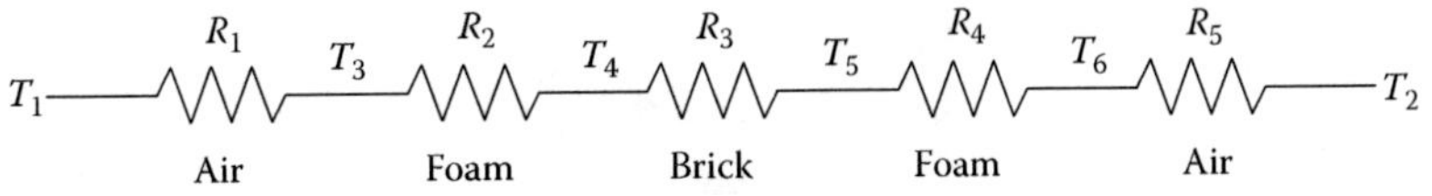

FIGURE 7.30 The equivalent thermal circuit for the heat transfer system in Figure 7.28.

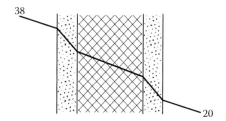

FIGURE 7.31 Temperature distribution through the insulated wall.

With the given temperatures T_1 and T_2, we have $T_3 = 35.61°C$, $T_4 = 31.40°C$, $T_5 = 26.61°C$, and $T_6 = 22.40°C$. Figure 7.31 shows the temperature distribution through the wall.

Note that the temperatures shown in Figure 7.31 are the values when the heat transfer process reaches steady state.

7.3.4 Modeling of Heat Transfer Systems

The mathematical model of a thermal system is often complicated because of the complex temperature distribution throughout the system. To simplify analysis, in this section, we discuss how to obtain a lumped-parameter model, which may approximate the gross system dynamics. The validity of this lumped-parameter assumption can be checked using the so-called Biot number.

Consider a solid body submerged in hot fluid as shown in Figure 7.32, where T_f is the fluid temperature, T_w is the temperature at the wall of the object, and T is the temperature of an arbitrary point inside the object. There are two modes of heat transfer involved, conduction within the body and convection between the fluid and the body. The heat flow rates for the two different modes can be approximated to have the same magnitude,

$$\frac{kA}{L}(T_w - T) \approx hA(T_f - T_w), \tag{7.56}$$

where h is the heat transfer coefficient, k is the thermal conductivity for the material of the object, and L is the relevant length between the wall and the point. The ratio of the temperature differences caused by the different modes of heat transfer is

$$\frac{T_w - T}{T_f - T_w} \approx \frac{hL}{k}. \tag{7.57}$$

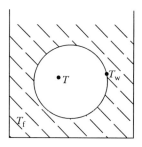

FIGURE 7.32 A solid object submerged in a fluid.

Fluid and Thermal Systems

Note that the temperature difference within the object can be negligible if the object is thin or small enough. The criterion for determining a solid being thin or small is based on the Biot number, which is defined as

$$Bi = \frac{hL_c}{k}, \tag{7.58}$$

where L_c is the characteristic length of the solid object. It is defined as the volume of the body divided by the surface area of the body,

$$L_c = \frac{V_{body}}{A_{surface}}. \tag{7.59}$$

For a body whose Biot number is much less than one, that is, $Bi < 0.1$, the interior of the body may be assumed to have the same temperature.

Example 7.7: Temperature Dynamics of a Heated Object

Consider a steel sphere with a radius of $r = 0.01$ m submerged in a hot water bath with a heat transfer coefficient of $h = 350$ W/(m²·°C). For steel, the density is $\rho = 7850$ kg/m³, the specific heat capacity is $c = 440$ J/(kg°·C), and the thermal conductivity is $k = 43$ W/(m·°C). The temperature of the water T_f is 100°C and the initial temperature of the sphere T_0 is 25°C.

  a. Determine whether the sphere's temperature can be considered uniform.
  b. Derive the differential equation relating the sphere's temperature $T(t)$ and the water's temperature T_f.
  c. Using the differential equation obtained in Part (b), construct a Simulink block diagram to find the sphere's temperature $T(t)$.

Solution

a. The characteristic length of the sphere is

$$L_c = \frac{V_{body}}{A_{surface}} = \frac{4/3\pi r^3}{4\pi r^2} = \frac{1}{3}r = \frac{0.01}{3}.$$

The Biot number of the steel sphere is

$$Bi = \frac{hL_c}{k} = \frac{350(0.01)}{43(3)} = 2.71 \times 10^{-2} < 0.1.$$

Thus, the sphere can be treated as a lump-temperature system, and its temperature can be considered uniform within the body.

b. The dynamic model of the sphere's temperature can be derived using the law of conservation of energy,

$$\frac{dU}{dt} = q_{hi} - q_{ho}.$$

Note that $U = mcT = \rho VcT$, and we have

$$\frac{dU}{dt} = \frac{d}{dt}(\rho VcT) = \rho Vc\frac{dT}{dt}.$$

The heat flow rate into the body is

$$q_{hi} = \frac{T_f - T}{R}$$

and the heat flow rate out of the body is $q_{ho} = 0$. Thus, the differential equation of the system is

$$\rho Vc(dT/dt) = (T_f - T)/R.$$

Introducing the expression for the thermal capacitance $\rho V c = C$, gives

$$RC\frac{dT}{dt} + T = T_f.$$

The thermal capacitance of the sphere is

$$C = \rho V c = 7850 \left(\frac{4}{3}\right)(\pi)(0.01)^3(440) = 14.47 \, J/°C,$$

and the thermal resistance due to convection is

$$R = \frac{1}{hA} = \frac{1}{350(4)(\pi)(0.01)^2} = 2.27 \, °C \, s/J.$$

Thus, the dynamic model of the sphere's temperature is

$$32.85\frac{dT}{dt} + T = T_f.$$

c. The Simulink block diagram for the thermal system can be constructed in the same way done for the pneumatic system in Example 7.2. Given $T_f = 100°C$, solving for the highest derivative of the output T gives

$$\frac{dT}{dt} = \frac{1}{32.85}(100 - T),$$

which can be represented using the block diagram shown in Figure 7.33. Double-click on the Integrator block and define the initial temperature of the sphere to be 25°C. Run the simulation. The results can be plotted as shown in Figure 7.34.

Example 7.7 shows that the temperature dynamics of a thermal system can be expressed in terms of thermal capacitance and thermal resistance. For multiple thermal capacitances, we apply the law of conservation of energy to each thermal capacitance.

Example 7.8: Temperature Dynamics of a House with a Heater

The room shown in Figure 7.35 has a heater with heat flow rate input q_0. The thermal capacitances of the heater and the room air are C_1 and C_2, respectively. The thermal resistances of the heater–air interface and the room wall-ambient air interface are R_1 and R_2, respectively. The temperatures of the heater and the room air are T_1 and T_2, respectively. The temperature outside the room is T_0, which is assumed to be constant.

a. Derive the differential equations relating the temperatures T_1, T_2, the input q_0, and the outside temperature T_0.

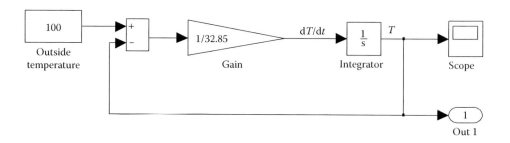

FIGURE 7.33 A Simulink block diagram representing the thermal system in Example 7.7.

Fluid and Thermal Systems

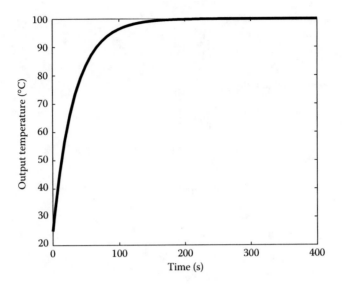

FIGURE 7.34 Temperature output $T(t)$ of the thermal system in Example 7.7.

b. Using the differential equations obtained in Part (a), determine the state-space form of the system. Assume the temperatures T_1 and T_2 as the outputs.

Solution

a. Applying the law of conservation of energy to the heater, we have

$$\frac{dU}{dt} = q_{hi} - q_{ho},$$

where

$$\frac{dU}{dt} = C_1 \frac{dT_1}{dt},$$

$$q_{hi} = q_0,$$

$$q_{ho} = \frac{T_1 - T_2}{R_1}.$$

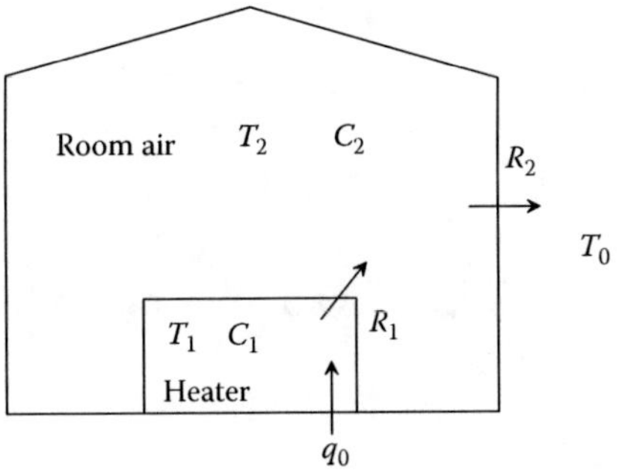

FIGURE 7.35 A room with a heater.

Substituting these expressions gives

$$C_1 \frac{dT_1}{dt} = q_0 - \frac{T_1 - T_2}{R_1},$$

which can be rearranged into

$$C_1 \frac{dT_1}{dt} + \frac{1}{R_1}T_1 - \frac{1}{R_1}T_2 = q_0.$$

Applying the law of conservation of energy to the room air, we have

$$\frac{dU}{dt} = q_{hi} - q_{ho},$$

where

$$\frac{dU}{dt} = C_2 \frac{dT_2}{dt},$$
$$q_{hi} = \frac{T_1 - T_2}{R_1},$$
$$q_{ho} = \frac{T_2 - T_0}{R_2}.$$

Substituting these expressions gives

$$C_2 \frac{dT_2}{dt} = \frac{T_1 - T_2}{R_1} - \frac{T_2 - T_0}{R_2},$$

which can be rearranged into

$$C_2 \frac{dT_2}{dt} - \frac{1}{R_1}T_1 + \left(\frac{1}{R_1} + \frac{1}{R_2}\right)T_2 = \frac{1}{R_2}T_0.$$

The system of differential equations can be written in second-order matrix form as follows:

$$\begin{bmatrix} C_1 & 0 \\ 0 & C_2 \end{bmatrix} \begin{Bmatrix} \frac{dT_1}{dt} \\ \frac{dT_2}{dt} \end{Bmatrix} + \begin{bmatrix} \frac{1}{R_1} & -\frac{1}{R_1} \\ -\frac{1}{R_1} & \left(\frac{1}{R_1} + \frac{1}{R_2}\right) \end{bmatrix} \begin{Bmatrix} T_1 \\ T_2 \end{Bmatrix} = \begin{Bmatrix} q_0 \\ \frac{1}{R_2}T_0 \end{Bmatrix}.$$

b. To represent a thermal system in the state-space form, the temperature of each thermal capacitance is often chosen as a state variable. As specified, the state, the input, and the output are

$$\mathbf{x} = \begin{Bmatrix} x_1 \\ x_2 \end{Bmatrix} = \begin{Bmatrix} T_1 \\ T_2 \end{Bmatrix}, \quad \mathbf{u} = \begin{Bmatrix} q_0 \\ T_0 \end{Bmatrix}, \quad \mathbf{y} = \begin{Bmatrix} T_1 \\ T_2 \end{Bmatrix}.$$

The state-variable equations are

$$\dot{x}_1 = \frac{dT_1}{dt} = -\frac{1}{R_1 C_1}T_1 + \frac{1}{R_1 C_1}T_2 + \frac{1}{C_1}q_0,$$
$$\dot{x}_2 = \frac{dT_2}{dt} = \frac{1}{R_1 C_2}T_1 - \left(\frac{1}{R_1 C_2} + \frac{1}{R_2 C_2}\right)T_2 + \frac{1}{R_2 C_2}T_0,$$

or in matrix form

$$\begin{Bmatrix} \dot{x}_1 \\ \dot{x}_2 \end{Bmatrix} = \begin{bmatrix} -\frac{1}{R_1 C_1} & \frac{1}{R_1 C_1} \\ \frac{1}{R_1 C_2} & -\left(\frac{1}{R_1 C_2} + \frac{1}{R_2 C_2}\right) \end{bmatrix} \begin{Bmatrix} x_1 \\ x_2 \end{Bmatrix} + \begin{bmatrix} \frac{1}{C_1} & 0 \\ 0 & \frac{1}{R_2 C_2} \end{bmatrix} \begin{Bmatrix} u_1 \\ u_2 \end{Bmatrix}.$$

The output equation is

$$\begin{Bmatrix} y_1 \\ y_2 \end{Bmatrix} = \begin{bmatrix} 1 & 0 \\ 0 & 1 \end{bmatrix} \begin{Bmatrix} x_1 \\ x_2 \end{Bmatrix} + \begin{bmatrix} 0 & 0 \\ 0 & 0 \end{bmatrix} \begin{Bmatrix} u_1 \\ u_2 \end{Bmatrix}.$$

PROBLEM SET 7.3

1. Consider heat transfer through a brick wall as shown in Figure 7.36. The temperatures of the inner and the outer surfaces of the wall are 20°C and 5°C, respectively. The thermal conductivity of the brick is $k = 0.5 \cdot W/(m \cdot °C)$. The thickness of the brick wall is 0.25 m. Determine the heat flow rate through the wall.

2. Consider heat transfer through a double-pane window as shown in Figure 7.37a. Two layers of glass with thermal conductivity k_1 are separated by a layer of stagnant air with thermal conductivity k_2. The inner surface of the window is at temperature T_1 and exposed to room air with heat transfer coefficient h_1. The outer surface of the wall is at temperature T_2 and exposed to air with heat transfer coefficient h_2. Assume that $k_1 = 0.95\,W/(m \cdot °C)$, $k_2 = 0.0285\,W/(m \cdot °C)$, $h_1 = h_2 = 10\,W/(m^2 \cdot °C)$, $T_1 = 20°C$, and $T_2 = 35°C$. The thickness of each glass layer is 5 mm, the thickness of the air layer is 10 mm, and the cross-sectional area of the window is $1.5\,m^2$.

   a. Determine the heat flow rate through the double-pane window.
   b. Determine the temperature distribution through the double-pane window.
   c. Repeat Parts (a) and (b) for the single-pane glass window shown in Figure 7.37b.

3. The junction of a thermocouple can be approximated as a sphere with a radius of 1 mm. As shown in Figure 7.38, the thermocouple is submerged in a hot water bath to measure the temperature. For the junction, the density is $\rho = 8800\,kg/m^3$, the specific heat capacity is $c = 420\,J/(kg \cdot °C)$, and the thermal conductivity is $k = 52\,W/(m \cdot °C)$. The

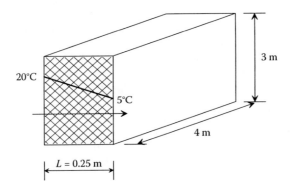

FIGURE 7.36 Problem 1.

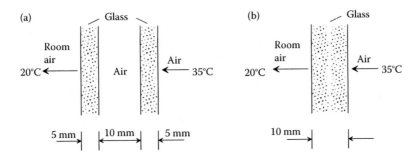

FIGURE 7.37 Problem 2. (a) Double-pane window and (b) single-pane window.

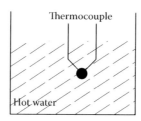

FIGURE 7.38 Problem 3.

temperature of the water T_f is 100°C and the initial temperature of the sphere T_0 is 25°C. The heat transfer coefficient between the liquid and the junction is $h = 340$ W/(m²·°C).
   a. Determine if the junction's temperature can be considered uniform.
   b. Derive the differential equation relating the junction's temperature $T(t)$ and the water's temperature T_f.
   c. ◂ Using the differential equation obtained in Part (b), construct a Simulink block diagram to find out how long it will take the thermocouple to read 99% of the water's temperature.
4. A chicken is taken out of the oven at a uniform temperature of 150°C and is left out in the open air at the room temperature of 25°C. Assume that the chicken can be approximated as a lumped model. The estimated parameters are: mass $m = 2$ kg, heat transfer surface area $A = 0.32$ m², specific heat capacity $c = 3220$ J/(kg·°C), and heat transfer coefficient $h = 15$ W/(m²·°C).
   a. Derive the differential equation relating the chicken's temperature $T(t)$ and the room temperature.
   b. ◂ Using the differential equation obtained in Part (a), construct a Simulink block diagram. Assume that the chicken can be served only if its temperature is above 80°C. Can it be left at the room temperature of 25°C for one hour?
5. The furnace shown in Figure 7.39 has an electric heater with heat flow rate input q_0. A pack of wood is being heated in the furnace. The thermal capacitances of the wood and the air inside the furnace are C_1 and C_2, respectively. The thermal resistances of the air–wood interface and the wall–ambient air interface are R_1 and R_2, respectively. The temperatures of the wood and the air inside the furnace are T_1 and T_2, respectively. The temperature outside the furnace is T_0, which is assumed to be constant.
   a. Derive the differential equations relating the temperatures T_1, T_2, the input q_0, and the outside temperature T_0.
   b. Using the differential equations obtained in Part (a), determine the state-space form of the system. Assume the temperatures T_1 and T_2 as the outputs.

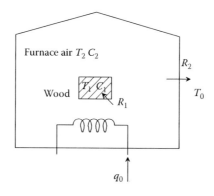

FIGURE 7.39 Problem 5.

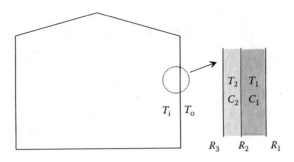

FIGURE 7.40 Problem 6.

6. As shown in Figure 7.40, the wall of a room consists of two layers, for which the thermal capacitances are C_1 and C_2, respectively. Assume that the temperatures in both layers are uniform and they are T_1 and T_2, respectively. The temperatures inside and outside the room are T_i and T_o, respectively. Both layers exchange heat by convection with air and the thermal resistances are R_1 and R_3, respectively. The thermal resistance of the interface between the layers is R_2.
   a. Derive the differential equations for this system.
   b. Using the differential equations obtained in Part (a), determine the state-space form of the system. Assume the temperatures T_1 and T_2 as the outputs.

7.4 SUMMARY

This chapter was devoted to the modeling of fluid and thermal systems. For each of them, we first introduced the concepts of capacitance and resistance. It is useful to think of fluid and thermal systems as electrical circuits. Along with the basic elements, the conservation of mass and the conservation of energy are the main laws used to develop mathematical models of fluid and thermal systems, respectively.

Fluid systems can be divided into pneumatics and hydraulics. A pneumatic system is one in which the fluid is compressible. At low pressure and moderate or high temperature, real gases may be approximated as ideal gases to simplify calculations for pneumatic systems. A hydraulic system is one in which the fluid is incompressible. Most liquids are generally considered incompressible, and this approximation greatly simplifies the modeling of hydraulic systems. A general category of hydraulic systems is liquid-level systems.

Fluid capacitance is the relation between the stored fluid mass and the resulting pressure caused by the stored mass,

$$C = \frac{dm}{dp}.$$

The pneumatic capacitance of a container of constant volume V is defined as

$$C = \frac{V}{nR_gT},$$

where the value of n depends on the type of thermodynamic process.

For a tank of cross-sectional area A with a liquid of height h, the hydraulic capacitance of the tank is defined as

$$C = \frac{A(h)}{g}.$$

When a fluid flows through a valve, a pipe, or an orifice, the fluid meets resistance and there is a drop in the pressure of the fluid. The pressure difference is associated with the mass flow rate q_m in a nonlinear relationship. Near a reference operating point, linearization can be performed to obtain the linearized resistance,

$$R = \frac{\Delta p}{\Delta q_m}.$$

To obtain a simple model of a fluid system, each mass storage element can be represented using a capacitance element and each valve can be represented using a resistance element. The differential equation of the system can be derived by applying the law of conservation of mass:

$$\frac{dm}{dt} = q_{mi} - q_{mo},$$

where the mass flow rate into or out of the system can be related to the resistance at the inlet or the outlet of the system, respectively. The resulting mathematical model may adequately describe the dynamics of the real system.

A thermal system is one that involves the transfer of heat from one object to another. For an object, the thermal capacitance is defined as the ratio of the change in heat flow to the change in the object's temperature,

$$C = \frac{dQ}{dT}.$$

For incompressible liquids and solids,

$$C = mc = \rho V c,$$

where c is the specific heat capacity.

The thermal resistance for heat transfer is defined as the ratio of the change in temperature difference to the change in heat flow rate,

$$R = \frac{dT}{dq_h}.$$

For simple one-dimensional conduction, Fourier's law gives

$$q_h = kA \frac{\Delta T}{L} = kA \frac{T_1 - T_2}{L}$$

and the thermal resistance for conduction is

$$R = \frac{L}{kA}.$$

For convective heat transfer, Newton's law of cooling gives

$$q_h = hA\Delta T = hA(T_s - T_{env})$$

and the thermal resistance for convection is

$$R = \frac{1}{hA}.$$

The mathematical model of a thermal system is often complicated because of the complex temperature distribution throughout the system. To simplify analysis, a lumped-parameter model may be used to approximate the gross system dynamics. The validity of this lumped-parameter assumption can be checked using the so-called Biot number,

$$Bi = \frac{hL_c}{k},$$

Fluid and Thermal Systems

where L_c is the characteristic length of the solid object,

$$L_c = \frac{V_{body}}{A_{surface}}.$$

For a body whose Biot number is much less than one, that is, $Bi < 0.1$, the interior of the body may be assumed to have the same temperature. Then, the dynamic model of a heat transfer system can be derived using the law of conservation of energy,

$$\frac{dU}{dt} = q_{hi} - q_{ho}.$$

REVIEW PROBLEMS

1. Dry air at a constant temperature of T passes through a valve into a rigid spherical container (see Figure 7.41). The pressure at the inlet is p_i. The linear resistances of the two valves at the inlet and the outlet are R_1 and R_2, respectively. Assume that the process is isothermal.
   a. Develop a mathematical model of the pressure p in the container.
   b. Denote the volume flow rate at the outlet as q_o. Determine the transfer function relating p_i and p for this pneumatic system if $q_o = 0$.
2. A single-tank liquid-level system is shown in Figure 7.42, where water flows into the tank at a volume flow rate q_i and out of the tank through two valves at points 1 and 2. The linear resistances of the two valves are R_1 and R_2, respectively. Assuming $h > h_1$, derive the differential equation relating the liquid height h and the volume flow rate q_i at the inlet. The cross-sectional area of the tank A is constant. The density ρ of the liquid is constant.
3. Consider the two-tank liquid-level system shown in Figure 7.43. The liquid is pumped into tanks 1 and 2 through valves of linear resistances R_1 and R_2, respectively. The pressure of the fluid increases by Δp when crossing the pump. The cross-sectional

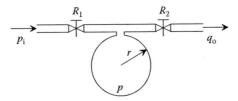

FIGURE 7.41 Problem 1.

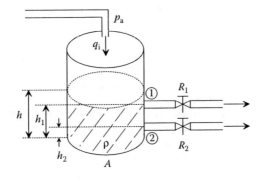

FIGURE 7.42 Problem 2.

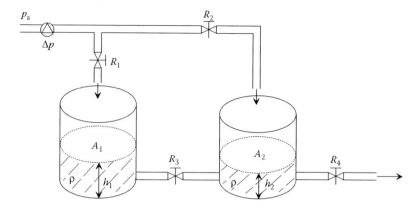

FIGURE 7.43 Problem 3.

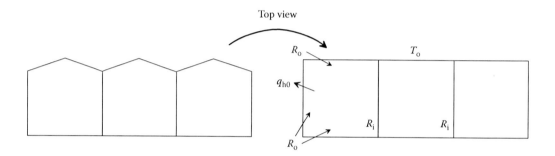

FIGURE 7.44 Problem 5.

areas of the two tanks are A_1 and A_2, respectively. The liquid flows from tank 1 to tank 2 through a valve of linear resistance R_3 and leaves tank 2 through a valve of linear resistance R_4. The density ρ of the liquid is constant. Derive the differential equations in terms of the liquid heights h_1 and h_2. Write the equations in second-order matrix form.

4. A watermelon is taken out of the refrigerator at a uniform temperature of 5°C and is exposed to 27°C air. Assume that the watermelon can be approximated as a sphere and the temperature of the watermelon is uniform. The estimated parameters are: density of watermelon $\rho = 120 \, \text{kg/m}^3$, diameter of the watermelon $D = 40$ cm, specific heat capacity $c = 4200 \, \text{J/(kg·°C)}$, and heat transfer coefficient $h = 15 \, \text{W/(m}^2\text{·°C)}$.
   a. Derive the differential equation relating the watermelon's temperature $T(t)$ and the air temperature.
   b. Using the differential equation obtained in Part (a), construct a Simulink block diagram. How long will it take before the watermelon is warmed up to 20°C?

5. For a three-room house shown in Figure 7.44, all rooms are perfect square and have the same dimensions. An air conditioner produces an equal amount of heat flow q_{ho} out of each room. The temperature outside the house is T_o. Assume that there is no heat flow through the floors or ceilings. The thermal resistances through the inner walls and the outer walls are R_i and R_o, respectively. The thermal capacitance of each room is C. Derive the differential equations for this system.

8 System Response

In Chapters 5 through 7 we learned how to derive the mathematical models of dynamic systems. The different forms of system model representation were introduced in Chapter 4. This chapter deals with the response analysis of dynamic systems, mainly transient response and frequency response. Systematic techniques to solve the state equation will also be presented. For the most part, linear dynamic systems are considered in this chapter. Treatment of nonlinear systems is presented in the final section of the chapter.

Types of Response

Consider an nth-order dynamic system defined by the differential equation

$$x^{(n)} + a_1 x^{(n-1)} + \cdots + a_{n-1}\dot{x} + a_n x = f(t), \tag{8.1}$$

where the coefficients $a_1, a_2, \ldots, a_n$ are constants, $x(t)$ is the dependent variable, t denotes time, and $f(t)$ is the input known as the forcing function. As seen in Chapter 2, the solution $x(t)$, the total response, is composed of the complementary (homogeneous) solution $x_c(t)$ and the particular solution $x_p(t)$, that is, $x(t) = x_c(t) + x_p(t)$. Note that $x_c(t)$ is the solution of Equation 8.1 when $f(t) = 0$, and $x_p(t)$ depends on the nature of the forcing function $f(t)$. The complementary solution $x_c(t)$ is called the natural response (free response) because it represents the natural behavior of the system in the absence of the input. The particular solution $x_p(t)$ is known as the forced response of the system.

Transient Response and Steady-State Response

The total response $x(t)$ can also be decomposed into transient response $x_{tr}(t)$ and steady-state response $x_{ss}(t)$, that is, $x(t) = x_{tr}(t) + x_{ss}(t)$. The transient response consists of those terms in $x(t)$ that decay to zero as $t \to \infty$. The part of the response $x(t)$ that remains after the transient terms have vanished is called the steady-state response.

Example 8.1: Transient and Steady-State Responses

The mathematical model of a dynamic system is defined by

$$\ddot{x} + x = 2e^{-t}, \quad x(0) = 1, \dot{x}(0) = 0.$$

Find the system's total response, and identify the transient and steady-state responses.

Solution

The complementary solution is readily found as $x_c(t) = c_1 \cos t + c_2 \sin t$. The particular solution is found via the method of undetermined coefficients (Chapter 2) as $x_p(t) = e^{-t}$. Therefore,

$$x(t) = c_1 \cos t + c_2 \sin t + e^{-t}.$$

Initial conditions yield $c_1 = 0$, $c_2 = 1$. The total response is then formed as

$$x(t) = \sin t + e^{-t}.$$

Consequently, $x_{tr}(t) = e^{-t}$ and $x_{ss}(t) = \sin t$.

8.1 TRANSIENT RESPONSE OF FIRST-ORDER SYSTEMS

Linear, first-order dynamic systems are described by

$$\tau \dot{x} + x = f(t), \quad \tau = \text{const} > 0, \quad x(0) = x_0, \tag{8.2}$$

where τ is called the time constant and gives a measure of how fast the response reaches steady state. Note that upon division by τ, Equation 8.2 agrees with the general form of Equation 8.1. To perform the transient-response analysis, we first take the Laplace transform of Equation 8.2, as

$$\tau[sX(s) - x_0] + X(s) = F(s) \xrightarrow[\text{and rearrange}]{\text{Simplify}} X(s) = \frac{\tau}{\tau s + 1} x_0 + \frac{1}{\tau s + 1} F(s).$$

Subsequently, inverse Laplace transformation yields

$$x(t) = \mathcal{L}^{-1}\left\{\frac{\tau}{\tau s + 1}\right\} x_0 + \mathcal{L}^{-1}\left\{\frac{1}{\tau s + 1} F(s)\right\}.$$

The first inverse Laplace term is simply

$$\mathcal{L}^{-1}\left\{\frac{\tau}{\tau s + 1}\right\} = \mathcal{L}^{-1}\left\{\frac{1}{s + 1/\tau}\right\} = e^{-t/\tau}.$$

Therefore, the total response can be expressed as

$$x(t) = e^{-t/\tau} x_0 + \mathcal{L}^{-1}\left\{\frac{1}{\tau s + 1} F(s)\right\}. \tag{8.3}$$

The first term on the right side represents the system response to initial condition (excitation), while the second term gives the response to the input. In this section, we will study the response behavior of the system when subjected to specific types of input such as step and ramp functions. The system's free response is discussed first.

8.1.1 Free Response of First-Order Systems

Free response is defined as the response to the initial condition only, hence given by the first term in Equation 8.3,

$$x(t) = e^{-t/\tau} x_0. \tag{8.4}$$

Example 8.2: Free Response

Suppose that a first-order system is described by

$$2\dot{x} + 3x = 0, \quad x(0) = 2.$$

The response $x(t)$ is clearly a free (natural) response as there is no forcing function present. Rewriting the equation as $(2/3)\dot{x} + x = 0$ reveals $\tau = 2/3$. With this, and noting $x(0) = 2$, Equation 8.4 yields

$$x(t) = 2e^{-3t/2}.$$

It is readily seen that the smaller the time constant, the faster the response reaches equilibrium.

System Response

8.1.2 Step Response of First-Order Systems

Step response refers to $x(t)$ in Equation 8.2 when $f(t) = Au(t)$, where $u(t)$ is the unit-step and A is a constant magnitude (Section 2.3). Inserting $F(s) = A/s$ in Equation 8.3, we find

$$x(t) = e^{-t/\tau}x_0 + \mathcal{L}^{-1}\left\{\frac{A}{s(\tau s + 1)}\right\}.$$

Using partial-fraction expansion, it can be shown that

$$\mathcal{L}^{-1}\left\{\frac{A}{s(\tau s + 1)}\right\} = A\left[1 - e^{-t/\tau}\right].$$

Therefore, the step response is described by

$$x(t) = e^{-t/\tau}x_0 + A\left[1 - e^{-t/\tau}\right]. \tag{8.5}$$

Since $\tau > 0$, we have $x(t) \to A$ as $t \to \infty$. In other words, the step response $x(t)$ has a steady-state value of A.

Example 8.3: Step Response of an RL Circuit

The RL circuit shown in Figure 8.1 is subjected to zero initial current. Determine the unit-step response of the circuit, and examine the role of the time constant.

Solution

Using Kirchhoff's voltage law (KVL), the mathematical model of the circuit is written as

$$L\frac{di}{dt} + Ri = v_a(t), \quad i(0) = 0.$$

Division of the differential equation by R yields

$$\frac{L}{R}\frac{di}{dt} + i = \frac{1}{R}v_a(t) \Rightarrow \underbrace{\tau = \frac{L}{R}}_{\text{time constant}}.$$

Comparing this equation with Equation 8.2, we find $f(t) = (1/R)v_a(t)$. But since $v_a(t) = u(t)$, the unit-step function, $f(t) = (1/R)u(t)$ as if $f(t)$ is a step function with magnitude $1/R$. As a result, using $A = 1/R$ and $x_0 = 0$ in Equation 8.5, we find

$$i(t) = \frac{1}{R}\left[1 - e^{-t/\tau}\right] = \frac{1}{R}\left[1 - e^{-(R/L)t}\right].$$

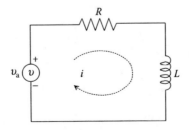

FIGURE 8.1 RL circuit.

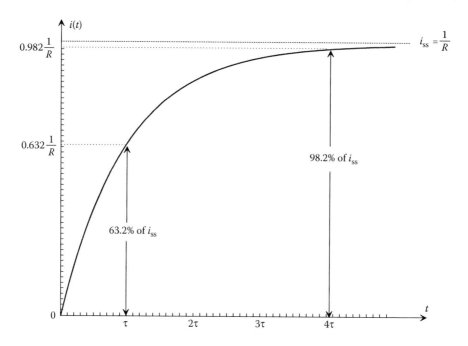

FIGURE 8.2 Unit-step response of an RL circuit (Example 8.3).

It is clear that $i(t)$ reaches a steady-state value of $1/R$ after a sufficiently long time (Figure 8.2); hence the steady-state current is $i_{ss} = 1/R$. We examine the role of τ as follows. After one time constant (at $t = \tau$),

$$i(\tau) = \frac{1}{R}(1 - e^{-1}) = 0.632 \left(\frac{1}{R}\right) = 0.632 i_{ss}.$$

This means that 63.2% of the steady-state current is recovered after one time constant. Similarly, the percentage of recovery at $t = 2\tau$ and $t = 3\tau$ may be calculated. At $t = 4\tau$,

$$i(4\tau) = \frac{1}{R}(1 - e^{-4}) = 0.982 \left(\frac{1}{R}\right) = 0.982 i_{ss}.$$

Therefore, after four time constants the response is within 2% of the steady-state value. This essentially serves as the settling time for the response curve. Settling time is one of the four transient-response specifications of second-order systems and plays a central role in the control of such systems (see Chapter 10).

8.1.3 Ramp Response of First-Order Systems

Ramp response refers to $x(t)$ in Equation 8.2 when $f(t) = Au_r(t)$, where $u_r(t)$ is the unit-ramp and A is a constant slope (Section 2.3). Inserting $F(s) = A/s^2$ in Equation 8.3, we find

$$x(t) = e^{-t/\tau} x_0 + \mathcal{L}^{-1}\left\{\frac{A}{s^2(\tau s + 1)}\right\}.$$

But

$$\mathcal{L}^{-1}\left\{\frac{A}{s^2(\tau s + 1)}\right\} = A\left[t - \tau(1 - e^{-t/\tau})\right].$$

Therefore, the ramp response is given by

$$x(t) = e^{-t/\tau} x_0 + A\left[t - \tau(1 - e^{-t/\tau})\right]. \tag{8.6}$$

System Response

8.1.3.1 Steady-State Error

For simplicity, assume that the initial condition is zero so that Equation 8.6 reads

$$x(t) = A\left[t - \tau(1 - e^{-t/\tau})\right].$$

The error between the ramp input and the ramp response is

$$e(t) = At - A\left[t - \tau(1 - e^{-t/\tau})\right] = A\tau(1 - e^{-t/\tau}).$$

As $t \to \infty$, this error approaches $A\tau$ so that the steady-state error is

$$e_{ss} = A\tau.$$

The ramp input, the ramp response, and the associated steady-state error are shown in Figure 8.3.

PROBLEM SET 8.1

1. Find the unit-ramp response of the RL circuit considered in Example 8.3.
2. The equation of motion for the rotational system in Figure 8.4 is derived as

$$I_0\ddot{\theta} + B\dot{\theta} = T(t).$$

   a. By introducing the angular velocity ($\omega = \dot{\theta}$), rewrite the above as a first-order equation.
   b. Assuming $\omega(0) = 0$, find the system response $\omega(t)$ to a step input of magnitude 2.

3. The governing equation for the single-tank, liquid-level system in Figure 8.5 is derived as

$$\frac{RA}{g}\dot{h} + h = \frac{R}{g}q_i(t).$$

   a. Assuming $h(0) = h_0$, find the system response $h(t)$ to a step input of magnitude Q_i.
   b. Find the steady-state response.

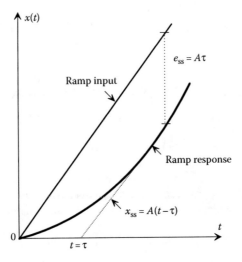

FIGURE 8.3 Ramp response of a first-order system.

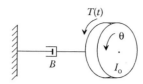

FIGURE 8.4 Problem 2.

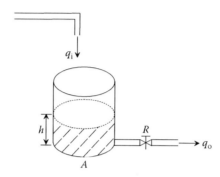

FIGURE 8.5 Problem 3.

4. For the system in Problem 3, find the response to a ramp input of slope $\frac{3}{2}$, assuming $h_0 = 0$.
5. A first-order dynamic system is governed by

$$3\dot{v} + v = F(t), \quad v(0) = 1.$$

   a. Find the response $v(t)$ when $F(t) = 2u_r(t)$.
   b. Determine the steady-state response v_{ss}.
6. A system is modeled as

$$3\dot{v} + 2v = u(t), \quad v(0) = 0,$$

   where $u(t)$ is the unit-step function.
   a. What is v_{ss}?
   b. How long does it take for the response to reach within 2% of its steady-state value?

8.2 TRANSIENT RESPONSE OF SECOND-ORDER SYSTEMS

Linear, second-order dynamic systems are mathematically modeled as

$$\ddot{x} + 2\zeta\omega_n\dot{x} + \omega_n^2 x = f(t), \quad x(0) = x_0, \; \dot{x}(0) = \dot{x}_0, \qquad (8.7)$$

where ζ is the damping ratio and ω_n is the (undamped) natural frequency, in rad/s. Although Equation 8.7 represents the model for any second-order dynamic system, it is best understood when it is viewed in relation to a mechanical system. To that end, consider the mass–spring–damper system in Figure 8.6.

The system's (undamped) natural frequency is defined as

$$\omega_n = \sqrt{\frac{k}{m}} \text{ rad/s}.$$

System Response

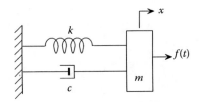

FIGURE 8.6 A mass–spring–damper system.

The damping ratio is defined as the ratio of the actual damping c and the critical damping $c_{cr} = 2\sqrt{mk}$, that is,

$$\zeta = \frac{c}{c_{cr}} = \frac{c}{2\sqrt{mk}}.$$

The system's equation of motion is derived as (Chapter 5)

$$m\ddot{x} + c\dot{x} + kx = f(t) \overset{\text{Divide by } m}{\Rightarrow} \ddot{x} + \frac{c}{m}\dot{x} + \frac{k}{m}x = \frac{1}{m}f(t).$$

Noting that

$$\frac{k}{m} = \omega_n^2, \quad \frac{c}{m} = 2\frac{c}{2\sqrt{mk}}\sqrt{\frac{k}{m}} = 2\zeta\omega_n,$$

the equation of motion can be expressed as that in Equation 8.7, with the force $(1/m)f(t)$ renamed $f(t)$. We now perform the transient-response analysis of second-order systems as follows. Taking the Laplace transform of Equation 8.7 and using the initial conditions, we find

$$[s^2 X(s) - sx_0 - \dot{x}_0] + 2\zeta\omega_n[sX(s) - x_0] + \omega_n^2 X(s) = F(s).$$

Collecting like terms and solving for $X(s)$ yields

$$X(s) = \frac{(s + 2\zeta\omega_n)x_0 + \dot{x}_0}{s^2 + 2\zeta\omega_n s + \omega_n^2} + \frac{1}{s^2 + 2\zeta\omega_n s + \omega_n^2}F(s).$$

Finally,

$$x(t) = \mathcal{L}^{-1}\left\{\frac{(s + 2\zeta\omega_n)x_0 + \dot{x}_0}{s^2 + 2\zeta\omega_n s + \omega_n^2}\right\} + \mathcal{L}^{-1}\left\{\frac{1}{s^2 + 2\zeta\omega_n s + \omega_n^2}F(s)\right\}. \tag{8.8}$$

The first term on the right side represents the system response to initial conditions, while the second term gives the response to the input.

8.2.1 Free Response of Second-Order Systems

Free response is the response to initial conditions only, defined by the first term in Equation 8.8:

$$x(t) = \mathcal{L}^{-1}\left\{\frac{(s + 2\zeta\omega_n)x_0 + \dot{x}_0}{s^2 + 2\zeta\omega_n s + \omega_n^2}\right\}. \tag{8.9}$$

This inverse Laplace transform depends on the nature of the poles (Section 2.3), that is, the roots of the characteristic equation

$$s^2 + 2\zeta\omega_n s + \omega_n^2 = 0.$$

Solution of this equation yields the poles

$$s = -\zeta\omega_n \pm \sqrt{(\zeta\omega_n)^2 - \omega_n^2} = -\zeta\omega_n \pm \omega_n\sqrt{\zeta^2 - 1}, \tag{8.10}$$

which depend on the value or range of values of ζ. The system is undamped if $\zeta = 0$, underdamped if $0 < \zeta < 1$, critically damped if $\zeta = 1$, and overdamped if $\zeta > 1$. The free response analysis for all these damping cases is conducted as follows.

Case (1) Undamped ($\zeta = 0$)
Inserting $\zeta = 0$ in Equation 8.9, we find

$$x(t) = \mathcal{L}^{-1}\left\{\frac{sx_0 + \dot{x}_0}{s^2 + \omega_n^2}\right\} = x_0 \cos \omega_n t + \frac{\dot{x}_0}{\omega_n}\sin \omega_n t. \tag{8.11}$$

Case (2) Underdamped ($0 < \zeta < 1$)
In this case, rewrite

$$\begin{aligned} s^2 + 2\zeta\omega_n s + \omega_n^2 &= (s + \zeta\omega_n)^2 - (\zeta\omega_n)^2 + \omega_n^2 \\ &= (s + \zeta\omega_n)^2 + \omega_n^2(1 - \zeta^2) \\ &= (s + \zeta\omega_n)^2 + \omega_d^2, \end{aligned}$$

where $\omega_d = \omega_n\sqrt{1 - \zeta^2}$ is the (damped) natural frequency. With this, Equation 8.9 yields

$$x(t) = \mathcal{L}^{-1}\left\{\frac{(s + 2\zeta\omega_n)x_0 + \dot{x}_0}{(s + \zeta\omega_n)^2 + \omega_d^2}\right\} = e^{-\zeta\omega_n t}\left[x_0 \cos \omega_d t + \frac{\zeta\omega_n x_0 + \dot{x}_0}{\omega_d}\sin \omega_d t\right]. \tag{8.12}$$

Case (3) Critically damped ($\zeta = 1$)
Using $\zeta = 1$ in Equation 8.9 leads to

$$x(t) = \mathcal{L}^{-1}\left\{\frac{(s + 2\omega_n)x_0 + \dot{x}_0}{(s + \omega_n)^2}\right\} = e^{-\omega_n t}\left[x_0 + (\omega_n x_0 + \dot{x}_0)t\right]. \tag{8.13}$$

Case (4) Overdamped ($\zeta > 1$)
In this case, the two poles are real and distinct as in Equation 8.10. Let

$$s_1 = -\zeta\omega_n + \omega_n\sqrt{\zeta^2 - 1}, \quad s_2 = -\zeta\omega_n - \omega_n\sqrt{\zeta^2 - 1}.$$

Then,

$$x(t) = \mathcal{L}^{-1}\left\{\frac{(s + 2\zeta\omega_n)x_0 + \dot{x}_0}{s^2 + 2\zeta\omega_n s + \omega_n^2}\right\} = \frac{-s_2 x_0 + \dot{x}_0}{s_1 - s_2}e^{s_1 t} - \frac{-s_1 x_0 + \dot{x}_0}{s_1 - s_2}e^{s_2 t}. \tag{8.14}$$

Note that besides the response for the undamped case which is oscillatory, all other responses eventually stabilize at zero. Also note that the response for the underdamped case exhibits decaying oscillations.

8.2.1.1 Initial Response in MATLAB®

The `initial` command calculates the free response of a state-space model (Section 4.2) with an initial condition on the states:

$$\dot{\mathbf{x}} = \mathbf{A}\mathbf{x}, \quad \mathbf{x}(0) = \mathbf{x}_0,$$
$$\mathbf{y} = \mathbf{C}\mathbf{x}.$$

Then `initial(sys,x0)` plots the response of system `sys` to an initial condition `x0`. The duration of simulation is determined automatically to reflect adequately the response transients.

Example 8.4: Free Response

Consider

$$\ddot{x} + 3\dot{x} + 2x = 0, \quad x(0) = 0, \; \dot{x}(0) = 1.$$

System Response

a. Determine $x(t)$.
b. Find and plot the free response in MATLAB.

SOLUTION

a. Comparing with Equation 8.7, it is readily seen that

$$\begin{aligned} \omega_n^2 &= 2 \\ 2\zeta\omega_n &= 3 \end{aligned} \Rightarrow \begin{aligned} \omega_n &= \sqrt{2} \text{ rad/s} \\ \zeta &= \frac{3}{2\sqrt{2}} > 1, \end{aligned}$$

so the system is overdamped. The characteristic equation $s^2 + 3s + 2 = 0$ has roots $s_1 = -1$ and $s_2 = -2$ so that $s_1 - s_2 = 1$. Inserting these and the given initial conditions into Equation 8.14,

$$x(t) = e^{-t} - e^{-2t}.$$

b. Choosing the state variables as $x_1 = x$ and $x_2 = \dot{x}$, the state-variable equations are

$$\begin{cases} \dot{x}_1 = x_2, \\ \dot{x}_2 = -2x_1 - 3x_2, \end{cases}$$

so that

$$\mathbf{A} = \begin{bmatrix} 0 & 1 \\ -2 & -3 \end{bmatrix}.$$

Since it is $x_1 = x$ that we are looking for, we set $\mathbf{C} = \begin{bmatrix} 1 & 0 \end{bmatrix}$. The following code will generate the plot of x versus t.

```
>> A = [0 1;-2 -3];   C = [1 0]; x0 = [0;1];
>> sys = ss(A,[],C,[]);
>> initial(sys,x0)   % Figure 8.7
```

Note that $t = 7(\text{sec})$ has been automatically determined by the `initial` command.

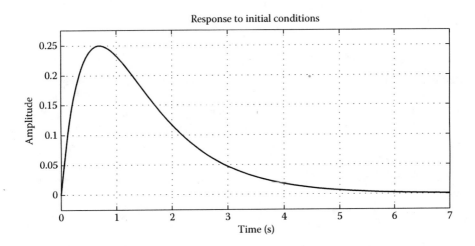

FIGURE 8.7 Initial response (Example 8.4).

8.2.2 Impulse Response of Second-Order Systems

Impulse response refers to $x(t)$ in Equation 8.7 when the input is $f(t) = A\delta(t)$, where $\delta(t)$ is the unit-impulse and A is a constant magnitude. Using $F(s) = A$ in Equation 8.8, the response is

$$x(t) = \mathcal{L}^{-1}\left\{\frac{(s + 2\zeta\omega_n)x_0 + \dot{x}_0 + A}{s^2 + 2\zeta\omega_n s + \omega_n^2}\right\}. \tag{8.15}$$

As in free response analysis, different damping cases are studied separately as follows.

Case (1) *Undamped* ($\zeta = 0$)
Inserting $\zeta = 0$ in Equation 8.15, we find

$$x(t) = \mathcal{L}^{-1}\left\{\frac{sx_0 + \dot{x}_0 + A}{s^2 + \omega_n^2}\right\} = x_0 \cos\omega_n t + \frac{\dot{x}_0 + A}{\omega_n}\sin\omega_n t. \tag{8.16}$$

Case (2) *Underdamped* ($0 < \zeta < 1$)

$$x(t) = \mathcal{L}^{-1}\left\{\frac{(s + 2\zeta\omega_n)x_0 + \dot{x}_0 + A}{(s + \zeta\omega_n)^2 + \omega_d^2}\right\} = e^{-\zeta\omega_n t}\left[x_0 \cos\omega_d t + \frac{\zeta\omega_n x_0 + \dot{x}_0 + A}{\omega_d}\sin\omega_d t\right]. \tag{8.17}$$

Case (3) *Critically damped* ($\zeta = 1$)

$$x(t) = \mathcal{L}^{-1}\left\{\frac{(s + 2\omega_n)x_0 + \dot{x}_0 + A}{(s + \omega_n)^2}\right\} = e^{-\omega_n t}[x_0 + (\omega_n x_0 + \dot{x}_0 + A)t]. \tag{8.18}$$

Case (4) *Overdamped* ($\zeta > 1$)

$$x(t) = \mathcal{L}^{-1}\left\{\frac{(s + 2\zeta\omega_n)x_0 + \dot{x}_0 + A}{s^2 + 2\zeta\omega_n s + \omega_n^2}\right\} = \frac{-s_2 x_0 + \dot{x}_0 + A}{s_1 - s_2}e^{s_1 t} - \frac{-s_1 x_0 + \dot{x}_0 + A}{s_1 - s_2}e^{s_2 t}, \tag{8.19}$$

where

$$s_1 = -\zeta\omega_n + \omega_n\sqrt{\zeta^2 - 1}, \quad s_2 = -\zeta\omega_n - \omega_n\sqrt{\zeta^2 - 1}.$$

Once again, as it was the case with free response of second-order systems, the impulse response for the undamped case is oscillatory. In each of the other cases, the response will stabilize at zero, while the underdamped impulse response exhibits decaying oscillations.

8.2.2.1 Impulse Response in MATLAB

The command `impulse(sys)` returns the unit-impulse response of the linear time-invariant (LTI) model `sys` created with either transfer function (TF) or state-space (ss), with the assumption of zero initial conditions. For multi-input models, independent impulse commands are applied to each input channel. The time range and number of points are chosen automatically. For nonzero initial conditions, it is best to use Equations 8.16 through 8.19.

Example 8.5: Impulse Response

Consider

$$\ddot{x} + 2\dot{x} + 2x = 3\delta(t), \quad x(0) = 0, \, \dot{x}(0) = 0.$$

a. Determine the response $x(t)$.
b. Find and plot the response in MATLAB.

System Response

Solution

a. Comparison with Equation 8.7 reveals that

$$\begin{matrix} \omega_n^2 = 2 \\ 2\zeta\omega_n = 2 \end{matrix} \Rightarrow \begin{matrix} \omega_n = \sqrt{2} \text{ rad/s} \\ \zeta = \dfrac{1}{\sqrt{2}} < 1 \end{matrix}$$

so that the system is underdamped. Also, $\omega_d = \omega_n\sqrt{1-\zeta^2} = 1$ rad/s. By Equation 8.17,

$$x(t) = 3e^{-t}\sin t.$$

b. Since the initial conditions are zero, it is appropriate to use the `impulse` command in MATLAB. We will create our system using the transfer function. Because the `impulse` command returns the unit-impulse response, we must define the transfer function as

$$\frac{3}{s^2 + 2s + 2}.$$

The following code will generate and plot the impulse response.

```
>> n = 3; d = [1 2 2];
>> sys = tf(n,d);
>> impulse(sys)    % Figure 8.8
```

8.2.3 Step Response of Second-Order Systems

Impulse response refers to $x(t)$ in Equation 8.7 when the input is $f(t) = Au(t)$, where $u(t)$ is the unit-step and A is a constant magnitude. Using $F(s) = A/s$ in Equation 8.8, the response is

$$x(t) = \mathcal{L}^{-1}\left\{\frac{(s+2\zeta\omega_n)x_0 + \dot{x}_0}{s^2 + 2\zeta\omega_n s + \omega_n^2}\right\} + \mathcal{L}^{-1}\left\{\frac{A}{s(s^2 + 2\zeta\omega_n s + \omega_n^2)}\right\}. \tag{8.20}$$

As always, different cases of damping are considered separately as follows.

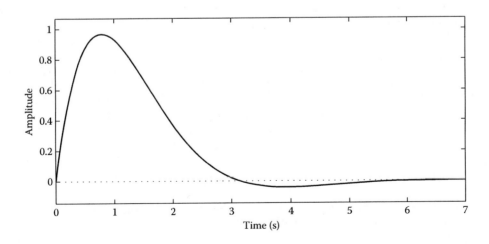

FIGURE 8.8 Impulse response (Example 8.5).

Case (1) *Undamped* ($\zeta = 0$)
Inserting $\zeta = 0$ in Equation 8.20 yields

$$x(t) = \mathcal{L}^{-1}\left\{\frac{sx_0 + \dot{x}_0}{s^2 + \omega_n^2}\right\} + \mathcal{L}^{-1}\left\{\frac{A}{s(s^2 + \omega_n^2)}\right\}$$

$$= x_0 \cos \omega_n t + \frac{\dot{x}_0}{\omega_n}\sin \omega_n t + \frac{A}{\omega_n^2}(1 - \cos \omega_n t). \tag{8.21}$$

Case (2) *Underdamped* ($0 < \zeta < 1$)

$$x(t) = \mathcal{L}^{-1}\left\{\frac{(s + 2\zeta\omega_n)x_0 + \dot{x}_0}{(s + \zeta\omega_n)^2 + \omega_d^2}\right\} + \mathcal{L}^{-1}\left\{\frac{A}{s\left[(s + \zeta\omega_n)^2 + \omega_d^2\right]}\right\}$$

$$= e^{-\zeta\omega_n t}\left[x_0 \cos \omega_d t + \frac{\zeta\omega_n x_0 + \dot{x}_0}{\omega_d}\sin \omega_d t\right]$$

$$+ \frac{A}{\omega_n^2}\left[1 - e^{-\zeta\omega_n t}\left(\cos \omega_d t + \frac{\zeta}{\sqrt{1-\zeta^2}}\sin \omega_d t\right)\right]. \tag{8.22}$$

Case (3) *Critically damped* ($\zeta = 1$)

$$x(t) = \mathcal{L}^{-1}\left\{\frac{(s + 2\omega_n)x_0 + \dot{x}_0}{(s + \omega_n)^2}\right\} + \mathcal{L}^{-1}\left\{\frac{A}{s(s + \omega_n)^2}\right\}$$

$$= e^{-\omega_n t}\left[x_0 + (\omega_n x_0 + \dot{x}_0)t\right] + \frac{A}{\omega_n^2}\left[1 - e^{-\omega_n t}(1 + \omega_n t)\right]. \tag{8.23}$$

Case (4) *Overdamped* ($\zeta > 1$)

$$x(t) = \mathcal{L}^{-1}\left\{\frac{(s + 2\zeta\omega_n)x_0 + \dot{x}_0}{s^2 + 2\zeta\omega_n s + \omega_n^2}\right\} + \mathcal{L}^{-1}\left\{\frac{A}{s(s^2 + 2\zeta\omega_n s + \omega_n^2)}\right\}$$

$$= \frac{-s_2 x_0 + \dot{x}_0}{s_1 - s_2}e^{s_1 t} - \frac{-s_1 x_0 + \dot{x}_0}{s_1 - s_2}e^{s_2 t}$$

$$+ \frac{A}{\omega_n^2}\left[1 + \frac{\omega_n^2}{s_1 - s_2}\left(\frac{1}{s_1}e^{s_1 t} - \frac{1}{s_2}e^{s_2 t}\right)\right]. \tag{8.24}$$

Besides the undamped case where the response oscillates, it is readily seen that for each of the other cases of damping, the step response settles at A/ω_n^2.

8.2.3.1 Step Response in MATLAB

The command `step(sys)` returns the unit-step response of the LTI model `sys` created with either transfer function (`tf`) or state-space (`ss`), with the assumption of zero initial conditions. For multi-input models, independent step commands are applied to each input channel. The time range and number of points are chosen automatically. For nonzero initial conditions, it is best to use Equations 8.21 through 8.24.

Example 8.6: Step Response

The mechanical system in Figure 8.9 is subjected to an applied force $f(t) = 10u(t)$ N, where $u(t)$ is the unit-step function. Assume zero initial displacement and velocity.

a. Determine the system response, the block displacement $x(t)$.
b. Find and plot the response in MATLAB.

System Response

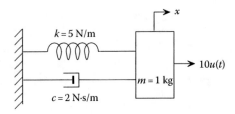

FIGURE 8.9 Mechanical system (Example 8.6).

SOLUTION

a. Using the given physical parameter values, the equation of motion is written as

$$\ddot{x} + 2\dot{x} + 5x = 10u(t), \quad x(0) = 0, \; \dot{x}(0) = 0.$$

Comparing with Equation 8.7,

$$\begin{aligned}\omega_n^2 &= 5 \\ 2\zeta\omega_n &= 2\end{aligned} \quad \Rightarrow \quad \begin{aligned}\omega_n &= \sqrt{5}\,\text{rad/s}, \\ \zeta &= \frac{1}{\sqrt{5}} < 1.\end{aligned}$$

This implies the system is underdamped and $\omega_d = \omega_n\sqrt{1-\zeta^2} = 2$ rad/s. Using Equation 8.22, with zero initial conditions and $A = 10$, we find the response as

$$x(t) = 2\left[1 - e^{-t}\left(\cos 2t + \frac{1}{2}\sin 2t\right)\right] \text{m}.$$

Note that the steady-state value of response is 2, which agrees with $A/\omega_n^2 = 10/5 = 2$.

b.

Since the **step** command returns the unit-step response, we will define the transfer function as

$$\frac{10}{s^2 + 2s + 5}.$$

Subsequently, the following code will generate the desired plot.

```
>> n = 10;
>> d = [1 2 5];
>> sys = tf(n,d);
>> step(sys)   % Figure 8.10
```

Note that, as expected, the step response has a steady-state value of $A/\omega_n^2 = 2$.

8.2.4 RESPONSE TO ARBITRARY INPUTS

The command `lsim(sys,u,t,x0)` plots the time response of the LTI model `sys` to the arbitrary input signal described by `u` and `t`. The time vector `t` consists of regularly spaced time samples and `u` is a matrix with as many columns as inputs and whose ith row specifies the input value at time `t(i)`. Vector `x0` is the initial state vector at `t(1)` (for state-space models only). It is set to zero when omitted.

Example 8.7: Response to a Sinusoidal Input

A dynamic system is modeled as

$$\ddot{x} + 4\dot{x} + 2x = \frac{1}{2}\sin t, \quad x(0) = 0, \; \dot{x}(0) = -1.$$

Plot the response $x(t)$ for $0 \le t \le 10$ s using the `lsim` command.

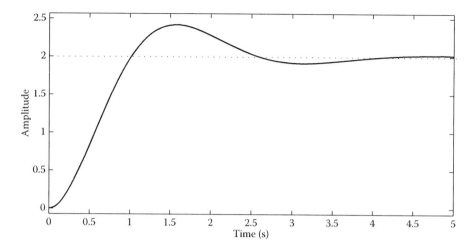

FIGURE 8.10 Step response of the mechanical system in Example 8.6.

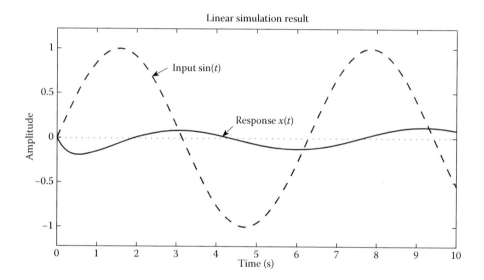

FIGURE 8.11 Simulation result (Example 8.7).

Solution

We first obtain the state-space model as

$$\begin{cases} \dot{\mathbf{x}} = \begin{bmatrix} 0 & 1 \\ -2 & -4 \end{bmatrix} \mathbf{x} + \begin{bmatrix} 0 \\ \frac{1}{2} \end{bmatrix} u, \quad \mathbf{x} = \begin{Bmatrix} x_1 \\ x_2 \end{Bmatrix}, \quad u = \sin t, \\ y = \begin{bmatrix} 1 & 0 \end{bmatrix} \mathbf{x}. \end{cases}$$

Note that the output matrix is $\begin{bmatrix} 1 & 0 \end{bmatrix}$ since $x = x_1$ is to be plotted. The following code will generate the desired plot.

```
>> A=[0 1;-2 -4]; B=[0;0.5]; C=[1 0]; D=0;
```

System Response

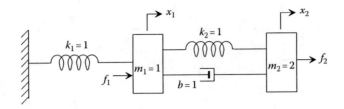

FIGURE 8.12 System in Example 8.8.

```
>> x0=[0;-1];
>> t=0:0.01:10;
>> u=sin(t);
>> lsim(sys,u,t,x0)   % Figure 8.11
```

Example 8.8: MIMO Systems

Consider the translational mechanical system shown in Figure 8.12. The equations of motion are (see Example 4.7, Section 4.2)

$$\begin{cases} \ddot{x}_1 + x_1 - (x_2 - x_1) - (\dot{x}_2 - \dot{x}_1) = f_1, \\ 2\ddot{x}_2 + (x_2 - x_1) + (\dot{x}_2 - \dot{x}_1) = f_2. \end{cases}$$

Selecting state variables $x_1 = x_1$, $x_2 = x_2$, $x_3 = \dot{x}_1$, $x_4 = \dot{x}_2$, the system's state equation is formed as

$$\dot{x} = \begin{bmatrix} 0 & 0 & 1 & 0 \\ 0 & 0 & 0 & 1 \\ -2 & 1 & -1 & 1 \\ \frac{1}{2} & -\frac{1}{2} & \frac{1}{2} & -\frac{1}{2} \end{bmatrix} x + \begin{bmatrix} 0 & 0 \\ 0 & 0 \\ 1 & 0 \\ 0 & \frac{1}{2} \end{bmatrix} u, \quad x = \begin{Bmatrix} x_1 \\ x_2 \\ x_3 \\ x_4 \end{Bmatrix}, \quad u = \begin{Bmatrix} f_1 \\ f_2 \end{Bmatrix}.$$

We will investigate the response $x_1(t)$ when both inputs f_1 and f_2 are unit-step functions and initial conditions are zero. The following code produces a pair of plots, one showing the contribution of f_1 to x_1 (labeled x_{11}), the other reflecting the contribution of f_2 to x_1 (labeled x_{12}). Since the inputs are step functions, we can indeed use the `step` command in MATLAB rather than the `lsim` command.

```
>> A=[0 0 1 0;0 0 0 1;-2 1 -1 1;1/2 -1/2 1/2 -1/2];B=[0 0;0 0;1 0;0 1/2];
>> C=[1 0 0 0];D=0;
>> sys=ss(A,B,C,D);
>> step(sys)   % Figure 8.13
```

Subsequently, the superposition of x_{11} and x_{12} yields the response x_1, that is, $x_1 = x_{11} + x_{12}$. In order to generate and store the numerical values of x_{11} and x_{12}, we execute the following code.

```
>> A=[0 0 1 0;0 0 0 1;-2 1 -1 1;1/2 -1/2 1/2 -1/2];
>> B1=[0;0;1;0];   % Input is f1 only
>> B2=[0;0;0;1/2]; % Input is f2 only
>> C=[1 0 0 0];D=0;
>> sys1=ss(A,B1,C,D);
>> sys2=ss(A,B2,C,D);
>> [x11,t]=step(sys1,150);  % Contribution of f1 to x1
>> [x12,t]=step(sys2,150);  % Contribution of f2 to x1
>> x1=x11+x12;   % Superimpose to obtain response x1
>> plot(t,x1)    % Figure 8.14
```

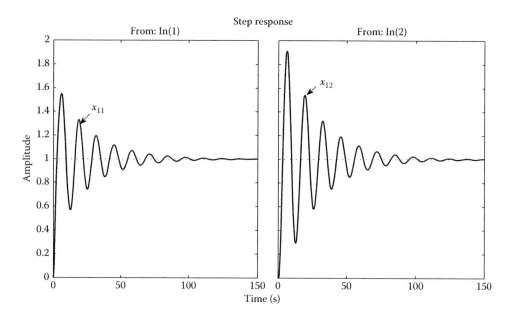

FIGURE 8.13 Contributions of the two inputs to x_1 (Example 8.8).

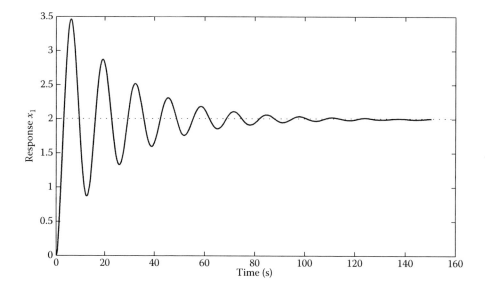

FIGURE 8.14 Time variations of response x_1 (Example 8.8).

Example 8.9: MIMO System

Suppose that the mechanical system in Example 8.8 is subjected to $f_1 = e^{-t}\sin t$, $f_2 = 0$, and initial conditions $x_1(0) = \frac{1}{2}$, $x_2(0) = 0 = \dot{x}_1(0) = \dot{x}_2(0)$. Plot x_1 versus $0 \leq t \leq 30\,\text{s}$.

Solution

```
>> A=[0 0 1 0;0 0 0 1;-2 1 -1 1;1/2 -1/2 1/2 -1/2];
```

System Response

```
>> B1=[0;0;1;0];C=[1 0 0 0];D=0;
>> t=linspace(0,30);
>> x0=[1/2;0;0;0];
>> sys=ss(A,B1,C,D);
>> u=exp(-t).*sin(t);
>> lsim(sys,u,t,x0)    % Figure 8.15
```

PROBLEM SET 8.2

1. A second-order dynamic system is modeled as

$$3\ddot{x} + 2\dot{x} + 2x = f(t),$$

where $f(t)$ is a prescribed function. Determine the undamped natural frequency and the damping ratio. If underdamped, also find the damped natural frequency.

2. Show that the free response of an overdamped, second-order system given by Equation 8.14 approaches zero as $t \to \infty$.

   In Problems 3 through 6, for each given system,
   a. Identify the damping type and find the free response.
   b. Find and plot the free response in MATLAB.

3. $4\ddot{x} + 4\dot{x} + x = 0$, $x(0) = -1$, $\dot{x}(0) = 0$
4. $\ddot{x} + 3\dot{x} + 9x = 0$, $x(0) = 0$, $\dot{x}(0) = \frac{1}{2}$
5. $2\ddot{x} + 5\dot{x} + 3x = 0$, $x(0) = 1$, $\dot{x}(0) = \frac{1}{2}$
6. $\ddot{x} + 2\dot{x} + x = 0$, $x(0) = 1$, $\dot{x}(0) = -1$
7. Show that the impulse response of an underdamped, second-order system given by Equation 8.17 approaches zero as $t \to \infty$.
8. Assuming zero initial conditions, show that the response of an underdamped, second-order system to an impulsive input $\delta(t-a)$, where $a = \text{const} > 0$, is described by

$$\left[\frac{1}{\omega_d} e^{-\zeta\omega_n(t-a)} \sin\omega_d(t-a) \right] u(t-a),$$

where $u(t)$ is the unit-step function. (*Hint*: Use the "shift on t-axis" property of Laplace transformation, Section 2.3.)

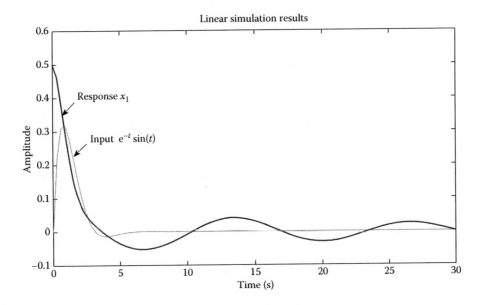

FIGURE 8.15 Response x_1 in Example 8.9.

9. A dynamic system is modeled as

$$\ddot{x} + 2\dot{x} + 2x = 3\delta(t-1).$$

Assuming zero initial conditions, use the result of Problem 8 to determine the response $x(t)$.

10. ◢ Consider a mass–spring–damper system such as that in Figure 8.6 of this section with

$$m = 4 \text{ kg}, \quad c = 4 \text{ N·s/m}, \quad k = 5 \text{ N/m}.$$

Find and plot its response when subjected to an applied force $\delta(t)$ and zero initial displacement and velocity.

11. ◢ Consider two scenarios pertaining to a mass–spring–damper system (Figure 8.6). In the first case, assume that the elements are

$$m = 1 \text{ kg}, \quad c = 1 \text{ N·s/m}, \quad k = 1 \text{ N/m}.$$

In the second case, suppose that the stiffness coefficient k is increased to 2 N/m. Assuming zero initial conditions, plot (in a single figure) the response x to a unit-impulse force for both cases and discuss the results.

12. ◢ For a mass–spring–damper system (Figure 8.6), consider the following two cases:
- $m = 1 \text{ kg}, \quad c = 1 \text{ N·s/m}, \quad k = 1 \text{ N/m}$
- $m = 1 \text{ kg}, \quad c = 2 \text{ N·s/m}, \quad k = 1 \text{ N/m}$

Assuming zero initial conditions, plot (in a single figure) the response x to a unit-impulse force for both cases and discuss the results.

13. The governing equation (via KVL) for the RLC circuit in Figure 8.16 is derived as

$$L\frac{di}{dt} + Ri + \frac{1}{C}\int i\,dt = v_a(t).$$

Noting that the electric charge q and current i are related via $i = \dot{q}$, rewrite the above equation as a second-order differential equation in q and subsequently determine the system's natural frequency and damping ratio.

14. Reconsider the circuit in Problem 13 with $L = 4 \text{ H}, R = 4\Omega$, and $C = \frac{1}{2} \text{ F}$.
    a. Write the governing equation in terms of the electric charge q.
    b. ◢ Find and plot q and i versus t (same figure) when the applied voltage v_a is a unit-impulse and initial conditions are zero.

15. The model of a dynamic system is described by

$$4\ddot{x} + 12\dot{x} + 10x = u(t), \quad u(t) = \text{unit-step},$$

subjected to zero initial conditions.

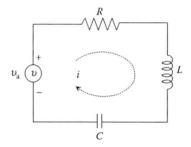

FIGURE 8.16 Problem 13.

System Response

a. What is the steady-state value x_{ss}?
b. ▲ Plot the response $x(t)$ and verify the value of x_{ss} from Part (a).

16. Repeat Problem 15 for
$$\ddot{x} + \dot{x} + 4x = 3u(t).$$

17. The mathematical model of a dynamic system is derived as
$$9\ddot{x} + 12\dot{x} + 4x = 2u(t), \quad x(0) = 0, \; \dot{x}(0) = 1.$$

   a. Find the analytical expression for response $x(t)$ and determine x_{ss}.
   b. ▲ Plot $x(t)$ of Part (a) versus $0 \le t \le 20$ s.

18. A dynamic system is modeled as
$$9\ddot{x} + 6\dot{x} + x = 4u(t), \quad x(0) = 1, \; \dot{x}(0) = 0.$$

   a. Find the analytical expression for response $x(t)$ and determine x_{ss}.
   b. ▲ Plot $x(t)$ of Part (a) versus $0 \le t \le 30$ s.

19. ▲ A dynamic system model is described by
$$3\ddot{x} + \dot{x} + 2x = e^{-t/2}, \quad x(0) = 0.5, \; \dot{x}(0) = 0.$$

   Plot the response $x(t)$ for $0 \le t \le 20$ s using the `lsim` command. Discuss the results in relation to the stability of the system at hand. (See Problem Set 4.2.)

20. ▲ Repeat Problem 10 for
$$10\ddot{x} + \dot{x} + 3x = 3e^{-t}\sin t, \quad x(0) = 0, \; \dot{x}(0) = \frac{1}{4}.$$

21. ▲ Repeat Example 8.8 for the case when x_2 is the output.
22. ▲ Repeat Example 8.8 when the system is subjected to initial conditions
$$x_1(0) = 0, \quad x_2(0) = 1, \quad \dot{x}_1(0) = 1, \quad \dot{x}_2(0) = 0.$$

23. ▲ The mechanical system in Example 8.8 is subjected to external forces $f_1 = e^{-t/2}\sin t$ and $f_2 = u(t) = $ unit-step, as well as zero initial conditions. Plot the time variations of x_1.

24. ▲ The system in Example 8.8 is subjected to inputs $f_1 = e^{-t/2}$, $f_2 = u(t) = $ unit-step, and initial conditions $x_1(0) = 1$, $x_2(0) = 0$, $\dot{x}_1(0) = 0$, $\dot{x}_2(0) = 1$. Plot the response x_1 versus t.

25. ▲ Repeat Problem 24 for the case when $\dot{x}_1$ is the response.
26. ▲ Repeat Problem 23 for the case when the response is $\dot{x}_2$.

8.3 FREQUENCY RESPONSE

Suppose that an LTI system is subjected to a sinusoidal input, characterized by its amplitude and forcing frequency. The system response will then contain two portions: one that vibrates at the natural frequency of the system and another that follows the forcing frequency. If damping is present, the portion vibrating at the system's natural frequency will eventually die out, as discussed earlier in this chapter. Therefore, the response at steady state is sinusoidal and it has the same frequency as the input (forcing frequency). The steady-state response to a sinusoidal input is known as the frequency response (Figure 8.17).

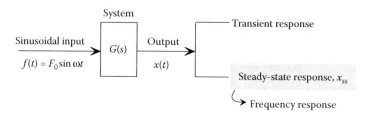

FIGURE 8.17 Frequency response of a linear system.

8.3.1 Frequency Response of Stable, Linear Systems

Consider a stable, LTI system described by its transfer function $G(s)$ as in Figure 8.17. The input is sinusoidal $f(t) = F_0 \sin \omega t$ with amplitude F_0 and (forcing) frequency ω. The frequency response is the steady-state portion of the response, denoted by x_{ss}, and is determined as follows.

Suppose that the transfer function is a ratio of two polynomials in s, that is,

$$G(s) = \frac{K(s+z_1)(s+z_2)\cdots(s+z_m)}{(s+p_1)(s+p_2)\cdots(s+p_n)}, \quad K = \text{const.} \tag{8.25}$$

Note that since x_{ss} is independent of the initial conditions, they are simply disregarded. Since $G(s) = X(s)/F(s)$, we have $X(s) = G(s)F(s)$, where $F(s) = F_0\omega/(s^2+\omega^2)$. If $G(s)$ has distinct poles only, then partial-fraction expansion gives

$$\begin{aligned} X(s) = G(s)F(s) &= \frac{K(s+z_1)(s+z_2)\cdots(s+z_m)}{(s+p_1)(s+p_2)\cdots(s+p_n)} \cdot \frac{F_0\omega}{s^2+\omega^2} \\ &= \frac{c_1}{s+p_1} + \frac{c_2}{s+p_2} + \cdots + \frac{c_n}{s+p_n} + \frac{d}{s+j\omega} + \frac{\bar{d}}{s-j\omega}, \end{aligned} \tag{8.26}$$

where $c_i (i = 1, 2, \ldots, n)$ and d are constants and $\bar{d}$ is the complex conjugate of d. Inverse Laplace transformation yields

$$x(t) = c_1 e^{-p_1 t} + c_2 e^{-p_2 t} + \cdots + c_n e^{-p_n t} + d e^{-j\omega t} + \bar{d} e^{j\omega t}.$$

Because the system is assumed to be stable, the poles $-p_1, -p_2, \ldots, -p_n$ all have negative real parts. Therefore, except the last two terms, all terms on the right side die out at steady state. The same result comes about for the case of one or more multiple poles. If a typical pole $-p_k$ has multiplicity q, then $x(t)$ will contain terms $t^r e^{-p_k t}$ ($r = 0, 1, \ldots, q-1$) so that they too will die out at steady state. Therefore, regardless of the multiplicity of the poles, we have

$$x_{ss}(t) = d e^{-j\omega t} + \bar{d} e^{j\omega t}. \tag{8.27}$$

The constants d and $\bar{d}$ can be evaluated from Equation 8.26 via

$$d = (s+j\omega)X(s)|_{s=-j\omega} = \left[(s+j\omega)G(s)\frac{F_0\omega}{s^2+\omega^2}\right]_{s=-j\omega} = -\frac{F_0}{2j}G(-j\omega),$$

$$\bar{d} = (s-j\omega)X(s)|_{s=j\omega} = \left[(s-j\omega)G(s)\frac{F_0\omega}{s^2+\omega^2}\right]_{s=j\omega} = \frac{F_0}{2j}G(j\omega).$$

It is readily seen that $\bar{d}$ and d are indeed complex conjugates. The quantity $G(j\omega)$ is known as the frequency response function (FRF) and is obtained by replacing s with $j\omega$ in $G(s)$. Since $G(j\omega)$ is

System Response

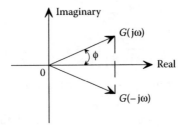

FIGURE 8.18 FRF and its conjugate.

generally complex, it can be expressed in polar form as (Figure 8.18)

$$G(j\omega) = |G(j\omega)|e^{j\phi}, \quad \phi = \angle G(j\omega) = \tan^{-1}\left\{\frac{\text{imaginary part of } G(j\omega)}{\text{real part of } G(j\omega)}\right\}.$$

Similarly,

$$G(-j\omega) = |G(-j\omega)|e^{-j\phi} = |G(j\omega)|e^{-j\phi}.$$

Using these in the expressions for d and $\bar{d}$, and inserting the results into Equation 8.27, we find

$$x_{ss}(t) = -\frac{F_0}{2j}|G(j\omega)|e^{-j\phi}e^{-j\omega t} + \frac{F_0}{2j}|G(j\omega)|e^{j\phi}e^{j\omega t}$$

$$= F_0|G(j\omega)|\frac{e^{j(\omega t + \phi)} - e^{-j(\omega t + \phi)}}{2j} = F_0|G(j\omega)|\sin(\omega t + \phi).$$

In summary, if the input of a stable, LTI system with transfer function $G(s)$ is $F_0 \sin \omega t$, the system's frequency response is

$$x_{ss}(t) = F_0|G(j\omega)|\sin(\omega t + \phi), \tag{8.28}$$

where $|G(j\omega)|$ and ϕ denote, respectively, the magnitude and phase of the FRF $G(j\omega)$.

8.3.1.1 Frequency Response of First-Order Systems

Linear, first-order systems are described by (see Equation 8.2)

$$\tau \dot{x} + x = f(t), \quad \tau = \text{time constant} > 0.$$

The frequency response x_{ss}, when the system is subjected to $f(t) = F_0 \sin \omega t$, is obtained as follows. The system's transfer function is

$$G(s) = \frac{1}{\tau s + 1} \stackrel{\text{FRF}}{\Rightarrow} G(j\omega) = \frac{1}{1 + \tau \omega j}.$$

The magnitude of the FRF is simply

$$|G(j\omega)| = \frac{1}{\sqrt{1 + (\tau\omega)^2}}.$$

To calculate the phase, however, it is advised that $G(j\omega)$ be expressed in standard rectangular form so that its location in the complex plane is determined. That is,

$$G(j\omega) = \frac{1}{1 + \tau\omega j} = \frac{1 - \tau\omega j}{1 + (\tau\omega)^2} = \frac{1}{1 + (\tau\omega)^2} - j\frac{\tau\omega}{1 + (\tau\omega)^2}.$$

This indicates that $G(j\omega)$ is located in the fourth quadrant. The phase is given by

$$\phi = \angle G(j\omega) = \tan^{-1}\left\{\frac{\text{imaginary part of } G(j\omega)}{\text{real part of } G(j\omega)}\right\} = \tan^{-1}(-\tau\omega) = -\tan^{-1}(\tau\omega).$$

Therefore, by Equation 8.28, the frequency response is found as

$$x_{ss}(t) = \frac{F_0}{\sqrt{1+(\tau\omega)^2}} \sin\left(\omega t - \tan^{-1}(\tau\omega)\right). \tag{8.29}$$

Example 8.10: Frequency Response of a First-Order System

Find the frequency response corresponding to

$$\dot{x} + 2x = 3\sin 2t.$$

SOLUTION

Rewrite in standard form as

$$\frac{1}{2}\dot{x} + x = \frac{3}{2}\sin 2t \;\Rightarrow\; \tau = \frac{1}{2}, \quad F_0 = \frac{3}{2}, \quad \omega = 2\,\text{rad/s}.$$

Thus

$$G(s) = \frac{1}{(1/2)s+1} \;\overset{\text{FRF}}{\underset{s=\omega j=2j}{\Rightarrow}}\; G(2j) = \frac{1}{1+j}.$$

It is clear that $|G(2j)| = 1/\sqrt{2}$. To find the phase, express $G(2j)$ in rectangular form as

$$G(2j) = \frac{1-j}{2} = \frac{1}{2} - \frac{1}{2}j.$$

This means that it is located in the fourth quadrant and $\phi = -\pi/4$ rad. Finally, by Equation 8.29, we have

$$x_{ss}(t) = \frac{3/2}{\sqrt{2}} \sin\left(2t - \frac{\pi}{4}\right) = \frac{3\sqrt{2}}{4} \sin\left(2t - \frac{\pi}{4}\right).$$

8.3.1.2 Frequency Response of Second-Order Systems

Frequency-response analysis of second-order systems is best understood when applied to a single-degree-of-freedom mechanical system. To that end, consider the mechanical system shown in Figure 8.19, where x is measured from the static equilibrium position and the applied force is $f(t) = F_0 \sin \omega t$. The frequency response x_{ss} is obtained as follows.

Because the equation of motion is $m\ddot{x} + b\dot{x} + kx = f(t)$, the system's transfer function is

$$G(s) = \frac{X(s)}{F(s)} = \frac{1}{ms^2 + bs + k} \;\overset{\text{FRF}}{\Rightarrow}\; G(j\omega) = \frac{1}{k - m\omega^2 + b\omega j}.$$

The magnitude and phase of the FRF are determined as

$$|G(j\omega)| = \frac{1}{\sqrt{(k-m\omega^2)^2 + (b\omega)^2}}, \quad \phi = \angle G(j\omega) = -\tan^{-1}\frac{b\omega}{k-m\omega^2}.$$

But from Section 8.2, we know $\omega_n = \sqrt{k/m}$ and $\zeta = b/(2\sqrt{mk})$ so that

$$\frac{k}{m} = \omega_n^2, \quad \frac{b}{k} = \frac{2\zeta}{\omega_n}.$$

System Response

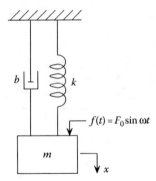

FIGURE 8.19 Second-order mechanical system.

Multiply and divide the fractions involved in $|G(j\omega)|$ and ϕ, use the above relations in the resulting expressions, and then insert into Equation 8.28 to obtain

$$x_{ss}(t) = \frac{x_{st}}{\sqrt{[1-(\omega/\omega_n)^2]^2 + (2\zeta\omega/\omega_n)^2}} \sin\left(\omega t - \tan^{-1}\frac{2\zeta\omega/\omega_n}{1-(\omega/\omega_n)^2}\right), \qquad (8.30)$$

where $x_{st} = F_0/k$ is the static deflection, and the dimensionless quantity ω/ω_n is the normalized frequency. If we let the amplitude of x_{ss} in Equation 8.30 be denoted by X, then

$$X = \frac{x_{st}}{\sqrt{[1-(\omega/\omega_n)^2]^2 + (2\zeta\omega/\omega_n)^2}} \Rightarrow \frac{X}{x_{st}} = \frac{1}{\sqrt{[1-(\omega/\omega_n)^2]^2 + (2\zeta\omega/\omega_n)^2}}.$$

The dimensionless ratio X/x_{st} is known as the normalized amplitude. Further details of the above findings, and in particular, applications in mechanical vibrations, will be presented in Section 9.2.

Example 8.11: Frequency Response of a Second-Order System

Find the frequency response of a mechanical system such as that in Figure 8.19 with the equation of motion

$$\ddot{x} + 2\dot{x} + 4x = 0.8\sin t.$$

Solution

We first observe that $F_0 = 0.8$, $m = 1$, $k = 4$, $b = 2$, $\omega = 1$ in consistent physical units. Thus

$$x_{st} = \frac{F_0}{k} = \frac{0.8}{4} = 0.2, \quad \omega_n = 2, \quad \zeta = \frac{1}{2}, \quad \frac{\omega}{\omega_n} = \frac{1}{2}, \quad 2\zeta\frac{\omega}{\omega_n} = \frac{1}{2}.$$

Substituting into Equation 8.30, and expressing the phase in radians, yields

$$x_{ss}(t) = 0.2219\sin(t - 0.5880).$$

8.3.2 Bode Diagram

An FRF may be represented by a pair of plots, one displaying the magnitude versus frequency, and the other the phase angle (in degrees) versus frequency. A very specific presentation of this pair of plots is known as a Bode diagram. For a given FRF $G(j\omega)$ with magnitude $|G(j\omega)|$ and phase angle ϕ, a Bode diagram consists of two plots: a curve of the logarithm of $|G(j\omega)|$ and a plot of the phase ϕ, both versus log ω. All logarithms are base 10. A common way to represent the logarithmic magnitude of $G(j\omega)$ is $20\log|G(j\omega)|$ with the unit of decibel, abbreviated dB. In a Bode diagram, the curves are sketched on semilog paper, using a linear scale for magnitude (in dB) and phase (in degrees) and a logarithmic scale for frequency (in rad/s). The logarithmic presentation described here allows the low- and high-frequency behavior of the transfer function to be shown in a single diagram. And this is very valuable because in practical situations, low-frequency characteristics of a system are most important. Bode diagrams play an important role in control systems design, as discussed in Section 10.6.

8.3.2.1 Plotting Bode Diagrams in MATLAB

The MATLAB command bode generates the Bode diagram for an LTI system. For a system sys, the command

```
>> bode(sys)
```

will plot the Bode diagram, comprised of a magnitude (dB) plot and a phase (degrees) plot. If the actual generated numerical values for magnitude and phase are desired, we define a range of frequency in the form of vector w, and use

```
>> [mag,phase,w] = bode(sys,w);
```

This returns matrices mag and phase but not a plot. Consider the following code:

```
>> w=linspace(0.1,100);    % Define frequency range (100 points default)
>> [mag,phase,w]=bode(sys,w);
>> for i=1:100,
m(i)=mag(1,1,i);    % Extract the magnitude column
p(i)=phase(1,1,i);  % Extract the phase column
end
>> subplot(2,1,1),semilogx(w,20*log10(m))  % Create the magnitude (dB) plot
>> subplot(2,1,2),semilogx(w,p)    % Complete Bode diagram
```

8.3.2.2 Bode Diagram of First-Order Systems

Consider a first-order system model in its standard form

$$\tau \dot{x} + x = f(t), \quad \tau > 0,$$

so that the transfer function is $G(s) = 1/(\tau s + 1)$ and the FRF is formed as

$$G(j\omega) = \frac{1}{1 + \tau \omega j}.$$

As shown earlier in this section, the magnitude and phase of the FRF are then

$$|G(j\omega)| = \frac{1}{\sqrt{1 + (\tau\omega)^2}}, \quad \phi = -\tan^{-1}(\tau\omega).$$

The Bode diagram is obtained as follows. By definition, the logarithmic magnitude is

$$20\log|G(j\omega)| = 20\log\frac{1}{\sqrt{1+(\tau\omega)^2}} = 20\log\left[1+(\tau\omega)^2\right]^{-1/2} = -10\log\left[1+(\tau\omega)^2\right] \text{ dB}.$$

System Response

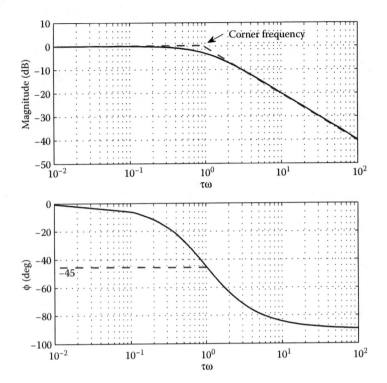

FIGURE 8.20 Bode diagram for $G(j\omega) = 1/(1 + \tau\omega j)$.

The plots of this quantity and the phase ϕ of the FRF versus $\tau\omega$ (logarithmic scale) will make up the Bode diagram shown in Figure 8.20. Since τ, the time constant, is in s and ω is in rad/s, $\tau\omega$ is dimensionless. The asymptotic approximations of magnitude (dB) for low- and high-frequency ranges intersect when $\tau\omega = 1$, so that $\omega = 1/\tau$ is the corner frequency. The phase at this frequency is $\phi = -\tan^{-1} 1 = -45°$. For high-frequency range, the asymptote of the magnitude has a slope of -20 dB/decade while the phase curve approaches $-90°$.

Example 8.12: Bode Diagram

Draw the Bode diagram for

$$G(s) = \frac{1}{s+3}.$$

Solution

We first rewrite $G(s)$ as

$$G(s) = \frac{1/3}{(1/3)s + 1} \quad \overset{\text{FRF}}{\Rightarrow} \quad G(j\omega) = \frac{1/3}{1 + (1/3)\omega j} = \frac{1}{3} \cdot \frac{1}{1 + (1/3)\omega j}.$$

This agrees with the standard form $1/(1 + \tau\omega j)$, with $\tau = 1/3$, except for the constant $1/3$ multiple. The magnitude is

$$|G(j\omega)| = \frac{1}{3} \cdot \frac{1}{\sqrt{1 + ((1/3)\omega)^2}} \quad \overset{\text{Logarithmic magnitude}}{\Rightarrow} \quad 20\log\frac{1}{3} + 20\log\frac{1}{\sqrt{1 + ((1/3)\omega)^2}} \text{ dB}.$$

The second term is the standard magnitude (dB) for first-order systems discussed earlier. Therefore, the complete magnitude (dB) for the present case is obtained by lowering the standard

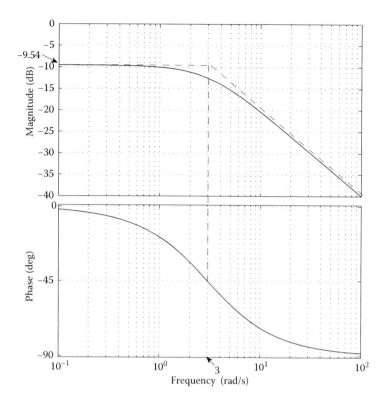

FIGURE 8.21 Bode diagram (Example 8.12).

magnitude curve by $20\log\frac{1}{3} = -9.54$ dB. The constant 1/3 does not affect the phase angle. The following code generates the Bode diagram.

```
>> sys = tf(1,[1 3]);
>> bode(sys)    % Figure 8.21
```

The same pair of plots can also be generated, as suggested earlier, by retaining the numerical values of magnitude and phase and then plotting them.

```
>> sys = tf(1,[1 3]);
>> w = linspace(0.1,100);
>> [mag,phase,w] = bode(sys,w);
>> for i=1:100,
m(i)=mag(1,1,i);
p(i)=phase(1,1,i);
end
>> subplot(2,1,1),semilogx(w,20*log10(m))
>> subplot(2,1,2),semilogx(w,p)
```

The corner frequency is $\omega = 1/\tau = 3$ rad/s. In the low-frequency range, the magnitude for the standard case is 0 dB as shown in Figure 8.20. Lowering that by $20\log\frac{1}{3} = -9.54$ dB yields the result depicted in Figure 8.21. The phase plot is clearly unchanged, as expected, compared with the standard case in Figure 8.20.

System Response

8.3.2.3 Bode Diagram of Second-Order Systems

Consider a second-order system in the standard (normalized) form

$$\ddot{x} + 2\zeta\omega_n \dot{x} + \omega_n^2 x = \omega_n^2 f(t), \tag{8.31}$$

so that the standard, second-order transfer function is obtained as

$$G(s) = \frac{\omega_n^2}{s^2 + 2\zeta\omega_n s + \omega_n^2}$$

and the FRF is subsequently formed as

$$G(j\omega) = \frac{\omega_n^2}{\omega_n^2 - \omega^2 + 2\zeta\omega_n \omega j} \stackrel{\text{Divide by } \omega_n^2}{=} \frac{1}{1 - (\omega/\omega_n)^2 + 2\zeta(\omega/\omega_n)j}.$$

As a result, the magnitude and phase of the FRF are calculated as

$$|G(j\omega)| = \frac{1}{\sqrt{[1-(\omega/\omega_n)^2]^2 + (2\zeta\omega/\omega_n)^2}}, \quad \phi = -\tan^{-1}\frac{2\zeta\omega/\omega_n}{1-(\omega/\omega_n)^2}. \tag{8.32}$$

The logarithmic magnitude is

$$20\log|G(j\omega)| = 20\log\frac{1}{\sqrt{[1-(\omega/\omega_n)^2]^2 + (2\zeta\omega/\omega_n)^2}} \text{ dB}.$$

For a fixed ζ, both the magnitude (dB) and phase angle ϕ are functions of the normalized frequency ω/ω_n and thus may be plotted versus ω/ω_n. Figure 8.22 shows several pairs of curves, each pair corresponding to a fixed ζ. It turns out that some magnitude curves attain a maximum peak, while others do not. To find the frequency at which a maximum peak occurs, we proceed as follows. Maximizing the logarithmic magnitude is equivalent to maximizing $|G(j\omega)|$, which is equivalent to minimizing the denominator of $|G(j\omega)|$, that is,

$$\sqrt{(1-r^2)^2 + (2\zeta r)^2},$$

where $r = \omega/\omega_n$.

It can be easily shown (Problem Set 8.3) that minimization of this quantity with respect to r leads to the critical value

$$r = \sqrt{1-2\zeta^2} \quad \Rightarrow \quad \frac{\omega}{\omega_n} = \sqrt{1-2\zeta^2}.$$

With this result, we define the resonant frequency ω_r as

$$\omega_r = \omega_n\sqrt{1-2\zeta^2}. \tag{8.33}$$

Based on this, the magnitude curves that attain a maximum peak correspond to $0 \leq \zeta \leq 0.707$, where 0.707 is an approximate value for $\sqrt{2}/2$. The maximum peak of the magnitude itself is then found as

$$|G(j\omega)|_{\max} = |G(j\omega)|_{\omega/\omega_n=\sqrt{1-2\zeta^2}} = \frac{1}{2\zeta\sqrt{1-\zeta^2}}. \tag{8.34}$$

For example, the maximum peak for the case of $\zeta = 0.3$ occurs at $\omega/\omega_n = \sqrt{1-2(0.3)^2} = 0.9055$ and

$$|G(j\omega)|_{\max} = \left.\frac{1}{2\zeta\sqrt{1-\zeta^2}}\right|_{\zeta=0.3} = 1.7471 \stackrel{\text{Logarithmic magnitude}}{\Rightarrow} 20\log(1.7471) = 4.8466\,\text{dB}.$$

All of these are readily verified via Figure 8.22.

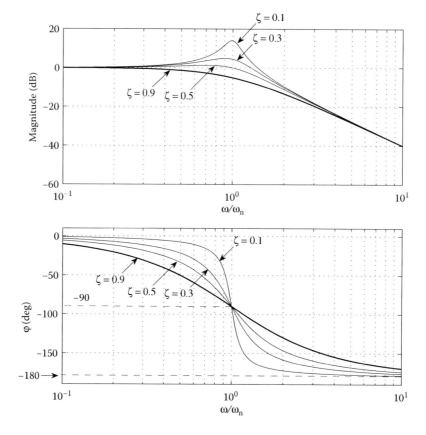

FIGURE 8.22 Bode diagram for $G(j\omega) = 1/\left[1 - (\omega/\omega_n)^2 + 2\zeta(\omega/\omega_n)j\right]$.

Example 8.13: Bode Diagram

Draw the Bode diagram for

$$G(s) = \frac{3}{s^2 + 2s + 4}.$$

Solution

Rewrite $G(s)$ to extract the standard, second-order transfer function, as

$$G(s) = \frac{3}{4} \cdot \underbrace{\frac{4}{s^2 + 2s + 4}}_{\text{Standard second-order TF}} \overset{\text{FRF}}{\Rightarrow} G(j\omega) = \frac{3}{4} \cdot \frac{4}{4 - \omega^2 + 2\omega j}.$$

The magnitude is then

$$|G(j\omega)| = \frac{3}{4} \cdot \frac{4}{\sqrt{(4 - \omega^2)^2 + (2\omega)^2}} \overset{\text{Logarithmic magnitude}}{\Rightarrow} 20\log\frac{3}{4} + 20\log\frac{4}{\sqrt{(4 - \omega^2)^2 + (2\omega)^2}} \, \text{dB}.$$

The second term is the standard magnitude (dB) for second-order systems. Therefore, the complete magnitude (dB) is obtained by lowering the standard magnitude curve by $20\log\frac{3}{4} = -2.50\,\text{dB}$. Once again, the constant 3/4 does not affect the phase angle. The following code generates the Bode diagram.

```
>> sys=tf([3],[1 2 2]);bode(sys)    % Figure 8.23
```

System Response

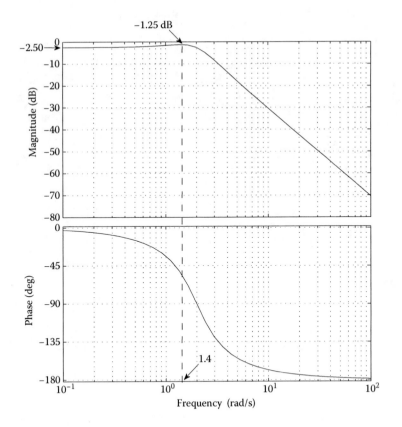

FIGURE 8.23 Bode diagram (Example 8.13).

Using the standard, second-order transfer function, we find $\omega_n = 2$ rad/s and $\zeta = 0.5$. Since $\zeta \leq 0.707$, the corresponding magnitude (dB) curve will have a maximum peak, which occurs at the resonant frequency

$$\omega_r = \omega_n \sqrt{1 - 2\zeta^2}\Big|_{\omega_n = 2, \zeta = 0.5} \cong 1.4 \text{ rad/s.}$$

For the standard form, the peak value is calculated as

$$|G(j\omega)|_{max} = \frac{1}{2\zeta\sqrt{1-\zeta^2}}\Big|_{\zeta = 0.5} = 1.16 \xrightarrow{\text{Logarithmic magnitude}} 20\log(1.16) = 1.25 \text{ dB.}$$

However, the magnitude curve has been lowered by -2.50 dB. Therefore, the maximum peak for our case is adjusted to $1.25 - 2.50 = -1.25$ dB. This is clearly indicated in Figure 8.23.

PROBLEM SET 8.3

In Problems 1 through 6 find the frequency response.

1. $3\dot{x} + 2x = 0.4 \sin t$
2. $\dot{x} + \frac{1}{2}x = 10 \sin 3t$
3. $\dot{y} + 3y = 6 \sin 2t$
4. $4\ddot{x} + 12\dot{x} + 13x = 15 \sin(t/2)$
5. $3\ddot{x} + 2\dot{x} + 9x = 0.9 \sin(t/3)$

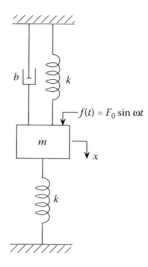

FIGURE 8.24 Problem 7.

6. $10\ddot{x} + 25\dot{x} + 400x = 10\sin 2t$
7. The equation of motion for the mechanical system shown in Figure 8.24 is

$$m\ddot{x} + b\dot{x} + 2kx = F_0 \sin \omega t,$$

where x is measured from the static equilibrium. Given that $m = 5\,\text{kg}$, $b = 20\,\text{N·s/m}$, $k = 200\,\text{N/m}$, $F_0 = 20\,\text{N}$, and $\omega = 2\,\text{rad/s}$, find the system's frequency response.
8. Repeat Problem 7 for $m = 10\,\text{kg}$, $b = 35\,\text{N·s/m}$, $k = 300\,\text{N/m}$, $F_0 = 10\,\text{N}$, and $\omega = 3\,\text{rad/s}$.
9. Consider the undamped mechanical system in Figure 8.25, where $m = 1\,\text{kg}$, $k = 50\,\text{N/m}$, $F_0 = 100\,\text{N}$, and $\omega = 2\,\text{rad/s}$. Assuming zero initial conditions, find the frequency response.
10. Show that the steady-state current in the RLC circuit shown in Figure 8.26 is described by

$$i_{ss}(t) = \frac{\omega C V_0}{\sqrt{(1 - LC\omega^2)^2 + (RC\omega)^2}} \cos\left(\omega t - \tan^{-1}\frac{RC\omega}{1 - LC\omega^2}\right).$$

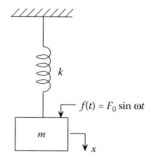

FIGURE 8.25 Problem 9.

System Response

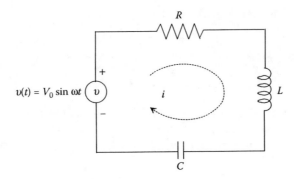

FIGURE 8.26 Problem 10.

11. For the transfer function given below, draw the Bode diagram and identify the corner frequency as well as the low-frequency magnitude (dB). See Example 8.12.

$$G(s) = \frac{2}{3s+1}.$$

12. Repeat Problem 11 for

$$G(s) = \frac{4}{2s+3}.$$

13. For the first-order system with FRF

$$G(j\omega) = \frac{1}{1+\tau\omega j},$$

show that the phase corresponding to the corner frequency is $-45°$.

14. Show that the magnitude $|G(j\omega)|$ in Equation 8.32 attains a maximum when

$$\frac{\omega}{\omega_n} = \sqrt{1-2\zeta^2}.$$

15. Show that the phase curves corresponding to all values of ζ in Figure 8.22 intersect at the critical point characterized by $\omega/\omega_n = 1$ and $\phi = -90°$.

16. For the second-order transfer function given below, draw the Bode diagram, identify the resonant frequency, find the magnitude (dB) peak, if applicable, and give the approximate low-frequency dB value. See Example 8.13.

$$G(s) = \frac{2}{s^2+2s+4}.$$

17. Repeat Problem 16 for

$$G(s) = \frac{17.28}{s^2+3.84s+5.76}.$$

18. Repeat Problem 16 for

$$G(s) = \frac{1}{1.8s^2+6.48s+28.80}.$$

8.4 SOLVING THE STATE EQUATION

In Section 4.2, we found out that using the state variables, the mathematical model of a linear dynamic system can be expressed in the form of the state equation

$$\dot{\mathbf{x}} = \mathbf{A}\mathbf{x} + \mathbf{B}\mathbf{u}, \quad \mathbf{x}(0) = \mathbf{x}_0, \tag{8.35}$$

where $\mathbf{x}(t)$ is the state vector, $\mathbf{A}$ is the state matrix, $\mathbf{B}$ is the input matrix, $\mathbf{u}(t)$ is the vector of the inputs, and $\mathbf{x}_0$ is the initial state vector. Note that matrix and vector sizes are such that all operations involved are valid. In this section, we will learn how to solve the state equation to derive the state vector.

8.4.1 Formal Solution of the State Equation

In order to solve Equation 8.35, we first need to recall the treatment of its scalar counterpart

$$\dot{x}(t) = ax(t) + bu(t), \quad x(0) = x_0.$$

The solution to this (initial-value) problem is obtained as

$$x(t) = e^{at}x_0 + \int_0^t e^{a(t-\tau)} bu(\tau)\, d\tau.$$

Now, inspired by this development, the formal solution of the state equation is expressed as

$$\mathbf{x}(t) = e^{\mathbf{A}t}\mathbf{x}_0 + \int_0^t e^{\mathbf{A}(t-\tau)} \mathbf{B}\mathbf{u}(\tau)\, d\tau. \tag{8.36}$$

The one quantity here that has yet to be defined is $e^{\mathbf{A}t}$, which we will call the matrix exponential of $\mathbf{A}$. Note that technically $e^{\mathbf{A}}$ is the matrix exponential, but $e^{\mathbf{A}t}$ is what is involved in dynamic systems analysis.

8.4.1.1 Matrix Exponential

When a is scalar, e^{at} is defined as

$$e^{at} = \sum_{k=0}^{\infty} \frac{1}{k!}(at)^k = 1 + ta + \frac{1}{2!}t^2 a^2 + \cdots.$$

This idea can be extended to the case involving a square matrix $\mathbf{A}_{n \times n}$, as

$$e^{\mathbf{A}t} = \sum_{k=0}^{\infty} \frac{1}{k!}(t\mathbf{A})^k = \mathbf{I} + t\mathbf{A} + \frac{1}{2!}t^2\mathbf{A}^2 + \cdots, \tag{8.37}$$

where $\mathbf{I}$ is the $n \times n$ identity matrix and $\mathbf{A}^k$ means $\mathbf{A}$ multiplying itself k times. Note that $e^{\mathbf{A}t}$ is an $n \times n$ matrix.

8.4.1.1.1 Properties of the Matrix Exponential

1. Given $\mathbf{A}_{n \times n}$ and scalars t and τ,

$$e^{\mathbf{A}(t+\tau)} = e^{\mathbf{A}t} e^{\mathbf{A}\tau},$$

which can be verified as follows.

$$e^{A(t+\tau)} = \sum_{k=0}^{\infty} \frac{1}{k!}A^k(t+\tau)^k = \sum_{k=0}^{\infty} \frac{1}{k!}A^k \sum_{m=0}^{\infty} \binom{k}{m} t^m \tau^{k-m} = \sum_{k=0}^{\infty} \frac{1}{k!}A^k \sum_{m=0}^{\infty} \frac{k!}{m!(k-m)!} t^m \tau^{k-m}$$

$$= \sum_{m=0}^{\infty} \sum_{k=m}^{\infty} A^k \frac{t^m \tau^{k-m}}{m!(k-m)!} \underset{k-m=r}{=} \sum_{m=0}^{\infty} \sum_{r=0}^{\infty} A^{r+m} \frac{t^m \tau^r}{m!r!}$$

$$= \left(\sum_{m=0}^{\infty} \frac{1}{m!} A^m t^m \right) \left(\sum_{r=0}^{\infty} \frac{1}{r!} A^r \tau^r \right) = e^{At} e^{A\tau}.$$

Here, $(t+\tau)^k = \sum_{m=0}^{\infty} \binom{k}{m} t^m \tau^{k-m}$ is the binomial series and $\binom{k}{m} = \frac{k!}{m!(k-m)!}$ is the binomial coefficient.

2. Since the series in Equation 8.37 is absolutely convergent for finite values of t, term-by-term differentiation of the series yields

$$\frac{d}{dt} e^{At} = A e^{At} = e^{At} A.$$

3. If $A_{n \times n}$ and $B_{n \times n}$ are similar matrices such that $S^{-1}AS = B$, then $At = S[Bt]S^{-1}$ and

$$e^{At} = S e^{Bt} S^{-1},$$

which is verified as follows.

$$e^{At} = I + tA + \frac{1}{2!} t^2 A^2 + \cdots = I + t SBS^{-1} + \frac{1}{2!} t^2 (SBS^{-1})(SBS^{-1}) + \cdots$$

$$= I + t SBS^{-1} + \frac{1}{2!} t^2 SB^2 S^{-1} + \cdots$$

$$= S \left[I + tB + \frac{1}{2!} t^2 B^2 + \cdots \right] S^{-1} = S e^{Bt} S^{-1}.$$

4. If matrix D is diagonal, say, $D = [d_{ii}]_{i=1,2,\ldots,n}$, then

$$e^{Dt} = \left[e^{d_{ii} t} \right]_{i=1,2,\ldots,n}.$$

We now present a systematic method to find e^{At}, where $A_{n \times n}$ has eigenvalues $\lambda_1, \lambda_2, \ldots, \lambda_n$ and linearly independent eigenvectors $v_1, v_2, \ldots, v_n$. We learned in Section 3.2 that the modal matrix $V = [v_1 \; v_2 \ldots v_n]$ diagonalizes A via

$$V^{-1} AV = D = [\lambda_i]_{i=1,2,\ldots,n} \;\Rightarrow\; A = VDV^{-1} \;\Rightarrow\; At = V[Dt]V^{-1}.$$

Then, by property 3 above, we find

$$e^{At} = V e^{Dt} V^{-1}. \tag{8.38}$$

Noting that the modal matrix V is known and e^{Dt} is easy to calculate (property 4), e^{At} will be readily available.

8.4.1.2 Formal Solution in MATLAB

The command expm calculates the matrix exponential. The integration portion of the formal solution is handled by the int command. For instance, the integral

$$\int_0^t g(\tau)\,d\tau$$

can be calculated as

int(g,tau,0,t)

Example 8.14: Matrix Exponential

Find e^{At}, where

$$A = \begin{bmatrix} 1 & 0 \\ -1 & -2 \end{bmatrix}.$$

SOLUTION

We first do this by using the expm command.

```
>> A=[1 0;-1 -2];
>> syms t
>> expm(A*t)
ans =
[                    exp(t),              0]
[ 1/3*exp(-2*t)-1/3*exp(t),      exp(-2*t)]
```

Next, we will confirm the result by following Equation 8.38.

```
% Find the modal matrix V and the diagonal matrix of the eigenvalues of A
>> [V,D]=eig(A)
V =
         0    0.9487
    1.0000   -0.3162
D =
    -2    0
     0    1
% Apply Eq. (8.38)
>> V*expm(D*t)*inv(V)
ans =
[                    exp(t),              0]
[ 1/3*exp(-2*t)-1/3*exp(t),      exp(-2*t)]
```

This confirms the previous result.

Example 8.15: Formal Solution

Find the formal solution of

$$\dot{x} = \begin{bmatrix} 1 & 0 \\ -1 & -2 \end{bmatrix} x + \begin{bmatrix} 1 \\ -2 \end{bmatrix} e^{-t}, \quad x_0 = \begin{Bmatrix} 0 \\ 1 \end{Bmatrix}.$$

System Response

Solution

```
>> syms t tau
>> A=[1 0;-1 -2]; B=[1;-2]; x0=[0;1];
% Eq. (8.36)
>> x=expm(A*t)*x0+int(expm(A*(t-tau))*B*exp((-tau),tau,0,t)
x =
                    1/2*exp(t)-1/2*exp(-t)
exp(-2*t)-1/6*(-10+exp(3*t)+9*exp(t))*exp(-2*t)
```

8.4.2 Solution of the State Equation via Laplace Transformation

We now solve the state equation using Laplace transformation, noting that (see Section 4.4)

$$\mathbf{x} = \begin{Bmatrix} x_1 \\ \cdot \\ \cdot \\ \cdot \\ x_n \end{Bmatrix} \xrightarrow{\text{Laplace transform}} \mathcal{L}\{\mathbf{x}\} = \mathbf{X}(s) = \begin{Bmatrix} \mathcal{L}\{x_1\} \\ \cdot \\ \cdot \\ \cdot \\ \mathcal{L}\{x_n\} \end{Bmatrix}.$$

Taking the Laplace transform of Equation 8.35, taking into account the initial state vector $\mathbf{x}_0$, yields

$$s\mathbf{X}(s) - \mathbf{x}_0 = \mathbf{A}\mathbf{X}(s) + \mathbf{B}\mathbf{U}(s) \xRightarrow{\text{Rearrange}} (s\mathbf{I} - \mathbf{A})\mathbf{X}(s) = \mathbf{x}_0 + \mathbf{B}\mathbf{U}(s).$$

Premultiply by $(s\mathbf{I} - \mathbf{A})^{-1}$ to obtain

$$\mathbf{X}(s) = (s\mathbf{I} - \mathbf{A})^{-1}\mathbf{x}_0 + (s\mathbf{I} - \mathbf{A})^{-1}\mathbf{B}\mathbf{U}(s),$$

so that

$$\mathbf{x}(t) = \mathcal{L}^{-1}\left\{(s\mathbf{I} - \mathbf{A})^{-1}\right\}\mathbf{x}_0 + \mathcal{L}^{-1}\left\{(s\mathbf{I} - \mathbf{A})^{-1}\mathbf{B}\mathbf{U}(s)\right\}. \tag{8.39}$$

We will manipulate the two terms in Equation 8.39. First note that

$$(s\mathbf{I} - \mathbf{A})^{-1} = \frac{1}{s}\mathbf{I} + \frac{1}{s^2}\mathbf{A} + \frac{1}{s^3}\mathbf{A}^2 + \cdots.$$

Therefore,

$$\mathcal{L}^{-1}\left\{(s\mathbf{I} - \mathbf{A})^{-1}\right\} = \mathcal{L}^{-1}\left\{\frac{1}{s}\mathbf{I} + \frac{1}{s^2}\mathbf{A} + \frac{1}{s^3}\mathbf{A}^2 + \cdots\right\} = \mathbf{I} + t\mathbf{A} + \frac{1}{2!}t^2\mathbf{A}^2 + \cdots \stackrel{\text{By Equation 8.37}}{=} e^{\mathbf{A}t}$$

This implies that the first term in Equation 8.39 is simply $e^{\mathbf{A}t}\mathbf{x}_0$. The second term may be worked out using the convolution integral (Section 2.3) as

$$\mathcal{L}^{-1}\left\{(s\mathbf{I} - \mathbf{A})^{-1}\mathbf{B}\mathbf{U}(s)\right\} = \int_0^t e^{\mathbf{A}(t-\tau)}\mathbf{B}\mathbf{u}(\tau)\,d\tau.$$

Combining the two terms yields

$$\mathbf{x}(t) = e^{\mathbf{A}t}\mathbf{x}_0 + \int_0^t e^{\mathbf{A}(t-\tau)}\mathbf{B}\mathbf{u}(\tau)d\tau,$$

which exactly agrees with the formal solution derived earlier.

Example 8.16: Laplace Transform Approach

Solve the state equation in Example 8.15 using the Laplace transform approach.

Solution

```
>> A=[1 0;-1 -2]; B=[1;-2]; x0=[0;1];
>> syms s t
>> U = laplace(exp(-t));
>> x = ilaplace(inv(s*eye(2)-A))*x0 + ilaplace(inv(s*eye(2)-A)*B*U)
x =
                        sinh(t)
8/3*exp(-2*t)-5/3*cosh(t)+4/3*sinh(t)
```

Using the definitions of $\sinh t$ and $\cosh t$, the above result matches that in Example 8.15.

8.4.3 Solution of the State Equation via State-Transition Matrix

The solution of the homogeneous state equation

$$\dot{\mathbf{x}} = \mathbf{A}\mathbf{x}, \quad \mathbf{x}(0) = \mathbf{x}_0$$

can be expressed as

$$\mathbf{x}(t) = \mathbf{\Phi}(t)\mathbf{x}_0, \tag{8.40}$$

so that $\mathbf{x}(t)$ is obtained by a transformation of the initial state vector $\mathbf{x}_0$. For this reason, $\mathbf{\Phi}(t)$ is called the state-transition matrix. We claim that $\mathbf{\Phi}(t)$ is the unique solution of

$$\dot{\mathbf{\Phi}} = \mathbf{A}\mathbf{\Phi}, \quad \mathbf{\Phi}(0) = \mathbf{I}$$

and prove it as follows. First, using Equation 8.40,

$$\mathbf{x}(0) = \mathbf{\Phi}(0)\mathbf{x}_0 \;\Rightarrow\; \mathbf{\Phi}(0) = \mathbf{I}.$$

Next, differentiation of Equation 8.40 with respect to t, and further manipulation, yields

$$\dot{\mathbf{x}}(t) = \dot{\mathbf{\Phi}}(t)\mathbf{x}_0 \stackrel{\dot{\mathbf{\Phi}}=\mathbf{A}\mathbf{\Phi}}{=} \mathbf{A}\mathbf{\Phi}(t)\mathbf{x}_0 \stackrel{\mathbf{\Phi}(t)\mathbf{x}_0=\mathbf{x}}{=} \mathbf{A}\mathbf{x}.$$

This completes the proof of the claim. Comparing Equation 8.40 with the homogeneous portions of Equations 8.39 and 8.36 reveals that

$$\mathbf{\Phi}(t) = \mathcal{L}^{-1}\left\{(s\mathbf{I} - \mathbf{A})^{-1}\right\} = e^{\mathbf{A}t}.$$

Based on this finding, the solution of the nonhomogeneous state equation

$$\dot{\mathbf{x}} = \mathbf{A}\mathbf{x} + \mathbf{B}\mathbf{u}, \quad \mathbf{x}(0) = \mathbf{x}_0$$

can be expressed as

$$\mathbf{x}(t) = \mathbf{\Phi}(t)\mathbf{x}_0 + \int_0^t \mathbf{\Phi}(t-\tau)\mathbf{B}\mathbf{u}(\tau)\,d\tau. \tag{8.41}$$

PROBLEM SET 8.4

In Problems 1 through 4, find $e^{\mathbf{A}t}$ using the inverse Laplace transform approach.

1. $\mathbf{A} = \begin{bmatrix} -1 & 2 \\ 2 & 2 \end{bmatrix}$

2. $\mathbf{A} = \begin{bmatrix} 2 & 4 \\ 4 & -4 \end{bmatrix}$

3. $\mathbf{A} = \begin{bmatrix} 0 & 0 & 1 \\ 1 & 0 & -3 \\ 0 & 1 & 3 \end{bmatrix}$

4. $\mathbf{A} = \begin{bmatrix} -1 & 1 & 2 \\ 0 & -1 & 3 \\ 0 & 0 & 2 \end{bmatrix}$

5. Consider the state equation

$$\dot{\mathbf{x}} = \begin{bmatrix} 2 & -1 \\ -4 & 2 \end{bmatrix} \mathbf{x} + \begin{bmatrix} 1 \\ -1 \end{bmatrix} u, \quad u = \text{unit-step function}, \mathbf{x}(0) = \begin{Bmatrix} 0 \\ -1 \end{Bmatrix}.$$

Determine the state vector via the formal-solution approach.

6. A system is described by its state-space form as

$$\begin{cases} \dot{\mathbf{x}} = \mathbf{A}\mathbf{x} + \mathbf{B}u, \\ y = \mathbf{C}\mathbf{x}, \end{cases} \quad \mathbf{A} = \begin{bmatrix} 0 & 1 \\ -4 & -4 \end{bmatrix}, \mathbf{B} = \begin{bmatrix} 1 \\ -2 \end{bmatrix}, \mathbf{C} = \begin{bmatrix} 1 & 0 \end{bmatrix}, u = \sin t, \mathbf{x}_0 = \begin{Bmatrix} 1 \\ -2 \end{Bmatrix}.$$

Find the response $y(t)$ using the formal-solution approach.

7. A dynamic system is described by

$$\begin{cases} \dot{\mathbf{x}} = \mathbf{A}\mathbf{x} + \mathbf{B}\mathbf{u}, \\ y = \mathbf{C}\mathbf{x}, \end{cases}, \quad \mathbf{A} = \begin{bmatrix} 1 & 0 & 2 \\ 0 & -1 & 3 \\ 0 & 0 & 2 \end{bmatrix}, \mathbf{B} = \begin{bmatrix} 0 & 0 \\ 1 & 0 \\ 0 & 1 \end{bmatrix}, \mathbf{C} = \begin{bmatrix} 0 & 0 & 1 \end{bmatrix},$$

$$\mathbf{u} = \begin{bmatrix} 1 \\ e^{-t} \end{bmatrix}, \mathbf{x}_0 = \begin{Bmatrix} 1 \\ 0 \\ 1 \end{Bmatrix}.$$

Use the formal-solution approach to determine the response $y(t)$.

8. In the state equation below, find $\mathbf{x}(t)$ using the Laplace transform approach.

$$\dot{\mathbf{x}} = \begin{bmatrix} 1 & 0 & 0 \\ 0 & 1 & 0 \\ -2 & 4 & -1 \end{bmatrix} \mathbf{x} + \begin{bmatrix} 0 \\ 0 \\ 2 \end{bmatrix} u, \quad u = \text{unit-step function}, \mathbf{x}(0) = \begin{Bmatrix} 0 \\ 1 \\ 0 \end{Bmatrix}.$$

9. In the state equation below, find $\mathbf{x}(t)$ using the Laplace transform approach.

$$\dot{\mathbf{x}} = \begin{bmatrix} 0 & 1 \\ 0 & -1 \end{bmatrix} \mathbf{x} + \begin{bmatrix} 0 \\ 1 \end{bmatrix} u, \quad u = \text{unit-ramp function}, \mathbf{x}(0) = \begin{Bmatrix} 1 \\ 0 \end{Bmatrix}.$$

10. Show that the state-transition matrix $\mathbf{\Phi}(t)$ satisfies $\mathbf{\Phi}^{-1}(t) = \mathbf{\Phi}(-t)$.

8.5 RESPONSE OF NONLINEAR SYSTEMS

In Section 4.6, we learned how to derive a linearized model of a nonlinear system, and that such approximation is reasonably accurate only in a small neighborhood of an operating point. To circumvent the obvious limitations created by this approach, one may tackle the nonlinear model directly, through a numerical method. Among the numerical methods designed for this purpose, the most commonly used is the very efficient and robust fourth-order Runge–Kutta (RK4) method. We know that the mathematical models of dynamic systems of any order can be expressed in terms of state-variable equations, which may be linear or nonlinear (see Section 4.2). Of course, in the present discussion we are more interested in the nonlinear case,

$$\begin{cases} \dot{x}_1 = f_1(x_1,\ldots,x_n; u_1,\ldots,u_m; t) \\ \dot{x}_2 = f_2(x_1,\ldots,x_n; u_1,\ldots,u_m; t) \\ \quad \vdots \\ \dot{x}_n = f_n(x_1,\ldots,x_n; u_1,\ldots,u_m; t) \end{cases} \tag{8.42}$$

which is conveniently written as (Equation 4.4, Section 4.2)

$$\dot{\mathbf{x}} = \mathbf{f}(\mathbf{x}, \mathbf{u}, t),$$

where

$$\mathbf{x} = \begin{Bmatrix} x_1 \\ \vdots \\ x_n \end{Bmatrix}_{n \times 1}, \quad \mathbf{u} = \begin{Bmatrix} u_1 \\ \vdots \\ u_m \end{Bmatrix}_{m \times 1}, \quad \mathbf{f} = \begin{Bmatrix} f_1 \\ \vdots \\ f_n \end{Bmatrix}_{n \times 1}.$$

The linear case results in the familiar state equation.

8.5.1 RK4 Method

To introduce the RK4, we reformulate the problem as

$$\dot{\mathbf{x}} = \mathbf{f}(t, \mathbf{x}), \quad \mathbf{x}(a) = \mathbf{x}_0, \quad a \le t \le b, \tag{8.43}$$

with all variables as defined previously. Define an integer $N > 0$ and let $h = (b-a)/N$ be the step size. The mesh points $t_i = a + ih$, $i = 0, 1, \ldots, N-1$, partition the interval $[a,b]$ into N subintervals. Knowing the initial state vector $\mathbf{x}_0$, the solution vector $\mathbf{x}_i$ at each of the subsequent mesh points t_i is obtained via

$$\mathbf{x}_{i+1} = \mathbf{x}_i + \frac{1}{6}[\mathbf{q}_1 + 2\mathbf{q}_2 + 2\mathbf{q}_3 + \mathbf{q}_4], \quad i = 0, 1, 2, \ldots, N-1, \tag{8.44}$$

where

$$\mathbf{q}_1 = h\mathbf{f}(t_i, \mathbf{x}_i),$$

$$\mathbf{q}_2 = h\mathbf{f}\left(t_i + \frac{1}{2}h, \mathbf{x}_i + \frac{1}{2}\mathbf{q}_1\right),$$

$$\mathbf{q}_3 = h\mathbf{f}\left(t_i + \frac{1}{2}h, \mathbf{x}_i + \frac{1}{2}\mathbf{q}_2\right),$$

$$\mathbf{q}_4 = h\mathbf{f}(t_i + h, \mathbf{x}_i + \mathbf{q}_3).$$

System Response

8.5.1.1 Running RK4 in MATLAB

The following user-defined function, called RK4system, numerically solves a system of state-variable equations subjected to a prescribed initial state over a specified time interval. Note that the function is also capable of handling a scalar initial-value problem.

```
function x = RK4system(f,t,x0)
%
%   RK4SYSTEM Uses the 4th-order Runge-Kutta (RK4) method to solve a system
%   of state-variable equations in the form xdot = f(t,x).
%
%       x = RK4system(f,t,x0) where
%
%           f is an inline function representing f(t,x).
%           t is a vector representing the mesh points.
%
%           x0 is the initial state vector.
%           x is the vector of approximate solutions at the mesh points.
%
x(:,1) = x0;
for n = 1:length(t)-1,
h = t(n+1) - t(n);
q1 = h*f(t(n), x(:,n));
q2 = h*f(t(n) + h/2, x(:,n) + q1/2);
q3 = h*f(t(n) + h/2, x(:,n) + q2/2);
q4 = h*f(t(n) + h, x(:,n) + q3);
x(:,n+1) = x(:,n) + 1/6*(q1 + 2*q2 + 2*q3 + q4);
end
```

Example 8.17: Nonlinear System

The system in Figure 8.27 contains a nonlinear spring identified by the spring force $f_s = x|x|$ and is subjected to a unit-step external force $u(t)$ and initial conditions $x(0) = 0$, $\dot{x}(0) = 1$.

   a. Derive the state-variable equations.
   b. Using RK4, find and plot the displacement $x(t)$ versus $0 \le t \le 10$ s.

Solution

a. The system's motion is described by

$$\ddot{x} + \dot{x} + x|x| = u(t), \quad x(0) = 0, \quad \dot{x}(0) = 1.$$

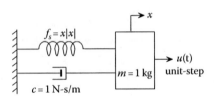

FIGURE 8.27 A nonlinear mechanical system.

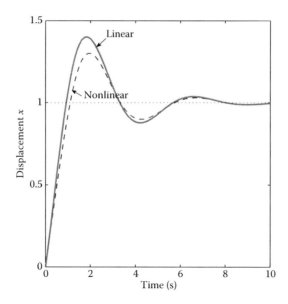

FIGURE 8.28 Nonlinear and linearized system responses, Example 8.17.

With state variables $x_1 = x$ and $x_2 = \dot{x}$, the nonlinear state-variable equations are obtained as

$$\begin{cases} \dot{x}_1 = x_2, & x_1(0) = 0, \\ \dot{x}_2 = -x_2 - x_1\,|x_1| + u(t), & x_2(0) = 1. \end{cases}$$

b. The problem is formulated as

$$\dot{\mathbf{x}} = \mathbf{f}(t, \mathbf{x}), \quad \mathbf{x} = \begin{Bmatrix} x_1 \\ x_2 \end{Bmatrix}, \quad \mathbf{f} = \begin{Bmatrix} x_2 \\ -x_2 - x_1\,|x_1| + 1 \end{Bmatrix}, \quad \mathbf{x}_0 = \begin{Bmatrix} 0 \\ 1 \end{Bmatrix}, \quad 0 \le t \le 10.$$

```
>> t=linspace(0,10);
>> x0=[0;1];
>> f=inline('[x(2,1);-x(2,1)-x(1,1)*abs(x(1,1))+1]','t','x');
>> x = RK4system(f, t, x0);
>> x1=x(1,:); % Extract x1 from the state vector
>> plot(t,x1) % Figure 8.28
```

Let us next compare these results with those from the linearized model. Based on the material in Section 4.6, the linearized model can be derived as

$$\ddot{x}_L + \dot{x}_L + 2x_L = 0, \quad x_L(0) = -1,\ \dot{x}_L(0) = 1.$$

Note that we have changed the notation from Δx (used in Section 4.6) to x_L here, with subscript L denoting the linear model. The relation between x_L and x (in the nonlinear system) is $x_L = x - 1$. Therefore, we need to first plot $x_L(t)$ and then raise the curve by one unit to obtain a variable compatible with x. The linearized model is formulated as

$$\dot{\mathbf{x}}_L = \mathbf{g}(t, \mathbf{x}_L), \quad \mathbf{x}_L = \begin{Bmatrix} x_{L1} \\ x_{L2} \end{Bmatrix}, \quad \mathbf{g} = \begin{Bmatrix} x_{L2} \\ -x_{L2} - 2x_{L1} \end{Bmatrix}, \quad \mathbf{x}_{L0} = \begin{Bmatrix} -1 \\ 1 \end{Bmatrix}, \quad 0 \le t \le 10,$$

where $x_{L1} = x_L$ and $x_{L2} = \dot{x}_L$.

System Response

```
>> hold on      % Hold Figure 8.28
>> xL0=[-1;1];
>> g=inline('[xL(2,1);-xL(2,1)-2*xL(1,1)]','t','xL');
>> xL = RK4system(g, t, xL0);
>> xL1=xL(1,:);   % Extract xL1 from the xL vector
>> plot(t,xL1+1)   % Add 1 to xL1 and complete Figure 8.28
```

We immediately observe that the nonlinear and linear response curves somewhat agree only in the neighborhood of $x = 1$, the operating point. This confirms that linearized models provide reasonable approximations in a close vicinity of an operating point.

PROBLEM SET 8.5

Solve each problem using RK4 in the indicated time interval and plot the response.

1. $\dddot{x} + 6\ddot{x} + 11\dot{x} + 6x = e^{-t}$, $x(0) = 0$, $\dot{x}(0) = 1$, $\ddot{x}(0) = -1$, $0 \leq t \leq 10$, plot $x(t)$
2. $\ddot{x} + \dot{x} + 2x^3 = 1$, $x(0) = -1$, $\dot{x}(0) = 0$, $0 \leq t \leq 10$, plot $x(t)$
3. $2\ddot{x} + \dot{x} + x|x| = 1 + \sin(t/2)$, $x(0) = \sqrt{2}$, $\dot{x}(0) = 1$, $0 \leq t \leq 15$, plot $x(t)$
4. $\ddot{x} + \dot{x} + x\sqrt{|x|} = 1 + \sin t$, $x(0) = 1$, $\dot{x}(0) = 0$, $0 \leq t \leq 5$, plot $x(t)$
5. $\begin{cases} \dot{x}_1 = x_2, \\ \dot{x}_2 = -x_1 - x_1|x_1| - x_2^3 - 2 + 0.1\sin t, \end{cases}$ $\begin{matrix} x_1(0) = 0 \\ x_2(0) = 1 \end{matrix}$, $0 \leq t \leq 5$, plot $x_1(t)$
6. $\begin{cases} \dot{x}_1 = -x_1 - 2x_2^3, \\ \dot{x}_2 = x_1 + 10 + 2\sin t, \end{cases}$ $\begin{matrix} x_1(0) = 0 \\ x_2(0) = -2 \end{matrix}$, $0 \leq t \leq 5$, plot $x_1(t)$
7. A nonlinear dynamic system model is derived as

$$\dot{x} + 2x^3 = u(t), \quad x(0) = 0, \, 0 \leq t \leq 3,$$

where $u(t)$ is the unit-step function. The linearized model is

$$\dot{x}_L + 6\left(\frac{1}{2}\right)^{2/3} x_L = 0, \quad x_L(0) = -\left(\frac{1}{2}\right)^{1/3}, \, 0 \leq t \leq 3.$$

The response $x(t)$ of the nonlinear system and $x_L(t)$ of the linear system are related through $x(t) = x_L(t) + (1/2)^{1/3}$. Using RK4, find and plot $x(t)$ and $x_L(t)$ versus $0 \leq t \leq 3$ s.

8. The model of a nonlinear dynamic system is

$$\begin{cases} \dot{x}_1 = -x_2 - x_1|x_1| - 3 + e^{-t}, \\ \dot{x}_2 = x_1 - x_2 - 3, \end{cases} \quad \begin{matrix} x_1(0) = -1 \\ x_2(0) = 1 \end{matrix}, \, 0 \leq t \leq 10.$$

The linearized model is

$$\begin{cases} \dot{x}_{L1} = -x_{L2} + e^{-t}, \\ \dot{x}_{L2} = x_{L1} - x_{L2}, \end{cases} \quad \begin{matrix} x_{L1}(0) = -1 \\ x_{L2}(0) = 4 \end{matrix}, \, 0 \leq t \leq 10.$$

The response $x_1(t)$ of the nonlinear system and $x_{L1}(t)$ of the linear system are related through $x_1(t) = x_{L1}(t)$. Using RK4, find and plot $x_1(t)$ and $x_{L1}(t)$ versus $0 \leq t \leq 10$ s.

9. Consider the nonlinear model

$$\dot{x} + \alpha x |x| = e^{-t/2} \sin t, \quad x(0) = 0, \, 0 \leq t \leq 10,$$

where α is a parameter. Find and plot $x(t)$ for $\alpha = 0.1, 1, 10$ versus $0 \leq t \leq 10$ s.

10. The equation of motion of an inverted pendulum mounted on a moving cart is derived as

$$\begin{cases} \ddot{x} + \dot{x} + x + 2\phi = 0, \\ \ddot{\phi} = 12\sin\phi + (x + 2\phi + \dot{x})\cos\phi, \end{cases}$$

where x and ϕ denote the cart position and the angular displacement (in radians) of the pendulum from the vertical, respectively. Assume that the initial conditions are

$$x(0) = 0, \quad \dot{x}(0) = 0, \quad \phi(0) = 0.01\,\text{rad}, \quad \dot{\phi}(0) = 0.$$

a. Using state variables $x_1 = x$, $x_2 = \phi$, $x_3 = \dot{x}$, $x_4 = \dot{\phi}$, obtain the state-variable equations.
b. Using RK4, find and plot $\phi(t)$ versus $0 \leq t \leq 10\,\text{s}$.

8.6 SUMMARY

The total response of a dynamic system is comprised of transient response and steady-state response. The terms in the response that eventually decay to zero form the transient response. The part of the response that remains after the transient terms have vanished is called the steady-state response. Linear, first-order systems are modeled as

$$\tau \dot{x} + x = f(t), \quad \tau = \text{const} > 0, \; x(0) = x_0,$$

where τ is the time constant that provides a measure of how quickly the system reaches steady state. Linear, second-order systems are modeled as

$$\ddot{x} + 2\zeta\omega_n \dot{x} + \omega_n^2 x = f(t), \quad x(0) = x_0, \; \dot{x}(0) = \dot{x}_0,$$

where ζ and ω_n are the damping ratio and (undamped) natural frequency, respectively. The steady-state response to a sinusoidal input is called the frequency response. The frequency response of a system with transfer function $G(s)$ to input $F_0 \sin \omega t$ is

$$F_0 |G(j\omega)| \sin(\omega t + \phi), \quad \phi = \angle G(j\omega), \; G(j\omega) = \text{FRF}.$$

A Bode diagram consists of two curves versus frequency: $20\log|G(j\omega)|$ with the unit of decibel (dB) and the phase ϕ (in degrees). The curves are sketched on semilog paper, using a linear scale for magnitude (in dB) and phase (in degrees) and a logarithmic scale for frequency (in rad/s). The formal solution of the state equation

$$\dot{\mathbf{x}} = \mathbf{A}\mathbf{x} + \mathbf{B}\mathbf{u}, \quad \mathbf{x}(0) = \mathbf{x}_0$$

can be expressed as

$$\mathbf{x}(t) = e^{\mathbf{A}t}\mathbf{x}_0 + \int_0^t e^{\mathbf{A}(t-\tau)} \mathbf{B}\mathbf{u}(\tau)\,d\tau,$$

where $e^{\mathbf{A}t}$ is the matrix exponential of $\mathbf{A}$. The RK4 method is designed to numerically solve any set of state-variable equations (linear or nonlinear)

$$\dot{\mathbf{x}} = \mathbf{f}(t, \mathbf{x}), \quad \mathbf{x}(a) = \mathbf{x}_0, \quad a \leq t \leq b,$$

where $\mathbf{x}_0$ is the initial state vector and $[a, b]$ is the interval in which the system is to be solved. Given an integer $N > 0$, choose the step size as $h = (b-a)/N$ so that the mesh points are $t_i = a + ih$, $i = 0, 1, \ldots, N-1$. Then, the solution vector $\mathbf{x}_i$ at each of the mesh points t_i is obtained via

$$\mathbf{x}_{i+1} = \mathbf{x}_i + \frac{1}{6}[\mathbf{q}_1 + 2\mathbf{q}_2 + 2\mathbf{q}_3 + \mathbf{q}_4], \quad i = 0, 1, 2, \ldots, N-1,$$

where

$$q_1 = hf(t_i, x_i),$$

$$q_2 = hf\left(t_i + \frac{1}{2}h, x_i + \frac{1}{2}q_1\right),$$

$$q_3 = hf\left(t_i + \frac{1}{2}h, x_i + \frac{1}{2}q_2\right),$$

$$q_4 = hf(t_i + h, x_i + q_3).$$

REVIEW PROBLEMS

1. Find the total response $y(t)$ for the system described by

$$2\dot{y} + \frac{1}{2}y = 2u(t) + u_r(t), \quad y(0) = 0,$$

where $u(t)$ and $u_r(t)$ denote the unit-step and unit-ramp functions, respectively.

2. A first-order system is modeled as

$$\frac{1}{3}\dot{y} + 2y = 3u_r(t), \quad y(0) = 0.$$

Determine the steady-state error between the input and the system response.

3. The model of a second-order system is given as

$$4\ddot{x} + 4\dot{x} + 2x = 0, \quad x(0) = 0, \dot{x}(0) = \frac{1}{2}.$$

a. Identify the damping type and find the analytical expression of the free response.
b. ◢ Find and plot the free response using the `initial` command in MATLAB.

4. A second-order dynamic system is modeled as

$$9\ddot{x} + 6\dot{x} + 5x = 2\delta(t), \quad x(0) = 0, \dot{x}(0) = \frac{2}{3}.$$

a. Find the analytical expression for the response $x(t)$.
b. ◢ Plot the expression obtained in (a) versus $0 \leq t \leq 18$ in MATLAB.

5. ◢ Using the `step` and `initial` commands, plot the response $x(t)$ of the following system versus $0 \leq t \leq 12$ s.

$$\ddot{x} + \dot{x} + x = 5u(t), \quad x(0) = 1, \dot{x}(0) = 0.$$

6. A dynamic system model is given as

$$\begin{cases} \ddot{x}_1 + \dot{x}_1 - 2(x_2 - x_1) = f_1(t), \\ \dot{x}_2 + x_2 + 2(x_2 - x_1) = f_2(t), \end{cases}$$

subjected to zero initial conditions. Suppose that x_2 is the system output.
a. Find the state-space form.
b. ◢ Assuming f_1 and f_2 are unit-step functions, plot x_{21} and x_{22} (contributions of f_1 and f_2 to x_2) as well as x_2 versus $0 \leq t \leq 8$ in the same graph.

7. The governing equations for a system are derived as

$$\begin{cases} \ddot{x}_1 + \dot{x}_1 + 2x_1 - 2x_2 = f(t), \\ \dot{x}_2 - 2x_1 + 3x_2 = 0, \end{cases}$$

subjected to initial conditions $x_1(0) = 0$, $x_2(0) = 0.5$, and $\dot{x}_1(0) = 0$. Assume that x_2 is the output.
a. Find the state-space form.
b. ◢ Assuming $f(t) = e^{-t/2} \cos t$, plot the response x_2 versus $0 \leq t \leq 10$ using the lsim command.

8. ◢ Consider the mechanical system in Example 8.8 (Section 8.2, Figure 8.12). Suppose that the physical parameter values are unchanged but the forces are $f_1(t) = u(t) =$ unit-step and $f_2(t) = e^{-t} \cos t$. Assuming zero initial conditions, plot displacement x_2 versus $0 \leq t \leq 150$.

9. Find the frequency response for a system described by

$$\frac{1}{3}\dot{x} + 2x = 10 \sin\left(\frac{t}{2}\right).$$

10. The equation of motion of a mechanical system is given as

$$\ddot{x} + 2\dot{x} + 16x = 32 \sin t.$$

Determine the system's frequency response.

11. ◢ Consider the transfer function

$$G(s) = \frac{3}{(1/2)s + 1}.$$

Draw the Bode diagram and insert the following information into the graph: the corner frequency (ω_c), the low-frequency approximate magnitude (dB), and the magnitude (dB) corresponding to ω_c.

12. ◢ Consider the second-order transfer function

$$G(s) = \frac{0.8}{s^2 + 1.4s + 4}.$$

Draw the Bode diagram and insert the following information into the graph: the low-frequency approximate magnitude (dB), the resonant frequency (ω_r), and the peak magnitude (dB) at ω_r.

13. ◢ The nonlinear state-variable equations for a dynamic system are derived as

$$\begin{cases} \dot{x}_1 = -x_2 + 0.1 \sin t, & x_1(0) = 0, \\ \dot{x}_2 = x_1^3 - 3x_2 + 8, & x_2(0) = 1. \end{cases}$$

Solve using RK4, and plot $x_1(t)$ and $x_2(t)$ versus $0 \leq t \leq 10$ in the same figure.

14. ◢ Consider

$$\begin{cases} \dot{x}_1 = -x_1 |x_1| + ax_2 - 1 + \cos t, & x_1(0) = 2, \\ \dot{x}_2 = -x_1 - x_2 - 1, & x_2(0) = -1, \end{cases}$$

where a is a parameter. Using RK4, solve the system for $a = 1$, $a = 2$, and plot the two resulting $x_2(t)$ versus $0 \leq t \leq 10$ in the same graph.

9 Introduction to Vibrations

Vibration can be regarded as a subset of dynamics, in which a system subjected to restoring forces oscillates about an equilibrium position. The restoring forces are due to elasticity or due to gravity. Two different types of excitations, initial excitations and external excitations, cause a system to vibrate. The vibration of a system caused by nonzero initial excitations, including initial displacements and/or initial velocities, is known as free vibration. The vibration of a system caused by externally applied forces is known as forced vibration.

In this chapter, we first extend the knowledge learned in Chapters 5 and 8 to the analysis of free vibration and forced vibration of single- or two-degree-of-freedom systems. Section 9.1 introduces the free vibration of Coulomb damped systems, in which energy is dissipated via dry friction. Section 9.2 considers the forced vibration of systems with rotating eccentric masses and systems with harmonically moving supports. Sections 9.1 and 9.2 also cover topics such as logarithmic decrement and bandwidth, both of which can be used to estimate the widely used viscous damping model. To reduce the effects of undesired vibration, Section 9.3 discusses the design of vibration suppression systems, including vibration isolators and vibration absorbers. For multi-degree-of-freedom systems, it is convenient to use the matrix-based method to perform vibration analysis. In Section 9.4, the concepts of the eigenvalue problem, natural modes, and orthogonality of modes are presented by means of matrix algebra. The modal analysis method is developed and used to obtain the response to initial or harmonic excitations. The chapter concludes with coverage of vibration measurement technology. Section 9.5 introduces the hardware available for vibration testing and the methods used to identify system parameters, such as natural frequencies and damping ratios.

9.1 FREE VIBRATION

Consider a single-degree-of-freedom viscously damped system subjected to nonzero initial excitations. Its governing differential equation is given by

$$m\ddot{x}(t) + b\dot{x}(t) + kx(t) = 0 \tag{9.1}$$

or

$$\ddot{x}(t) + 2\zeta\omega_n\dot{x}(t) + \omega_n^2 x(t) = 0, \tag{9.2}$$

with $x(0) = x_0$ and $\dot{x}(0) = v_0$. As indicated in Section 8.2, the nature of the system response to initial excitations depends on the value of the damping ratio ζ. For an overdamped system, $\zeta > 1$, the response represents aperiodic decay, which is also true for a critically damped system, $\zeta = 1$. For an underdamped system, $0 < \zeta < 1$, the response represents oscillatory decay. For an undamped system, $\zeta = 0$, the response represents harmonic oscillation with natural frequency ω_n.

Viscous damping is a widely used damping model, but not the only one. Damping is a very complex phenomenon and a wide range of damping models can be found in the literature. In this section, we first discuss the measurement of the viscous damping ratio ζ based on transient time response plots and then introduce another damping model, Coulomb damping.

9.1.1 Logarithmic Decrement

Unlike mass m and spring stiffness k, both of which can easily be measured with static tests, the viscous damping coefficient b has to be measured with a dynamic test. A usual way is to use the free response of the whole system to measure the damping ratio ζ and then determine the damping coefficient b by using $b = 2\zeta\sqrt{mk}$.

As discussed in Section 8.2, the free response of an underdamped single-degree-of-freedom system is

$$x(t) = e^{-\zeta\omega_n t}\left(x_0 \cos\omega_d t + \frac{\zeta\omega_n x_0 + v_0}{\omega_d}\sin\omega_d t\right), \tag{9.3}$$

where the damped natural frequency is $\omega_d = \omega_n\sqrt{1-\zeta^2}$. Equation 9.3 can also be written as (see Section 2.2)

$$x(t) = Ae^{-\zeta\omega_n t}\cos(\omega_d t - \phi). \tag{9.4}$$

The amplitude A and the phase ϕ are given by

$$A = \sqrt{x_0^2 + \left(\frac{\zeta\omega_n x_0 + v_0}{\omega_d}\right)^2} \tag{9.5}$$

and

$$\phi = \tan^{-1}\frac{\zeta\omega_n x_0 + v_0}{\omega_d x_0}. \tag{9.6}$$

Figure 9.1 shows the system response to initial excitations, where $T = 2\pi/\omega_d$ is the period of damped oscillation.

Note that the peak drops after one cycle of vibration, and the ratio between the first and the second peaks is

$$\frac{x_1}{x_2} = \frac{x(t_1)}{x(t_2)} = \frac{Ae^{-\zeta\omega_n t_1}\cos(\omega_d t_1 - \phi)}{Ae^{-\zeta\omega_n t_2}\cos(\omega_d t_2 - \phi)} = \frac{e^{-\zeta\omega_n t_1}\cos(\omega_d t_1 - \phi)}{e^{-\zeta\omega_n (t_1+T)}\cos[\omega_d(t_1+T) - \phi]}. \tag{9.7}$$

Since $\omega_d T = 2\pi$, we have $\zeta\omega_n T = \zeta\omega_n(2\pi/\omega_d) = 2\pi\zeta/\sqrt{1-\zeta^2}$ and $\cos[\omega_d(t_1+T) - \phi] = \cos(\omega_d t_1 + 2\pi - \phi) = \cos(\omega_d t_1 - \phi)$. Thus, Equation 9.7 reduces to

$$\frac{x_1}{x_2} = e^{2\pi\zeta/\sqrt{1-\zeta^2}}. \tag{9.8}$$

Taking the natural logarithm of both sides of Equation 9.8 yields

$$\delta = \ln\frac{x_1}{x_2} = \frac{2\pi\zeta}{\sqrt{1-\zeta^2}}, \tag{9.9}$$

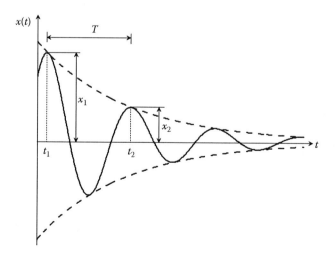

FIGURE 9.1 Free response of an underdamped single-degree-of-freedom system to initial excitations.

Introduction to Vibrations

where δ is called the logarithmic decrement. It turns out that the ratio of any two consecutive displacement peaks gives the same result as Equation 9.8,

$$\frac{x_1}{x_2} = \frac{x_2}{x_3} = \cdots = \frac{x_n}{x_{n+1}} = e^{2\pi\zeta/\sqrt{1-\zeta^2}}. \tag{9.10}$$

Thus, the logarithmic decrement δ can be determined by measuring any two consecutive displacement peaks. The damping ratio ζ can be solved from Equation 9.9 as

$$\zeta = \frac{\delta}{\sqrt{(2\pi)^2 + \delta^2}}, \tag{9.11}$$

For a more accurate estimation, the damping ratio ζ can be determined by measuring the displacements of two peaks separated by a number of periods instead of two consecutive peaks. If we denote the peak displacements at time t_1 and t_{n+1} as x_1 and x_{n+1}, respectively, where $t_{n+1} = t_1 + nT$, then the ratio between the two peak displacements can be written as

$$\frac{x_1}{x_{n+1}} = \frac{x_1}{x_2}\frac{x_2}{x_3}\cdots\frac{x_n}{x_{n+1}}. \tag{9.12}$$

Taking the natural logarithm of both sides of Equation 9.12, we have

$$\ln\frac{x_1}{x_{n+1}} = \ln\left(\frac{x_1}{x_2}\frac{x_2}{x_3}\cdots\frac{x_n}{x_{n+1}}\right) = \ln\frac{x_1}{x_2} + \ln\frac{x_2}{x_3} + \cdots + \ln\frac{x_n}{x_{n+1}} = n\delta, \tag{9.13}$$

which yields the logarithmic decrement as

$$\delta = \frac{1}{n}\ln\frac{x_1}{x_{n+1}}. \tag{9.14}$$

Substituting Equation 9.14 into Equation 9.11 gives the damping ratio. If the damping is very small, that is, $\zeta \ll 1$ and $\sqrt{1-\zeta^2} \approx 1$, then Equation 9.9 gives

$$\zeta \approx \frac{\delta}{2\pi}. \tag{9.15}$$

Example 9.1: Logarithmic Decrement

A vibrating system consisting of a mass of 2.5 kg and a spring of stiffness 1750 N/m is viscously damped. The ratio of any two consecutive amplitudes is 1/0.6.

a. Determine the logarithmic decrement δ.
b. Determine the exact value of the damping ratio ζ.
c. Determine the damping coefficient b.
d. Assuming small damping, recalculate the damping ratio ζ and determine the percent error.

Solution

a. The logarithmic decrement can be determined by measuring the displacements of any two consecutive peaks, and it is solved as

$$\delta = \ln\frac{x_j}{x_{j+1}} = \ln\frac{1}{0.6} = 0.5108.$$

b. The exact value of the damping ratio is

$$\zeta = \frac{\delta}{\sqrt{(2\pi)^2 + \delta^2}} = \frac{0.5108}{\sqrt{(2\pi)^2 + (0.5108)^2}} = 0.0810.$$

c. The viscous damping coefficient is

$$b = 2\zeta\sqrt{mk} = 2(0.0810)\sqrt{2.5(1750)} = 10.7153 \, \text{N·s/m}.$$

d. For small damping,

$$\zeta \approx \frac{\delta}{2\pi} = \frac{0.5108}{2\pi} = 0.0813$$

and the percent error is 0.33%. Note that the error is close to zero and thus the assumption of small damping is valid.

9.1.2 Coulomb Damping

In the previous chapters the viscous damping model was considered, which is widely used in vibration. The viscous damping force is linearly dependent on velocity, and this makes analysis easy. However, it is not the only damping model. Coulomb damping is another type of damping in which energy is dissipated via dry friction.

Figure 9.2 shows a mass–spring system subject to Coulomb damping, where N is the normal force and F_f is the dry friction force. Note that the friction force remains constant in magnitude, which is $\mu_k mg$, where μ_k is the kinetic friction coefficient, and is opposite in direction to the motion or the velocity $\dot{x}$. Introducing the signum function $\text{sgn}(\cdot)$,

$$\text{sgn}(\dot{x}) = \begin{cases} 1, & \dot{x} > 0, \\ -1, & \dot{x} < 0, \end{cases} \tag{9.16}$$

the friction force can be expressed as

$$\mathbf{F}_f = -F_f \text{sgn}(\dot{x}) = \begin{cases} -F_f, & \dot{x} > 0, \\ F_f, & \dot{x} < 0, \end{cases} \tag{9.17}$$

where $F_f = \mu_k mg$. Equation 9.17 implies that the friction force points to the left if the mass moves to the right and points to the right if the mass moves to the left.

Using Equation 9.17, we can write the dynamics equation of motion as

$$m\ddot{x} + F_f \text{sgn}(\dot{x}) + kx = 0, \tag{9.18}$$

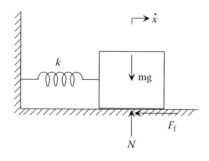

FIGURE 9.2 A mass–spring system subjected to Coulomb damping.

Introduction to Vibrations

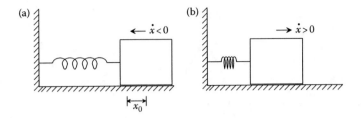

FIGURE 9.3 Motion of the mass with (a) negative velocity and (b) positive velocity.

which is a nonlinear equation that can be separated into two linear equations,

$$m\ddot{x} + F_f + kx = 0, \quad \dot{x} > 0, \tag{9.19}$$

$$m\ddot{x} - F_f + kx = 0, \quad \dot{x} < 0. \tag{9.20}$$

Without loss of generality, let us assume the initial conditions to be $x(0) = x_0 > 0$ and $\dot{x}(0) = v_0 = 0$. Due to the restoring spring force, the mass first moves from right to left as shown in Figure 9.3a. The velocity is negative. The dynamics of the system is expressed by Equation 9.20, which can be rewritten as

$$m\ddot{x} + k(x - \Delta) = 0, \tag{9.21}$$

where $\Delta = F_f/k$. For a given system, the mass m, spring stiffness k, and kinetic friction coefficient μ_k are all constants. Thus, Δ is a constant and $\ddot{\Delta} = 0$. Equation 9.21 can then be rewritten as

$$m(\ddot{x} - \ddot{\Delta}) + k(x - \Delta) = 0. \tag{9.22}$$

Note that Equation 9.22 has the same format as $m\ddot{x} + kx = 0$ if we replace $x - \Delta$ with x. Therefore, the free response of the system can be determined by using Equations 9.4 through 9.6 and neglecting damping. If we let $x' = x - \Delta$, the initial conditions can be expressed as $x'(0) = x(0) - \Delta = x_0 - \Delta$ and $\dot{x}'(0) = \dot{x}(0) - \dot{\Delta} = 0$. Substituting the initial conditions into Equations 9.4 through 9.6 yields

$$x(t) - \Delta = (x_0 - \Delta)\cos \omega_n t \tag{9.23}$$

or

$$x(t) = (x_0 - \Delta)\cos \omega_n t + \Delta. \tag{9.24}$$

When the spring reaches maximum compression as shown in Figure 9.3b, the velocity of the mass reduces to zero, that is, $\dot{x}(t) = -(x_0 - \Delta)\omega_n \sin \omega_n t = 0$, which yields $t = \pi/\omega_n = T/2$. The corresponding displacement is $-(x_0 - 2\Delta)$. The mass then starts to move from left to right and the velocity becomes positive. Thus, Equation 9.24 is only valid for $0 \leq t < T/2$.

For $t \geq T/2$, the dynamics of the system is expressed by Equation 9.19, which can be rewritten as

$$m\ddot{x} + k(x + \Delta) = 0 \tag{9.25}$$

or

$$m(\ddot{x} + \ddot{\Delta}) + k(x + \Delta) = 0. \tag{9.26}$$

The response of the system can also be determined by using Equations 9.4 through 9.6 and neglecting damping. Let $x' = x + \Delta$. Note that the initial values of x and $\dot{x}$ in Equation 9.26 are the displacement and the velocity at time $T/2$, respectively, that is, $x(T/2) = -(x_0 - 2\Delta)$ and $\dot{x}(T/2) = 0$. As a result, $x'(0) = -(x_0 - 3\Delta)$ and $\dot{x}'(0) = 0$. The solution of Equation 9.26 is

$$x(t) + \Delta = (x_0 - 3\Delta)\cos \omega_n t \tag{9.27}$$

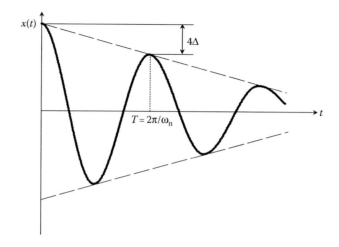

FIGURE 9.4 Free response of a mass–spring system subject to Coulomb damping.

or

$$x(t) = (x_0 - 3\Delta)\cos\omega_n t - \Delta. \tag{9.28}$$

When the spring reaches maximum elongation, the velocity of the mass reduces to zero once again. By differentiating Equation 9.28 and equating it to zero, we obtain the time corresponding to the maximum elongation, that is, T. The displacement at $t = T$ is $x_0 - 4\Delta$. After this point, the motion reverses in direction and the mass moves from right to left. Thus, Equation 9.28 is only valid for $T/2 \leq t < T$.

The above discussion gives the displacement magnitudes at $t = 0$, $T/2$, and T, which are x_0, $x_0 - 2\Delta$, and $x_0 - 4\Delta$, respectively. It can be concluded that the displacement magnitude is reduced by 2Δ after every half-cycle. This process is repeated as the mass oscillates back and forth about its equilibrium position. The motion stops when the displacement is not large enough for the restoring spring force to overcome the static friction force. The free response of the mass–spring system with Coulomb damping is shown in Figure 9.4. Different from viscously damped systems, the envelope of the response for a vibrating system with Coulomb damping is a straight line instead of an exponential decay curve, as shown in Figure 9.1.

Example 9.2: Coulomb Damping

A Coulomb damped vibrating system as shown in Figure 9.2 consists of a mass of 8 kg and a spring of stiffness 6000 N/m. The kinetic friction coefficient μ_k is 0.15. The initial conditions are $x_0 = 0.025$ m and $v_0 = 0$ m/s.

  a. Determine the decay per cycle.
  b. Determine the position when the oscillation stops.

Solution

a. The magnitude of the Coulomb damping force is

$$F_f = \mu_k mg = 0.15(8)(9.81) = 11.7720 \text{ N}$$

and

$$\Delta = \frac{F_f}{k} = \frac{11.7720}{6000} = 1.9620 \times 10^{-3} \text{ m}.$$

Introduction to Vibrations

Thus, the decay per cycle is

$$4\Delta = 4(1.9620 \times 10^{-3}) = 7.8480 \times 10^{-3} \text{ m}.$$

b. Note that the displacement magnitude is reduced by 2Δ after every half-cycle. The motion stops at the end of the half-cycle for which the displacement magnitude is smaller than 2Δ. This can be expressed mathematically as

$$x_0 - n(2\Delta) < 2\Delta,$$

where n denotes the number of half-cycles before stopping. Solving for n and substituting the appropriate values, we have

$$n > \frac{x_0}{2\Delta} - 1 = \frac{0.025}{2(1.9620 \times 10^{-3})} - 1 = 5.37.$$

The smallest integer satisfying the inequality is $n = 6$. Thus, the oscillation stops after six half-cycles, or three cycles, with

$$x(t = 6T/2) = x_0 - 6(2\Delta) = 0.025 - 12(1.9620 \times 10^{-3}) = 1.4560 \times 10^{-3} \text{ m}.$$

PROBLEM SET 9.1

1. A lightly damped single-degree-of-freedom system is subjected to free vibration. The response of the system is shown in Figure 9.5. Estimate the value of the viscous damping ratio ζ.
2. An underdamped single-degree-of-freedom vibrating system is viscously damped. It is observed that the maximum displacement amplitude during the second cycle is 70% of the first. Calculate the damping ratio ζ and determine the maximum displacement amplitude during the fourth cycle as a fraction of the first.
3. Figure 9.6 is the free response of a single-degree-of-freedom system subjected to Coulomb damping. The parameters of the system include the mass, $m = 40$ kg, and the spring stiffness, $k = 1500$ N/m. Estimate the value of the kinetic friction coefficient μ_k.
4. For a Coulomb damped system, it is observed that the first three consecutive maximum displacement amplitudes x_0, x_1, and x_2 for a free vibration are 15 cm, 12.55 cm, and 10.10 cm, respectively. The time duration between any two of these amplitudes is 0.7 s.
   a. Determine the value of the kinetic friction coefficient μ_k.
   b. Determine the position when the oscillation stops.

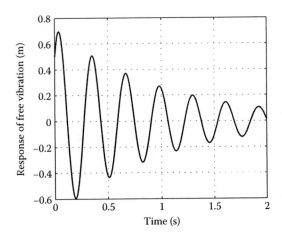

FIGURE 9.5 Problem 1.

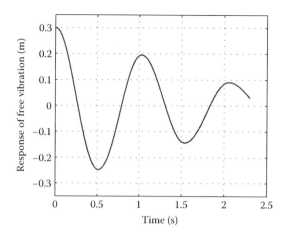

FIGURE 9.6 Problem 3.

9.2 FORCED VIBRATION

The vibration of a system caused by externally applied forces is known as forced vibration. A very important class of external excitations involves harmonic forces. Consider a single-degree-of-freedom viscously damped system subjected to a harmonic excitation, $f(t) = F_0 \sin \omega t$. Its governing differential equation is given by

$$m\ddot{x}(t) + b\dot{x}(t) + kx(t) = F_0 \sin \omega t \tag{9.29}$$

or

$$\ddot{x}(t) + 2\zeta\omega_n\dot{x}(t) + \omega_n^2 x(t) = \omega_n^2 \frac{F_0}{k} \sin \omega t. \tag{9.30}$$

As discussed in Section 8.3, the steady-state response of the system subjected to the harmonic excitation $f(t)$ may be obtained using the frequency response, which is a very important concept in vibration. The steady-state response of the system described by Equation 9.29 or 9.30 is

$$x(t) = X \sin(\omega t + \phi), \tag{9.31}$$

where the amplitude and the phase angle are given by

$$X = \frac{F_0/k}{\sqrt{[1 - (\omega/\omega_n)^2]^2 + [2\zeta(\omega/\omega_n)]^2}} \tag{9.32}$$

and

$$\phi = -\tan^{-1} \frac{2\zeta(\omega/\omega_n)}{1 - (\omega/\omega_n)^2}. \tag{9.33}$$

Note that the value of X depends on the driving frequency ω, and X is called the dynamic amplitude. The term F_0/k in Equation 9.32 has units of displacement and is known as the static deflection. If we use x_{st} to denote the static deflection, we can obtain the nondimensional ratio

$$\frac{X}{x_{st}} = \frac{1}{\sqrt{[1 - (\omega/\omega_n)^2]^2 + [2\zeta(\omega/\omega_n)]^2}}. \tag{9.34}$$

When compared to other excitations, more information on the response to the steady-state harmonic excitations can be extracted by means of the frequency domain technique than the time

Introduction to Vibrations

domain technique. In this section, we first discuss how to measure the viscous damping coefficient ζ based on frequency response plots. Systems subjected to two types of harmonic excitations, including rotating unbalance and base excitation, are then introduced and their responses are determined using the frequency response method.

9.2.1 Half-Power Bandwidth

Figure 9.7 shows the frequency response of the nondimensional ratio X/x_{st} near the natural frequency for a viscously damped system. As proven in Section 8.3, the maximum peak occurs at $\omega/\omega_n = \sqrt{1 - 2\zeta^2}$. If the system is lightly damped, the peak occurs in the immediate neighborhood of $\omega/\omega_n = 1$, as shown in Figure 9.7. From Equation 9.34, the corresponding value of the peak amplitude can be approximated as

$$\left(\frac{X}{x_{st}}\right)_{max} = Q \approx \frac{1}{2\zeta} \qquad (9.35)$$

for small damping. The symbol Q introduced in Equation 9.35 is known as the quality factor or Q factor, which is usually used in electrical engineering applications and is related to the amplitude at resonance. Thus, the damping ratio ζ can be solved from Equation 9.35 as

$$\zeta \approx \frac{1}{2Q}, \qquad (9.36)$$

which can be used as a quick way of estimating the viscous damping ratio by measuring the peak amplitude Q.

For a more accurate estimation, the damping ratio ζ can be determined by measuring the frequencies at two half-power points instead of the peak amplitude. Note that for a viscously damped system subjected to a harmonic force, the velocity response is

$$\dot{x}(t) = \omega X \cos(\omega t + \phi), \qquad (9.37)$$

which is obtained from Equation 9.31 by taking the time derivative. The maximum kinetic energy is

$$T_{max} = \frac{1}{2}m\omega^2 X_{max}^2 = \frac{1}{2}m\omega^2 x_{st}^2 Q^2, \qquad (9.38)$$

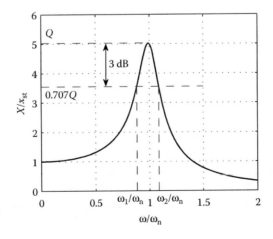

FIGURE 9.7 Magnitude of the frequency response for a viscously damped system.

which is proportional to the square of Q. Half-power points occur at frequencies where the power drops to half its maximum level. As shown in Figure 9.7, points 1 and 2, at which the amplitude falls to $Q/\sqrt{2} \approx 0.707Q$, or drops by 3 dB down from the peak, are half-power points.

Let $v = \omega/\omega_n$. At the half-power points, combining Equations 9.34 and 9.35 gives

$$\frac{1}{\sqrt{(1-v^2)^2 + (2\zeta v)^2}} \approx \frac{1}{\sqrt{2}} \frac{1}{2\zeta} \tag{9.39}$$

or

$$v^4 - 2(1 - 2\zeta^2)v^2 + (1 - 8\zeta^2) = 0, \tag{9.40}$$

which has the solutions

$$v^2 = (1 - 2\zeta^2) \pm 2\zeta\sqrt{1 + \zeta^2}. \tag{9.41}$$

If the system is lightly damped, ζ^2 can be ignored and Equation 9.41 can be approximated as

$$v^2 = 1 \pm 2\zeta \tag{9.42}$$

or

$$\left(\frac{\omega_1}{\omega_n}\right)^2 = 1 - 2\zeta, \tag{9.43}$$

$$\left(\frac{\omega_2}{\omega_n}\right)^2 = 1 + 2\zeta. \tag{9.44}$$

Subtracting Equation 9.43 from 9.44 gives

$$\frac{\omega_2^2 - \omega_1^2}{\omega_n^2} = 4\zeta. \tag{9.45}$$

In general, the natural frequency ω_n is between the half-power points and $\omega_n \approx \frac{1}{2}(\omega_1 + \omega_2)$ for light damping. Thus, we have

$$\frac{\omega_2 - \omega_1}{\omega_n} \approx 2\zeta \tag{9.46}$$

or

$$\zeta \approx \frac{\Delta\omega}{2\omega_n}, \tag{9.47}$$

where $\Delta\omega = \omega_2 - \omega_1$ is referred to as the bandwidth of the system.

Example 9.3: Half-Power Bandwidth

A viscously damped system consisting of a mass of 2.5 kg and a spring of stiffness 2250 N/m is subjected to a harmonic force excitation. The frequency ratios ω_1/ω_n and ω_2/ω_n at half-power points are observed to be 0.9064 and 1.0731, respectively. Assume that the system is lightly damped.

  a. Determine the bandwidth of the system.
  b. Determine the damping ratio ζ.

Solution

  a. Noticing the undamped natural frequency is $\omega_n = \sqrt{k/m} = \sqrt{2250/2.5} = 30\,\text{rad/s}$, the bandwidth of the system is

$$\Delta\omega = \omega_n \left(\frac{\omega_2 - \omega_1}{\omega_n}\right) = 30(1.0731 - 0.9064) = 5.00\,\text{rad/s},$$

Introduction to Vibrations

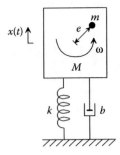

FIGURE 9.8 A system with rotating unbalance.

b. For small damping, the damping ratio is

$$\zeta \approx \frac{\Delta\omega}{2\omega_n} = \frac{5.00}{2(30)} = 0.08.$$

9.2.2 Rotating Unbalance

Unbalance in rotating machines is a common source of harmonic excitation. The unbalance is caused by the rotating part, for which the mass center does not coincide with the center of rotation. Figure 9.8 shows a system with an unbalanced mass m rotating at a constant angular velocity of ω. The distance between the unbalanced mass and the center of rotation is e, which represents the eccentricity. The mass of the entire system is M and its motion is assumed to be constrained along the vertical direction only. The entire system can therefore be considered as a single-degree-of-freedom system.

Choose the static equilibrium position of the entire system as the coordinate origin. Denote the mass of the unrotating part as $M - m$ and the corresponding vertical displacement as the generalized coordinate x. Thus, the vertical displacement of the rotating unbalanced mass is $x + e\sin\omega t$. The free-body diagrams of $M - m$ and m are shown in Figure 9.9.

Applying Newton's second law to $M - m$ and m along the vertical direction yields

$$(M - m)\ddot{x} = -F_V - kx - b\dot{x}, \qquad (9.48)$$

$$m\frac{d^2}{dt^2}(x + e\sin\omega t) = F_V, \qquad (9.49)$$

where F_V is the vertical component of the internal force between the rotating and the unrotating parts. Combining Equations 9.48 and 9.49, and eliminating F_V, gives the differential equation of the

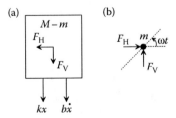

FIGURE 9.9 Free-body diagram of the system shown in Figure 9.8: (a) unrotating part $M - m$ and (b) rotating unbalance m.

system with rotating unbalance,

$$(M-m)\ddot{x} + m\frac{d^2}{dt^2}(x + e\sin\omega t) + b\dot{x} + kx = 0. \quad (9.50)$$

Differentiating $x + e\sin\omega t$ with respect to time twice, we have

$$M\ddot{x} + b\dot{x} + kx = me\omega^2 \sin\omega t, \quad (9.51)$$

which implies that the effect of a rotating unbalance mass is to exert a harmonic force $me\omega^2 \sin\omega t$ on the system.

Note that Equation 9.51 is similar to Equation 9.29, except that the magnitude of the harmonic force is $me\omega^2$ instead of F_0. Thus, with this modification, we can obtain the steady-state solution of the system using Equations 9.31 through 9.33. The dynamic amplitude is given by

$$X = X(\omega) = \frac{me\omega^2/k}{\sqrt{\left[1-(\omega/\omega_n)^2\right]^2 + [2\zeta(\omega/\omega_n)]^2}}, \quad (9.52)$$

where $\omega_n = \sqrt{k/M}$. Replacing k in Equation 9.52 with $M\omega_n^2$ yields

$$X = X(\omega) = \frac{me(\omega/\omega_n)^2/M}{\sqrt{\left[1-(\omega/\omega_n)^2\right]^2 + [2\zeta(\omega/\omega_n)]^2}}. \quad (9.53)$$

For a system with rotating unbalance, the nondimensional ratio

$$\frac{MX}{me} = \frac{(\omega/\omega_n)^2}{\sqrt{[1-(\omega/\omega_n)^2]^2 + [2\zeta(\omega/\omega_n)]^2}} \quad (9.54)$$

is usually plotted versus ω/ω_n. Figure 9.10 shows the magnitude of the frequency response for a system with rotating unbalance, where the vertical axis is $MX/(me)$. For low-speed rotations, that is, $\omega/\omega_n \ll 1$, the magnitude of the response $MX/(me)$ is very small and thus the vibration amplitude X is close to zero. Resonance peaks occur in the neighborhood of $\omega/\omega_n = 1$. For high-speed rotations, that is, $\omega/\omega_n \gg 1$, the magnitude of the response $MX/(me)$ tends to 1. This implies that the dynamic amplitude X is me/M, which is constant regardless of the driving frequency and the amount of damping.

Example 9.4: Rotating Unbalance

An industrial machine of mass $M = 450$ kg is supported by a spring with a static deflection $x_{st} = 0.5$ cm. If the machine has a rotating unbalance $me = 0.25$ kg·m, determine the dynamic amplitude X at 1200 rpm. Assume damping to be negligible.

Solution

The stiffness of the spring is

$$k = \frac{Mg}{x_{st}} = \frac{450(9.81)}{0.5 \times 10^{-2}} = 882,900 \text{ N/m}$$

and the natural frequency of the system is

$$\omega_n = \sqrt{\frac{k}{M}} = \sqrt{\frac{882,900}{450}} = 44.29 \text{ rad/s}.$$

Introduction to Vibrations

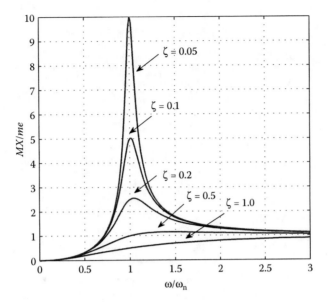

FIGURE 9.10 Magnitude of the frequency response for a system with rotating unbalance.

Given the driving frequency of 1200 rpm, the frequency ratio is

$$\frac{\omega}{\omega_n} = \frac{1200(2\pi)}{60(44.29)} = 2.84.$$

The dynamic amplitude can be solved from Equation 9.53 with damping neglected,

$$X = \frac{(me/M)(\omega/\omega_n)^2}{|1-(\omega/\omega_n)^2|} = \frac{(0.25/450)(2.84^2)}{|1-2.84^2|} = 6.34 \times 10^{-4} \, \text{m}.$$

9.2.3 Harmonic Base Excitation

Many applications in vibration involve systems with displacement as the input. Examples include a machine placed on a foundation undergoing vibration and a vehicle traveling on a wavy road. Assume that each system mentioned can be modeled as a single-degree-of-freedom system as shown in Figure 9.11, where x and y are the displacements of the mass and the base, respectively. Applying Newton's second law gives the differential equation of motion as

$$m\ddot{x}(t) + b\dot{x}(t) + kx(t) = b\dot{y}(t) + ky(t) \tag{9.55}$$

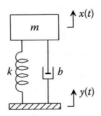

FIGURE 9.11 A single-degree-of-freedom system undergoing base excitation.

or
$$\ddot{x}(t) + 2\zeta\omega_n \dot{x}(t) + \omega_n^2 x(t) = 2\zeta\omega_n \dot{y}(t) + \omega_n^2 y(t). \tag{9.56}$$

As discussed in Section 8.3, the system's frequency response function is

$$G(j\omega) = \frac{X(j\omega)}{Y(j\omega)} = \frac{j2\zeta\omega_n\omega + \omega_n^2}{-\omega^2 + j2\zeta\omega_n\omega + \omega_n^2} = \frac{1 + j2\zeta\omega/\omega_n}{1 - (\omega/\omega_n)^2 + j2\zeta\omega/\omega_n}, \tag{9.57}$$

where the magnitude and phase of the frequency response function are

$$|G(j\omega)| = \frac{\sqrt{1 + (2\zeta\omega/\omega_n)^2}}{\sqrt{\left[1 - (\omega/\omega_n)^2\right]^2 + (2\zeta\omega/\omega_n)^2}} \tag{9.58}$$

and

$$\phi = \angle G(j\omega) = -\tan^{-1} \frac{2\zeta(\omega/\omega_n)^3}{1 - (\omega/\omega_n)^2 + (2\zeta\omega/\omega_n)^2}. \tag{9.59}$$

Assume that the motion of the base is harmonic, for example, $y(t) = Y_0 \sin\omega t$. Then, the steady-state response of the system is also harmonic, $x(t) = X \sin(\omega t + \phi)$, where the dynamic amplitude X is given by

$$X = Y_0 |G(j\omega)| = Y_0 \frac{\sqrt{1 + (2\zeta\omega/\omega_n)^2}}{\sqrt{\left[1 - (\omega/\omega_n)^2\right]^2 + (2\zeta\omega/\omega_n)^2}}. \tag{9.60}$$

Thus, the nondimensional ratio between the dynamic amplitude X and the amplitude of the base displacement is

$$\frac{X}{Y_0} = \frac{\sqrt{1 + (2\zeta\omega/\omega_n)^2}}{\sqrt{\left[1 - (\omega/\omega_n)^2\right]^2 + (2\zeta\omega/\omega_n)^2}}, \tag{9.61}$$

which is known as the transmissibility. Figure 9.12 shows the magnitude of the frequency response for a system with harmonic excitation, where the vertical axis is X/Y_0. Note that $X/Y_0 = 1$ for all curves when the frequency ratio $\omega/\omega_n = \sqrt{2}$. This implies that the response has the same magnitude as the excitation. The response is amplified for $\omega/\omega_n < \sqrt{2}$ and is reduced for $\omega/\omega_n > \sqrt{2}$.

Example 9.5: Harmonic Base Excitation

Precision instruments must be placed on rubber mounts, which act as springs and dampers, to reduce the effects of base vibration. Consider a precision instrument of mass 110 kg mounted on a rubber block. For the entire assembly, the spring stiffness is 280 kN/m and the damping ratio is 0.10. Assume that the base undergoes vibration, and the displacement of the base is expressed as $y(t) = Y_0 \sin\omega t$. Determine the dynamic amplitude of the system if the acceleration amplitude of the base excitation is 0.15 m/s^2 and the excitation frequency is 20 Hz.

Solution

The expression for the acceleration of the base is

$$\ddot{y}(t) = -\omega^2 Y_0 \sin\omega t,$$

where $\omega^2 Y_0 = 0.15$ m/s^2. For an excitation frequency of 20 Hz, the displacement amplitude of the base is

$$Y_0 = \frac{0.15}{\omega^2} = \frac{0.15}{[20(2\pi)]^2} = 9.50 \times 10^{-6} \text{ m}.$$

Introduction to Vibrations

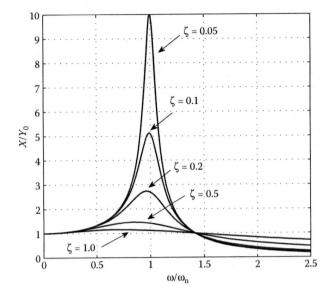

FIGURE 9.12 Magnitude of the frequency response for a system with harmonic base excitation.

The frequency ratio is

$$\frac{\omega}{\omega_n} = \frac{20(2\pi)}{\sqrt{280{,}000/110}} = 2.49$$

and the transmissibility can be calculated from Equation 9.61,

$$\frac{X}{Y_0} = \frac{\sqrt{1 + [2(0.1)(2.49)]^2}}{\sqrt{(1 - 2.49^2)^2 + (2(0.1)(2.49))^2}} = 0.21.$$

Thus, the dynamic amplitude of the system is $X = 0.21(9.50 \times 10^{-6}) = 2.00 \times 10^{-6}$ m.

PROBLEM SET 9.2

1. Figure 9.13 shows the experimental data for the frequency response of a single-degree-of-freedom system. The peak amplitude is 0.611 mm. Estimate the damping ratio ζ of the system.
2. A viscously damped single-degree-of-freedom system is subjected to a harmonic force excitation. It is observed that the peak amplitude Q of the frequency response X/x_{st} is 32 when the driving frequency ω is 10 rad/s.
   a. Determine the damping ratio ζ.
   b. Determine the bandwidth of the system.
3. An electric motor of mass $M = 500$ kg is mounted on a simply supported beam with negligible mass. Assume that the supporting beam is equivalent to a spring of stiffness $k = 4500$ kN/m and the damping ratio of the system is 0.1. The unbalance in the rotor of the motor is $me = 0.5$ kg·m. Determine the dynamic amplitude X of the motor when it runs at a speed of 950 rpm.
4. Tires must be balanced so that no periodic forces develop during operation. Figure 9.14 shows a tire with an eccentric mass because of uneven wear. The parameters are given as follows: the mass of the tire $M = 11.75$ kg, the unbalanced mass $m = 0.1$ kg, the radius of the tire $r = 22.5$ cm, and the eccentric distance $e = 15$ cm. Assume that the stiffness of the tire is 120 kN/m. Neglect the damping of the system. Determine the

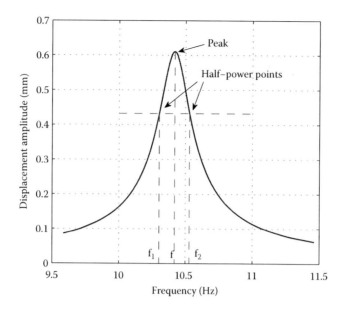

FIGURE 9.13 Problem 1.

amplitude of the steady-state response of the tire caused by mass unbalance when the car moves at 100 km/h.

5. A vehicle of mass 3000 kg travels on a rough road, as shown in Figure 9.15. Assume that the surface of the road can be approximated as a sine wave $y = Y_0 \sin(2\pi vt/L)$, where $Y_0 = 0.01$ m and $L = 10$ m. The suspension of the vehicle is modeled as a spring–damper system with $k = 50$ kN/m and $b = 2000$ N·s/m. Determine the dynamic amplitude X of the vehicle when it moves at a speed of (a) 20 km/h and (b) 100 km/h.

6. Many vibration measuring instruments consist of a case containing a mass–damper–spring system, as shown in Figure 9.16. The displacement of the mass relative to the case is measured electrically. Denote the displacement of the mass, the displacement of the case, and the displacement of the mass relative to the case as $x(t)$, $y(t)$, and $z(t)$, respectively, where $z(t) = x(t) - y(t)$. Assume harmonic excitation, $y(t) = Y_0 \sin \omega t$.
   a. Show that the amplitude Z is given by
   $$Z = \frac{(\omega/\omega_n)^2}{\sqrt{[1-(\omega/\omega_n)^2]^2 + (2\zeta\omega/\omega_n)^2}} Y_0.$$
   b. Using MATLAB®, write an m-file to plot the frequency response Z/Y_0 versus ω/ω_n for different values of damping ratio: $\zeta = 0.1, 0.25, 0.5, 0.7,$ and 1.0.

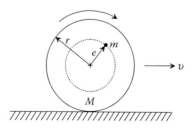

FIGURE 9.14 Problem 4.

Introduction to Vibrations

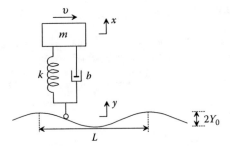

FIGURE 9.15 Problem 5.

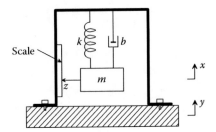

FIGURE 9.16 Problem 6.

9.3 VIBRATION SUPPRESSIONS

Vibrations are undesirable in most cases, particularly in cars, machining tools, precision instruments, buildings in an active seismic zone, and so on. To protect these systems and enhance their life, it is necessary to reduce vibration. This can be achieved with vibration isolators or vibration absorbers.

9.3.1 VIBRATION ISOLATORS

In order to isolate an object from the source of vibrations, two types of vibration isolation systems are used, passive and active. Passive vibration isolation systems consist of springs and dampers. Active vibration isolation systems contain (along with the springs) piezoelectric accelerometers, electromagnetic actuators, and control circuits. The topic of active vibration isolation is beyond the scope of this text, and therefore will not be discussed.

A vibration isolation system attempts either to protect delicate equipment from vibration transmitted to it from its support system or to prevent the vibratory force generated by a machine from being transmitted to its surroundings. The essence of these two objectives is the same. The concept of transmissibility introduced in Section 9.2 can be used for either displacement isolation design or force isolation design.

As shown in Equation 9.61, for a system placed on a support undergoing harmonic vibration, the dynamic response of the system depends on the natural frequency ω_n and the damping ratio ζ. This fact is also shown in Figure 9.12. When the natural frequency ω_n is much less than the excitation frequency ω, more specifically, $\omega/\omega_n > \sqrt{2}$, the displacement transmissibility X/Y_0 is less than 1. This implies that the magnitude of the system response is reduced. Thus, it is desirable to design a vibration isolator such that the natural frequency of the entire assembly is within the region of $\omega/\omega_n > \sqrt{2}$. This can be achieved by placing the system on a spring and/or damper system. Note that the value of damping should not be too large because the displacement transmissibility increases when the damping ratio increases for $\omega/\omega_n > \sqrt{2}$.

Example 9.6: Displacement Isolation

Consider a machine of mass 4000 kg mounted on a vibration isolator consisting of four parallel-connected springs. Assume that the base undergoes vibration, and the displacement of the base is expressed as $y(t) = 0.001 \sin \pi t$. Determine the stiffness of each spring if the maximum allowable displacement amplitude of the machine is 0.0001 m. Assume that all the springs are identical and damping is negligible.

SOLUTION

The displacement transmissibility is $X/Y_0 = 0.0001/0.001 = 0.1$. From Equation 9.61, the transmissibility with damping neglected is

$$\frac{X}{Y_0} = \frac{1}{\left|1-(\omega/\omega_n)^2\right|} = 0.10,$$

from which the frequency ratio ω/ω_n is solved to be $\sqrt{11}$. For the excitation frequency $\omega = \pi$ rad/s, the natural frequency is

$$\omega_n = \frac{\pi}{\sqrt{11}} = 0.95 \text{ rad/s}.$$

Thus, the equivalent spring stiffness of the system is

$$k_e = \omega_n^2 m = 0.95^2(4000) = 3610 \text{ N/m}.$$

For each spring, the stiffness is $k = k_e/4 = 902.50$ N/m.

For a machine placed on a rigid base, the machine itself is a vibration source. The vibratory force generated by the machine will be transmitted to the base and then affect the surrounding equipment. To reduce the damaging effect of the vibratory machine on its surroundings, it is necessary to isolate it from the base.

Figure 9.17a shows a machine placed on a rigid foundation through a spring and damper system. The machine is subjected to a harmonic excitation force $f(t)$. Recall that the displacement response $x(t)$ to harmonic excitation is given by Equations 9.31 through 9.33. Differentiating Equation 9.31 with respect to time yields the velocity response as given in Equation 9.37, where X is the amplitude of the displacement and the amplitude of the velocity is ωX. Moreover, the velocity leads the displacement by the phase angle $\pi/2$ because $\cos(\omega t + \phi)$ in Equation 9.37 can be expressed as $\sin(\omega t + \phi + \pi/2)$.

From the free-body diagram in Figure 9.17b, it is clear that the force transmitted to the base includes two parts, the spring force kx and the damping force $b\dot{x}$. The amplitudes of the spring force and the damping force are kX and $b\omega X$, respectively. The force vectors are shown in Figure 9.17c.

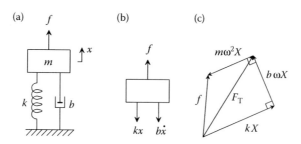

FIGURE 9.17 Force isolation: (a) physical system, (b) free-body diagram, and (c) force vector diagram.

Introduction to Vibrations

The angle between the two force vectors is $\pi/2$. Thus, the amplitude of the force transmitted to the base is

$$F_T = \sqrt{(kX)^2 + (b\omega X)^2}. \tag{9.62}$$

Note that $b/k = (b/m)(m/k) = 2\zeta\omega_n/\omega_n^2 = 2\zeta/\omega_n$. Then, Equation 9.62 can be rewritten as

$$F_T = kX\sqrt{1 + (2\zeta\omega/\omega_n)^2}. \tag{9.63}$$

Substituting Equation 9.32 into Equation 9.63 yields

$$F_T = F_0 \frac{\sqrt{1 + (2\zeta\omega/\omega_n)^2}}{\sqrt{\left[1 - (\omega/\omega_n)^2\right]^2 + [2\zeta(\omega/\omega_n)]^2}}. \tag{9.64}$$

The nondimensional ratio F_T/F_0 is the transmissibility given by Equation 9.61, and it is a measure of the force transmitted to the base. Thus, the plot of F_T/F_0 versus ω/ω_n is the same as the plot of X/Y_0 versus ω/ω_n, as shown in Figure 9.12. When the frequency ratio $\omega/\omega_n = \sqrt{2}$, $F_T/F_0 = 1$ and the excitation force is fully transmitted to the base. For $\omega/\omega_n > \sqrt{2}$, the force transmissibility F_T/F_0 is less than 1 and the force transmitted reduces with increasing excitation frequency ω.

Example 9.7: Force Isolation

A rotating machine of mass 2000 kg is mounted on an isolator block of stiffness 500 kN/m and damping ratio $\zeta = 0.1$. The machine is subjected to a harmonic disturbance force, for which the frequency is the same as the rotational speed of the machine. Assuming that 20% of the disturbance force is transmitted to the base, determine the rotational speed of the machine.

Solution

From Equation 9.64, we have

$$\frac{F_T}{F_0} = \frac{\sqrt{1 + [2(0.1)(\omega/\omega_n)]^2}}{\sqrt{\left[1 - (\omega/\omega_n)^2\right]^2 + [2(0.1)(\omega/\omega_n)]^2}} = 0.2.$$

Solving for the frequency ratio yields $\omega/\omega_n = 2.57$. The natural frequency of the system is $\omega_n = \sqrt{500,000/2000} = 15.81$ rad/s. Thus, the excitation frequency is

$$\omega = 2.57(15.81) = 40.63 \text{ rad/s},$$

which is also the rotational speed of the machine.

9.3.2 Vibration Absorbers

As discussed in Section 8.3, for a single-degree-of-freedom system subjected to harmonic excitation, violent vibration is induced when the excitation frequency is close to the natural frequency of the system. To protect the system, we can change either the mass or the spring stiffness so that the natural frequency is not too close to the excitation frequency. However, this may not always be possible. To solve this issue, a vibration absorber consisting of a second mass and spring can be added to the system to protect the original single-degree-of-freedom system from harmonic excitation.

Consider the two-degree-of-freedom system as shown in Figure 9.18, where m_1 and k_1 are the mass and the spring stiffness of the primary system, and m_2 and k_2 are the mass and the spring stiffness of the absorber. A harmonic force $F_1 \sin \omega t$ is applied to the primary system, which would undergo violent vibrations if the absorber is not installed.

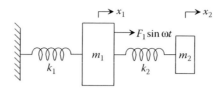

FIGURE 9.18 A vibration absorber.

For the combined system, the equations of motion in matrix form are

$$\begin{bmatrix} m_1 & 0 \\ 0 & m_2 \end{bmatrix} \begin{Bmatrix} \ddot{x}_1 \\ \ddot{x}_2 \end{Bmatrix} + \begin{bmatrix} k_1 + k_2 & -k_2 \\ -k_2 & k_2 \end{bmatrix} \begin{Bmatrix} x_1 \\ x_2 \end{Bmatrix} = \begin{Bmatrix} F_1 \\ 0 \end{Bmatrix} \sin \omega t. \quad (9.65)$$

Because the system is undamped, the steady-state response to a sinusoidal input is still sinusoidal, which has the same frequency as the excitation frequency and a zero phase angle. We can express the steady-state response as

$$\begin{Bmatrix} x_1(t) \\ x_2(t) \end{Bmatrix} = \begin{Bmatrix} X_1 \\ X_2 \end{Bmatrix} \sin \omega t. \quad (9.66)$$

Substituting Equation 9.66 into Equation 9.65 yields

$$\begin{bmatrix} k_1 + k_2 - m_1 \omega^2 & -k_2 \\ -k_2 & k_2 - m_2 \omega^2 \end{bmatrix} \begin{Bmatrix} X_1 \\ X_2 \end{Bmatrix} = \begin{Bmatrix} F_1 \\ 0 \end{Bmatrix}. \quad (9.67)$$

Solving for X_1 and X_2 gives

$$\begin{Bmatrix} X_1 \\ X_2 \end{Bmatrix} = \frac{1}{\Delta(\omega)} \begin{bmatrix} k_2 - m_2 \omega^2 & k_2 \\ k_2 & k_1 + k_2 - m_1 \omega^2 \end{bmatrix} \begin{Bmatrix} F_1 \\ 0 \end{Bmatrix}, \quad (9.68)$$

where

$$\Delta(\omega) = (k_2 - m_2 \omega^2)(k_1 + k_2 - m_1 \omega^2) - k_2^2. \quad (9.69)$$

Thus, the amplitudes are

$$X_1 = \frac{(k_2 - m_2 \omega^2) F_1}{(k_2 - m_2 \omega^2)(k_1 + k_2 - m_1 \omega^2) - k_2^2}, \quad (9.70)$$

$$X_2 = \frac{k_2 F_1}{(k_2 - m_2 \omega^2)(k_1 + k_2 - m_1 \omega^2) - k_2^2}. \quad (9.71)$$

Dividing the numerators and denominators in Equations 9.70 and 9.71 by $k_1 k_2$, we obtain

$$X_1 = \frac{\left[1 - (m_2/k_2)\omega^2\right](F_1/k_1)}{\left[1 - (m_2/k_2)\omega^2\right]\left[1 + k_2/k_1 - (m_1/k_1)\omega^2\right] - (k_2/k_1)}, \quad (9.72)$$

$$X_2 = \frac{F_1/k_1}{\left[1 - (m_2/k_2)\omega^2\right]\left[1 + k_2/k_1 - (m_1/k_1)\omega^2\right] - (k_2/k_1)}. \quad (9.73)$$

From Equation 9.72, it is clear that the amplitude X_1 of the primary system reduces to zero when the excitation frequency ω equals $\sqrt{k_2/m_2}$. Correspondingly, the amplitude X_2 of the absorber becomes

$$X_2 = -\frac{F_1}{k_2}. \quad (9.74)$$

Introduction to Vibrations

where the negative sign indicates that the absorber moves opposite to the direction of the force F_1. Thus, the force exerted on the primary mass m_1 by the spring of the absorber is

$$k_2 x_2 = k_2 X_2 \sin \omega t = -F_1 \sin \omega t, \tag{9.75}$$

which exactly balances the externally applied force $F_1 \sin \omega t$.

Let us denote the natural frequency of the original single-degree-of-freedom system alone as $\omega_1 = \sqrt{k_1/m_1}$ and the natural frequency of the absorber alone as $\omega_2 = \sqrt{k_2/m_2}$. To determine the parameters of the absorber, we introduce the following notations, $x_{st} = F_1/k_1$, $\mu = m_2/m_1$, and $\nu = \omega_2/\omega_1$, where x_{st} is the static deflection of the original system, μ is the mass ratio, and ν is the natural frequency ratio. Then, Equations 9.72 and 9.73 can be rewritten as

$$X_1 = \frac{\left[1 - (\omega/\omega_2)^2\right] x_{st}}{\left[1 - (\omega/\omega_2)^2\right]\left[1 + k_2/k_1 - (\omega/\omega_1)^2\right] - (k_2/k_1)}, \tag{9.76}$$

$$X_2 = \frac{x_{st}}{\left[1 - (\omega/\omega_2)^2\right]\left[1 + k_2/k_1 - (\omega/\omega_1)^2\right] - (k_2/k_1)}. \tag{9.77}$$

Note that $\omega/\omega_2 = (\omega/\omega_1)(1/\nu)$ and $k_2/k_1 = (m_2/m_1)(k_2/m_2)(m_1/k_1) = \mu \nu^2$. Thus,

$$X_1 = \frac{\left[\nu^2 - (\omega/\omega_1)^2\right] x_{st}}{\left[\nu^2 - (\omega/\omega_1)^2\right]\left[1 + \mu \nu^2 - (\omega/\omega_1)^2\right] - \mu \nu^4}, \tag{9.78}$$

$$X_2 = \frac{x_{st}}{\left[\nu^2 - (\omega/\omega_1)^2\right]\left[1 + \mu \nu^2 - (\omega/\omega_1)^2\right] - \mu \nu^4}. \tag{9.79}$$

From Equations 9.78 and 9.79, it is clear that the values of X_1 and X_2 depend on the mass ratio μ and the natural frequency ratio ν. Note that the natural frequency ratio ν must be very close to one. This is due to the fact that adding the absorber aims to alleviate the vibration of the original system at resonance, that is, $\omega = \omega_1$, and the absorber can only be effective when its own natural frequency is the same as the excitation frequency, that is, $\omega_2 = \omega$. Thus, we have $\omega_2 = \omega_1$ or $\nu = 1$. If ν is very close to one, the motion of m_1 is not zero, but its amplitude is still very small.

Figures 9.19 shows two curves, X_1/x_{st} versus ω/ω_1 and X_2/x_{st} versus ω/ω_1 for $\mu = 0.2$ and $\nu = 1$. Note that the horizontal axis can be replaced by ω/ω_2 because $\nu = 1$ or $\omega_2 = \omega_1$. As observed from Figure 9.19a, when the excitation frequency ω is in the immediate neighborhood of ω_2, the amplitude X_1 is very small. However, when the excitation frequency ω shifts slightly away from ω_2, the amplitude X_1 increases significantly. This shows that the absorber is only effective when the excitation frequency ω is close to ω_2.

It should be pointed out that one disadvantage of the vibration absorber is that two new resonant frequencies in the neighborhood of the excitation frequency are created as seen from Figure 9.20. Since the values of X_1/x_{st} and X_2/x_{st} change dramatically, the vertical axis in Figure 9.20 is given in the unit of dB, which can be used to conveniently represent very large or small numbers. The details on how to determine natural frequencies for multi-degree-of-freedom systems will be discussed in the next section.

Example 9.8: Vibration Absorber Design

A rotating machine has a natural frequency of 1.5 Hz. Due to a rotating unbalanced mass, the machine is subjected to a harmonic disturbance force. The force has a frequency of 1.6 Hz and an amplitude of 15 N. Design a vibration absorber assuming that the maximum allowable displacement of the absorber is 0.1 cm.

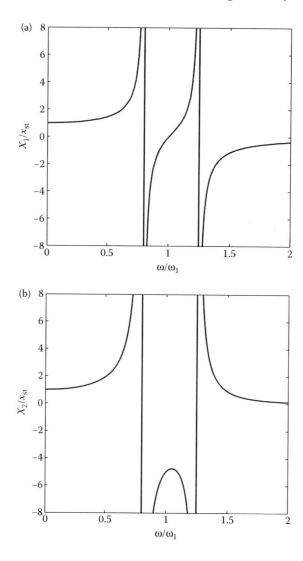

FIGURE 9.19 Frequency response curves: (a) main mass and (b) absorber mass.

Solution

The natural frequency of the absorber is required to be the same as the disturbance frequency,

$$\omega_2 = \omega = 1.6\,\text{Hz} = 3.2\pi\,\text{rad/s}.$$

The maximum allowable displacement amplitude of the absorber is 0.1 cm. From Equation 9.74, considering the magnitude only, the spring stiffness of the absorber can be obtained as

$$k_2 = \frac{F_1}{X_2} = \frac{15}{0.1 \times 10^{-2}} = 15{,}000\,\text{N/m}.$$

Thus, the mass of the absorber is

$$m_2 = \frac{k_2}{\omega_2^2} = \frac{15{,}000}{(3.2\pi)^2} = 148.42\,\text{kg}.$$

Introduction to Vibrations

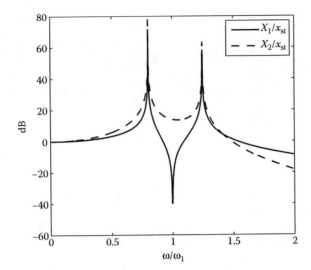

FIGURE 9.20 Frequency response curves (in dB) for the main mass and the absorber mass.

PROBLEM SET 9.3

1. A machine of mass 50 kg is mounted on a rubber isolator to protect it from the ground's vibration caused by the operation of other machines nearby. Assume that the ground vibrates at 10 Hz. Determine the stiffness of the rubber isolation spring if only 10% of the ground's motion is transmitted to the machine.
   a. Neglect damping.
   b. Assume that the damping coefficient of the rubber isolator is 0.1.
2. Consider a 10-kg instrument placed on a floor that vibrates with a frequency of 3000 rpm and an amplitude of 2 mm due to nearby machinery. A vibration isolator is designed to protect the instrument from the vibration of the floor. Assume that the damping ratio of the isolator is 0.05 and the maximum allowable acceleration amplitude of the instrument is 2 g.
   a. Determine the stiffness of the isolator.
   b. Determine the maximum displacement amplitude of the instrument.
3. Reconsider Problem 5 in Problem Set 9.2. Determine the maximum force transmitted to the vehicle when it moves at a speed of (a) 20 km/h and (b) 100 km/h.
4. Consider the single-degree-of-freedom system shown in Figure 9.8. The excitation force due to the rotating unbalance is $me\omega^2 \sin\omega t$. Assume that the system has the following parameters: $m = 5$ kg, $M = 100$ kg, $e = 0.15$ m, $k = 5000$ N/m, and $b = 200$ N·s/m.
   a. Determine the force transmitted to the support when the system runs at the rotating speed $\omega = 2\omega_n$.
   b. Determine the force transmissibility.
   c. The spring is replaced in order to decrease the force transmissibility to 20%. Determine the stiffness of the new spring.
5. Reconsider Problem 3 in Problem Set 9.2. Design a vibration absorber assuming that the maximum allowable displacement of the absorber is 0.5 cm.
6. A rotating machine has a mass of 6 kg and a natural frequency of 5 Hz. Due to a rotating unbalanced mass, the machine is subjected to a harmonic disturbance force $me\omega^2 \sin\omega t$. When the machine operates at a frequency of 3.5 Hz, the amplitude of the disturbance force is 40 N.
   a. Design a vibration absorber assuming that the maximum allowable displacement of the absorber is 5 cm.
   b. Using MATLAB, write an m-file to plot X_1/me versus ω.

9.4 MODAL ANALYSIS

The systems discussed in the previous sections are mainly single-degree-of-freedom systems, for which the vibration can be studied using elementary methods presented in Chapter 8. However, for a multi-degree-of-freedom system, more advanced mathematical tools are required to solve the equations of motion due to coordinate coupling. In this section, we first introduce concepts, such as the eigenvalue problem, natural modes, and orthogonality of modes. Then, we develop modal analysis in a rigorous manner to decouple coordinates and use it to obtain the response to initial excitations or external forces. All derivations in this section are presented in matrix form.

9.4.1 Eigenvalue Problem

Consider a three-degree-of-freedom mass–spring system as shown in Figure 9.21, where the motion of the system is described by the displacement coordinates x_1, x_2, and x_3. In the absence of damping and external forces, the system undergoes undamped free vibration, and the differential equations of motion in matrix form are

$$\begin{bmatrix} m_1 & 0 & 0 \\ 0 & m_2 & 0 \\ 0 & 0 & m_3 \end{bmatrix} \begin{Bmatrix} \ddot{x}_1 \\ \ddot{x}_2 \\ \ddot{x}_3 \end{Bmatrix} + \begin{bmatrix} k_1+k_2 & -k_2 & 0 \\ -k_2 & k_2+k_3 & -k_3 \\ 0 & -k_3 & k_3 \end{bmatrix} \begin{Bmatrix} x_1 \\ x_2 \\ x_3 \end{Bmatrix} = \begin{Bmatrix} 0 \\ 0 \\ 0 \end{Bmatrix}. \qquad (9.80)$$

Note that the mass matrix is diagonal and the stiffness matrix is not. Thus, Equation 9.80 represents a set of three simultaneous, or coupled, second-order differential equations. It is generally difficult to obtain the analytical solution of Equation 9.80 because of coordinate coupling. If we can find a coordinate transformation such that the mass and stiffness matrices are both diagonalized, then the system dynamics can be decoupled into a set of independent second-order differential equations in the form of $m\ddot{x} + kx = 0$, which can be easily solved as a single-degree-of-freedom system. Such a transformation matrix does exist, which is not unique. One can be found by solving the eigenvalue problem.

The eigenvalue problem is a problem associated with free, undamped vibration, for which the differential equation in the general matrix form is given by

$$\mathbf{M}\ddot{\mathbf{x}} + \mathbf{K}\mathbf{x} = \mathbf{0}. \qquad (9.81)$$

The solution to Equation 9.81 is harmonic. Assume that the solution is in the form

$$\mathbf{x} = \mathbf{X}e^{j\omega t}, \qquad (9.82)$$

where ω is the frequency of the harmonic motion. Note that $\ddot{\mathbf{x}} = -\omega^2 \mathbf{X}e^{j\omega t}$. Thus, Equation 9.81 becomes

$$(\mathbf{K} - \omega^2 \mathbf{M})\mathbf{X}e^{j\omega t} = \mathbf{0}. \qquad (9.83)$$

Canceling the nonzero term $e^{j\omega t}$ gives

$$(\mathbf{K} - \omega^2 \mathbf{M})\mathbf{X} = \mathbf{0} \qquad (9.84)$$

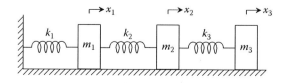

FIGURE 9.21 A three-degree-of-freedom mass–spring system.

Introduction to Vibrations

or
$$KX = \omega^2 MX. \tag{9.85}$$

Equation 9.84 or 9.85 represents the eigenvalue problem associated with matrices M and K. In particular, it is known as the algebraic eigenvalue problem, which can be solved in a similar way as solving the eigenvalue problem associated with a matrix A as discussed in Section 3.2. First, Equation 9.84 has a nontrivial solution $X \neq 0$ if and only if the coefficient matrix is singular,

$$|K - \omega^2 M| = 0. \tag{9.86}$$

Equation 9.86 is known as the characteristic equation or frequency equation. For an n-degree-of-freedom system, the determinant $|K - \omega^2 M|$ is a polynomial of degree n in ω^2. The n roots of Equation 9.86 are referred to as eigenvalues and denoted by $\omega_1^2, \omega_2^2, \ldots,$ and ω_n^2. Once the eigenvalues are identified, each eigenvector corresponding to each of the eigenvalues can be obtained by solving Equation 9.84 or 9.85. The n eigenvectors are denoted by $X_1, X_2, \ldots,$ and X_n.

For a vibration system, the eigenvalues and eigenvectors associated with the eigenvalue problem defined in Equation 9.84 or 9.85 have significant physical meanings. The square roots of the eigenvalues are the system's natural frequencies ω_r, where $r = 1, 2, \ldots, n$. The natural frequencies are usually arranged in increasing order of magnitude, that is, $\omega_1 \leq \omega_2 \leq \cdots \leq \omega_n$. The lowest frequency ω_1 is referred to as the fundamental frequency, which is extremely important for many practical problems. The eigenvectors are referred to as modal vectors. Each modal vector represents physically the shape of a normal mode, a certain pattern of motion in which all masses move harmonically with the same natural frequency associated with this modal vector. The following example shows how to solve the eigenvalue problem and describes the physical significance of the eigenvalues and eigenvectors.

Example 9.9: Natural Frequencies and Modal Vectors

Consider the three-degree-of-freedom system shown in Figure 9.21. Assume that $m_1 = m_2 = m_3 = m$, $k_1 = 3k$, $k_2 = 2k$, and $k_3 = k$.

a. Solve the associated eigenvalue problem by hand.
b. ◢ Solve the associated eigenvalue problem using MATLAB. Without loss of generality, assume $m = 1$ kg and $k = 1$ N/m.

Solution

a. The differential equations for the system in Figure 9.21 are given by Equation 9.80. Thus, the mass and stiffness matrices are

$$M = \begin{bmatrix} m & 0 & 0 \\ 0 & m & 0 \\ 0 & 0 & m \end{bmatrix}, \quad K = \begin{bmatrix} 5k & -2k & 0 \\ -2k & 3k & -k \\ 0 & -k & k \end{bmatrix}.$$

Substituting M and K into the frequency equation defined by Equation 9.86, we have

$$\begin{vmatrix} 5k - \omega^2 m & -2k & 0 \\ -2k & 3k - \omega^2 m & -k \\ 0 & -k & k - \omega^2 m \end{vmatrix} = 0.$$

To simplify the calculation, divide the above equation by k and let $\beta = \omega^2 m/k$. Thus, the determinant is

$$\begin{vmatrix} 5 - \beta & -2 & 0 \\ -2 & 3 - \beta & -1 \\ 0 & -1 & 1 - \beta \end{vmatrix} = (5 - \beta)[(3 - \beta)(1 - \beta) - 1] - (-2)[(-2)(1 - \beta)] = 0,$$

which reduces to
$$\beta^3 - 9\beta^2 + 18\beta - 6 = 0.$$

The three roots are
$$\beta_1 = 0.4158, \quad \beta_2 = 2.2943, \quad \beta_3 = 6.2899,$$

which gives the three eigenvalues
$$\omega_1^2 = 0.4158 \frac{k}{m}, \quad \omega_2^2 = 2.2943 \frac{k}{m}, \quad \omega_3^2 = 6.2899 \frac{k}{m}.$$

Thus, the three natural frequencies are
$$\omega_1 = 0.6448 \sqrt{\frac{k}{m}}, \quad \omega_2 = 1.5147 \sqrt{\frac{k}{m}}, \quad \omega_3 = 2.5080 \sqrt{\frac{k}{m}}.$$

To determine the eigenvectors or modal vectors, we insert ω_r^2 ($r = 1, 2, 3$) into Equation 9.84. For $r = 1$, we have
$$\left(\mathbf{K} - \omega_1^2 \mathbf{M} \right) \mathbf{X}_1 = \mathbf{0}.$$

Note that the modal vectors are 3×1 column vectors for a three-degree-of-freedom system. The above equation can be written as
$$\begin{bmatrix} 5k - \omega_1^2 m & -2k & 0 \\ -2k & 3k - \omega_1^2 m & -k \\ 0 & -k & k - \omega_1^2 m \end{bmatrix} \begin{bmatrix} X_{11} \\ X_{12} \\ X_{13} \end{bmatrix} = 0.$$

Again, to simplify the calculation, we use β_1 to replace $\omega_1^2 m/k$ and write the equations in scalar form
$$(5 - \beta_1)X_{11} - 2X_{12} = 0,$$
$$-2X_{11} + (3 - \beta_1)X_{12} - X_{13} = 0,$$
$$-X_{12} + (1 - \beta_1)X_{13} = 0.$$

Note that combining the first and the third equations by canceling X_{12} gives
$$\frac{X_{13}}{X_{11}} = \frac{5 - \beta_1}{2 - 2\beta_1}.$$

Then, expressing X_{13} in the second equation in terms of X_{11} yields
$$\frac{X_{12}}{X_{11}} = \frac{9 - 5\beta_1}{(2 - 2\beta_1)(3 - \beta_1)}.$$

If we assign one element in $\mathbf{X}_1$ an arbitrary value, the other two can be determined uniquely. This implies that we have three equations and two unknowns. Assuming $X_{11} = 1$ leads to $X_{13} = 3.9235$ and $X_{12} = 2.2922$. Thus, the first modal vector is
$$\mathbf{X}_1 = \begin{bmatrix} 1 \\ 2.2922 \\ 3.9235 \end{bmatrix}.$$

In a similar way, we can find the other two modal vectors.
$$\mathbf{X}_2 = \begin{bmatrix} 1 \\ 1.3529 \\ -1.0452 \end{bmatrix} \quad \mathbf{X}_3 = \begin{bmatrix} 1 \\ -0.6450 \\ 0.1219 \end{bmatrix}.$$

Here, we set the magnitude of the first component in all modal vectors as 1. The modal vectors can also be normalized so that the largest component in magnitude is equal to 1.

$$\mathbf{X}_1 = \begin{bmatrix} 0.2549 \\ 0.5842 \\ 1 \end{bmatrix} \quad \mathbf{X}_2 = \begin{bmatrix} 0.7392 \\ 1 \\ -0.7726 \end{bmatrix} \quad \mathbf{X}_3 = \begin{bmatrix} 1 \\ -0.6450 \\ 0.1219 \end{bmatrix}.$$

The modal vectors represent physically the shape of the natural modes as shown in Figure 9.22. In mode 1, all masses move in the same direction and oscillate with a frequency of $\omega_1 = 0.6448\sqrt{k/m}$. In mode 2, all masses oscillate with a frequency of $\omega_2 = 1.5147\sqrt{k/m}$. Masses 1 and 2 move in the same direction, while mass 3 moves in the opposite direction. The direction change is also implied by the one sign change in the second mode $\mathbf{X}_2$. In mode 3, there are two sign changes in $\mathbf{X}_3$. Masses 1 and 3 move in the same direction, while mass 2 moves in the opposite direction. All masses oscillate with a frequency of $\omega_3 = 2.5080\sqrt{k/m}$.

b. Given the mass and stiffness matrices, it is easy to use MATLAB to solve the associated algebraic eigenvalue problem. The following is the MATLAB session.

```
>> M = eye(3);
>> K = [5 -2 0; -2 3 -1; 0 -1 1];
>> [V,D] = eig(K,M);
```

The eig command returns two matrices named $\mathbf{V}$ and $\mathbf{D}$,

$$\mathbf{V} = \begin{bmatrix} 0.2149 & -0.5049 & -0.8360 \\ 0.4927 & -0.6831 & 0.5392 \\ 0.8433 & 0.5277 & -0.1019 \end{bmatrix} \quad \mathbf{D} = \begin{bmatrix} 0.4158 & 0 & 0 \\ 0 & 2.2943 & 0 \\ 0 & 0 & 6.2899 \end{bmatrix},$$

where $\mathbf{D}$ is a diagonal matrix containing the eigenvalues and the columns in matrix $\mathbf{V}$ are the corresponding eigenvectors. Note that the modal vectors returned by the MATLAB command eig are normalized in a way so that the magnitude of each vector is equal to 1. For example, the first column in matrix $\mathbf{V}$ is the first modal vector, and we have

$$\sqrt{0.2149^2 + 0.4927^2 + 0.8433^2} = 1.$$

Comparing the eigenvectors obtained by hand and by MATLAB, we observe that the ratios between the elements are the same. For example, for mode 1, $X_{11}/X_{13} = 0.2149/0.8433 = 0.2548$ and $X_{12}/X_{13} = 0.4927/0.8433 = 0.5843$. Due to round-off error, the ratios are slightly different from that obtained in Part (a). This implies that the shape of modes is unique but the amplitude is not.

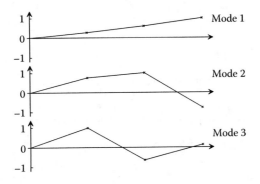

FIGURE 9.22 Natural modes of the three-degree-of-freedom system shown in Figure 9.21.

9.4.2 Orthogonality of Modes

The natural modes possess an important property known as orthogonality. To show this concept, let us consider an n-degree-of-freedom system. Following the same procedure as in Example 9.9, we can obtain n natural frequencies and n modal vectors by solving the eigenvalue problem. Denote any two solutions of the eigenvalue problem as ω_r^2, $\mathbf{X}_r$, and ω_s^2, $\mathbf{X}_s$. Both pairs should satisfy Equation 9.85, that is,

$$\mathbf{KX}_r = \omega_r^2 \mathbf{MX}_r, \tag{9.87}$$

$$\mathbf{KX}_s = \omega_s^2 \mathbf{MX}_s. \tag{9.88}$$

Premultiplying both sides of Equation 9.87 by $\mathbf{X}_s^T$ and Equation 9.88 by $\mathbf{X}_r^T$ gives

$$\mathbf{X}_s^T \mathbf{KX}_r = \omega_r^2 \mathbf{X}_s^T \mathbf{MX}_r, \tag{9.89}$$

$$\mathbf{X}_r^T \mathbf{KX}_s = \omega_s^2 \mathbf{X}_r^T \mathbf{MX}_s. \tag{9.90}$$

Taking the transpose of Equation 9.90 on both sides yields

$$\mathbf{X}_s^T \mathbf{K}^T \mathbf{X}_r = \omega_s^2 \mathbf{X}_s^T \mathbf{M}^T \mathbf{X}_r. \tag{9.91}$$

Recall that the mass and stiffness matrices are symmetric, $\mathbf{M} = \mathbf{M}^T$ and $\mathbf{K} = \mathbf{K}^T$. Then, Equation 9.91 reduces to

$$\mathbf{X}_s^T \mathbf{KX}_r = \omega_s^2 \mathbf{X}_s^T \mathbf{MX}_r. \tag{9.92}$$

Subtracting Equation 9.92 from Equation 9.89, we have

$$\left(\omega_r^2 - \omega_s^2\right) \mathbf{X}_s^T \mathbf{MX}_r = 0. \tag{9.93}$$

For two distinct modes, the frequencies are different, $\omega_r^2 \neq \omega_s^2$. Thus, Equation 9.93 is satisfied if and only if

$$\mathbf{X}_s^T \mathbf{MX}_r = 0, \quad r \neq s. \tag{9.94}$$

Substituting Equation 9.94 into Equation 9.89 gives

$$\mathbf{X}_s^T \mathbf{KX}_r = 0, \quad r \neq s. \tag{9.95}$$

Equations 9.94 and 9.95 represent the orthogonality relation for two distinct modal vectors $\mathbf{X}_r$ and $\mathbf{X}_s$. They are orthogonal with respect to the mass matrix $\mathbf{M}$ as well as the stiffness matrix $\mathbf{K}$. If $r = s$, then the values of $\mathbf{X}_r^T \mathbf{MX}_r$ and $\mathbf{X}_r^T \mathbf{KX}_r$ are nonzero, and their values depend on the method used to normalize the modal vectors. A convenient scheme is to normalize $\mathbf{X}_r$ such that

$$\mathbf{X}_r^T \mathbf{MX}_r = 1. \tag{9.96}$$

The modal vector $\mathbf{X}_r$ is then called orthonormal. Premultiplying both sides of Equation 9.87 by $\mathbf{X}_r^T$ yields

$$\mathbf{X}_r^T \mathbf{KX}_r = \omega_r^2. \tag{9.97}$$

Example 9.10: Normalization of Modal Vectors

Consider the three-degree-of-freedom system discussed in Example 9.9. Normalize the modal vectors $\mathbf{X}_1$, $\mathbf{X}_2$, and $\mathbf{X}_3$ obtained in Part (a) to a set of orthonormal modal vectors satisfying Equations 9.96 and 9.97. Assume $m = 1$ kg and $k = 1$ N/m.

Introduction to Vibrations

SOLUTION

Note that regardless of how the modal vectors are normalized, they result in the same mode shape but scaled in magnitude. So, we can write the first modal vector in the form

$$\tilde{\mathbf{X}}_1 = \alpha \mathbf{X}_1,$$

where $\tilde{\mathbf{X}}_1$ is the modal vector after normalization and α is a nonzero scaling constant. Substituting $\tilde{\mathbf{X}}_1$ into Equation 9.96 gives

$$\tilde{\mathbf{X}}_1^T \mathbf{M} \tilde{\mathbf{X}}_1 = (\alpha \mathbf{X}_1)^T \mathbf{M} (\alpha \mathbf{X}_1) = \alpha^2 \mathbf{X}_1^T \mathbf{M} \mathbf{X}_1,$$

which should be equal to 1 if $\tilde{\mathbf{X}}_1$ is orthonormal. Inserting $\mathbf{X}_1 = \{0.2549 \; 0.5842 \; 1\}^T$ and $\mathbf{M}$ obtained in Example 9.9 yields

$$\alpha^2 \mathbf{X}_1^T \mathbf{M} \mathbf{X}_1 = \alpha^2 \begin{bmatrix} 0.2549 \\ 0.5842 \\ 1 \end{bmatrix}^T \begin{bmatrix} 1 & 0 & 0 \\ 0 & 1 & 0 \\ 0 & 0 & 1 \end{bmatrix} \begin{bmatrix} 0.2549 \\ 0.5842 \\ 1 \end{bmatrix} = 1.4063 \alpha^2 = 1.$$

Thus, $\alpha = 0.8433$ and

$$\tilde{\mathbf{X}}_1 = \alpha \mathbf{X}_1 = \begin{bmatrix} 0.2150 \\ 0.4927 \\ 0.8433 \end{bmatrix}.$$

Neglecting round-off error, note that $\tilde{\mathbf{X}}_1$ is the same as the first column in matrix $\mathbf{V}$ obtained in Example 9.9, Part (b). Following the same procedure yields the other two orthonormal modal vectors, $\tilde{\mathbf{X}}_2$ and $\tilde{\mathbf{X}}_3$,

$$\tilde{\mathbf{X}}_2 = 0.6831 \mathbf{X}_2 = \begin{bmatrix} 0.5049 \\ 0.6831 \\ -0.5278 \end{bmatrix} \quad \tilde{\mathbf{X}}_3 = 0.8360 \mathbf{X}_3 = \begin{bmatrix} 0.8360 \\ -0.5392 \\ 0.1019 \end{bmatrix},$$

which corresponds to the second and third columns in matrix $\mathbf{V}$ with all elements multiplied by -1. Thus, the MATLAB command `eig` returns a set of orthonormal modal vectors.

The modal vectors are generally placed side by side to formulate a matrix. For example, $\boldsymbol{\Phi} = [\tilde{\mathbf{X}}_1 \; \cdots \; \tilde{\mathbf{X}}_n]$, which is defined as a modal matrix. If the modal vectors are orthonormal, then we have

$$\boldsymbol{\Phi}^T \mathbf{M} \boldsymbol{\Phi} = \begin{bmatrix} \tilde{\mathbf{X}}_1^T \\ \vdots \\ \tilde{\mathbf{X}}_n^T \end{bmatrix} \mathbf{M} [\tilde{\mathbf{X}}_1 \; \cdots \; \tilde{\mathbf{X}}_n] = \begin{bmatrix} \tilde{\mathbf{X}}_1^T \mathbf{M} \tilde{\mathbf{X}}_1 & \cdots & \tilde{\mathbf{X}}_1^T \mathbf{M} \tilde{\mathbf{X}}_n \\ \vdots & \ddots & \vdots \\ \tilde{\mathbf{X}}_n^T \mathbf{M} \tilde{\mathbf{X}}_1 & \cdots & \tilde{\mathbf{X}}_n^T \mathbf{M} \tilde{\mathbf{X}}_n \end{bmatrix}. \quad (9.98)$$

Due to the orthogonality of modal vectors as defined in Equation 9.94, all the off-diagonal entries are zero. Also, the modal vectors from $\tilde{\mathbf{X}}_1$ to $\tilde{\mathbf{X}}_n$ are orthonormal. Following Equation 9.96, each entry along the diagonal should be 1. Thus,

$$\boldsymbol{\Phi}^T \mathbf{M} \boldsymbol{\Phi} = \mathbf{I}_n. \quad (9.99)$$

Similarly, we have

$$\boldsymbol{\Phi}^T \mathbf{K} \boldsymbol{\Phi} = \begin{bmatrix} \tilde{\mathbf{X}}_1^T \\ \vdots \\ \tilde{\mathbf{X}}_n^T \end{bmatrix} \mathbf{K} [\tilde{\mathbf{X}}_1 \; \cdots \; \tilde{\mathbf{X}}_n] = \begin{bmatrix} \tilde{\mathbf{X}}_1^T \mathbf{K} \tilde{\mathbf{X}}_1 & \cdots & \tilde{\mathbf{X}}_1^T \mathbf{K} \tilde{\mathbf{X}}_n \\ \vdots & \ddots & \vdots \\ \tilde{\mathbf{X}}_n^T \mathbf{K} \tilde{\mathbf{X}}_1 & \cdots & \tilde{\mathbf{X}}_n^T \mathbf{K} \tilde{\mathbf{X}}_n \end{bmatrix}. \quad (9.100)$$

Following Equations 9.95 and 9.97 gives

$$\Phi^T K \Phi = \begin{bmatrix} \omega_1^2 & 0 & 0 \\ 0 & \ddots & 0 \\ 0 & 0 & \omega_n^2 \end{bmatrix} = \Omega. \tag{9.101}$$

Equations 9.99 and 9.101 are very useful for decoupling equations of motion and obtaining responses of vibration systems.

9.4.3 Response to Initial Excitations

Now let us consider the free, undamped vibration problem described by Equation 9.81, where the initial conditions are $\mathbf{x}(0)$ and $\dot{\mathbf{x}}(0)$. Solving the associated eigenvalue problem yields a set of natural frequencies, $\omega_1, \omega_2, \ldots,$ and ω_n, and a set of orthonormal modal vectors, $\tilde{\mathbf{X}}_1, \tilde{\mathbf{X}}_2, \ldots,$ and $\tilde{\mathbf{X}}_n$. To decouple the equations of motion, we introduce a coordinate transformation

$$\mathbf{x} = \Phi \mathbf{q}, \tag{9.102}$$

where $\Phi = \begin{bmatrix} \tilde{\mathbf{X}}_1 & \tilde{\mathbf{X}}_2 & \cdots & \tilde{\mathbf{X}}_n \end{bmatrix}$ is the modal matrix and $\mathbf{q}$ is known as the vector of modal coordinates. Substituting Equation 9.102 into Equation 9.81 and premultiplying by Φ^T on both sides gives

$$\Phi^T M \Phi \ddot{\mathbf{q}} + \Phi^T K \Phi \mathbf{q} = \mathbf{0}. \tag{9.103}$$

Due to the orthogonality of modal vectors given by Equations 9.99 and 9.101, Equation 9.103 reduces to

$$\ddot{\mathbf{q}} + \Omega \mathbf{q} = \mathbf{0}, \tag{9.104}$$

which can be written as n independent modal equations

$$\ddot{q}_r + \omega_r^2 q_r = 0, \quad r = 1, 2, \ldots, n \tag{9.105}$$

Obviously, modal equations are analogous to differential equations of single-degree-of-freedom systems, for which the response was given in Section 8.2. Following the discussion in Chapter 8, the solutions to Equation 9.105 are

$$q_r(t) = q_r(0) \cos \omega_r t + \frac{\dot{q}_r(0)}{\omega_r} \sin \omega_r t, \quad r = 1, 2, \ldots, n, \tag{9.106}$$

where $q_r(0)$ and $\dot{q}_r(0)$ are the initial displacements and initial velocities of modal coordinates. Note that the initial conditions given are associated with the physical coordinates $\mathbf{x}$. To transform the initial conditions, we use Equation 9.102, which gives $\mathbf{q} = \Phi^{-1} \mathbf{x}$. Because of the property of orthonormal modal vectors, $\Phi^T M \Phi = \mathbf{I}_n$, we have $\Phi^{-1} = \Phi^T M$. Thus, the initial modal displacements and velocities are

$$\mathbf{q}(0) = \Phi^T M \mathbf{x}(0), \tag{9.107}$$

$$\dot{\mathbf{q}}(0) = \Phi^T M \dot{\mathbf{x}}(0). \tag{9.108}$$

Introduction to Vibrations

Expanding the modal matrix yields

$$\begin{Bmatrix} q_1(0) \\ \vdots \\ q_n(0) \end{Bmatrix} = \begin{bmatrix} \tilde{\mathbf{X}}_1^T \\ \vdots \\ \tilde{\mathbf{X}}_n^T \end{bmatrix} \mathbf{M}\mathbf{x}(0) = \begin{Bmatrix} \tilde{\mathbf{X}}_1^T \mathbf{M}\mathbf{x}(0) \\ \vdots \\ \tilde{\mathbf{X}}_n^T \mathbf{M}\mathbf{x}(0) \end{Bmatrix}, \quad (9.109)$$

$$\begin{Bmatrix} \dot{q}_1(0) \\ \vdots \\ \dot{q}_n(0) \end{Bmatrix} = \begin{bmatrix} \tilde{\mathbf{X}}_1^T \\ \vdots \\ \tilde{\mathbf{X}}_n^T \end{bmatrix} \mathbf{M}\dot{\mathbf{x}}(0) = \begin{Bmatrix} \tilde{\mathbf{X}}_1^T \mathbf{M}\dot{\mathbf{x}}(0) \\ \vdots \\ \tilde{\mathbf{X}}_n^T \mathbf{M}\dot{\mathbf{x}}(0) \end{Bmatrix}, \quad (9.110)$$

or

$$q_r(0) = \tilde{\mathbf{X}}_r^T \mathbf{M}\mathbf{x}(0), \quad r = 1, 2, \ldots, n, \quad (9.111)$$

$$\dot{q}_r(0) = \tilde{\mathbf{X}}_r^T \mathbf{M}\dot{\mathbf{x}}(0), \quad r = 1, 2, \ldots, n. \quad (9.112)$$

Then, the modal responses are

$$q_r(t) = \tilde{\mathbf{X}}_r^T \mathbf{M}\mathbf{x}(0) \cos\omega_r t + \frac{1}{\omega_r} \tilde{\mathbf{X}}_r^T \mathbf{M}\dot{\mathbf{x}}(0) \sin\omega_r t, \quad r = 1, 2, \ldots, n. \quad (9.113)$$

Finally, applying Equation 9.102 gives the responses of n-degree-of-freedom systems to initial excitations

$$\mathbf{x}(t) = \begin{bmatrix} \tilde{\mathbf{X}}_1 & \cdots & \tilde{\mathbf{X}}_n \end{bmatrix} \begin{Bmatrix} q_1(t) \\ \vdots \\ q_n(t) \end{Bmatrix} = \sum_{r=1}^n \tilde{\mathbf{X}}_r q_r(t) = \sum_{r=1}^n q_r(t) \tilde{\mathbf{X}}_r \quad (9.114)$$

Equation 9.114 represents the so-called expansion theorem, by which the solution $\mathbf{x}(t)$ can be regarded as a superposition of the normal modes $\tilde{\mathbf{X}}_r(t)$ ($r = 1, 2, \ldots, n$). The modal responses $q_r(t)$ represent the contributions of the particular configuration $\tilde{\mathbf{X}}_r(t)$ to the total solution.

The approach using the orthogonality properties of the modal matrix to obtain a set of simultaneous independent equations is known as modal analysis. The basic steps to obtain the solutions using modal analysis are summarized as follows.

Step 1: Solve the eigenvalue problem and obtain the natural frequencies and modal vectors.
Step 2: Normalize the modal vectors to obtain orthonormality.
Step 3: Determine the modal responses and combine them into the response of the original system using the expansion theorem.

Example 9.11: Response to Initial Excitation by Modal Analysis

Consider the three-degree-of-freedom system discussed in Example 9.9. Assume $m = 1$ kg and $k = 1$ N/m. Using the natural frequencies obtained in Example 9.9 and the orthonormal modal vectors obtained in Example 9.10, determine the response of the system subjected to initial excitations $\mathbf{x}(0) = [0 \ 0 \ 0.01]^T$ and $\dot{\mathbf{x}}(0) = [0 \ 0 \ 0]^T$.

Solution

In Examples 9.9 and 9.10, we obtained the three natural frequencies

$$\omega_1 = 0.6448 \, \text{rad/s}, \quad \omega_2 = 1.5147 \, \text{rad/s}, \quad \omega_3 = 2.5080 \, \text{rad/s}$$

and the three orthonormal modal vectors

$$\tilde{\mathbf{X}}_1 = \begin{bmatrix} 0.2150 \\ 0.4927 \\ 0.8433 \end{bmatrix}, \quad \tilde{\mathbf{X}}_2 = \begin{bmatrix} 0.5049 \\ 0.6831 \\ -0.5278 \end{bmatrix}, \quad \tilde{\mathbf{X}}_3 = \begin{bmatrix} 0.8360 \\ -0.5392 \\ 0.1019 \end{bmatrix}.$$

Applying Equation 9.113 gives the modal responses

$$q_1(t) = [0.2150 \; 0.4927 \; 0.8433] \begin{bmatrix} 1 & 0 & 0 \\ 0 & 1 & 0 \\ 0 & 0 & 1 \end{bmatrix} \begin{Bmatrix} 0 \\ 0 \\ 0.01 \end{Bmatrix} \cos(0.6448t) = 0.0084 \cos(0.6448t),$$

$$q_2(t) = [0.5049 \; 0.6831 \; -0.5278] \begin{bmatrix} 1 & 0 & 0 \\ 0 & 1 & 0 \\ 0 & 0 & 1 \end{bmatrix} \begin{Bmatrix} 0 \\ 0 \\ 0.01 \end{Bmatrix} \cos(1.5147t) = -0.0053 \cos(1.5147t),$$

$$q_3(t) = [0.8360 \; -0.5392 \; 0.1019] \begin{bmatrix} 1 & 0 & 0 \\ 0 & 1 & 0 \\ 0 & 0 & 1 \end{bmatrix} \begin{Bmatrix} 0 \\ 0 \\ 0.01 \end{Bmatrix} \cos(2.5080t) = 0.0010 \cos(2.5080t).$$

Inserting the modal responses into Equation 9.114, we obtain the response of the system to the given initial excitations.

$$\mathbf{x}(t) = 0.0084 \cos(0.6448t) \begin{bmatrix} 0.2150 \\ 0.4927 \\ 0.8433 \end{bmatrix} - 0.0053 \cos(1.5147t) \begin{bmatrix} 0.5049 \\ 0.6831 \\ -0.5278 \end{bmatrix}$$

$$+ 0.0010 \cos(2.5080t) \begin{bmatrix} 0.8360 \\ -0.5392 \\ 0.1019 \end{bmatrix}$$

$$= \begin{bmatrix} 0.0018 \\ 0.0041 \\ 0.0071 \end{bmatrix} \cos(0.6448t) + \begin{bmatrix} -0.0027 \\ -0.0036 \\ 0.0028 \end{bmatrix} \cos(1.5147t) + \begin{bmatrix} 0.0084 \\ -0.0054 \\ 0.0010 \end{bmatrix} \cos(2.5080t).$$

9.4.4 Response to Harmonic Excitations

The essence of modal analysis is to determine the response of an n-degree-of-freedom system by decomposing it into n single-degree-of-freedom systems, determining the response of each single-degree-of-freedom system, and combining the individual responses into the response of the original system. The previous subsection showed how one can obtain the free response of an undamped system using modal analysis, which can also be used to find the response of an undamped system to harmonic excitations. In this case, the equations of motion in matrix form are

$$\mathbf{M}\ddot{\mathbf{x}} + \mathbf{K}\mathbf{x} = \mathbf{f}, \tag{9.115}$$

where $\mathbf{f}$ is an $n \times 1$ force vector. Following the basic steps of modal analysis given in Subsection 9.4.3, we first need to solve the eigenvalue problem given by $(\mathbf{K} - \omega^2 \mathbf{M})\mathbf{X} = \mathbf{0}$ and obtain the solutions ω_r^2 and $\mathbf{X}_r$ ($r = 1, 2, \ldots, n$). The n modal vectors are then normalized and arranged into the $n \times n$ modal matrix $\mathbf{\Phi} = [\tilde{\mathbf{X}}_1 \; \cdots \; \tilde{\mathbf{X}}_n]$, which satisfies the orthonormality relations $\mathbf{\Phi}^T \mathbf{M} \mathbf{\Phi} = \mathbf{I}_n$ and $\mathbf{\Phi}^T \mathbf{K} \mathbf{\Phi} = \mathbf{\Omega}$. Using the expansion theorem, the solution to Equation 9.115 is a linear combination of

Introduction to Vibrations

the modal vectors, $\mathbf{x} = \sum_{r=1}^{n} q_r(t)\tilde{\mathbf{X}}_r = \mathbf{\Phi q}$. Substituting it into Equation 9.115 and premultiplying the result by $\mathbf{\Phi}^T$ gives

$$\mathbf{\Phi}^T \mathbf{M} \mathbf{\Phi} \ddot{\mathbf{q}} + \mathbf{\Phi}^T \mathbf{K} \mathbf{\Phi} \mathbf{q} = \mathbf{\Phi}^T \mathbf{f}. \tag{9.116}$$

Combining the orthonormality relations, we have

$$\ddot{\mathbf{q}} + \mathbf{\Omega} \mathbf{q} = \mathbf{N}, \tag{9.117}$$

where $\mathbf{N} = \mathbf{\Phi}^T \mathbf{f}$. Note that

$$\mathbf{\Phi}^T \mathbf{f} = \begin{bmatrix} \tilde{\mathbf{X}}_1^T \\ \vdots \\ \tilde{\mathbf{X}}_n^T \end{bmatrix} \mathbf{f} = \begin{Bmatrix} \tilde{\mathbf{X}}_1^T \mathbf{f} \\ \vdots \\ \tilde{\mathbf{X}}_n^T \mathbf{f} \end{Bmatrix} = \begin{Bmatrix} N_1 \\ \vdots \\ N_n \end{Bmatrix}, \tag{9.118}$$

where the entries $\tilde{\mathbf{X}}_r^T \mathbf{f}$ ($r = 1, 2, \ldots, n$) are the products of the $1 \times n$ row vectors $\tilde{\mathbf{X}}_r^T$ and the $n \times 1$ column vector $\mathbf{f}$, resulting in scalars N_r. They are known as modal forces. Equation 9.117 is equivalent to a set of n independent modal equations

$$\ddot{q}_r + \omega_r^2 q_r = N_r, \quad r = 1, 2, \ldots, n. \tag{9.119}$$

We consider harmonic excitations, without loss of generality, which have the form

$$\mathbf{f} = \mathbf{f}_0 \sin \omega t. \tag{9.120}$$

Thus, the modal forces N_r can be written as $\tilde{\mathbf{X}}_r^T \mathbf{f}_0 \sin \omega t$ ($r = 1, 2, \ldots, n$). The solutions to Equation 9.119 can be obtained using the analogy with the response of undamped single-degree-of-freedom systems to harmonic excitations. Applying Equations 9.31 and 9.32 with $\zeta = 0$, we have

$$q_r(t) = \frac{\tilde{\mathbf{X}}_r^T \mathbf{f}_0}{\omega_r^2 - \omega^2} \sin \omega t, \quad r = 1, 2, \ldots, n. \tag{9.121}$$

Thus, the steady-state response of the original system is

$$\mathbf{x} = \sum_{r=1}^{n} q_r(t) \tilde{\mathbf{X}}_r = \sum_{r=1}^{n} \frac{\tilde{\mathbf{X}}_r^T \mathbf{f}_0 \tilde{\mathbf{X}}_r}{\omega_r^2 - \omega^2} \sin \omega t. \tag{9.122}$$

Example 9.12: Response to Harmonic Excitation by Modal Analysis

Consider the three-degree-of-freedom system discussed in Example 9.9. Assume $m = 1$ kg and $k = 1$ N/m. Using the natural frequencies obtained in Example 9.9 and the orthonormal modal vectors obtained in Example 9.10, determine the response of the system if a harmonic force $F(t) = 3 \sin 2t$ is applied to mass 3.

SOLUTION

The force vector can be written as

$$\mathbf{f} = \begin{Bmatrix} 0 \\ 0 \\ F(t) \end{Bmatrix} = \begin{Bmatrix} 0 \\ 0 \\ 3 \sin 2t \end{Bmatrix} = \begin{Bmatrix} 0 \\ 0 \\ 3 \end{Bmatrix} \sin 2t.$$

From Examples 9.9 and 9.10, the three natural frequencies are

$$\omega_1 = 0.6448 \, \text{rad/s}, \quad \omega_2 = 1.5147 \, \text{rad/s}, \quad \omega_3 = 2.5080 \, \text{rad/s}$$

and the three orthonormal modal vectors

$$\tilde{\mathbf{X}}_1 = \begin{bmatrix} 0.2150 \\ 0.4927 \\ 0.8433 \end{bmatrix}, \quad \tilde{\mathbf{X}}_2 = \begin{bmatrix} 0.5049 \\ 0.6831 \\ -0.5278 \end{bmatrix}, \quad \tilde{\mathbf{X}}_3 = \begin{bmatrix} 0.8360 \\ -0.5392 \\ 0.1019 \end{bmatrix}.$$

Applying Equation 9.121 gives the modal responses

$$q_1(t) = \frac{[0.2150 \; 0.4927 \; 0.8433] \begin{Bmatrix} 0 \\ 0 \\ 3 \end{Bmatrix}}{0.6448^2 - 3^2} \sin 3t = -0.2948 \sin 3t,$$

$$q_2(t) = \frac{[0.5049 \; 0.6831 \; -0.5278] \begin{Bmatrix} 0 \\ 0 \\ 3 \end{Bmatrix}}{1.5147^2 - 3^2} \sin 3t = 0.2361 \sin 3t,$$

$$q_3(t) = \frac{[0.8360 \; -0.5392 \; 0.1019] \begin{Bmatrix} 0 \\ 0 \\ 3 \end{Bmatrix}}{2.5080^2 - 3^2} \sin 3t = -0.1128 \sin 3t.$$

Inserting the modal responses into Equation 9.122, we obtain the response of the system to the given harmonic excitation.

$$\mathbf{x}(t) = -0.2948 \sin 3t \begin{bmatrix} 0.2150 \\ 0.4927 \\ 0.8433 \end{bmatrix} + 0.2361 \sin 3t \begin{bmatrix} 0.5049 \\ 0.6831 \\ -0.5277 \end{bmatrix} - 0.1128 \sin 3t \begin{bmatrix} 0.8360 \\ -0.5392 \\ 0.1019 \end{bmatrix}$$

$$= \begin{Bmatrix} -0.0385 \\ 0.0768 \\ -0.3848 \end{Bmatrix} \sin 3t.$$

PROBLEM SET 9.4

1. Consider the two-degree-of-freedom mass–spring system shown in Figure 9.23. Assume that $m_1 = m_2 = m$, $k_1 = k$, and $k_2 = 2k$, where $m = 5$ kg and $k = 2000$ N/m.
   a. Derive the equations of motion and write them in matrix form.
   b. Solve the associated eigenvalue problem by hand. Plot the two modes and explain the nature of the mode shapes.
   c. ◢ Solve the associated eigenvalue problem using MATLAB.
2. A three-story building can be modeled as a three-degree-of-freedom system as shown in Figure 9.24, where the horizontal members are rigid and the columns are massless beams acting as springs. Assume that $m_1 = 1500$ kg, $m_2 = 3000$ kg, $m_3 = 4500$ kg, $k_1 = 400$ kN/m, $k_2 = 800$ kN/m, and $k_3 = 1200$ kN/m.
   a. Derive the differential equations for the horizontal motion of the masses.
   b. Solve the associated eigenvalue problem by hand. Plot the three modes and explain the nature of the mode shapes.
   c. ◢ Solve the associated eigenvalue problem using MATLAB.
3. Consider the two-degree-of-freedom mass–spring system in Figure 9.23. Normalize the modal vectors $\mathbf{X}_1$ and $\mathbf{X}_2$ obtained in Part (b) of Problem 1 to a set of orthonormal modal vectors satisfying Equations 9.96 and 9.97.

Introduction to Vibrations

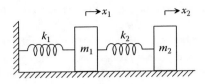

FIGURE 9.23 Problem 1.

4. Consider the three-story building system in Figure 9.24. Normalize the modal vectors X_1, X_2, and X_3 obtained in Part (b) of Problem 2 to a set of orthonormal modal vectors satisfying Equations 9.96 and 9.97.
5. Consider the two-degree-of-freedom mass–spring system in Figure 9.23. Determine the response of the system subjected to initial excitations $\mathbf{x}(0) = [0\ 0]^T$ and $\dot{\mathbf{x}}(0) = [1\ 0]^T$ by modal analysis. Note that the natural frequencies were solved in Problem 1 and the modal vectors were normalized in Problem 3.
6. Consider the three-story building system in Figure 9.24. Determine the response of the system subjected to initial excitations $\mathbf{x}(0) = [0.01\ 0\ 0]^T$ and $\dot{\mathbf{x}}(0) = [0\ 0\ 0]^T$ by modal analysis. Note that the natural frequencies were solved in Problem 2 and the modal vectors were normalized in Problem 4.
7. Consider the two-degree-of-freedom mass–spring system in Figure 9.23. Determine the response of the system if a harmonic force $F(t) = 2\cos t$ is applied to mass 2. Note that the natural frequencies were solved in Problem 1 and the modal vectors were normalized in Problem 3.
8. Consider the three-story building system in Figure 9.24. Determine the response of the system if a harmonic force $F(t) = 0.15\sin 0.15t$ is applied to the top story. Note that the natural frequencies were solved in Problem 2 and the modal vectors were normalized in Problem 4.
9. Solve Problem 5 using MATLAB.
10. Solve Problem 6 using MATLAB.
11. Solve Problem 7 using MATLAB.
12. Solve Problem 8 using MATLAB.

9.5 VIBRATION MEASUREMENT AND ANALYSIS

Models are necessary for designing dynamic systems and understanding system dynamics. For many vibration systems, theoretical models are too difficult to develop. In such a case, we can resort

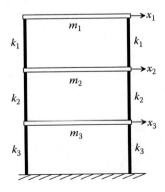

FIGURE 9.24 Problem 2.

to experimental models, which can be obtained using experimental modal techniques, specifically the method known as frequency response function testing. There are two phases needed to obtain an experimental model of a vibration system. The first phase is the measurement phase, where the frequency response functions of the system are measured. The second phase is the analysis phase, where system parameters are estimated from the measured frequency response functions. In this section, we first introduce the measurement methods for acquiring frequency response data and then discuss the parameter estimation methods for extracting system properties.

9.5.1 Vibration Measurement

As presented in Chapter 8, a frequency response function is a transfer function, expressed in the frequency domain instead of the s domain. For a vibration system, a frequency response function describes the system response to an external excitation force as a function of frequency. The response may be displacement, velocity, or acceleration. In order to experimentally obtain the frequency response function of a vibration system, the input excitation and output response must be measured simultaneously. Both the excitation and response are obtained in the frequency domain by operating fast Fourier transform, and the frequency response function is the ratio of the response to the excitation.

Figure 9.25 shows a diagram of the basic components for frequency response function measurement. An actuator, or so-called exciter, is used to apply force to the system under test. Sensors, or so-called transducers, are used to measure the force and responses. A data acquisition system is used to acquire and process the signals from the sensors. A computer with analysis software provides measurement functions such as windowing, averaging, and fast Fourier transform computation.

The first step in the measurement process involves selecting an excitation function along with an excitation system. The excitation function is the mathematical signal used for the input. The excitation system is the physical mechanism used to provide the signal. Two of the general categories of excitation functions are steady-state and transient. For example, a sine function is a steady-state signal and an impulse function is a transient signal. Two of the most common excitation mechanisms are a shaker and an impact hammer.

Figure 9.26 shows an electromagnetic shaker, which is one of the most common shakers for vibration testing. With the electromagnetic shaker, a force is generated by an alternating current that drives a magnetic coil. Such a shaker can generate a variety of time-varying forces, such as a sinusoidal force with a constant frequency and a swept sine function with gradually increased frequency but constant amplitude. The maximum frequency range and the maximum force level depend on the size of the shaker. Smaller shakers have a higher frequency range and a lower force level. When using a shaker for excitation, the shaker should be physically mounted on the system via a force transducer.

An impact hammer is another common excitation mechanism, which is used to apply impulse. As shown in Figure 9.27, the impact hammer has a transducer at the tip for measuring the impact force. If the system being tested is struck by the hammer quickly, the resulting excitation force

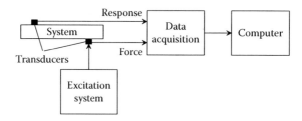

FIGURE 9.25 Basic components for frequency response function measurement.

Introduction to Vibrations

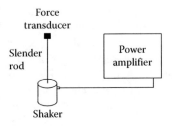

FIGURE 9.26 An electromagnetic shaker with power amplifier.

resembles an impulse. Since the impact hammer does not have to be attached to the system under test, this technique is relatively simple to implement. However, it is difficult to obtain consistent results. The frequency content and the pulse duration depend on the material of the tip. The harder the tip, the shorter the pulse duration included and thus the higher the frequency content measured.

The second step in the measurement process is to select the transducers for sensing force and response. The piezoelectric type is the most widely used for vibration testing. The piezoelectric transducer is an electromechanical sensor that generates an electrical output when subjected to vibration. The response measured is usually acceleration. This particular type of response transducer is called an accelerometer. For some measurements, it may be too small or too hot to attach an accelerometer to the system being tested. A laser Doppler vibrometer can then be used to make noncontact vibration measurements.

With the selected excitation mechanism and transducers, the frequency response function can be obtained in several different ways. For example, if a shaker is used, then harmonic excitation is applied to the system under test and the resulting harmonic response is measured. This type of test is referred to as sine wave testing. The frequency range is covered either by stepping from one frequency to the next or by slowly sweeping the frequency continuously. In both cases, the measurement time should be long enough to allow steady-state conditions to be attained. If an impact hammer is used, then impulsive excitation is applied to the system under test and the resulting transient response is measured.

9.5.2 System Identification

After having acquired frequency response data via vibration testing, the next major step is to identify the system parameters, more specifically, modal parameters. The basic information that can be determined from frequency response functions includes natural frequencies, the damping ratio associated with each mode, and mode shapes. The discussions in this subsection are only limited to the identification of natural frequencies and damping ratios.

To obtain an accurate estimation, it is important to understand the relationships between frequency response functions and their individual modal parameters. The basics of a single-degree-of-freedom dynamic system form the basis for parameter estimation techniques. As discussed in

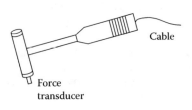

FIGURE 9.27 An impact hammer for vibration testing.

Section 8.3, the frequency response function is a complex quantity, which can be presented in terms of magnitude and phase versus frequency. Another method of presenting the frequency response data is to plot the real part and the imaginary part versus frequency. Denote the Fourier transforms of displacement response and excitation force as $X(j\omega)$ and $F(j\omega)$, respectively. The expression of the frequency response function $X(j\omega)/F(j\omega)$ was given in Section 8.3. Figure 9.28 shows the frequency response presented in different forms, where the real and imaginary parts are

$$\text{Re}\left(\frac{X(j\omega)}{F(j\omega)}\right) = \frac{k - m\omega^2}{\sqrt{(k - m\omega^2)^2 + (b\omega)^2}} = \frac{1 - (\omega/\omega_n)^2}{\sqrt{\left[1 - (\omega/\omega_n)^2\right]^2 + (2\zeta\omega/\omega_n)^2}}, \quad (9.123)$$

$$\text{Im}\left(\frac{X(j\omega)}{F(j\omega)}\right) = \frac{-b\omega}{\sqrt{(k - m\omega^2)^2 + (b\omega)^2}} = \frac{-2\zeta\omega/\omega_n}{\sqrt{\left[1 - (\omega/\omega_n)^2\right]^2 + (2\zeta\omega/\omega_n)^2}}. \quad (9.124)$$

Assume that the system is lightly damped. As shown in Figure 9.28, when the excitation frequency approaches the natural frequency, $\omega = \omega_n$, the system resonates, and the magnitude of the frequency response function reaches its maximum. This conclusion was proved in Section 8.3. Also, at resonance, the real part is equal to zero, and the imaginary part reaches a peak. The former observation can be easily proved by setting $\omega = \omega_n$ in Equation 9.123, and the latter will be left as an exercise for the reader. It was discussed earlier in Section 9.2 that the damping ratio ζ can be estimated by the half-power bandwidth method.

In reality, most dynamic systems cannot be simplified as ideal single-degree-of-freedom systems. As presented in Section 9.4, an n-degree-of-freedom system has n modes. For systems with lightly damped and well-separated modes, as shown in Figure 9.29a, the natural frequencies and damping ratios can be estimated using the single-mode method. The basic assumption for single-mode approximation is that in the vicinity of a resonance, the response is due primarily to that single mode. Just like a single-degree-of-freedom system, the natural frequency associated with that single mode can be estimated from the frequency response data by observing the frequency at which any of the following trends occur:

- The magnitude of the frequency response is a maximum
- The real part of the frequency response is zero
- The imaginary part of the frequency response is a maximum or minimum

The damping ratio associated with that single mode can be estimated using the half-power bandwidth method. For systems with heavily damped and closely spaced modes, as shown in Figure 9.29b, the adjacent modes can affect each other significantly. In general, it will be necessary to implement a multiple-mode method to more accurately identify the modal parameters of these types of systems.

PROBLEM SET 9.5

1. For a single-degree-of-freedom system, denote the Fourier transforms of the displacement response and excitation force as $X(j\omega)$ and $F(j\omega)$, respectively. The expression of the imaginary part of the frequency response function $X(j\omega)/F(j\omega)$ is given by Equation 9.124. Prove that the imaginary part reaches a peak at resonance.

2. Accelerations are often measured in vibration testing. For a single-degree-of-freedom system, denote the Fourier transforms of the acceleration response and excitation force as $A(j\omega)$ and $F(j\omega)$, respectively. The expression of the frequency response function $A(j\omega)/F(j\omega)$ is given by

$$\frac{A(j\omega)}{F(j\omega)} = \frac{-\omega^2}{k - m\omega^2 + jb\omega}.$$

Introduction to Vibrations

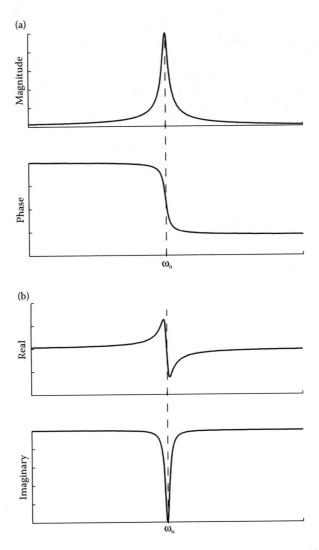

FIGURE 9.28 Frequency response of a single-degree-of-freedom system represented in terms of (a) magnitude and phase, and (b) real and imaginary.

Using MATLAB, write an m-file to plot the magnitude, phase, real part, and imaginary part of the frequency response versus ω/ω_n. Assume that $m = 5$ kg $b = 30$ N·s/m, and $k = 2000$ N/m.

3. Rods, beams, plates, and so on are continuous systems, which have an infinite number of degrees of freedom and an infinite number of modes. For simplicity, assume that a cantilever beam is approximated as a single-degree-of-freedom mass–damper–spring system, for which the natural frequency is close to the first mode of the beam. The parameters of the cantilever beam are length $L = 0.5$ m, width $b = 0.025$ m, thickness $h = 0.005$ m, density $\rho = 7850$ kg/m^3, and Young's modulus $E = 210 \times 10^9$ N/m^2.

   a. It is known that the equivalent mass for the beam is $m_{eq} = m/3$, where m is the actual mass of the beam. Determine the equivalent stiffness k_{eq} for the beam. Calculate the natural frequency of the equivalent single-degree-of-freedom system.
   b. Figure 9.30 is the measured frequency response of the cantilever beam for the first mode. Determine the natural frequency and the damping ratio based on the given information in the plot.

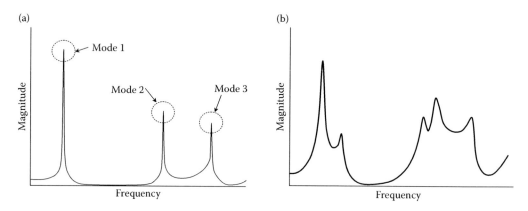

FIGURE 9.29 Frequency response of a multi-degree-of-freedom system with (a) light damping and (b) heavy damping.

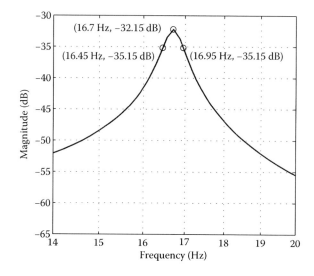

FIGURE 9.30 Problem 3.

   c. Compare the frequencies obtained in Parts (a) and (b). What is the error if the cantilever beam is approximated as a single-degree-of-freedom system?
4. Figure 9.31 shows the magnitude of an experimentally determined frequency response. Estimate the degree-of-freedom of the system and its natural frequencies.

9.6 SUMMARY

This chapter presented an introduction to vibrations. Two different types of excitations, initial excitations and external excitations, cause a system to vibrate. The vibration of a system caused by nonzero initial displacements and/or initial velocities is known as free vibration. The vibration of a system caused by externally applied forces is known as forced vibration.

The nature of system response to initial or external excitations depends on the system's damping, which is a very complex phenomenon. Logarithmic decrement and half-power bandwidth are

Introduction to Vibrations

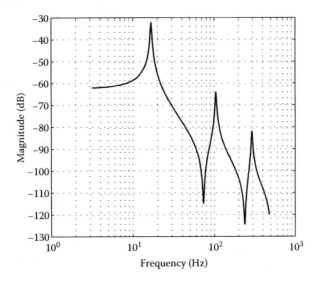

FIGURE 9.31 Problem 4.

two commonly used methods for measuring the damping of a vibration system. These two damping identification methods are only valid for the viscous damping model, which is widely used in vibration.

In the logarithmic decrement method, the free response to initial excitations is measured. The damping ratio ζ is estimated using the displacements of two consecutive peaks x_1 and x_2. The value of the damping ratio ζ is given by

$$\zeta = \frac{\delta}{\sqrt{(2\pi)^2 + \delta^2}},$$

where δ is called the logarithmic decrement and $\delta = \ln(x_1/x_2)$. For a more accurate estimation, the damping ratio ζ can be determined by measuring the displacements of two peaks separated by a number of periods instead of two consecutive peaks.

In the half-power bandwidth method, the frequency response of the system is measured. The damping ratio ζ is estimated using a peak in the magnitude curve of the frequency response and two half-power points, which are 3 dB down from the peak. The value of the damping ratio ζ is given by

$$\zeta \approx \frac{\omega_2 - \omega_1}{2\omega_n},$$

where ω_1 and ω_2 are the frequencies at the two half-power points. For light damping, the natural frequency ω_n corresponds to the frequency at the peak or is approximated as $\frac{1}{2}(\omega_1 + \omega_2)$.

Frequency response is a very important concept in vibration. For harmonic excitations, more information on the steady-state response can be extracted by means of the frequency domain technique than the time domain technique. As presented in Chapter 8, the steady-state response of a single-degree-of-freedom mass–damper–spring system to harmonic excitation, for example, $f(t) = F_0 \sin \omega t$, is still harmonic with a frequency that is the same as the excitation frequency. Denote the frequency response function of the system as $G(j\omega)$; then the steady-state response is $x(t) = X \sin(\omega t + \phi)$, where $X = F_0 |G(j\omega)|$ and $\phi = \angle G(j\omega)$. The nondimensional ratio between the dynamic amplitude X and the static deflection x_{st} is given by

$$\frac{X}{x_{\text{st}}} = \frac{1}{\sqrt{\left[1 - (\omega/\omega_n)^2\right]^2 + \left[2\zeta(\omega/\omega_n)\right]^2}}.$$

TABLE 9.1
Summarized Results for Rotating Unbalance and Harmonic Base Excitation

Rotating Unbalance

$M\ddot{x} + b\dot{x} + kx = me\omega^2 \sin \omega t$,

where the effect of a rotating unbalance mass is to exert a harmonic force $me\omega^2 \sin \omega t$

$$G(j\omega) = \frac{1/k}{1 - (\omega/\omega_n)^2 + j2\zeta\omega/\omega_n},$$

$x(t) = X \sin(\omega t + \phi)$,

where

$$X = \frac{me\omega^2/k}{\sqrt{\left[1 - (\omega/\omega_n)^2\right]^2 + [2\zeta(\omega/\omega_n)]^2}},$$

$$\phi = -\tan^{-1} \frac{2\zeta(\omega/\omega_n)}{1 - (\omega/\omega_n)^2},$$

$$\frac{MX}{me} = \frac{(\omega/\omega_n)^2}{\sqrt{\left[1 - (\omega/\omega_n)^2\right]^2 + [2\zeta(\omega/\omega_n)]^2}}.$$

Harmonic Base Excitation

$\ddot{x}(t) + 2\zeta\omega_n\dot{x}(t) + \omega_n^2 x(t) = 2\zeta\omega_n\dot{y}(t) + \omega_n^2 y(t)$,

where the motion of the base is harmonic, for example, $y(t) = Y_0 \sin \omega t$

$$G(j\omega) = \frac{1 + j2\zeta\omega/\omega_n}{1 - (\omega/\omega_n)^2 + j2\zeta\omega/\omega_n},$$

$x(t) = X \sin(\omega t + \phi)$,

where

$$X = Y_0 \frac{\sqrt{1 + (2\zeta\omega/\omega_n)^2}}{\sqrt{\left[1 - (\omega/\omega_n)^2\right]^2 + (2\zeta\omega/\omega_n)^2}},$$

$$\phi = -\tan^{-1} \frac{2\zeta(\omega/\omega_n)^3}{1 - (\omega/\omega_n)^2 + (2\zeta\omega/\omega_n)^2},$$

$$\frac{X}{Y_0} = \frac{\sqrt{1 + (2\zeta\omega/\omega_n)^2}}{\sqrt{\left[1 - (\omega/\omega_n)^2\right]^2 + (2\zeta\omega/\omega_n)^2}}.$$

Rotating eccentric masses and harmonically moving supports are two common harmonic excitation sources in engineering applications. The steady-state responses to these two excitations can be obtained using the same pattern as the general case. The results are summarized in Table 9.1. The magnitude of the nondimensional ratios MX/me or X/Y_0 versus the driving frequency can be plotted using MATLAB. These plots provide significant information for harmonic responses.

Vibrations are undesirable in many systems. The reduction of vibration can be achieved through vibration isolators or vibration absorbers. A vibration isolation system attempts either to protect delicate equipment from vibration transmitted to it from its support system or to prevent the vibratory force generated by a machine from being transmitted to its surroundings. The essence of these two objectives is the same. The concept of displacement transmissibility X/Y_0 can be used for displacement isolation design, while force transmissibility F_T/F_0 can be used for force isolation design, where $X/Y_0 = F_T/F_0$.

To prevent harmonic resonance for a single-degree-of-freedom system, it is not always possible to prevent the natural frequency from being close to the driving frequency by changing either the mass or the spring stiffness. To address this issue, a vibration absorber consisting of a second mass and spring can be added to the system and protect the original single-degree-of-freedom system from harmonic excitation. The vibration of the original mass can be reduced to zero, provided that the natural frequency of the absorber is the same as the driving frequency. One disadvantage of the vibration absorber is that two new resonant frequencies are created.

For multi-degree-of-freedom systems, more advanced mathematical tools are required to solve the equations of motion due to coordinate coupling. Modal analysis is an approach that utilizes the orthogonality of modal vectors to reduce the equations of motion to a set of independent second-order differential equations for modal coordinates. The basic steps in obtaining the response of an n-degree-of-freedom undamped system to initial excitations or external forces using modal analysis are summarized as follows.

TABLE 9.2
Responses to Initial Excitations or External Harmonic Excitations Using Modal Analysis

	Initial Excitations	External Harmonic Excitations
Equations of motion	$\mathbf{M}\ddot{\mathbf{x}} + \mathbf{K}\mathbf{x} = \mathbf{0}$	$\mathbf{M}\ddot{\mathbf{x}} + \mathbf{K}\mathbf{x} = \mathbf{f}_0 \sin \omega t$
Modal equations	$\ddot{q}_r + \omega_r^2 q_r = 0, \quad r = 1,2,\ldots,n$	$\ddot{q}_r + \omega_r^2 q_r = N_r, \quad r = 1,2,\ldots,n$ where the modal forces are $N_r = \tilde{\mathbf{X}}_r^T \mathbf{f}_0 \sin \omega t$
Modal responses	$q_r(t) = q_r(0)\cos\omega_r t + \dfrac{\dot{q}_r(0)}{\omega_r}\sin\omega_r t,$ where the modal initial conditions are $q_r(0) = \tilde{\mathbf{X}}_r^T \mathbf{M}\mathbf{x}(0),$ $\dot{q}_r(0) = \tilde{\mathbf{X}}_r^T \mathbf{M}\dot{\mathbf{x}}(0)$	$q_r(t) = \dfrac{\tilde{\mathbf{X}}_r^T \mathbf{f}_0}{\omega_r^2 - \omega^2}\sin\omega t$
System responses	$\mathbf{x}(t) = \sum_{r=1}^{n} q_r(t)\tilde{\mathbf{X}}_r$	$\mathbf{x}(t) = \sum_{r=1}^{n} q_r(t)\tilde{\mathbf{X}}_r$

Step 1: Solve the eigenvalue problem associated with the mass and stiffness matrices, that is,

$$(\mathbf{K} - \omega^2 \mathbf{M})\mathbf{X} = \mathbf{0},$$

and obtain the natural frequencies, which are the square roots of eigenvalues $\omega_1^2, \omega_2^2, \ldots,$ and ω_n^2, and the modal vectors, which are eigenvectors $\mathbf{X}_1, \mathbf{X}_2, \ldots,$ and $\mathbf{X}_n$. The natural frequencies are usually arranged in increasing order of magnitude, that is, $\omega_1 \leq \omega_2 \leq \cdots \leq \omega_n$. The modal vectors represent the shape of the normal modes physically.

Step 2: Normalize each modal vector to satisfy the relations $\tilde{\mathbf{X}}_r^T \mathbf{M}\tilde{\mathbf{X}}_r = 1$ and $\tilde{\mathbf{X}}_r^T \mathbf{K}\tilde{\mathbf{X}}_r = \omega_r^2$, where $r = 1,2,\ldots,n$. Then, the orthonormal modal matrix $\mathbf{\Phi} = \begin{bmatrix} \tilde{\mathbf{X}}_1 & \cdots & \tilde{\mathbf{X}}_n \end{bmatrix}$ yields

$$\mathbf{\Phi}^T \mathbf{M} \mathbf{\Phi} = \mathbf{I}_n, \quad \mathbf{\Phi}^T \mathbf{K} \mathbf{\Phi} = \mathbf{\Omega},$$

where $\mathbf{\Omega}$ is a diagonal matrix of the natural frequencies squared. Introducing the modal coordinate vector $\mathbf{q}$ and the coordinate transformation $\mathbf{x} = \mathbf{\Phi}\mathbf{q}$, we can decouple the original equations as n independent modal equations.

Step 3: Determine the modal responses and combine them into the response of the original system using the expansion theorem. Table 9.2 lists the responses to initial excitations or external harmonic excitations.

For many vibration systems, theoretical models are too difficult to develop. In such a case, experimental modal techniques can be used to obtain experimental models. There are two phases needed to obtain an experimental model of a vibration system. The first phase is the measurement phase, where the frequency response functions of the system are measured. Two of the most common excitation mechanisms are the shaker and the impact hammer. The second phase is the analysis phase, where the system parameters are estimated from the measured frequency response functions. The basic information that can be determined from the frequency response functions includes natural frequencies, the damping ratio associated with each mode, and mode shapes.

REVIEW PROBLEMS

1. Consider the single-degree-of-freedom system shown in Figure 5.39.
   a. Determine the undamped natural frequency ω_n, the damping ratio ζ, and the damped natural frequency ω_d.
   b. Assume $f(t) = 0$. Find the free vibration response of the system subjected to the initial conditions $x(0) = 0$ and $\dot{x}(0) = 0.5\,\text{m/s}$.
   c. ◆ Write a MATLAB m-file to plot the system's response obtained in Part (b).
   d. ◆ Using the differential equation of motion in Part (b), construct a Simulink® block diagram to find the response of the system.

2. Repeat Problem 1 for the system shown in Figure 5.64, where $m = 1.2\,\text{kg}$, $L = 0.6\,\text{m}$, $k = 100\,\text{N/m}$, and $B = 0.5\,\text{N·s/m}$. Assume $f(t) = 0$ and small angles θ with initial conditions $\theta(0) = 5°$ and $\dot{\theta}(0) = 30°/\text{s}$.

3. Consider the single-degree-of-freedom system shown in Figure 5.39.
   a. Assume $f(t) = 5\sin 4t$ and initial conditions $x(0) = 0$ and $\dot{x}(0) = 0$. Determine the dynamic amplitude X and the static deflection x_{st} of the system.
   b. Find the steady-state response of the system. Determine the period of the response.
   c. ◆ Write a MATLAB m-file to plot the system's response obtained in Part (b).
   d. ◆ Assume that the driving frequency varies from 0 to 10 rad/s. Write a MATLAB m-file to plot the nondimensional ratio X/x_{st} versus the driving frequency ω.

4. Repeat Problem 3 for the system shown in Figure 5.64, where $m = 1.2\,\text{kg}$, $L = 0.6\,\text{m}$, $k = 100\,\text{N/m}$, and $B = 0.5\,\text{N·s/m}$. Assume $f(t) = 10\sin 25t$ and small angles θ with zero initial conditions. For Part (d), assume that the excitation frequency varies from 0 rad/s to 40 rad/s.

5. A machine can be considered as a rigid mass with a rotating unbalanced mass. To reduce the vibration, the machine is mounted on a support system with a stiffness of 10 kN/m. Assume that the total mass $M = 10$ kg and the unbalanced mass $m = 0.5$ kg. Determine the damping ratio of the support system so that the vibration amplitude will not exceed 10% of the rotating mass's eccentricity when the machine operates at a speed of 350 rpm.

6. As shown in Figure 9.32, a machine of a mass $M = 150$ kg is mounted on a simply supported steel beam with negligible mass. The parameters of the beam are given as follows: Young's modulus of the beam $E = 210$ GPa, length $L = 2$ m, width $b = 0.75$ m, and thickness $h = 3$ cm. The rotating unbalance in the machine is $me = 0.015$ kg·m. Assuming that the machine runs at a speed of 2000 rpm and neglecting the damping,
   a. Determine the dynamic amplitude X of the machine.
   b. Determine the amplitude of the force transmitted to the ground.

7. A 110-kg machine is placed on a floor that vibrates with a frequency of 20 Hz. The maximum acceleration of the floor is $15\,\text{cm/s}^2$. A vibration isolator consisting of four parallel-connected springs is designed to protect the machine from the vibration of the floor. Assume that the damping ratio of the isolator is 0.1 and the maximum allowable acceleration is $2.25\,\text{cm/s}^2$. Determine the stiffness of each spring.

8. Consider the single-degree-of-freedom system shown in Figure 9.17a, where $m = 10$ kg. When the mass is in equilibrium, the static deformation of the spring is 2.5 mm. When the mass is allowed to vibrate freely, the maximum displacement amplitude during the fifth cycle is 20% of the first. Assume that a harmonic excitation force is applied to the system.

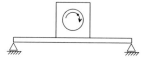

FIGURE 9.32 Problem 6.

Introduction to Vibrations

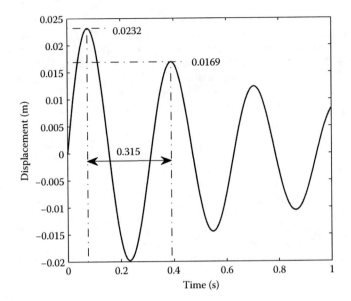

FIGURE 9.33 Problem 13.

a. Determine the minimum allowable driving frequency if the force transmitted to the ground is less than the excitation force.
b. If the allowable force transmissibility is 20%, determine the stiffness of the spring.

9. Consider the vibration absorber shown in Figure 9.18. It is known that two new resonant frequencies in the neighborhood of the excitation frequency are created. Denote the two new natural frequencies as ω_{n1} and ω_{n2}. Assume $\nu = 1$. Determine ω_{n1}/ω_1 and ω_{n2}/ω_1 for the following cases: $\mu = 0.05, 0.1, 0.15, 0.2$, and 0.5.

10. A 1-Mg machine is supported by a spring, which is deformed by 5 cm when the machine is in equilibrium. A harmonic force of magnitude 500 N is applied to the machine

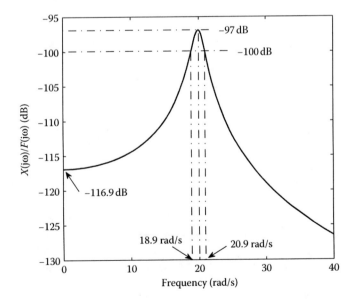

FIGURE 9.34 Problem 14.

and induces resonance. Design a vibration absorber assuming that the maximum allowable displacement of the absorber is 2 mm. What is the value of the mass ratio μ?

11. Consider the two-degree-of-freedom rotational mass–spring system in Example 5.9. Assume $2I_1 = I_2 = 2I = 0.008$ kg·m^2 and $K_1 = K_2 = K = 4150$ N·m/rad.
    a. Use the modal analysis approach to determine the response of the system to the initial excitations $\theta_1(0) = 0.5$ rad, $\theta_2(0) = 1$ rad, and $\dot{\theta}_1(0) = \dot{\theta}_2(0) = 0$ rad/s.
    b. ◢ Write a MATLAB m-file and plot the response of the system.

12. Consider the two-degree-of-freedom rotational mass–spring system in Example 5.9. Assume $2I_1 = I_2 = 2I = 0.008$ kg·m^2 and $K_1 = K_2 = K = 4150$ N·m/rad.
    a. Use modal analysis to determine the response of the system to the harmonic excitation $\tau = T \sin \omega t = 10 \sin 600 t$.
    b. ◢ Write a MATLAB m-file and plot the frequency response of the system.

13. A single-degree-of-freedom system undergoes free vibration. Figure 9.33 is the recorded displacement response of the first three cycles. The mass of the system is known to be 1750 kg. Determine the stiffness k and the damping b of the system.

14. The frequency response of a single-degree-of-freedom system is shown in Figure 9.34. Determine the system's parameters including the mass m, the stiffness k, and the damping b.

10 Introduction to Feedback Control Systems

Control deals with the modeling of a variety of dynamic systems and the design of controllers that will cause these systems to behave in a desired manner. In Chapters 5 through 7 we discussed how to derive a mathematical model for a dynamic system, which can be mechanical, electrical, fluid, or thermal. In this chapter, we focus on how to design a controller for a dynamic system based on its mathematical model. Basic concepts and terminologies in control are first introduced in Section 10.1, such as feedback, open-loop control, closed-loop control, regulators, and servos. In general, there are two main reasons why control is needed. One is to maintain the system stability, and the other is to improve the system performance. Section 10.2 covers how to determine the stability of a system and how to define the performance in either time domain or frequency domain. Section 10.3 discusses the advantages of feedback control, which is utilized in most of the cases. Following the overview of feedback in Section 10.3, the classical structure of proportional, integral, and derivative control is introduced in Section 10.4. Finally, three different control design methods based on root locus, Bode plot, and state variable feedback are presented in Sections 10.5 through 10.7, respectively.

10.1 BASIC CONCEPTS AND TERMINOLOGIES

Control is the process of manipulating, manually or automatically, the input of a dynamic system so that the system output will behave as desired. If the output signal is measured and fed back for use in computing the input signal, the system is called feedback control. A familiar example is the cruise control of an automobile. To maintain a constant vehicle speed set by the driver, the actual speed of the vehicle is measured by the speedometer and fed back to the controller, which adjusts the engine's throttle position. Then, the engine torque is changed accordingly, which influences the actual vehicle speed.

To analyze and/or design a feedback control system, a block diagram is usually drawn to show the major components and their interconnections in graphical form. Figure 10.1 is a general block diagram of an elementary feedback control. The essential components of this feedback control include a system we want to control, a controller we need to design, an actuator used to drive the controlled system, and a sensor used to measure the system output. The connecting lines in the block diagram carry signals. As shown in Figure 10.1, the important signals in this feedback control system include the output, the control signal, and the reference.

In the example of the automobile's cruise control, the controlled system is the autobody, whose output is the speed. The speedometer, which acts as a sensor, measures the vehicle speed. The measured speed is fed back and compared with the desired speed, which is the reference signal. Based on the error between the measured and the reference signals, the controller computes the control signal, which is the engine's throttle position in our case. The engine is the actuator, and the torque provided by it is applied to the autobody and influences the vehicle speed.

Any control system must have these four essential components. Generally, the controlled system and the actuator are intimately connected, and they are combined as one component called the plant. There are two other signals shown in Figure 10.1, disturbance and sensor noise. Both of them are undesired system inputs that adversely affect the performance of a system.

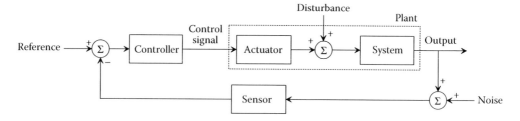

FIGURE 10.1 Block diagram of an elementary feedback control.

Example 10.1: Block Diagram of a Feedback Control System

Consider the electromechanical system described in Problem 3 of Problem Set 6.4. It consists of a cart that moves along a linear track and a DC motor that drives the cart. An encoder is included to measure the position of the cart. Assume that a controller is designed to control the position of the cart. Draw a block diagram for this feedback control system. Clearly label essential components and signals.

Solution

Note that the essential components of this feedback control are the cart (the controlled system), the controller, the DC motor (the actuator), and the encoder (the sensor). The corresponding block diagram is shown in Figure 10.2. The actual position of the cart is the output, the desired position is the reference, and the voltage applied to the DC motor is the control signal.

Transfer functions are usually used to represent the mathematical model of each block in a block diagram. The input and output signals of each block are also expressed in the Laplace domain, and they are denoted by capital letters. If the disturbance and noise signals are negligible, then the general block diagram of a feedback control system given in Figure 10.1 can be redrawn as shown in Figure 10.3, where $G(s)$ represents the dynamics of the plant, $C(s)$ is the controller, $H(s)$ is the sensor, $U(s)$ is the control signal, $Y(s)$ is the actual system output, $Y_m(s)$ is the measured output, and $R(s)$ is the reference. The difference between the reference and the feedback is defined as the error signal,

$$E(s) = R(s) - Y_m(s). \tag{10.1}$$

For an ideal sensor, its output is exactly the same as its input, that is, $H(s) = 1$. Therefore, we have $Y_m(s) = Y(s)$ and

$$E(s) = R(s) - Y(s). \tag{10.2}$$

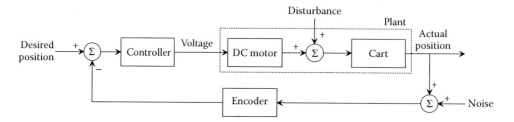

FIGURE 10.2 Block diagram of cart position control.

Introduction to Feedback Control Systems

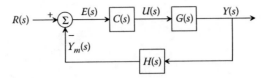

FIGURE 10.3 General block diagram of a feedback control system with transfer function representation.

A control system with feedback is also called closed-loop control. If the feedback is subtracted, it is called negative feedback, whereas the feedback that is added is called positive feedback. The negative feedback is usually required for system stability, while the positive feedback tends to make the system unstable. Unlike feedback control, open-loop control does not use the measured output to compute the control signal. The advantages of closed-loop control over open-loop control will be discussed in Section 10.3.

Example 10.2: Closed-Loop Transfer Function

Reconsider the cart position control system in Example 10.1. The transfer functions of the plant (combining the cart and the DC motor), the controller, and the sensor are

$$G(s) = \frac{3.778}{s^2 + 16.883s}, \quad C(s) = 85, \quad H(s) = 1.$$

Derive the closed-loop transfer functions $Y(s)/R(s)$ and $E(s)/R(s)$.

SOLUTION

Using the result presented in Section 4.5, the equivalent transfer function for a negative feedback control system is

$$\frac{Y(s)}{R(s)} = \frac{C(s)G(s)}{1 + C(s)G(s)H(s)}.$$

Substituting the given transfer functions results in

$$\frac{Y(s)}{R(s)} = \frac{85(3.778/(s^2 + 16.883s))}{1 + 85(3.778/(s^2 + 16.883s))} = \frac{321.13}{s^2 + 16.883s + 321.13}.$$

Since $E(s) = R(s) - Y(s)$, we have

$$\frac{E(s)}{R(s)} = \frac{R(s) - Y(s)}{R(s)} = 1 - \frac{Y(s)}{R(s)}.$$

Inserting $Y(s)/R(s)$ gives

$$\frac{E(s)}{R(s)} = 1 - \frac{321.13}{s^2 + 16.883s + 321.13} = \frac{s^2 + 16.883s}{s^2 + 16.883s + 321.13}.$$

Note that the system in Example 10.1 is designed to track a reference signal. This type of control system is called a tracking or servo system, where the reference signal usually varies with time. If the reference signal is constant, usually zero, and the system is designed to hold an output steady against unknown disturbances, then this type of control system is known as a regulator.

PROBLEM SET 10.1

1. Draw a block diagram for the feedback control of a liquid-level system, which consists of a valve with a control knob (0–100%) and a liquid-level sensor. Clearly label essential components and signals.

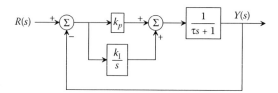

FIGURE 10.4 Problem 3.

2. Draw a block diagram for the feedback control of a single-link robot arm system, which consists of a DC motor to produce the driving force and an encoder to measure the joint angle. Clearly label essential components and signals.
3. Determine the transfer function between $R(s)$ and $Y(s)$ in Figure 10.4.
4. Determine the transfer function between $R(s)$ and $Y(s)$ in Figure 10.5.
5. ◢ Consider the control system in Example 10.2. Build a Simulink block diagram to simulate reference tracking control, where the signal $R(s)$ is a sine wave with a magnitude of 0.1 m and a frequency of 2 rad/s. Show the actual position response and the reference signal in the same scope.
6. Reconsider the control system in Example 10.2.
   a. Convert the transfer function $G(s) = Y(s)/U(s)$ to a differential equation of $y(t)$.
   b. ◢ Using the differential equation obtained in Part (a) to represent the plant, build a Simulink block diagram to simulate regulation control, where the reference signal $R(s)$ is zero. Assume that the initial conditions are $y(0) = 0.1$ m and $\dot{y}(0) = 0$ m/s.

10.2 STABILITY AND PERFORMANCE

Stability and performance are two important subjects in control. Generally, before designing a controller for a dynamic system, control designers check the stability and performance of the uncontrolled system. Then they come up with reasonable control design objectives from the perspective of the stability and performance. After a controller is designed, the stability and performance of the closed-loop control system are verified to meet the design objectives. In this section, we introduce the stability condition, time-domain performance specifications, and frequency-domain performance specifications.

10.2.1 STABILITY OF LINEAR TIME-INVARIANT SYSTEMS

Intuitively, a system is stable if its transient response decays and is unstable if it diverges. Thus, we can determine the stability of a system by solving and plotting the transient response of the system. However, this is not the only way to determine the system stability.

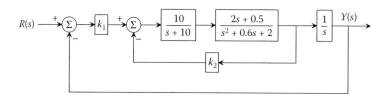

FIGURE 10.5 Problem 4.

Introduction to Feedback Control Systems

Consider a linear time-invariant system, whose transfer function is given by

$$G(s) = \frac{b(s)}{a(s)} = \frac{b_0 s^m + b_1 s^{m-1} + \cdots + b_{m-1} s + b_m}{s^n + a_1 s^{n-1} + \cdots + a_{n-1} s + a_n}. \tag{10.3}$$

Assume that the numerator and denominator polynomials $b(s)$ and $a(s)$, respectively, have no common factors. Setting the denominator of the transfer function equal to zero leads to the characteristic equation

$$a(s) = s^n + a_1 s^{n-1} + \cdots + a_{n-1} s + a_n = 0. \tag{10.4}$$

The roots of the characteristic equation are called the poles (see Section 2.3), which are complex and can be defined in terms of real and imaginary parts, $p = \sigma + j\omega$. A complex s-plane is usually sketched to show pole locations.

The following is the condition for stability: a linear time-invariant system is said to be stable if all the poles of its transfer function have negative real parts and is unstable otherwise. In terms of the pole locations in the s-plane, a linear time-invariant system is stable if all the poles of the system are inside the left-half s-plane, that is, $\sigma < 0$. If the system has any poles, even only one pole, in the right-half s-plane, that is, $\sigma > 0$, then the system is unstable. Thus, the imaginary $j\omega$-axis is the stability boundary. If the system has nonrepeated poles on the $j\omega$-axis, that is, $\sigma = 0$, then the system is marginally stable. If the system has repeated poles on the $j\omega$-axis, then the system is unstable. Note that we can solve for another set of roots by setting the numerator of the transfer function equal to zero. They are called zeros, which are not related to the stability of the system.

Essentially, the transient response of a linear time-invariant system is associated with its pole locations in the s-plane. For example, consider a first-order system whose transfer function is $Y(s)/U(s) = 1/(s-p)$. Assume that the input is an impulse function. We can obtain the response $y(t)$ by applying the inverse Laplace transform, $y(t) = e^{pt}$. The transient response e^{pt} approaches zero if and only if the real part of the pole p is negative. This simple example explains the reason why poles can be used to determine stability.

It is not an easy task to solve for the roots of a high-order characteristic equation by hand. Routh's stability criterion is a method for obtaining information about pole locations without solving for the poles. Consider the characteristic equation given in Equation 10.4. Routh's stability criterion consists of two conditions:

- A necessary (but not sufficient) condition for stability is that all the coefficients of the characteristic polynomial are positive.
- A necessary and sufficient condition for stability is that all the elements in the first column of the Routh array are positive.

If any of the coefficients in the characteristic polynomial is negative, we can conclude that the system is unstable by applying the first condition. If all of the coefficients in the characteristic polynomial are positive, then we have to check the second condition by constructing a Routh array as shown in Equation 10.5. For an n-th order characteristic polynomial, the Routh array has $n + 1$ rows. The first two rows are obtained by arranging the coefficients of the characteristic polynomial, beginning with the first and second coefficients and followed by the even-numbered and odd-numbered coefficients, respectively. Starting from the third row, the elements are formed from the two previous rows using determinants, with two elements in the first column and the other two elements from successive columns. Equation 10.6 shows how to compute the elements in the third and fourth rows. The rest of the rows can be obtained in the same manner as rows 3 and 4.

$$
\begin{array}{c|cccc}
s^n & : & 1 & a_2 & a_4 & a_6 & \cdots \\
s^{n-1} & : & a_1 & a_3 & a_5 & a_7 & \cdots \\
s^{n-2} & : & b_1 & b_2 & b_3 & \cdots \\
s^{n-3} & : & c_1 & c_2 & c_3 & \cdots \\
\vdots & & \vdots & \vdots & \vdots & \vdots \\
s^2 & : & * & * \\
s^1 & : & * \\
s^0 & : & *
\end{array}
\tag{10.5}
$$

$$
b_1 = \frac{-\det\begin{bmatrix} 1 & a_2 \\ a_1 & a_3 \end{bmatrix}}{a_1} = \frac{a_1 a_2 - a_3}{a_1}, \quad c_1 = \frac{-\det\begin{bmatrix} a_1 & a_3 \\ b_1 & b_2 \end{bmatrix}}{b_1} = \frac{b_1 a_3 - a_1 b_2}{b_1},
$$

$$
b_2 = \frac{-\det\begin{bmatrix} 1 & a_4 \\ a_1 & a_5 \end{bmatrix}}{a_1} = \frac{a_1 a_4 - a_5}{a_1}, \quad c_2 = \frac{-\det\begin{bmatrix} a_1 & a_5 \\ b_1 & b_3 \end{bmatrix}}{b_1} = \frac{b_1 a_5 - a_1 b_3}{b_1}, \tag{10.6}
$$

$$
b_3 = \frac{-\det\begin{bmatrix} 1 & a_6 \\ a_1 & a_7 \end{bmatrix}}{a_1} = \frac{a_1 a_6 - a_7}{a_1}, \quad c_3 = \frac{-\det\begin{bmatrix} a_1 & a_7 \\ b_1 & b_4 \end{bmatrix}}{b_1} = \frac{b_1 a_7 - a_1 b_4}{b_1}.
$$

Example 10.3: Stability

The transfer function of a dynamic system is given by

$$G(s) = \frac{s+4}{s^5 + s^4 + 3s^3 + 8s^2 + 4s + 5}.$$

Determine the stability of the system

a. Using Routh's stability criterion without solving for the poles of the system.
b. ◀ Using MATLAB® to solve for the poles.

SOLUTION

a. The characteristic equation is $s^5 + s^4 + 3s^3 + 8s^2 + 4s + 5 = 0$. The Routh array can be formed as follows.

$$
\begin{array}{r|lll}
s^5: & 1 & 3 & 4 \\
s^4: & 1 & 8 & 5 \\
s^3: & -5 = \dfrac{1(3)-8}{1} & -1 = \dfrac{1(4)-5}{1} & 0 \\
s^2: & 7.8 = \dfrac{(-5)(8)-1(-1)}{-5} & 5 = \dfrac{(-5)(5)-0}{-5} & 0 \\
s^1: & 2.21 = \dfrac{(7.8)(-1)-(-5)(5)}{7.8} & 0 \\
s^0: & 5
\end{array}
$$

Since the elements in the first column of the Routh array are not all positive, we conclude that the system is unstable.

Introduction to Feedback Control Systems

b. One of two MATLAB commands can be used to solve for the poles, `pole` or `roots`.

```
>> num = [1 4];
>> den = [1 1 3 8 4 5];
>> sys = tf(num,den);
>> pole(sys)
```

The command `pole` returns the poles of the system: $0.55 \pm 1.90j$, $-0.18 \pm 0.84j$, and -1.73. There are two poles that have positive real parts, and thus the system is unstable. We can also use the command `roots` to solve for the roots of the characteristic equation.

```
>> roots(den)
```

Routh's stability criterion was especially useful before the availability of mathematical and scientific computing software, like MATLAB. However, it is still useful for determining the ranges of system parameters for stability (see Example 10.5 in Section 10.3). It should be pointed out that the study of stability discussed here is limited to only linear time-invariant systems. The study of stability for nonlinear and time-varying systems is very complex and is beyond the scope of this text.

10.2.2 Time-Domain Performance Specifications

Performance specifications are certain requirements associated with the response of the system. In the time domain, the requirements are usually given for the step response. Consider a second-order system whose transfer function is given by

$$\frac{Y(s)}{U(s)} = \frac{\omega_n^2}{s^2 + 2\zeta\omega_n s + \omega_n^2}. \tag{10.7}$$

Figure 10.6 shows a unit-step response, where the vertical axis is normalized so that the steady-state value is equal to 1. Four quantities are defined to specify the performance of the system, and they are rise time t_r, overshoot M_p, peak time t_p, and settling time t_s.

The rise time t_r is the time from $y = 0.1$ to $y = 0.9$. The shorter the rise time, the faster the system reaches the vicinity of the steady-state value. The approximated relationship between t_r and

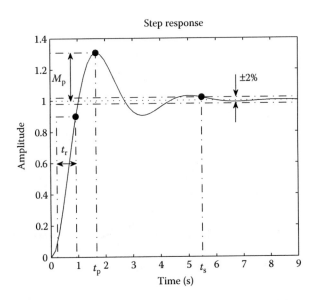

FIGURE 10.6 A unit-step response with time-domain performance specifications indicated.

the natural frequency ω_n is given by

$$t_r \approx \frac{1.8}{\omega_n}. \tag{10.8}$$

The overshoot M_p is the maximum amount of the response of the system exceeding the steady-state value divided by the steady-state value. It is usually expressed as a percentage value. The peak time t_p is the time it takes the system to reach the maximum value. As discussed in Section 8.2, assuming zero initial conditions, the unit-step response for the second-order system in Equation 10.7 is

$$y(t) = 1 - e^{-\zeta\omega_n t}\left(\cos\omega_d t + \frac{\zeta}{\sqrt{1-\zeta^2}}\sin\omega_d t\right). \tag{10.9}$$

Differentiating $y(t)$ with respect to t and setting it equal to zero yields the peak time

$$t_p = \frac{\pi}{\omega_d}. \tag{10.10}$$

Substituting Equation 10.10 into Equation 10.9 gives the value of y at the peak time. The overshoot M_p can be determined by computing $y(t_p) - 1$,

$$M_p = e^{-\pi\zeta/\sqrt{1-\zeta^2}}. \tag{10.11}$$

The detailed derivation will be left to the reader as an exercise. As shown in Equation 10.11, the overshoot of the step response is related to the damping of the system. Figure 10.7 is the plot of M_p versus ζ. The larger the damping ratio, the smaller the overshoot.

The setting time t_s is the time required for the transient to decay to a small value such that $y(t)$ almost reaches the steady-state value. It is observed from Equation 10.9 that the value of t_s is determined by the decaying exponential component $e^{-\zeta\omega_n t}$. For example, the settling time when the decay exponential reaches 2% can be determined by solving $e^{-\zeta\omega_n t_s} = 0.02$ for t_s. Thus, we have

$$t_s \approx \frac{3.9}{\zeta\omega_n}. \tag{10.12}$$

There are other small values used to compute the settling time, such as 1% and 5%.

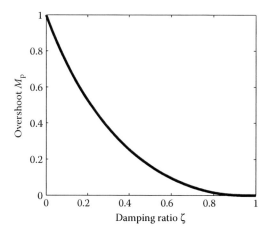

FIGURE 10.7 Plot of overshoot versus damping ratio for a second-order system.

Introduction to Feedback Control Systems

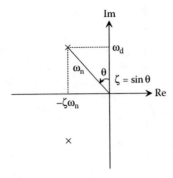

FIGURE 10.8 Relationship between a pair of complex conjugate poles and system parameters.

As indicated by Equation 10.8 and Equations 10.10 through 10.12, the time-domain performance specifications are related to the system parameters, specifically the undamped natural frequency ω_n and the damping ratio ζ. From Equation 10.7, the poles of a second-order system are also related to ω_n and ζ,

$$p_{1,2} = -\zeta\omega_n \pm j\omega_n\sqrt{1-\zeta^2} = -\zeta\omega_n \pm j\omega_d, \tag{10.13}$$

which is a pair of complex conjugate poles for $0 \leq \zeta < 1$. Figure 10.8 shows the plot of the poles in the s-plane. Considering either one of them, the magnitude of the complex pole is

$$\sqrt{\text{Re}^2 + \text{Im}^2} = \sqrt{(-\zeta\omega_n)^2 + \omega_n^2(1-\zeta^2)} = \omega_n \tag{10.14}$$

and the angle between the pole and the imaginary axis satisfies

$$\sin\theta = \frac{|\text{Re}|}{\sqrt{\text{Re}^2 + \text{Im}^2}} = \frac{\zeta\omega_n}{\omega_n} = \zeta. \tag{10.15}$$

Thus, the time-domain performance specifications are associated with the pole locations. In control design, one or more of these requirements are often specified to determine the allowable region for the poles in the s-plane. The three plots in Figure 10.9 show the regions graphed based on the transient requirements t_r, M_p, and t_s, respectively.

Example 10.4: Time-Domain Performance Specifications

Consider a second-order system whose transfer function is given by Equation 10.7. Assume that the requirements for the system unit-step response are $t_r \leq 0.1$ s and $M_p \leq 20\%$. Sketch the allowable region for the poles in the s-plane.

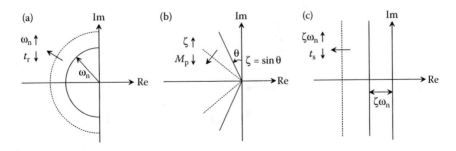

FIGURE 10.9 Regions graphed based on (a) rise time, (b) overshoot, and (c) settling time.

Solution

Equation 10.8 indicates that

$$\omega_n \approx \frac{1.8}{t_r} \geq \frac{1.8}{0.1} = 18 \text{ rad/s}.$$

To express the damping ratio in terms of the overshoot, we take the natural logarithm of both sides of Equation 10.11,

$$\ln M_p = -\frac{\pi \zeta}{\sqrt{1-\zeta^2}}.$$

Squaring both sides and solving for ζ,

$$\zeta = \frac{|\ln M_p|}{\sqrt{\pi^2 + (\ln M_p)^2}}.$$

As shown in both Figures 10.7 and 10.9b, ζ is mono decreasing with respect to M_p. That is, ζ will increase if M_p decreases and vice versa. Thus, for $M_p \leq 20\%$, we have

$$\zeta \geq \frac{|\ln 0.2|}{\sqrt{\pi^2 + (\ln 0.2)^2}} = 0.46.$$

The corresponding angle θ is

$$\theta = \sin^{-1}\zeta \geq \sin^{-1} 0.46 = 27°.$$

The area to the left of the gray boundary shown in Figure 10.10 is the allowable region for the poles in the s-plane so that the two performance requirements are met.

Note that Equation 10.8 and Equations 10.10 through 10.12 are derived based on the assumption that the system has no zeros and has two complex poles. Thus, they are not precise design formulas for any system. However, they can be used as qualitative guides to provide a starting point for design. The transient time response of the system is often checked to verify the time-domain specifications after control design.

10.2.3 Frequency-Domain Performance Specifications

System performance can also be specified in terms of frequency response. Figure 10.11 illustrates the ideal frequency response magnitude of a closed-loop control system. Two specifications are defined, and they are bandwidth ω_{BW} and resonant peak M_r.

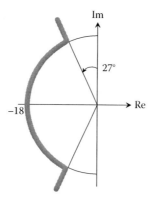

FIGURE 10.10 Allowable region of the poles in the s-plane.

Introduction to Feedback Control Systems

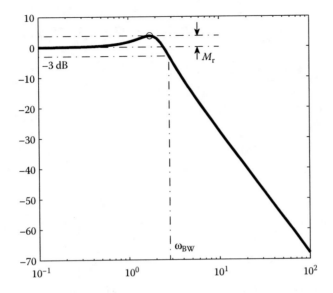

FIGURE 10.11 An ideal frequency response with performance specifications indicated.

Assume that the transfer function of the closed-loop system is $Y(s)/R(s)$, where $Y(s)$ is the system output and $R(s)$ is the reference input. The bandwidth is defined as the frequency at which the magnitude of the closed-loop transfer function crosses -3 dB or 0.707. Recall that the steady-state response of a linear system to sinusoidal excitations is called the system's frequency response. As shown in Figure 10.11, if the excitation frequency is lower than ω_{BW}, the magnitude $|Y(s)/R(s)|$ is close to 1 or 0 dB. This indicates that the system output follows the reference input. If the excitation frequency is higher than ω_{BW}, the magnitude $|Y(s)/R(s)|$ drops to a small value, and the system output no longer follows the reference input. The higher the bandwidth, the faster the reference signal the system can follow. Thus, the bandwidth is a measure of the speed of the response.

The resonant peak M_r was introduced previously in Section 8.3. It is defined as the maximum value of the magnitude of the frequency response. As shown in Section 8.3, the resonant peak is related to the damping of the system. The smaller the damping, the higher the resonant peak. Compared with the time-domain performance specifications, the resonant peak M_r is similar to overshoot M_p, both of which are related to the damping ratio ζ, while the bandwidth ω_{BW} is similar to the rise time t_r, both of which are related to the natural frequency ω_n.

PROBLEM SET 10.2

1. The transfer function of a dynamic system is given by

$$G(s) = \frac{5s + 10}{s^4 + 3s^3 + 2s^2 + 5s}.$$

   Suppose that a negative unity feedback is applied to this open-loop system.
   a. Using Routh's stability criterion, determine the stability of the resulting closed-loop system.
   b. Using MATLAB, solve for the poles of the resulting closed-loop system and verify the result obtained in Part (a).
2. Repeat Problem 1 for the dynamic system with a transfer function given by

$$G(s) = \frac{2s^2 + s + 1}{3s^5 + s^4 + 2s^3 + 5s^2 + s + 1}.$$

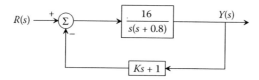

FIGURE 10.12 Problem 6.

3. The unit-step response for a second-order system $Y(s)/U(s) = \omega_n^2/(s^2 + 2\zeta\omega_n s + \omega_n^2)$ is given by

$$y(t) = 1 - e^{-\zeta\omega_n t}\left(\cos\omega_d t + \frac{\zeta}{\sqrt{1-\zeta^2}}\sin\omega_d t\right).$$

Prove that the relationship between the overshoot M_p and the damping ratio is

$$M_p = e^{-\pi\zeta/\sqrt{1-\zeta^2}}.$$

4. Consider a second-order system $Y(s)/U(s) = \omega_n^2/(s^2 + 2\zeta\omega_n s + \omega_n^2)$. Determine the relationship between the settling time t_s and the system parameters, that is, ω_n and ζ, when the system's unit-step response reaches and remains
   a. Within a 1% band of the steady-state value.
   b. Within a 5% band of the steady-state value.
5. Consider a second-order system $Y(s)/U(s) = \omega_n^2/(s^2 + 2\zeta\omega_n s + \omega_n^2)$. Sketch the allowable region of the poles in the s-plane if the requirements for the system's unit-step response are $M_p \leq 5\%$ and 1% settling time $t_s \leq 0.1$ s.
6. Figure 10.12 shows a negative feedback control system.
   a. Find the closed-loop transfer function $Y(s)/R(s)$.
   b. Determine the value of K such that the damping ratio of the closed-loop system is 0.5.
   c. Compute the rise time, overshoot, and 1% settling time in the unit-step response for the system.
7. ▲ Consider a second-order system $Y(s)/U(s) = \omega_n^2/(s^2 + 2\zeta\omega_n s + \omega_n^2)$. Write a MATLAB m-file to plot the magnitude of the system's frequency response function for the following cases: $\omega_n = 2$ rad/s and $\zeta = 0.01, 0.1, 0.5,$ and 1. Summarize the effects of the damping ratio on the frequency-domain performance.
8. ▲ Repeat Problem 7 for the following cases: $\zeta = 0.1$ and $\omega_n = 1, 2,$ and 6 rad/s. Summarize the effects of the natural frequency on the frequency-domain performance.

10.3 BENEFITS OF FEEDBACK CONTROL

As introduced in Section 10.1, a closed-loop controller uses feedback to control the output of a plant. The input to a plant has an effect on its output, which is measured with a sensor and fed back to the controller, and then the computed control signal is used as the input to the plant, closing the loop. Feedback control has the following advantages over open-loop control, such as stabilization, disturbance rejection, improved reference tracking performance, and reduced sensitivity to parameter variations. In this section, we show these benefits one by one using MATLAB and/or Simulink®. The controller in each discussion and example is assumed to be a gain denoted by K.

10.3.1 STABILIZATION

Consider a plant represented by a transfer function $G(s) = b(s)/a(s)$, where $a(s)$ and $b(s)$ are the denominator and numerator polynomials, respectively. Assume that $G(s)$ is unstable. This assumption implies that not all the poles, or the roots of the characteristic equation $a(s) = 0$, have negative

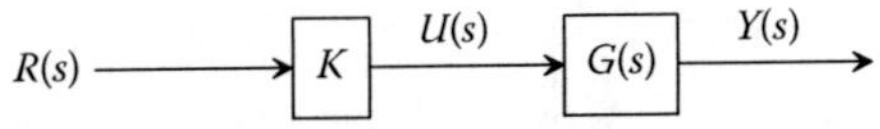

Introduction to Feedback Control Systems

FIGURE 10.13 Open-loop control.

real parts. If an open-loop control system is implemented as shown in Figure 10.13, where the output is $Y(s)$ and the input is $R(s)$, then the transfer function $Y(s)/R(s)$ is

$$\frac{Y(s)}{R(s)} = KG(s) = \frac{Kb(s)}{a(s)}. \tag{10.16}$$

It is obvious that the open-loop control system has the same poles as the plant $G(s)$, and thus it is still unstable.

Now, we use a negative feedback control system. Figure 10.14 shows the corresponding block diagram. The closed-loop transfer function $Y(s)/R(s)$ is

$$\frac{Y(s)}{R(s)} = \frac{KG(s)}{1 + KG(s)} = \frac{Kb(s)}{a(s) + Kb(s)}. \tag{10.17}$$

The characteristic equation of the closed-loop system is $a(s) + Kb(s) = 0$. Properly choosing the control gain K can make the closed-loop system stable. Routh's stability criterion introduced in Section 10.2 is a way to determine the range of K such that the closed-loop control system is stable.

Example 10.5: Stabilization Using Feedback

Consider an unstable plant

$$G(s) = \frac{s+2}{s^3 + 4s^2 - 5s},$$

with feedback control as shown in Figure 10.14.

a. Using Routh's stability criterion, determine the range of the control gain K for which the closed-loop system is stable.

b. Use MATLAB commands to find the unit-step responses for open-loop and closed-loop control. Assume that the control gain is $K = 20$. Compare the open- and closed-loop responses.

Solution

a. Solving the characteristic equation of $G(s)$ gives the poles 0, 1, and −5. The positive real pole, 1, indicates that the plant $G(s)$ is unstable. The closed-loop transfer function is

$$\frac{Y(s)}{R(s)} = \frac{K((s+2)/(s^3 + 4s^2 - 5s))}{1 + K((s+2)/(s^3 + 4s^2 - 5s))} = \frac{Ks + 2K}{s^3 + 4s^2 + (K-5)s + 2K}.$$

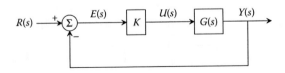

FIGURE 10.14 Closed-loop control.

The closed-loop characteristic equation is
$$s^3 + 4s^2 + (K-5)s + 2K = 0,$$
for which we can construct the Routh array

$$
\begin{array}{lll}
s^3: & 1 & K-5 \\
s^2: & 4 & 2K \\
s^1: & \dfrac{K}{2}-5 = \dfrac{4(K-5)-2K}{4} & 0 \\
s^0: & 2K &
\end{array}
$$

In order to make the closed-loop system stable, all elements in the first column of the Routh array must be positive. Therefore,
$$\begin{cases} \dfrac{K}{2} - 5 > 0, \\ 2K > 0, \end{cases}$$
which leads to $K > 10$.

b. For the open-loop control system as shown in Figure 10.13, the controller block is connected with the plant in series. The following is the MATLAB session.

```
>> G = tf([1 2],[1 4 -5 0]);
>> K = 20;
>> olp = K*G;
>> step(olp);
```

Figure 10.15 is the unit-step response of the open-loop control system, which is obviously unstable because the response diverges. Therefore, the unstable plant cannot be stabilized using open-loop control.

As introduced in Section 4.5, the MATLAB command `feedback` can be used to find the transfer function of the closed-loop system for the feedback control system in Figure 10.14.

```
>> clp = feedback(K*G,1);
>> step(clp);
```

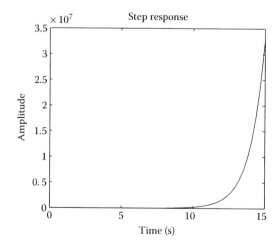

FIGURE 10.15 Unit-step response of the open-loop control system in Example 10.5.

Introduction to Feedback Control Systems

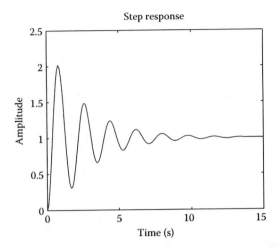

FIGURE 10.16 Unit-step response of the closed-loop control system in Example 10.5.

As shown in Figure 10.16, the unit-step response of the closed-loop control system converges and the unstable plant is stabilized.

10.3.2 Disturbance Rejection

To compare the capabilities of the open-loop and the closed-loop control for disturbance rejection, assume that a disturbance is an input to the plant $G(s)$. Figure 10.17 is the block diagram for the open-loop control, where

$$Y(s) = G(s)U(s) + G(s)D(s) = KG(s)R(s) + G(s)D(s). \quad (10.18)$$

Equation 10.18 shows that the output $Y(s)$ depends on the reference input $R(s)$ and the disturbance input $D(s)$. Letting $R(s) = 0$ yields the transfer function relating the disturbance $D(s)$ and the output $Y(s)$,

$$\frac{Y(s)}{D(s)} = G(s). \quad (10.19)$$

Note that K does not appear in Equation 10.19 and it has no control over the disturbance in the open-loop case.

Using closed-loop control as shown in Figure 10.18, we have

$$Y(s) = G(s)U(s) + G(s)D(s) = KG(s)[R(s) - Y(s)] + G(s)D(s). \quad (10.20)$$

Solving for $Y(s)$ gives

$$Y(s) = \frac{KG(s)}{1 + KG(s)}R(s) + \frac{G(s)}{1 + KG(s)}D(s). \quad (10.21)$$

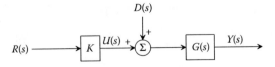

FIGURE 10.17 Open-loop control with disturbance input.

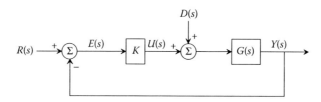

FIGURE 10.18 Closed-loop control with disturbance input.

The transfer function relating the disturbance $D(s)$ and the output $Y(s)$ is

$$\frac{Y(s)}{D(s)} = \frac{G(s)}{1 + KG(s)}, \tag{10.22}$$

which indicates that K in the closed-loop case has control over the disturbance.

Example 10.6: Disturbance Rejection Using Feedback

Consider a plant whose transfer function is given by

$$G(s) = \frac{4}{s + 10}.$$

a. Build a Simulink block diagram as shown in Figure 10.17 to simulate open-loop control with disturbance input. Assume that the controller is $K = 2.5$, the disturbance is a constant of -1, and the reference is a unit-step function with the step time at $t = 0$ s. Compare the steady-state values of the responses without and with the disturbance.

b. Build a Simulink block diagram as shown in Figure 10.18 to simulate closed-loop control with disturbance input. Assume that the controller is $K = 50$. The disturbance and the reference are the same as those in Part (a). Compare the steady-state values of the responses without and with the disturbance.

Solution

a. The Simulink block diagram of the open-loop control is shown in Figure 10.19, where the Constant block is used to represent the disturbance signal. We first set the Constant block to be 0 and run the simulation. Then, we change the Constant block to be -1 and rerun the simulation. The responses for those two cases are plotted in Figure 10.20. The steady-state value of the unit-step response is 1 without disturbance and 0.6 with disturbance. The error to the disturbance is 0.4.

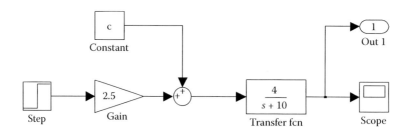

FIGURE 10.19 Simulink block diagram of open-loop control with disturbance input.

Introduction to Feedback Control Systems

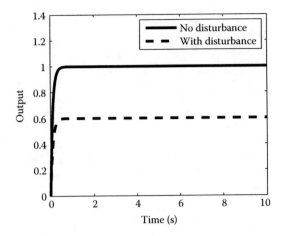

FIGURE 10.20 Unit-step responses of open-loop control without and with disturbance input.

b. The Simulink block diagram of the closed-loop control is shown in Figure 10.21. The responses are plotted in Figure 10.22. The steady-state value of the unit-step response is 0.95 without disturbance and 0.93 with disturbance. The error to the disturbance is 0.02.

Although the closed-loop control is not as good as the open-loop control in the absence of disturbance, the error resulting from a constant disturbance can be made smaller in a closed-loop feedback system when compared with an open-loop system.

10.3.3 Reference Tracking

As introduced in Section 10.1, there are two types of control systems, regulators and servos. The former are designed for disturbance rejection and the latter are designed for reference tracking. With feedback, a control system can achieve improved reference tracking performance. To show this, let us consider the closed-loop control case in Figure 10.18, where the disturbance $D(s)$ is now set as zero. As a result, the closed-loop transfer function of the system is

$$\frac{Y(s)}{R(s)} = \frac{KG(s)}{1 + KG(s)}. \tag{10.23}$$

A typical frequency response plot of the closed-loop system was given in Figure 10.11. As we discussed in Section 10.2, the output follows the reference input when $|Y(s)/R(s)| \approx 1$.

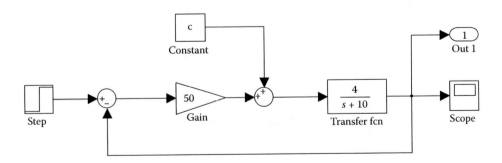

FIGURE 10.21 Simulink block diagram of closed-loop control with disturbance input.

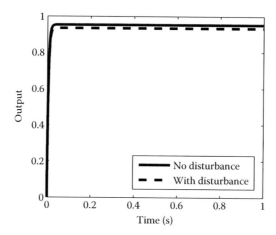

FIGURE 10.22 Unit-step responses of closed-loop control without and with disturbance input.

Equation 10.23 implies that $Y(s)$ is approximately equal to $R(s)$ if the magnitude $|KG(s)| \gg 1$. In general, this can be done by increasing the value of the control gain K. Thus, a large control gain can reduce the steady-state error of the response.

Example 10.7: Reference Tracking Using Feedback

Consider the closed-loop control system shown in Figure 10.23.

a. Using MATLAB, plot the unit-step responses of the system for the following values of K: 5, 50, and 500.
b. Compute the steady-state errors for the different values of K.

Solution

a. The following is the MATLAB m-script. The unit-step responses of the system for the three different values of the control gain are shown in Figure 10.24. We observe that the steady-state error becomes smaller when the control gain is increased.

```
K = [5 50 500];
G = tf([4],[1 10]);
for i = 1:length(K)
    clp = feedback(K(i)*G,1);
    step(clp);
    hold on;
end
```

b. The value of the steady-state error can be computed by applying the final value theorem.

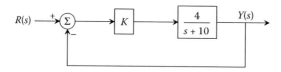

FIGURE 10.23 A reference tracking control system with feedback.

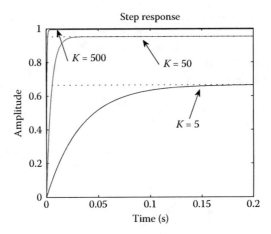

FIGURE 10.24 Unit-step responses for the system in Figure 10.23 with different values of K.

The closed-loop transfer function $E(s)/R(s)$ is

$$\frac{E(s)}{R(s)} = \frac{1}{1+KG(s)}.$$

For a unit-step input, we have

$$e_{ss} = \lim_{s\to 0} sE(s) = \lim_{s\to 0} sR(s)\frac{E(s)}{R(s)} = \lim_{s\to 0} s\frac{1}{s}\frac{1}{1+KG(s)} = \lim_{s\to 0}\frac{s+10}{s+10+4K} = \frac{10}{10+4K}.$$

The steady-state errors for $K = 5$, 50, and 500 are 0.3333, 0.0476, and 0.005, respectively. It should be pointed out that a large control gain may also result in unsatisfactory transient response and even destabilize the system. Details will be discussed in Section 10.4.

10.3.4 Sensitivity to Parameter Variations

For model-based control, a controller implemented for a practical system is designed based on a mathematical model of the system. Therefore, it is important to obtain a precise model of the plant for control design. However, modeling errors do exist due to uncertainties in system parameters. For example, the dynamic behavior of a mass–spring system depends on the values of the mass and stiffness. The values used for modeling might be different from the actual values due to inevitable measurement errors from the start or slight parameter changes caused by external effects. To maintain control performance, a controller should be insensitive to parameter changes.

Consider a function f that depends on a parameter a. Denote the change in the parameter as δa. If the change in f due to the parameter change is δf, then the sensitivity of the function f with respect to the parameter a is given by

$$S_a^f = \frac{\delta f/f}{\delta a/a}, \tag{10.24}$$

where the first-order variation δf is proportional to the derivative df/da and is given by

$$\delta f = \frac{df}{da}\delta a. \tag{10.25}$$

Thus, the sensitivity S_a^f can be written as

$$S_a^f = \frac{a}{f}\frac{df}{da}. \tag{10.26}$$

To compare the sensitivity of open-loop control with that of closed-loop control, we assume that one or more parameters in the plant change. Without loss of generality, the disturbance is assumed to be zero. For the open-loop control shown in Figure 10.17, denote the transfer function from the reference signal to the system output as $T_o(s)$, which is $KG(s)$. Applying Equation 10.26 gives the sensitivity of the open-loop control system $T_o(s)$ to the plant $G(s)$

$$S_G^{T_o} = \frac{G}{T_o}\frac{dT_o}{dG} = \frac{G}{KG}\frac{d(KG)}{dG} = \frac{G}{KG}\frac{K\,dG}{dG} = 1, \qquad (10.27)$$

which implies that open-loop control is very sensitive to the parameter variations in the plant. For example, a 5% error in the plant would yield the same percentage error in the open-loop transfer function.

For the closed-loop control case as shown in Figure 10.18, where the disturbance is still set as zero, denote the closed-loop transfer function as $T_c(s)$. Applying Equation 10.26 gives

$$S_G^{T_c} = \frac{G}{T_c}\frac{dT_c}{dG}. \qquad (10.28)$$

The closed-loop transfer function $T_c(s)$ is given by Equation 10.23. Thus, we have

$$\frac{dT_c}{dG} = \frac{d}{dG}\left(\frac{KG}{1+KG}\right) = \frac{K(1+KG) - KG(K)}{(1+KG)^2} = \frac{K}{(1+KG)^2}. \qquad (10.29)$$

Substituting it into Equation 10.28 yields

$$S_G^{T_c} = \frac{G(1+KG)}{KG}\frac{K}{(1+KG)^2} = \frac{1}{1+KG}, \qquad (10.30)$$

which can be made much less than 1 by adjusting the controller gain K. Thus, the overall transfer function in feedback control is less sensitive to variations in the plant gain compared with the one in open-loop control.

Example 10.8: Sensitivity to Parameter Variations

Consider a mass–damper–spring system $G(s) = Y(s)/U(s) = 1/(ms^2 + bs + k)$, where $m = 1$ kg, $b = 8$ N·s/m, and $k = 40$ N/m.

a. Assume that the mass–damper–spring system is controlled in an open-loop control system with a controller $K = 40$. Determine the steady-state value of the response to a unit-step input. If the spring stiffness is actually 50 N/m, recalculate the steady-state value of the response and determine the fractional change in the steady-state value.
b. Repeat Part (a) assuming that the mass–damper–spring system is controlled in a feedback control system with a controller $K = 2000$.

Solution

a. In the open-loop case, the steady-state value of the response to a unit-step input is

$$y_{ss} = \lim_{s\to 0} sR(s)\frac{Y(s)}{R(s)} = \lim_{s\to 0} s\frac{1}{s}KG(s) = \lim_{s\to 0}\frac{40}{s^2+8s+40} = 1.$$

If the spring stiffness is 50 N/m, then the steady-state value becomes

$$y_{ss} = \lim_{s\to 0} sR(s)\frac{Y(s)}{R(s)} = \lim_{s\to 0} s\frac{1}{s}KG(s) = \lim_{s\to 0}\frac{40}{s^2+8s+50} = 0.8.$$

Note that y_{ss} is reduced by 20%, which is the same as the error in k.

b. In the closed-loop case, the steady-state value of the response to a unit-step input is

$$y_{ss} = \lim_{s \to 0} sR(s)\frac{Y(s)}{R(s)} = \lim_{s \to 0} s\frac{1}{s}\frac{KG(s)}{1+KG(s)} = \lim_{s \to 0} \frac{2000}{s^2 + 8s + 2040} = 0.9804.$$

If the spring stiffness is 50 N/m, then the steady-state value becomes

$$y_{ss} = \lim_{s \to 0} sR(s)\frac{Y(s)}{R(s)} = \lim_{s \to 0} s\frac{1}{s}\frac{KG(s)}{1+KG(s)} = \lim_{s \to 0} \frac{2000}{s^2 + 8s + 2050} = 0.9756.$$

Note that y_{ss} is reduced by 0.49%, which is much less than the error in k.

PROBLEM SET 10.3

1. Consider the feedback system shown in Figure 10.25.
   a. Using Routh's stability criterion, determine the range of the control gain K for which the closed-loop system is stable.
   b. ◢ Use MATLAB commands to find the unit-step responses for open-loop and closed-loop control. Assume that the control gain is $K = 2$. Compare the open- and closed-loop responses.
2. Consider the feedback system shown in Figure 10.26. Using Routh's stability criterion, determine the range of the control gain K for which the closed-loop system is stable.
3. Reconsider Example 10.6. Using the final value theorem, verify the steady-state errors to a unit-step input for open-loop and closed-loop control without and with disturbance.
4. A stable system can be classified as a system type, which is defined as the degree of the polynomial for which the steady-state error is a nonzero finite constant. For instance, if the error to a step input, which is a polynomial of zero degree, is a nonzero finite constant, then such a system is called type 0, and so on. Consider the system in Figure 10.27.
   a. Compute the following steady-state errors: (1) for a unit-step reference input, (2) for a unit-ramp reference input $u = t$, and (3) for a parabolic reference input $u = 0.5t^2$.
   b. Determine the system type.
5. ◢ Reconsider Example 10.8. Build Simulink block diagrams to simulate open-loop and closed-loop control with parameter variations. Verify the steady-state response values y_{ss} obtained in Example 10.8.
6. Consider the feedback control system shown in Figure 10.28.
   a. Compute the sensitivity of the closed-loop transfer function to changes in the parameter τ.

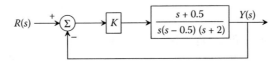

FIGURE 10.25 Problem 1.

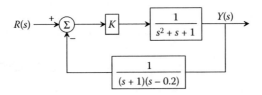

FIGURE 10.26 Problem 2.

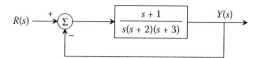

FIGURE 10.27 Problem 4.

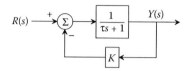

FIGURE 10.28 Problem 6.

b. Compute the sensitivity of the closed-loop transfer function to changes in the parameter K.

c. Assuming $\tau = 1$ and $K = 1$, use MATLAB to plot the magnitude of each of the sensitivity functions for $s = j\omega$. Use the logarithmic scale for the y-axis. Comment on the effect of parameter variations in τ and K for different driving frequencies ω.

10.4 PROPORTIONAL–INTEGRAL–DERIVATIVE CONTROL

A proportional–integral–derivative (PID) controller is a generic feedback control structure widely used in industries. The PID controller involves three terms. The proportional term determines the control signal based on the current error, the integral term determines the control signal based on the integral of error, and the derivative term determines the control signal based on the derivative of error. The expression of the PID controller in the time domain is given by

$$u(t) = k_\mathrm{P} e(t) + k_\mathrm{I} \int_0^t e(\tau)\,\mathrm{d}\tau + k_\mathrm{D} \frac{\mathrm{d}e(t)}{\mathrm{d}t}, \tag{10.31}$$

where k_P is the proportional control gain, k_I is the integral control gain, and k_D is the derivative control gain. Taking the Laplace transform of Equation 10.31 yields the transfer function of the PID controller

$$\frac{U(s)}{E(s)} = k_\mathrm{P} + \frac{k_\mathrm{I}}{s} + k_\mathrm{D} s. \tag{10.32}$$

Figure 10.29 shows a block diagram of feedback control using a PID controller denoted by $C(s)$. In this section, we discuss the advantages and disadvantages of PID control.

10.4.1 PROPORTIONAL CONTROL

For proportional feedback control, the controller transfer function is

$$\frac{U(s)}{E(s)} = k_\mathrm{P}. \tag{10.33}$$

Note that the controller structures in the previous sections are the most basic proportional feedback control. As discussed in Section 10.3, a high proportional control gain can result in smaller steady-state error. However, if k_P is made too large, the closed-loop system may have reduced damping and even become unstable.

Introduction to Feedback Control Systems

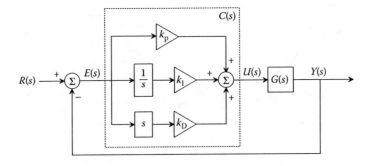

FIGURE 10.29 A block diagram of PID feedback control.

Example 10.9: Proportional Control

Consider the mass–damper–spring system in Example 10.8,

$$G(s) = \frac{1}{s^2 + 8s + 40}.$$

Use Simulink to build a block diagram for proportional feedback control. Find the unit-step responses for $k_p = 25$, 250, and 2000. Discuss the effects of the proportional feedback on the step response of the system.

Solution

The Simulink block diagram is shown in Figure 10.30. Note that Figure 10.30 also includes integral and derivative control loops, which will be used later in this section. Set k_I and k_D to be zero. Run the simulation for $k_p = 25$, 250, and 2000.

Figure 10.31 illustrates the effects of proportional feedback control. On the one hand, as k_p increases, the steady-state error to the unit-step input becomes smaller. Applying the final value theorem gives

$$e_{ss} = \lim_{s \to 0} sR(s)\frac{E(s)}{R(s)} = \lim_{s \to 0} s\frac{1}{s}\frac{1}{1 + k_p G(s)} = \frac{40}{40 + k_p}.$$

Substituting the three different proportional control gains yields $e_{ss} = 0.62$, 0.14, and 0.02, for $k_p = 25$, 250, and 2000, respectively. For this particular system, the steady-state error will approach zero as k_p increases. However, it will never be zero.

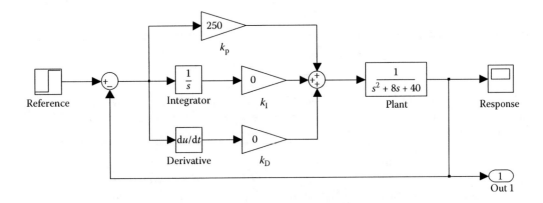

FIGURE 10.30 Simulink block diagram for PID feedback control.

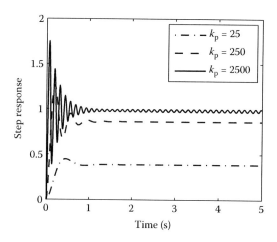

FIGURE 10.31 Responses of proportional control.

On the other hand, the larger proportional gain results in less satisfactory oscillatory response. This is caused by reduced damping. Note that the characteristic equation of the closed-loop system with proportional control is

$$s^2 + 8s + 40 + k_p = 0.$$

For a second-order system, the constant term and the coefficient of s in the characteristic equation are related to the natural frequency and the damping ratio of the system, that is,

$$8 = 2\zeta\omega_n, \quad 40 + k_p = \omega_n^2.$$

Obviously, if k_p is made large, the natural frequency is increased. However, the damping ratio becomes smaller. This leads to a faster response, but with a bigger overshoot and much more severe oscillations.

10.4.2 Proportional–Integral Control

As seen in Example 10.9, increasing the proportional gain k_p can reduce steady-state error, but cannot achieve zero steady-state error. Adding an integral term to the controller in Equation 10.33 results in a proportional–integral (PI) controller

$$\frac{U(s)}{E(s)} = k_p + \frac{k_I}{s}. \tag{10.34}$$

If PI control is used in the previous example, the steady-state error becomes

$$e_{ss} = \lim_{s \to 0} s \frac{1}{s} \frac{1}{1 + (k_p + (k_I/s))G(s)} = \lim_{s \to 0} \frac{s}{s + (k_p s + k_I)G(s)}. \tag{10.35}$$

Substituting the transfer function $G(s)$ given in Example 10.9, we can obtain the result by evaluating the limit in Equation 10.35. The system with PI control has a zero steady-state error. Thus, the primary reason to introduce the integral control is to reduce the steady-state error.

Example 10.10: PI Control

Use the Simulink block diagram built in Example 10.9 to find the unit-step responses for $k_I = 50$, 500, and 1550. Set $k_p = 250$ and $k_D = 0$. Discuss the effects of the integral term on the step response of the system.

Introduction to Feedback Control Systems

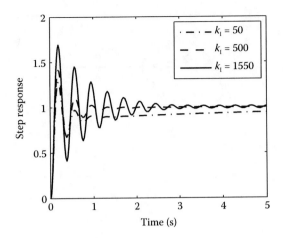

FIGURE 10.32 Responses of PI control.

SOLUTION

Figure 10.32 shows the unit-step responses of PI feedback control for $k_I = 50$, 500, and 1550. The steady-state error of the system will achieve zero for each case if the simulation time is long enough. However, a large integral control gain results in lightly damped oscillations.

10.4.3 PID CONTROL

The final term in a PID controller stands for derivative control. The complete three-term controller is given by Equation 10.32. The main reason to introduce the derivative control is to increase the damping and thus to improve the stability of the system.

Example 10.11: PID Control

Use the Simulink block diagram built in Example 10.9 to find the unit-step responses for $k_D = 1$, 10, and 50. Set $k_p = 250$ and $k_I = 500$. Discuss the effects of the derivative term on the step response of the system.

SOLUTION

Figure 10.33 shows the unit-step responses of PID feedback control for $k_D = 1$, 10, and 50. As k_D increases, the overshoot of the unit-step response becomes smaller. This implies that the damping of the feedback control system becomes larger. However, a large derivative control gain k_D leads to a slower response.

In summary, a larger proportional gain k_p results in a faster response and a smaller steady-state error. However, an excessively large proportional gain k_p leads to lightly damped oscillations and even instability. A larger integral gain k_I eliminates steady-state errors more quickly, but reduces damping and leads to a larger overshoot. A larger derivative control decreases the overshoot, but slows down the speed of the response.

For a PID feedback control system, if the closed-loop system is second-order with any two of these parameters in k_p, k_I, and k_D free, then these two parameters can be determined based on the stability and performance requirements.

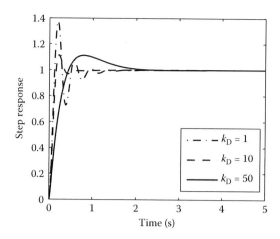

FIGURE 10.33 Responses of PID control.

Example 10.12: Proportional-Derivative (PD) Control of a DC-Motor-Driven Cart

Consider the feedback control system shown in Figure 10.34, where the plant is the DC-motor-driven cart given in Example 10.2. The input to the plant is the voltage applied to the DC motor, and the output is the position of the cart. Design a PD controller such that the maximum overshoot in the response to a unit-step reference input is less than 20%, and the rise time is less than 0.1 s.

Solution

The closed-loop transfer function $Y(s)/R(s)$ is

$$\frac{Y(s)}{R(s)} = \frac{C(s)G(s)}{1+C(s)G(s)} = \frac{(k_p + k_D s)(3.778/(s^2 + 16.883s))}{1 + (k_p + k_D s)(3.778/(s^2 + 16.883s))} = \frac{3.778 k_D s + 3.778 k_p}{s^2 + (16.883 + 3.778 k_D)s + 3.778 k_p},$$

which is a second-order system. The closed-loop characteristic equation is

$$s^2 + (16.883 + 3.778 k_D)s + 3.778 k_p = 0,$$

where the coefficient of s and the constant term are related to the natural frequency and the damping ratio of the closed-loop system, that is,

$$16.883 + 3.778 k_D = 2\zeta\omega_n$$
$$3.778 k_p = \omega_n^2.$$

Note that there are two requirements for the transient response of the closed-loop system, that is, $M_P < 20\%$ and $t_r < 0.1$ s. To satisfy these two requirements, a set of reasonable values of ω_n

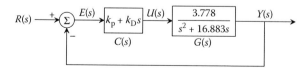

FIGURE 10.34 Block diagram of a feedback control system.

Introduction to Feedback Control Systems

and ζ can be approximated using the relationships given in Section 10.2.

$$t_r \approx \frac{1.8}{\omega_n} < 0.1\,\text{s},$$
$$M_p = e^{-\pi\zeta/\sqrt{1-\zeta^2}} < 20\%.$$

Refer to Example 10.4; these two equations indicate that

$$\omega_n > 18\,\text{rad/s},$$
$$\zeta > 0.46.$$

This region is shown in Figure 10.10. If the closed-loop poles are located anywhere to the left of the gray boundary, then the response to a unit-step reference input will meet the desired requirements. Choosing $\omega_n = 18.5$ rad/s and $\zeta = 0.5$ yields $k_p = 90.59$ and $k_D = 0.43$. Figure 10.35 shows the response of the closed-loop system to a unit-step reference input.

10.4.4 Ziegler–Nichols Tuning of PID Controllers

As mentioned earlier, all the methods considered in this chapter are model-based control, which requires a dynamic model of the plant to be available. To avoid this requirement, in the early 1940s, Ziegler and Nichols conducted numerous experiments and proposed two useful tuning methods for determining the PID control gains. The form of the PID controller used by Ziegler and Nichols is given by

$$C(s) = k_p \left(1 + \frac{1}{T_I s} + T_D s\right), \tag{10.36}$$

where T_I is the integral time and T_D is the derivative time. The gain parameters T_I and T_D are related to the parameters k_p, k_I, and k_D through $k_I = k_p/T_I$ and $k_D = k_p T_D$.

For the first method, a step response of the plant is measured. As shown in Figure 10.36, the S-shaped curve is characterized by two constants, lag time L and reaction rate R, which are determined by drawing a line tangent to the curve and finding the intersections of the tangent line with the

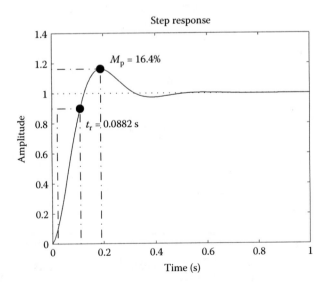

FIGURE 10.35 Unit-step response of the cart system with PD control $C(s) = 90.59 + 0.43s$.

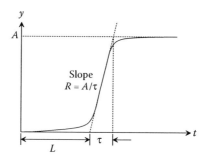

FIGURE 10.36 Reaction curve method.

TABLE 10.1
Ziegler–Nichols Tuning Based on Reaction Curve

Type of Controller	Optimum Gains
P	$k_p = 1/(RL)$
PI	$k_p = 0.9/(RL)$, $T_I = L/0.3$
PID	$k_p = 1.2/(RL)$, $T_I = 2L$, $T_D = 0.5L$

---

time axis and the steady-state level line. The values of k_p, T_I, and T_D can be set using the parameters L and R according to the rules in Table 10.1.

For the second method, known as ultimate sensitivity method, the frequency of the oscillations of the plant at the limit of stability is measured. To use this method, a proportional feedback control is applied to the plant and the proportional gain is increased until the closed-loop system becomes marginally stable. The corresponding proportional gain is defined as K_u, also called the ultimate gain. Figure 10.37 shows the response of a marginally stable system. It is a pure harmonic response, where the period of oscillation P_u can be measured, also known as the ultimate period. The values of k_p, T_I, and T_D can be set using the parameters K_u and P_u according to the rules in Table 10.2.

PROBLEM SET 10.4

1. Consider the feedback control system shown in Figure 10.38.
   a. Design a PD controller such that the closed-loop poles are at $p_{1,2} = -3 \pm 3\sqrt{2}j$.
   b. Use MATLAB to plot the unit-step response of the closed-loop system. Find the rise time, overshoot, peak time, and 1% settling time.

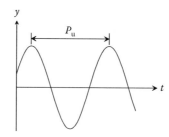

FIGURE 10.37 Response of a marginally stable system.

TABLE 10.2
Ziegler–Nichols Tuning Based on Ultimate Gain and Ultimate Period

Type of Controller	Optimum Gains
P	$k_p = 0.5 K_u$
PI	$k_p = 0.45 K_u$, $T_I = P_u/1.2$
PID	$k_p = 0.6 K_u$, $T_I = P_u/2$, $T_D = P_u/8$

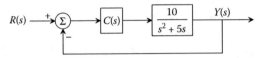

FIGURE 10.38 Problem 1.

2. Consider the feedback control system shown in Figure 10.39.
   a. Find the values for k_p and T_D such that the maximum overshoot in the response to a unit-step reference input is less than 10% and the 1% settling time is less than 0.5 s.
   b. Compute the steady-state error of the closed-loop system to a unit-step reference input.
   c. ▲ Verify the results in Parts (a) and (b) using MATLAB by plotting the unit-step response of the closed-loop system. If the maximum overshoot and the settling time exceed the requirements, make a fine tuning and reduce them to be approximately the specified values or less.
3. Consider the feedback control system shown in Figure 10.40a.
   a. If the desired closed-loop poles are located within the shaded regions shown in Figure 10.40b, determine the corresponding ranges of ω_n and ζ of the closed-loop system.
   b. Design a PI controller such that the closed-loop poles are at $p_{1,2} = -8 \pm 6j$.
   c. Compute the steady-state errors of the plant and the closed-loop system to a unit-step reference input.
   d. ▲ Verify the results in Part (c) using MATLAB by plotting the unit-step responses of the plant and the closed-loop system.
4. Consider the feedback control system shown in Figure 10.41.
   a. Find the values for k_p and T_I such that the maximum overshoot in the response to a unit-step reference input is less than 15% and the peak time is less than 0.25 s.
   b. ▲ Verify the results in Part (a) using MATLAB by plotting the unit-step response of the closed-loop system. If the maximum overshoot and the peak time exceed the requirements, make a fine tuning and reduce them to be approximately the specified values or less.

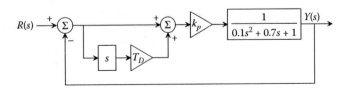

FIGURE 10.39 Problem 2.

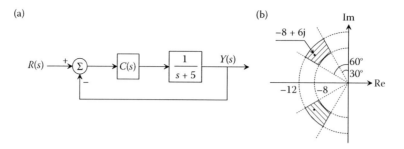

FIGURE 10.40 Problem 3. (a) Block diagram and (b) locations of closed-loop poles.

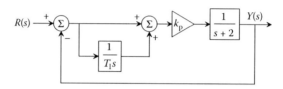

FIGURE 10.41 Problem 4.

5. The unit-step response of a plant is shown in Figure 10.42. The lag time L and the reaction rate R can be determined from the figure. Find the proportional, PI, and PID controller parameters using the Ziegler–Nichols reaction curve method.
6. Consider a proportional feedback control system. As shown in Figure 10.43, the closed-loop system becomes marginally stable when the proportional gain is 0.3. Find the proportional, PI, and PID controller parameters using the Ziegler–Nichols ultimate sensitivity method.

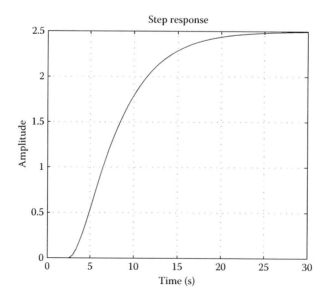

FIGURE 10.42 Problem 5.

Introduction to Feedback Control Systems

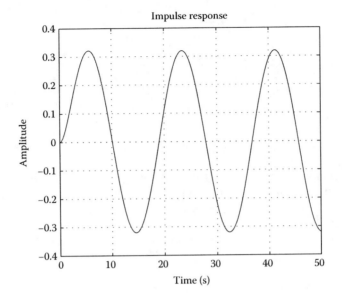

FIGURE 10.43 Problem 6.

10.5 ROOT LOCUS

Starting from this section, we discuss how to design a feedback controller to meet stability and performance requirements. Three different control design methods, including root locus, Bode plot, and state-space techniques, are introduced in Sections 10.5 through 10.7, respectively. To simplify the discussion, all controllers in those three sections are limited to proportional feedback control.

As presented in Section 10.2, the time-domain specifications, such as rise time t_r, overshoot M_p, peak time t_p, and settling time t_s, are related to the natural frequency ω_n and the damping ratio ζ, both of which can be used to express the pole locations of a second-order system in the s-plane. Thus, the dynamic response of a system can be influenced by changing the system's pole locations. Root locus developed by W. R. Evans in the late 1940s is a graphical design technique, which shows how changes in one of the system parameters will modify roots of the closed-loop characteristic equation, or the closed-loop poles, and thus change the dynamic response of the system. In this section, we first look at how to sketch the root locus of a feedback control system. Then, we discuss how to analyze stability and performance of the closed-loop system based on the root locus. Finally, we learn how to design a proportional feedback controller using the root locus technique.

10.5.1 ROOT LOCUS OF A BASIC FEEDBACK SYSTEM

Consider a basic feedback control system as shown in Figure 10.44, where $G(s)$ is the transfer function of the plant. The controller is assumed to be a proportional gain, $C(s) = K$. The closed-loop transfer function is

$$\frac{Y(s)}{R(s)} = \frac{C(s)G(s)}{1 + C(s)G(s)} = \frac{KG(s)}{1 + KG(s)} \tag{10.37}$$

and the characteristic equation is

$$1 + KG(s) = 0. \tag{10.38}$$

Note that the roots of the closed-loop characteristic equation are the poles of the closed-loop system. As observed from Equation 10.38, the closed-loop poles are affected by the value of K. When K

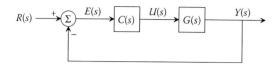

FIGURE 10.44 Block diagram of a basic feedback control system.

varies from 0 to ∞, the closed-loop poles will move around the s-plane and create a trajectory, or a locus of poles.

Intuitively, a root locus can be constructed by changing the value of K from 0 to ∞, solving the characteristic equation for the roots, and plotting the poles in the s-plane. However, it was difficult to obtain the poles for high-order systems before the availability of mathematical and scientific computing software. In the late 1940s, Evans developed rules to plot a locus without actually solving for the roots of the characteristic equation. The following example shows how to construct a root locus manually step by step.

Example 10.13: Root Locus Sketching

For the system in Figure 10.45, find the locus of closed-loop poles with respect to K.

Solution

Step 1: Express the closed-loop characteristic equation in the form of

$$1 + K \frac{b(s)}{a(s)} = 0. \tag{10.39}$$

From Figure 10.45, the closed-loop transfer function is

$$\frac{Y(s)}{R(s)} = \frac{C(s)G(s)}{1 + C(s)G(s)H(s)}.$$

Thus, the closed-loop characteristic equation is

$$1 + C(s)G(s)H(s) = 1 + K \frac{s+1}{(s+3)(s^2+2s+2)} \frac{1}{s} = 0,$$

which is in the form of $1 + Kb(s)/a(s) = 0$ with

$$a(s) = (s+3)(s^2+2s+2)s,$$
$$b(s) = s+1.$$

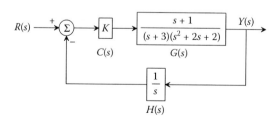

FIGURE 10.45 Block diagram for the feedback control in Example 10.13.

Introduction to Feedback Control Systems

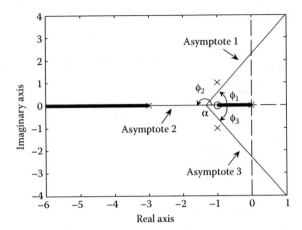

FIGURE 10.46 Real axis portions of the root locus and asymptotes for Example 10.13.

Denote $L(s) = b(s)/a(s)$, where $L(s)$ is the loop gain. Note that the roots of $a(s)$ are the poles of $L(s)$ and the roots of $b(s)$ are the zeros of $L(s)$. The number of poles determines the number of branches of the root locus.

RULE 1: Assume that $L(s)$ has n poles and m zeros. The n branches of the locus start at the poles of $L(s)$ and the m of these branches end at the zeros of $L(s)$.

Step 2: Draw the axes of the s-plane to a suitable scale and mark a cross symbol "×" for each pole of $L(s)$ and a circle symbol "o" for each zero of $L(s)$.

Solving $a(s) = 0$ for the poles gives

$$p_1 = 0, \quad p_2 = -3, \quad p_{3,4} = -1 \pm j.$$

Solving $b(s) = 0$ for the zero gives

$$z_1 = -1.$$

The locations of the four poles and one zero are shown in Figure 10.46.

Step 3: Find the real axis portions of the locus.

RULE 2: The portions of the root locus on the real axis are to the left of an odd number of poles and zeros.

This implies that a point on the real axis is part of the root locus if there is an odd number of poles and zeros to its right. As shown in Figure 10.46, there are two poles, 0 and -3, and one zero, -1, located on the real axis, which is divided into four segments, $(-\infty, -3)$, $(-3, -1)$, $(-1, 0)$, and $(0, +\infty)$. To demonstrate RULE 2, consider a point within $(-1, 0)$. There is one pole, 0, to its right, and since one is an odd number, this portion of the real axis is part of the root locus. Similarly, the segment $(-\infty, -3)$ is also part of the root locus because there are four poles and one zero, totally five, to its right. The thick solid lines in Figure 10.46 represent the portions of the root locus on the real axis.

Step 4: Draw the asymptotes for large values of K.

RULE 3: For large s and K, $n - m$ branches of the locus are asymptotic to lines radiating out from the point $s = \alpha$ on the real axis at an angle ϕ_l, where

$$\alpha = \frac{\sum \text{Re}(p_i) - \sum \text{Re}(z_i)}{n - m}, \tag{10.40}$$

$$\phi_l = \frac{180° + 360°(l-1)}{n - m}, \quad l = 1, 2, \ldots, n - m. \tag{10.41}$$

In our case, $n = 4$ and $m = 1$. Thus, there are three asymptotes, which radiate from a centroid α on the real axis

$$\alpha = \frac{(0 - 1 - 1 - 3) - (-1)}{3} = -\frac{4}{3},$$

with angles of

$$\phi_1 = \frac{180° + 360°(1 - 1)}{3} = 60°,$$

$$\phi_2 = \frac{180° + 360°(2 - 1)}{3} = 180°,$$

$$\phi_3 = \frac{180° + 360°(3 - 1)}{3} = 300° \text{ or } 60°.$$

The centroid and asymptotes are shown in Figure 10.46.

Step 5: Compute the departure and arrival angles.

RULE 4: The angle of departure of a branch of the locus from a pole is given by

$$\varphi_{dep} = \sum \psi_i - \sum \varphi_i - 180°, \quad (10.42)$$

where $\sum \varphi_i$ is the sum of the angles from this pole to the remaining poles and $\sum \psi_i$ is the sum of the angles from this pole to all the zeros. The angle of arrival of a branch at a zero is given by

$$\psi_{arr} = \sum \varphi_i - \sum \psi_i + 180°, \quad (10.43)$$

where $\sum \psi_i$ is the sum of the angles from this zero to the remaining zeros and $\sum \varphi_i$ is the sum of the angles from this zero to all the poles.

Note that Equations 10.42 and 10.43 are valid for nonrepeated poles and zeros. The formula used for computing the departure or arrival angles from a repeated pole or to a repeated zero can be found in control books.

As seen in Figure 10.46, two branches of the locus on the real axis have been completely drawn. One of them starts from the pole at 0 and ends at the zero at -1, and the other starts from the pole at -3 and ends at $-\infty$ by approaching the second asymptote. There is no need to compute the departure or arrival angles for the poles and the zero on the real axis. Note that there is a pair of complex conjugate poles at $-1 \pm j$, from each of which one branch of the locus starts. Selecting the pole at $-1 + j$ and applying Equation 10.42 gives

$$\varphi_{dep} = \sum \psi_i - \sum \varphi_i - 180° = \psi_1 - (\varphi_1 + \varphi_2 + \varphi_3) - 180°.$$

As sketched in Figure 10.47, $\psi_1 = 90°$, which is the angle of the line connecting the complex pole at $-1 + j$ and the zero with respect to the positive real axis. Similarly, we can determine the angles φ_i, which are $\varphi_1 = 135°$, $\varphi_2 = 90°$, and $\varphi_3 = \tan^{-1}(\frac{1}{2}) = 26.57°$. Thus,

$$\varphi_{dep} = 90° - (135° + 90° + 26.57°) - 180° = -341.57° = 18.43°.$$

The departure angle from the pole at $-1 - j$ is $-18.43°$ because the root locus is symmetric about the real axis.

Step 6: Determine the points where the root locus crosses the imaginary axis.

RULE 5: The root locus crosses the imaginary axis at points where the characteristic equation satisfies

$$1 + K\frac{b(j\omega)}{a(j\omega)} = 0 \quad (10.44)$$

or

$$a(j\omega) + Kb(j\omega) = 0. \quad (10.45)$$

Note that if a point on the imaginary axis is part of the root locus, then it must be a root of the characteristic equation given by Equation 10.39. The point on the imaginary axis can be expressed

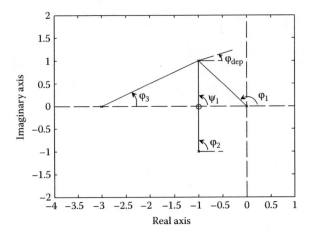

FIGURE 10.47 Departure angles for Example 10.13.

as $s = j\omega$ and substituting it into Equation 10.39 leads to Equation 10.44 or 10.45. Solving either of them yields the crossing $j\omega$-axis points and the corresponding values of K.

In this example, we have

$$s(s+3)(s^2+2s+2) + K(s+1) = 0.$$

Substituting $s = j\omega$ and rearranging the equation gives

$$(j\omega)^4 + 5(j\omega)^3 + 8(j\omega)^2 + 6j\omega + Kj\omega + K = 0.$$

Separating the real and imaginary terms results in

$$(\omega^4 - 8\omega^2 + K) + j(-5\omega^3 + 6\omega + K\omega) = 0,$$

which is equivalent to two equations

$$\begin{cases} \omega^4 - 8\omega^2 + K = 0, \\ -5\omega^3 + 6\omega + K\omega = 0. \end{cases}$$

Solving for ω, we obtain

$$\omega_1 = 0, \quad \omega_{2,3} = \pm 2.09, \quad \omega_{4,5} = \pm 1.17j,$$

where the last two roots are not valid because the physical meaning of ω is frequency and it is real. The corresponding values of K are

$$K_1 = 0, \quad K_{2,3} = 15.86.$$

Step 7: Complete the sketch.

As shown in Figure 10.48, two complex branches of the locus depart from the complex poles, cross the imaginary axis, enter the right-half s-plane, and end at infinity by approaching the asymptotes. Combining the complex branches with the real axis portions of the locus, we have the sketch completed.

Rules 1 through 5 can be used to roughly sketch a root locus. One more rule is available for computing the so-called break-in or break-away points, but this will not be covered in this text. With the availability of MATLAB, these rules are not necessary to plot a root locus. However, learning these rules can help understand classical control design techniques and evaluate the correctness of a computer-generated root locus.

The MATLAB command used to sketch a root locus is `rlocus`. The following is the MATLAB session to generate the root locus shown in Figure 10.48.

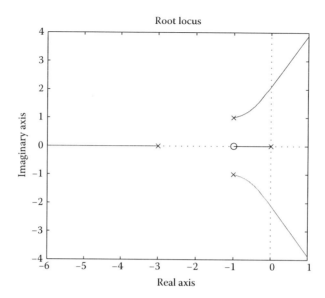

FIGURE 10.48 Root locus for Example 10.13.

```
>> num = [1 1];
>> den = conv([1 3 0],[1 2 2]);
>> sysL = tf(num,den);
>> rlocus(sysL);
```

where the loop gain $L(s)$ is defined in transfer function form and called by the command `rlocus`.

10.5.2 Analysis Using Root Locus

Using the root locus technique, it is very easy to determine the stability of a closed-loop system when the proportional gain K varies from 0 to ∞. For a particular value of K, if any of the poles are in the right-half s-plane, then the corresponding closed-loop system is unstable. If all of the poles are in the left-half s-plane, then the closed-loop system is stable. If any of the poles are on the imaginary axis and they are nonrepeated, then the closed-loop system is marginally stable.

From the root locus, we can also obtain information about the performance of a closed-loop system using the concept of a dominant pole. For a system with multiple poles, the pole closest to the origin is called the dominant pole. If the dominant poles are a pair of complex conjugate poles, the distance between either of them and the origin is associated with the natural frequency if the system is approximated as a second-order system. If the dominant pole is a real pole, the distance between the pole and the origin is associated with the time constant if the system is approximated as a first-order system. Both the natural frequency and the time constant determine the speed of transient response. The dominant pole has the slowest speed of response, and it dominates the effect of all other poles with higher frequencies or lower time constants.

Example 10.14: Analysis Using Root Locus

Refer to the root locus obtained in Example 10.13. Comment on the stability and performance of the closed-loop system when K varies from 0 to ∞.

Introduction to Feedback Control Systems

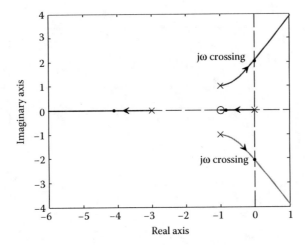

FIGURE 10.49 Root locus for Example 10.13 with poles highlighted when $K = 15.86$.

Solution

When $K = 0$, which corresponds to having no control, the root locus starts from the poles of the loop gain $L(s)$. As shown in Figure 10.49, there is one pole at 0. This implies that the system is marginally stable. As K increases, the four closed-loop poles move along four different branches of the root locus. For $0 < K < 15.86$, all of the poles are in the left-half s-plane, and thus the closed-loop system is stable. When $K = 15.86$, two complex poles, $\pm 2.09j$, appear on the imaginary axis, and thus the closed-loop system becomes marginally stable. For $K > 15.86$, the two complex branches of the root locus cross the imaginary axis and enter the right-half s-plane. The closed-loop system becomes unstable.

When K is small, the real pole close to the origin dominates the effect of all other poles, and thus the closed-loop system exhibits slow first-order response. For example, the closed-loop poles are -0.17, $-0.85 \pm 1.06j$, and -3.12 when $K = 1$. Figure 10.50 is the corresponding

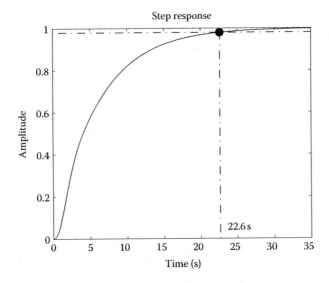

FIGURE 10.50 Unit-step response of the closed-loop system when $K = 1$.

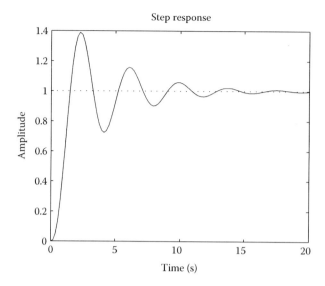

FIGURE 10.51 Unit-step response of the closed-loop system when $K = 8$.

closed-loop unit-step response, which is slow (2% settling time $t_s = 22.6$ s) and has no oscillation. As K increases, the closed-loop system will exhibit undamped oscillations due to the dominant complex poles. For example, the closed-loop poles are -0.77, -0.26 ± 1.65j, and -3.70 when $K = 8$. Figure 10.51 is the corresponding closed-loop unit-step response, where the oscillation is a feature of the complex poles.

Note that although the root locus is constructed based on the loop gain $L(s)$, it also gives information on the stability and performance of the closed-loop system varying with respect to the proportional gain K. This is what makes the root locus technique attractive.

10.5.3 CONTROL DESIGN USING ROOT LOCUS

As shown in Subsection 10.5.1, the root locus is a plot of all possible roots to the closed-loop characteristic equation $1 + KL(s) = 0$ for real positive values of K, which is generally the gain of a proportional controller. It is very easy to select K from a particular root locus so that the closed-loop system meets the performance specifications.

Example 10.15: Proportional Control Design Using Root Locus

Design a proportional controller for the cart system in Example 10.12 using the root locus technique.

SOLUTION

For proportional feedback control of the DC-motor-driven cart, the loop gain $L(s)$ is equal to the transfer function of the plant $G(s)$,

$$L(s) = G(s) = \frac{3.778}{s^2 + 16.883s}.$$

The root locus is plotted in Figure 10.52 using the MATLAB command `rlocus`. Note that the closed-loop system is a second-order system. Two closed-loop poles that are real appear for small values of K, specifically $K \leq 18.9$. When $K > 18.9$, the closed-loop system has a pair of complex

Introduction to Feedback Control Systems

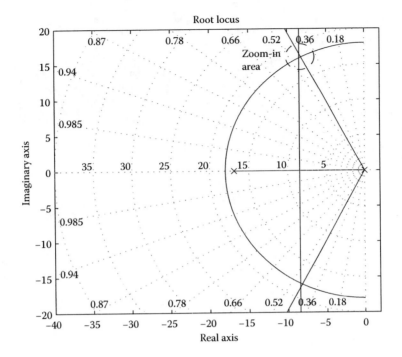

FIGURE 10.52 Root locus for Example 10.15 with grid lines.

conjugate poles, which move along the complex portion of the root locus as K varies between 18.9 and ∞.

To select the value of the proportional control gain that will meet the performance specifications, we can turn on the grid lines to the root locus using the command

```
>> grid on
```

As discussed in Subsection 10.2.2, the semicircles in Figure 10.52 indicate lines of constant natural frequencies ω_n and the diagonal lines indicate constant damping ratios ζ. In this example, we need an overshoot that is less than 20% (which implies $\zeta > 0.46$) and a rise time that is less than 0.1 s (which implies $\omega_n > 18$ rad/s). In Figure 10.52, the solid diagonal lines indicate pole locations with a damping ratio of about 0.46. In between these lines, the poles have $\zeta > 0.46$ and outside of the lines $\zeta < 0.46$. The solid semicircle is the locus of all poles with $\omega_n = 18$ rad/s, while those inside the semicircle have $\omega_n < 18$ rad/s, and those outside correspond to $\omega_n > 18$ rad/s. Thus, only the part of the root locus in between the two diagonal lines and outside of the semicircle is acceptable.

Figure 10.53 is the zoom-in of the desired region, where the vertical line is a part of the root locus. Left-clicking on the root locus, you will see the values of the pole and the gain that is required to place one of the closed-loop poles at that particular location. Holding down the left mouse button and moving the mouse along the root locus, you can see the values of the pole and the gain varying correspondingly.

Let us select $K = 88$. Figure 10.54 is the corresponding unit-step response of the closed-loop system with proportional feedback control. The closed-loop system meets the given specifications.

Note that the closed-loop poles cannot be placed arbitrarily in the s-plane with only a static proportional controller since the shape of the root locus is fixed for a given plant. A more useful design can be obtained by adding a pole or zero to the controller and making it a dynamic controller.

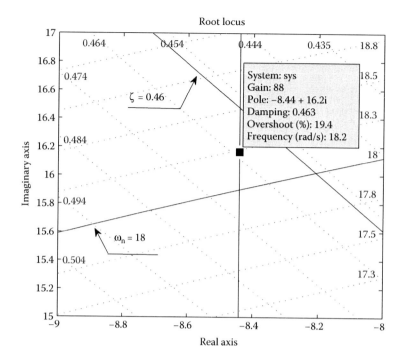

FIGURE 10.53 Zoom-in of the designed region in Example 10.15.

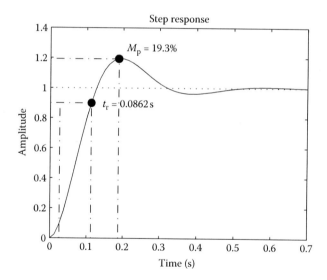

FIGURE 10.54 Unit-step response of the cart system with proportional feedback control $K = 88$.

This results in so-called lead or lag compensators, $C(s) = (s + z)/(s + p)$. The reader can refer to the control books for more details.

PROBLEM SET 10.5

1. Roughly sketch the root locus with respect to K for the equation of $1 + KL(s) = 0$ and the following choices for $L(s)$. Make sure to give the asymptotes, arrival and/or

departure angles, and points crossing the imaginary axis. Verify your results using MATLAB.

a. $L(s) = \dfrac{1}{(s+1)(s+2)}$

b. $L(s) = \dfrac{1}{(s+1)(s+2)(s+10)}$

c. $L(s) = \dfrac{s+5}{(s+1)(s+2)(s+10)}$

d. $L(s) = \dfrac{s(s+5)}{(s+1)(s+2)(s+10)}$

2. Repeat Problem 1 for the following choices for $L(s)$.

a. $L(s) = \dfrac{1}{s^2 + 2s + 2}$

b. $L(s) = \dfrac{1}{(s+4)(s^2 + 2s + 2)}$

c. $L(s) = \dfrac{s^2 + 2s + 5}{(s+4)(s^2 + 2s + 2)}$

d. $L(s) = \dfrac{s(s^2 + 2s + 5)}{(s+4)(s^2 + 2s + 2)}$

3. A control system is represented using the block diagram shown in Figure 10.55. Sketch the root locus with respect to the proportional control gain K. Determine all the values of K for which the closed-loop system is stable.

4. A control system is represented using the block diagram shown in Figure 10.56, where the parameter a is subjected to variations. Sketch the root locus with respect to the parameter a. Determine all the values of a for which the closed-loop system is stable.

5. Figure 10.57 shows the root locus of a unity negative feedback control system, where K is the proportional control gain.

   a. Determine the transfer function of the plant. Use MATLAB to plot the root locus based on your choice of the plant, compare it with the root locus shown in Figure 10.57, and check the accuracy of your plant.

   b. Give comments on the stability of the closed-loop system when K varies from 0 to ∞.

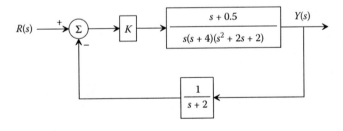

FIGURE 10.55 Problem 3.

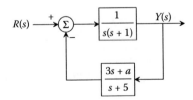

FIGURE 10.56 Problem 4.

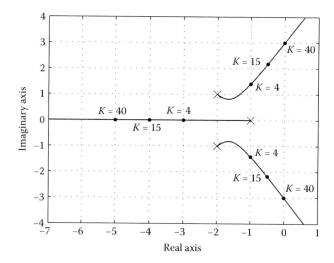

FIGURE 10.57 Problem 5.

    c. Give comments on the transient performance of the closed-loop system when $K = 4$, 15, and 40. Use MATLAB to plot the corresponding unit-step responses and verify your comments.

6. Figure 10.58 shows the root locus of a unity negative feedback control system, where K is the proportional control gain.
    a. Determine the transfer function of the plant. Use MATLAB to plot the root locus based on your choice of the plant, compare it with the root locus shown in Figure 10.58, and check the accuracy of your plant.
    b. Find the range of values of K for which the system has damped oscillatory response. What is the greatest value of K that can be used before pure harmonic oscillations occur? Also, what is the frequency of pure harmonic oscillations? Use MATLAB to plot the corresponding unit-step response and verify the accuracy of your computed frequency.

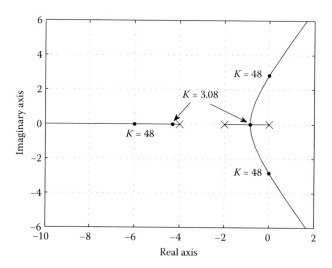

FIGURE 10.58 Problem 6.

Introduction to Feedback Control Systems

FIGURE 10.59 Problem 7.

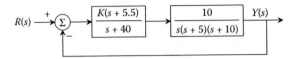

FIGURE 10.60 Problem 8.

7. Consider the feedback system shown in Figure 10.59.
   a. Find the locus of the closed-loop poles with respect to K.
   b. Find a value of K such that the maximum overshoot in the response to a unit-step reference input is less than 10%.
   c. Plot the unit-step response of the closed-loop system to verify the result for Part (b).
8. Consider the feedback system shown in Figure 10.60.
   a. Find the locus of the closed-loop poles with respect to K.
   b. Find a value of K such that the maximum overshoot in the response to a unit-step reference input is less than 20% and the 2% settling time is less than 1.1 s.
   c. Plot the unit-step response of the closed-loop system to verify the result in Part (b).

10.6 BODE PLOT

The Bode plot technique is widely used to display a frequency response function. It also gives useful information for analyzing and designing control systems. Stability criteria can be interpreted using the Bode plot and numerous control design techniques are based on the Bode plot. In Section 8.3, we introduced the concept of Bode plot and presented the Bode plot of the frequency response function for two fundamental systems, first-order and second-order. In this section, we first discuss how to use the Bode plot to display the frequency response function for a general dynamic system. Then, we learn how to determine stability based on the Bode plot. Finally, we show how to design a proportional feedback controller using the Bode plot technique.

10.6.1 BODE PLOT OF A BASIC FEEDBACK SYSTEM

Consider the basic feedback control system shown in Figure 10.61. The open-loop transfer function is $KG(s)$, which can be written in the form

$$KG(s) = K\frac{(s-z_1)(s-z_2)\cdots(s-z_m)}{(s-p_1)(s-p_2)\cdots(s-p_n)}, \qquad (10.46)$$

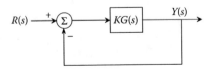

FIGURE 10.61 Simplified feedback control system.

which explicitly shows the poles and zeros. Replacing s with $j\omega$ yields the frequency response function

$$KG(j\omega) = K\frac{(j\omega - z_1)(j\omega - z_2)\cdots(j\omega - z_m)}{(j\omega - p_1)(j\omega - p_2)\cdots(j\omega - p_n)}, \qquad (10.47)$$

The frequency response function can be displayed using two curves: the Bode magnitude plot and the Bode phase plot. By definition, the magnitude of $KG(j\omega)$ in dB is

$$\begin{aligned}|KG(j\omega)|_{dB} &= 20\log_{10}|KG(j\omega)| \\ &= 20\log_{10}|K| + 20\log_{10}|j\omega - z_1| + 20\log_{10}|j\omega - z_2| + \cdots \\ &\quad - 20\log_{10}|j\omega - p_1| - 20\log_{10}|j\omega - p_2| - \cdots \end{aligned} \qquad (10.48)$$

and the phase of $KG(j\omega)$ is

$$\angle[KG(j\omega)] = \angle K + \angle(j\omega - z_1) + \angle(j\omega - z_2) + \cdots - \angle(j\omega - p_1) - \angle(j\omega - p_2) - \cdots. \qquad (10.49)$$

As shown in Equations 10.48 and 10.49, for a frequency response function, its magnitude in dB or its phase is the sum of the magnitudes or phases of simple terms, which are similar to each other. If we know how to draw the Bode plot for each individual term, then the composite curve can be obtained by combining all the terms together.

Depending on the locations of poles or zeros, there are four classes of basic terms:

1. Constant terms K (no pole or zero)
2. Integral or derivative terms $(j\omega)^{\pm n}$ (with pole(s) or zero(s) at the origin)
3. First-order terms $(j\omega\tau + 1)^{\pm 1}$ (with a real pole or zero at $-1/\tau$)
4. Second-order terms $[(j\omega/\omega_n)^2 + 2\zeta j\omega/\omega_n + 1]^{\pm 1}$ (with a pair of complex conjugate poles or zeros at $-\zeta\omega_n \pm j\omega_n\sqrt{1-\zeta^2}$)

Figure 10.62 is an example of a plot with a constant term $K = 10$. Since K is independent of the frequency, both magnitude and phase in the entire frequency region are horizontal lines. The magnitude in dB is

$$|K|_{dB} = 20\log_{10}|K| \qquad (10.50)$$

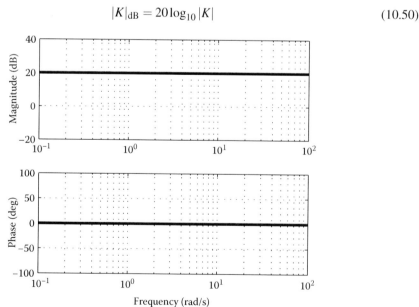

FIGURE 10.62 Bode plot for $K = 10$.

Introduction to Feedback Control Systems

and the phase is
$$\angle K = 0°. \tag{10.51}$$

For integral or derivative terms, the magnitude of $(j\omega)^{\pm n}$ in dB is

$$|(j\omega)^{\pm n}|_{dB} = 20\log_{10}|(j\omega)^{\pm n}| = \pm n \times 20\log_{10}|j\omega| = \pm n \times 20\log_{10}\omega. \tag{10.52}$$

Note that the magnitude plot is drawn using the logarithmic scale for the frequency, that is, $\log_{10}\omega$. Thus, the magnitude plot of an integral or derivative term is a straight line with a slope $\pm n \times 20$ dB/decade, which means that the magnitude will change by $\pm n \times 20$ dB as the frequency increases by a factor of 10. Geometrically, a straight line can be determined using its slope and one point that it crosses. By observing Equation 10.52, we can find that this line always crosses (1 rad/s, 0 dB) no matter what the value of n. Figure 10.63 shows Bode plots for an integral term $1/j\omega$ and a derivative term $j\omega$. Their magnitude plots intersect at (1 rad/s, 0 dB) and the slope is -20 dB/decade for $1/j\omega$ and $+20$ dB/decade for $j\omega$. The phase of $(j\omega)^{\pm n}$ is

$$\angle(j\omega)^{\pm n} = \pm n\angle(j\omega) = \pm n \times 90°, \tag{10.53}$$

which is independent of frequency. As shown in Figure 10.63, the phase is a horizontal line and it is $-90°$ for $1/j\omega$ and $+90°$ for $j\omega$.

The Bode plots of the frequency response function for a first-order or second-order system are shown in Figures 8.20 and 8.22. In this section, we discuss the plotting of asymptotes for more general cases as indicated in the third and fourth classes.

For the first-order terms $(j\omega\tau + 1)^{\pm 1}$, Figure 10.64 shows the magnitude plots with the asymptotes. Let us take $j\omega\tau + 1$ as an example. At low frequencies, $\omega\tau \ll 1$, we have $j\omega\tau + 1 \approx 1$. Thus, the magnitude approaches a horizontal line crossing 0 dB. At high frequencies, $\omega\tau \gg 1$, we have $j\omega\tau + 1 \approx j\omega\tau$, for which the magnitude is

$$20\log_{10}|j\omega\tau| = 20\log_{10}(\omega\tau) = 20\log_{10}\omega + 20\log_{10}\tau \approx 20\log_{10}\omega. \tag{10.54}$$

Note that τ is a finite number for a given frequency response function, and its effect on the magnitude can be ignored at very high frequencies. Thus, the magnitude approaches a straight line with a slope

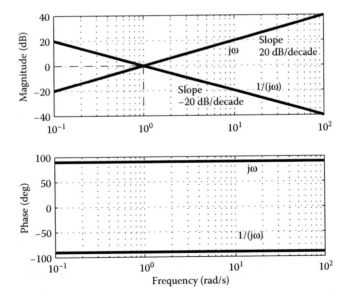

FIGURE 10.63 Bode plots for $1/j\omega$ and $j\omega$.

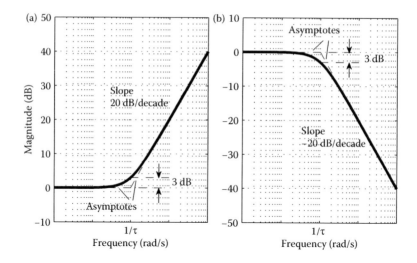

FIGURE 10.64 Magnitude plots for (a) $(j\omega\tau + 1)^{+1}$ and (b) $(j\omega\tau + 1)^{-1}$.

of 20 dB/decade. When $\omega\tau = 1$ or $\omega = 1/\tau$, the magnitude is

$$20\log_{10}|j + 1| = 20\log_{10}\sqrt{2} = 3\,\text{dB}. \tag{10.55}$$

The corresponding frequency $1/\tau$ is the corner frequency, where the slope of the asymptote changes from 0 to 20 dB/decade. The magnitude plot in the entire frequency region can be obtained by drawing a smooth curve following the asymptotes with 3 dB above the line at the corner frequency.

Figure 10.65 shows the asymptotes of the phase plots for the terms $(j\omega\tau + 1)^{\pm 1}$. Again, let us take $j\omega\tau + 1$ as an example. The phase can be approximated as $\angle 1 = 0°$ at low frequencies and $\angle(j\omega\tau) = 90°$ at high frequencies. The phase at the corner frequency $1/\tau$ is $\angle(j + 1) = 45°$.

The Bode plot of the second-order terms $[(j\omega/\omega_n)^2 + 2\zeta j\omega/\omega_n + 1]^{\pm 1}$ can be drawn in a similar way as the first-order terms. The asymptotes for magnitude and phase plots are shown in Figures 10.66 and 10.67, respectively. Different from the first-order terms, the corner frequency is

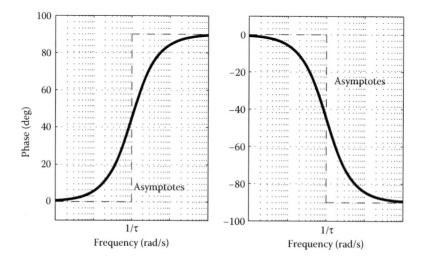

FIGURE 10.65 Phase plots for (a) $(j\omega\tau + 1)^{+1}$ and (b) $(j\omega\tau + 1)^{-1}$.

Introduction to Feedback Control Systems

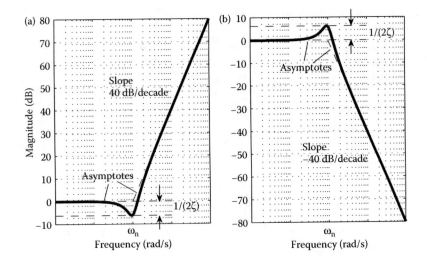

FIGURE 10.66 Magnitude plot for a second-order term in the (a) numerator and (b) denominator.

$\omega = \omega_n$, at which the magnitude changes slope from 0 to $+40$ dB/decade if the term is in the numerator or to -40 dB/decade if the term is in the denominator. Correspondingly, the phase changes from $0°$ to $+180°$ or $-180°$. The magnitude at the corner frequency ω_n greatly depends on the damping ratio ζ. A rough sketch can be made by noting whether the peak is either $1/(2\zeta)$ below or above the asymptotes.

For a general dynamic system, the frequency response function can always be written as the product of several basic terms. As seen in Equations 10.48 and 10.49, the composite magnitude curve and the phase curve are the sum of their respective individual curves. The following example shows how to obtain a quick sketch of the composite curve using the asymptotes of the basic terms.

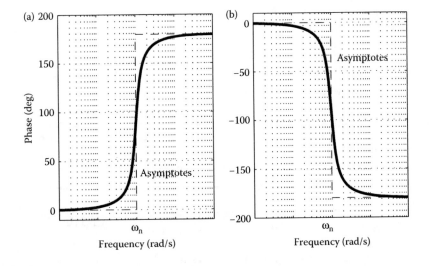

FIGURE 10.67 Phase plot for a second-order term in the (a) numerator and (b) denominator.

Example 10.16: Bode Plot Sketching

Plot the Bode magnitude and phase for the system with the transfer function

$$KG(s) = K\frac{s+2}{s(s^2+8s+400)},$$

where $K = 1$.

Solution

Step 1: Convert the transfer function to the frequency response function

$$KG(s) = \frac{j\omega+2}{j\omega[(j\omega)^2+8j\omega+400]} = \frac{0.005((j\omega/2)+1)}{j\omega[(j\omega/20)^2+2(0.2)(j\omega/20)+1]}.$$

Note that first-order and second-order terms are expressed in their corresponding basic forms, $j\omega\tau + 1$ and $(j\omega/\omega_n)^2 + 2\zeta j\omega/\omega_n + 1$.

Step 2: Identify the basic terms and the corner frequencies associated with first-order and second-order terms.
  The basic terms in this example are listed as follows:

1. One constant term 0.005
2. One integral term $1/j\omega$
3. One first-order term $j\omega/2 + 1$ in the numerator with $1/\tau = 2$ rad/s
4. One second-order term in the denominator with $\omega_n = 20$ rad/s and $\zeta = 0.2$

Step 3: Draw the asymptotes for the magnitude curve.
  We start with indicating the corner frequencies on the frequency axis. Then, the asymptote for the derivative term is drawn through the point (1 rad/s, 0 dB) with a slope of -20 dB/decade. The asymptote is extended until the first corner frequency 2 rad/s is met, which is associated with the first-order term in the numerator. At the first corner frequency, the slope increases by 20 dB/decade and changes to 0 dB/decade. We then continue extending the asymptote until the second corner frequency 20 rad/s is met, which is associated with the second-order term in the denominator. At the second corner frequency, the slope decreases by -40 dB/decade and changes from 0 dB/decade to -40 dB/decade. Finally, we consider the effect of the constant term 0.005 by sliding the asymptotes downward by 46 dB (i.e., $20\log_{10} 0.005 = -46$ dB). This completes the composite magnitude asymptotes.
  The approximate Bode magnitude plot can be obtained by drawing a smooth curve following the asymptotes. The magnitude is 3 dB above the asymptote at the first-order numerator corner frequency. A resonant peak is sketched at the second-order denominator corner frequency. Note that the associated damping ratio ζ is 0.2. Thus, the magnitude above the asymptote is about $1/0.2 = 5$ or 14 dB. Figure 10.68a shows the magnitude plot and the asymptotes.

Step 4: Draw the asymptotes for the phase curve.
  Following the same procedure as in Step 3, we can sketch the asymptotes for the composite phase plot. We start with sketching the asymptote for the derivative term with a horizontal line at $-90°$. The phase changes by $90°$ at the first-order numerator corner frequency 2 rad/s, and $-180°$ at the second-order denominator corner frequency 20 rad/s. The phase of a constant is $0°$ for any frequency, the constant is therefore not considered when sketching the phase plot. Figure 10.68b shows the phase plot and the asymptotes.

In summary, the asymptote for a composite magnitude or phase curve is plotted by starting with the $(j\omega)^{\pm n}$ term, and changing the slope or the phase at each corner frequency depending on whether the corner frequency is associated with a first-order or second-order term in the numerator or denominator. For first-order terms, the changes of slope and phase are $+20$ dB/decade and $+90°$, respectively, when in the numerator, and -20 dB/decade and $-90°$, respectively, when in

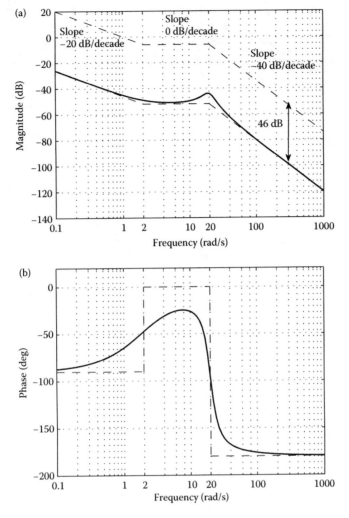

FIGURE 10.68 Bode plot for the system in Example 10.16: (a) magnitude plot with asymptotes and (b) phase plot with asymptotes.

the denominator. For second-order terms, the changes of slope and phase are $+40$ dB/decade and $+180°$, respectively, when in the numerator, and -40 dB/decade and $-180°$, respectively, when in the denominator. The asymptote for the magnitude is completed by shifting it up or down depending on the value of the constant term.

10.6.2 Analysis Using Bode Plot

Similar to the root locus technique, a Bode plot can be used to determine the stability of a closed-loop system without solving for the poles. Consider the proportional feedback control system as shown in Figure 10.61. The Bode plot for $K = 1$ is often drawn. Two margins can be read from the magnitude and phase plots. As shown in Figure 10.69, the gain margin (GM) is the amount of the gain that can be added before the magnitude curve reaches 0 dB at the frequency where the phase plot crosses $-180°$. The phase margin (PM) is the amount of the phase that can be subtracted before the phase curve reaches $-180°$ at the frequency where the magnitude plot crosses 0 dB.

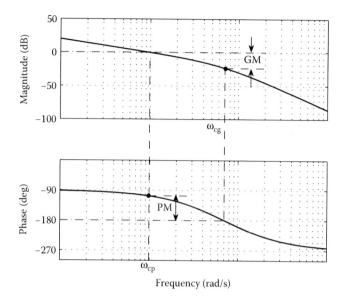

FIGURE 10.69 Gain margin and phase margin.

The stability criteria are given by

$$\begin{cases} \text{GM} > 0\,\text{dB} & \text{stable} \\ \text{GM} = 0\,\text{dB} & \text{marginally stable} \\ \text{GM} < 0\,\text{dB} & \text{unstable} \end{cases} \quad (10.56)$$

or

$$\begin{cases} \text{PM} > 0° & \text{stable} \\ \text{PM} = 0° & \text{marginally stable} \\ \text{PM} < 0° & \text{unstable} \end{cases} \quad (10.57)$$

Example 10.17: Stability Analysis Using a Bode Plot

Consider the feedback control system in Example 10.13, where K is assumed to be 1. Plot the Bode magnitude and phase curves using MATLAB. Give comments on the stability of the closed-loop system.

Solution

Note that the closed-loop characteristic equation is

$$1 + C(s)G(s)H(s) = 1 + K\frac{s+1}{(s+3)(s^2+2s+2)}\frac{1}{s} = 0.$$

The Bode plot is drawn based on the loop gain

$$L(s) = K\frac{s+1}{s(s+3)(s^2+2s+2)}.$$

The MATLAB command used to sketch the Bode plot is `bode`. Assume that $K = 1$. The following is the MATLAB session.

Introduction to Feedback Control Systems

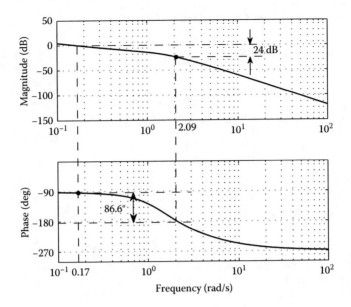

FIGURE 10.70 Bode plot for the system in Example 10.17.

```
>> num = [1 1];
>> den = conv([1 3 0],[1 2 2]);
>> sysL = tf(num,den);
>> bode(sysL);
```

As shown in Figure 10.70, the phase curve crosses $-180°$ at frequency 2.09 rad/s, and the corresponding magnitude is about -24 dB. This implies that a gain of 24 dB can be added before the magnitude plot reaches 0 dB at that frequency. Thus, the GM is 24 dB. The magnitude curve crosses 0 dB at frequency 0.17 rad/s, and the corresponding phase is $-93.4°$. This implies that a phase of $86.6°$ can be subtracted before the phase plot reaches $-180°$. Thus, the PM is $86.6°$. According to Equations 10.56 and 10.57, the closed-loop system with the proportional controller $K = 1$ is stable.

More information on stability can be extracted from the GM. Specifically, the GM when $K = 1$ gives the stability range of K for which the proportional feedback control system is stable. In our case, GM = 24 dB or 15.85, and this indicates that the closed-loop system is stable for $0 < K < 15.85$. We can also use the MATLAB command `margin` as follows:

```
>> [gm,pm,wcg,wcp] = margin(sysL)
```

which returns GM, PM, and the associated frequencies as defined in Figure 10.69. In our case, the GM returned by the command `margin` is 15.86, which is exactly the same as the one found using root locus in Example 10.14. Note that the stability range of K can be determined in this way only for systems that change from being stable to unstable as K increases.

10.6.3 CONTROL DESIGN USING BODE PLOT

Different from the root locus technique, which uses time-domain performance specifications, the Bode plot technique deals with control design in the frequency domain. The requirements are defined in terms of GM, PM, bandwidth, resonant peak, and so on. If a time-domain specification is given, it will usually be converted to one in the frequency domain.

Example 10.18: Proportional Control Design Using a Bode Plot

Design a proportional controller for the cart system in Example 10.12 using the Bode plot technique.

Solution

The Bode plot for the open-loop transfer funciton $KG(s)$, where

$$G(s) = \frac{3.778}{s^2 + 16.883s} \quad \text{and} \quad K = 1$$

is shown in Figure 10.71.

Note that the requirements are given as overshoot $M_p < 20\%$ and rise time $t_r < 0.1\,\text{s}$. They correspond to $\zeta > 0.46$ and $\omega_n > 18\,\text{rad/s}$. It is found that the relationship between the damping ratio and PM is

$$\text{PM} \approx 100\zeta, \quad (10.58)$$

which yields the requirement $\text{PM} > 46°$. In addition, the natural frequency ω_n is related to bandwidth, which is somewhat greater than the frequency when the Bode magnitude plot of $KG(s)$ crosses $-3\,\text{dB}$. Denote this crossover frequency as ω_c, and we have

$$\omega_c \leq \omega_{BW} \leq 2\omega_c. \quad (10.59)$$

The higher the crossover frequency, the higher the bandwidth and the natural frequency.

As shown in Figure 10.71, $\text{PM} = 89.2°$, which meets the requirement. However, the crossover frequency ω_c is only about 0.3 rad/s, which is too slow. We must adjust the value of the proportional control gain K in order to meet both requirements. Since the current PM is way above the requirement, let us decrease the PM and pick $\text{PM} = 50°$. Based on the definition of PM, this implies that the frequency at which the magnitude plot crosses 0 dB should be $-130°$. It is observed from Figure 10.71 that the frequency corresponding to $-130°$ is 14.4 rad/s, at which the magnitude is $-38.6\,\text{dB}$. To make the magnitude 0 dB, the magnitude plot should slide upward by 38.6 dB. This is the effect of multiplying a constant term of

$$10^{38.6/20} = 85.11,$$

which is the value of the proportional control gain K.

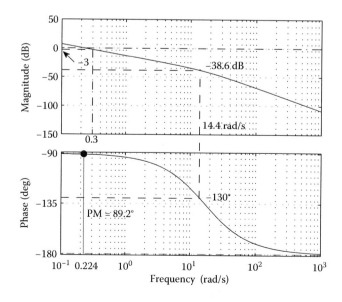

FIGURE 10.71 Bode plot for the system in Example 10.18 with $K = 1$.

Introduction to Feedback Control Systems

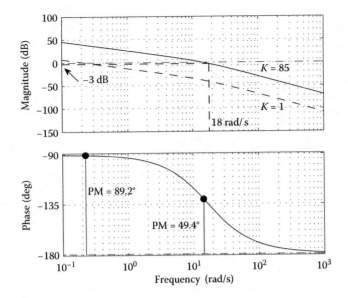

FIGURE 10.72 Bode plots for the system in Example 10.18 with $K = 85$ and $K = 1$.

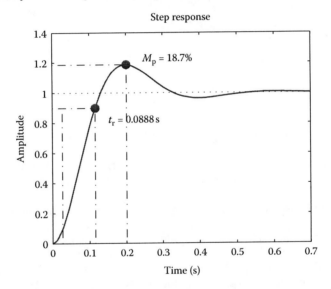

FIGURE 10.73 Unit-step response of the cart system with proportional feedback control $K = 85$.

Let us set K to be 85. The Bode plot of the open-loop transfer function $KG(s)$ with the new value of K is shown in Figure 10.72. The PM is 49.4° and the crossover frequency ω_c is 18 rad/s. The Bode plot with $K = 1$ is also shown in Figure 10.72. Comparing the two magnitude plots, we can find that the magnitude response corresponding to $K = 85$ is 38.6 dB above the one corresponding to $K = 1$, as designed.

The unit-step response of the closed-loop system is shown in Figure 10.73. The overshoot is 18.7% and the rise time is 0.0888 s. Both requirements are met.

PROBLEM SET 10.6

1. Sketch the asymptotes of the Bode plot magnitude and phase for the following open-loop transfer functions. Make sure to give the corner frequencies, slopes of the magnitude plot, and phase angles. Verify the results using MATLAB.

a. $G(s) = \dfrac{s+10}{s+1}$

b. $G(s) = \dfrac{s+10}{(s+1)(s+500)}$

c. $G(s) = \dfrac{s+1}{s(s+10)(s+500)}$

d. $G(s) = \dfrac{(s+10)^2}{s(s+1)(s+500)}$

2. Repeat Problem 1 for the following open-loop transfer functions.

   a. $G(s) = \dfrac{1}{s^2+2s+100}$

   b. $G(s) = \dfrac{1}{(s+0.5)(s^2+s+25)}$

   c. $G(s) = \dfrac{s^2+0.24s+144}{s^2+0.1s+25}$

   d. $G(s) = \dfrac{100(s^2+7s+49)}{(s+1)(s+5)(s+500)}$

3. ◢ For each of the following open-loop transfer functions, construct a Bode plot for $K = 1$ using the MATLAB command `bode`. Estimate the GM, PM, and their associated crossover frequencies from the plot. Verify the results using the MATLAB command `margin`. Determine the stability of the corresponding closed-loop system.

   a. $KG(s) = \dfrac{K}{s(s^2+2s+25)}$

   b. $KG(s) = \dfrac{K}{(0.01s+0.04)(s^2+2s+2)}$

   c. $KG(s) = K\dfrac{s+0.5}{s^2(s+20)}$

   d. $KG(s) = K\dfrac{1}{(s+10)(s+1)^2}$

4. ◢ Reconsider Problem 3. Each plant $G(s)$ is controlled by a proportional controller K via unity negative feedback. Determine the stability range of K by sliding the magnitude plot up or down until instability occurs. Verify the results by sketching a root locus.

5. Figure 10.74 shows the Bode plot for an open-loop transfer function $KG(s)$ with $K = 50$.
   a. Determine the stability of the closed-loop system with $K = 50$.
   b. Determine the value of K that would yield a PM of $50°$.

6. The Bode plot for an open-loop transfer function $KG(s)$ is shown in Figure 10.75.
   a. Determine the stability of the closed-loop system.
   b. Assume that the proportional control gain K is increased by a factor of 100. Will the closed-loop system still be stable with the new value of K?

7. ◢ Consider the unity negative feedback system shown in Figure 10.76.
   a. Use MATLAB to obtain the Bode plot $KG(s)$ for $K = 2$.
   b. Determine the stability of the closed-loop system when $K = 2$ using the stability margins.
   c. Determine the value of K that would yield a PM of $30°$.
   d. Verify the result obtained in Part (c) by using the MATLAB command `margin`.

8. ◢ Reconsider the feedback system in Figure 10.60. Using the Bode plot technique, find a value of K such that the maximum overshoot in the response to a unit-step reference input is less than 20% and the 2% settling time is less than 1.1 s. Plot the unit-step response of the closed-loop system to verify the result.

Introduction to Feedback Control Systems

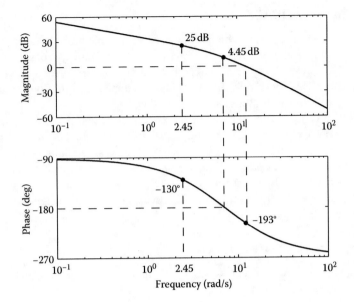

FIGURE 10.74 Problem 5.

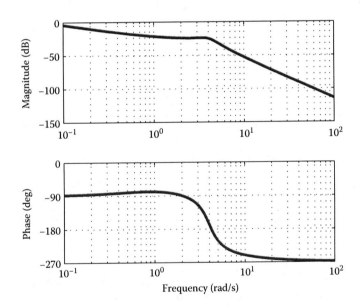

FIGURE 10.75 Problem 6.

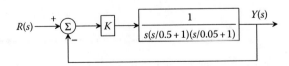

FIGURE 10.76 Problem 7.

10.7 FULL-STATE FEEDBACK

Unlike the root locus and Bode plot techniques, the state-space method works directly with mathematical models in state-space form instead of transfer function form. Often, use of the state-space method is referred to as modern control design, and use of transfer-function-based methods, such as the root locus and Bode plot, is referred to as classical control design. Compared with the techniques based on transfer functions, the state-space method provides a convenient and compact way to model and analyze systems with multiple inputs and multiple outputs. This is one of the advantages of state-space design because most practical systems have more than one control input and/or more than one measured output. In this section, we only discuss single-input–single-output systems to show the basic ideas of state-space design. We first present how to analyze stability if a system is given in state-space form. Two other important properties of dynamic systems, controllability and observability, are also briefly introduced. Then, we show how to design a full-state feedback controller using the pole placement method.

10.7.1 Analysis of State-Space Equations

Consider a linear dynamic system with single input, single output, and n states. The state-space representation is written in the form

$$\dot{\mathbf{x}} = \mathbf{A}\mathbf{x} + \mathbf{B}u,$$
$$y = \mathbf{C}\mathbf{x} + Du, \tag{10.60}$$

where both u and y are scalars. The stability characteristics of a dynamic system in state-space form can be studied from the eigenvalues of matrix $\mathbf{A}$. The determinant equation

$$|s\mathbf{I} - \mathbf{A}| = 0 \tag{10.61}$$

gives eigenvalues of the system matrix $\mathbf{A}$, which are the poles of the system. Equation 10.61 is known as the characteristic equation.

Example 10.19: Stability Analysis in State Space

a. Compute the poles of the system described by

$$\dot{\mathbf{x}} = \begin{bmatrix} 0 & 1 \\ 0 & -16.883 \end{bmatrix} \mathbf{x} + \begin{bmatrix} 0 \\ 3.778 \end{bmatrix} u,$$
$$y = \begin{bmatrix} 1 & 0 \end{bmatrix} \mathbf{x}.$$

b. Verify the results by converting the state-space representation to a transfer function and then solving for the poles of the transfer function.

Solution

a. The characteristic equation is

$$|s\mathbf{I} - \mathbf{A}| = \begin{vmatrix} s & -1 \\ 0 & s + 16.883 \end{vmatrix} = s^2 + 16.883s = 0,$$

which yields the poles $s_1 = 0$ and $s_2 = -16.883$.

b. As presented in Section 4.4, state-space equations for a single-input-single-output system can be converted to a transfer matrix using

$$G(s) = \mathbf{C}(s\mathbf{I} - \mathbf{A})^{-1}\mathbf{B} + D.$$

Introduction to Feedback Control Systems

Substituting the system matrices **A**, **B**, **C**, and **D**, which is 0 in this example, gives

$$G(s) = \begin{bmatrix} 1 & 0 \end{bmatrix} \begin{bmatrix} s & -1 \\ 0 & s+16.883 \end{bmatrix}^{-1} \begin{bmatrix} 0 \\ 3.778 \end{bmatrix} + 0 = \begin{bmatrix} 1 & 0 \end{bmatrix} \frac{\begin{bmatrix} s+16.883 & 1 \\ 0 & s \end{bmatrix}}{s^2 + 16.883s} \begin{bmatrix} 0 \\ 3.778 \end{bmatrix}$$

$$= \frac{3.778}{s^2 + 16.883s}.$$

The characteristic equation is $s^2 + 16.883s = 0$, which yields the poles at 0 and -16.883. The results agree with the poles obtained in Part (a).

Two other important properties for a control system are controllability and observability. Before we introduce their definitions, let us consider the following example.

Example 10.20: Controllability and Observability

Consider a dynamic system given by $G(s) = 2/(s+4)$, which can be written in the state-space form

$$\dot{x}_1 = -4x_1 + 2u,$$
$$y = x_1.$$

a. A new state is added and the state-space equation is

$$\dot{x}_1 = -4x_1 + 2u,$$
$$\dot{x}_2 = -x_2,$$
$$y = x_1 + 3x_2.$$

Determine the transfer function for this new dynamics model.

b. Determine the transfer function for another case

$$\dot{x}_1 = -4x_1 + 2u,$$
$$\dot{x}_2 = -x_2 + u,$$
$$y = x_1.$$

SOLUTION

a. The system matrices **A**, **B**, **C**, and **D** are

$$\mathbf{A} = \begin{bmatrix} -4 & 0 \\ 0 & -1 \end{bmatrix}, \quad \mathbf{B} = \begin{bmatrix} 2 \\ 0 \end{bmatrix}, \quad \mathbf{C} = \begin{bmatrix} 1 & 3 \end{bmatrix}, \quad D = 0.$$

The transfer function is

$$G(s) = \begin{bmatrix} 1 & 3 \end{bmatrix} \begin{bmatrix} s+4 & 0 \\ 0 & s+1 \end{bmatrix}^{-1} \begin{bmatrix} 2 \\ 0 \end{bmatrix} + 0 = \begin{bmatrix} 1 & 3 \end{bmatrix} \frac{\begin{bmatrix} s+1 & 0 \\ 0 & s+4 \end{bmatrix}}{(s+1)(s+4)} \begin{bmatrix} 2 \\ 0 \end{bmatrix}$$

$$= \frac{\begin{bmatrix} s+1 & 3(s+4) \end{bmatrix}}{(s+1)(s+4)} \begin{bmatrix} 2 \\ 0 \end{bmatrix} = \frac{2}{s+4}.$$

b. Similarly, we have

$$\mathbf{A} = \begin{bmatrix} -4 & 0 \\ 0 & -1 \end{bmatrix}, \quad \mathbf{B} = \begin{bmatrix} 2 \\ 1 \end{bmatrix}, \quad \mathbf{C} = \begin{bmatrix} 1 & 0 \end{bmatrix}, \quad D = 0$$

and

$$G(s) = \begin{bmatrix} 1 & 0 \end{bmatrix} \begin{bmatrix} s+4 & 0 \\ 0 & s+1 \end{bmatrix}^{-1} \begin{bmatrix} 2 \\ 1 \end{bmatrix} + 0 = \begin{bmatrix} 1 & 0 \end{bmatrix} \dfrac{\begin{bmatrix} s+1 & 0 \\ 0 & s+4 \end{bmatrix}}{(s+1)(s+4)} \begin{bmatrix} 2 \\ 1 \end{bmatrix}$$

$$= \begin{bmatrix} 1 & 0 \end{bmatrix} \dfrac{\begin{bmatrix} 2(s+1) \\ s+4 \end{bmatrix}}{(s+1)(s+4)} = \dfrac{2}{s+4}.$$

Note that the dynamic models in Parts (a) and (b) are both second-order systems, which have different state-space forms from the given first-order system. However, they end up with the same transfer function due to pole–zero cancellation. As seen in Part (a), the second state cannot be affected by the input matrix **B**,

$$G(s) = \dfrac{\begin{bmatrix} s+1 & 3(s+4) \end{bmatrix}}{(s+1)(s+4)} \begin{bmatrix} 2 \\ 0 \end{bmatrix}.$$

This implies that the second state is uncontrollable by the actuator defined by matrix **B**. Similarly, as seen in Part (b), the second state cannot be seen by the output matrix **C**,

$$G(s) = \begin{bmatrix} 1 & 0 \end{bmatrix} \dfrac{\begin{bmatrix} 2(s+1) \\ s+4 \end{bmatrix}}{(s+1)(s+4)}.$$

This implies that the second state is unobservable by the sensor defined by matrix **C**.

Now we introduce more rigorous definitions of controllability and observability. A system is controllable if there exists a control signal $u(t)$ that will take the state of the system from any initial state $\mathbf{x}_0$ to any desired final state $\mathbf{x}_f$ in a finite time interval. A system is observable if for any initial state $\mathbf{x}_0$ there is a finite time τ such that $\mathbf{x}_0$ can be determined from $u(t)$ and $y(t)$ for $0 \le t \le \tau$. An nth-order single-input–single-output system is controllable if and only if the square matrix given by

$$\mathbf{P} = \begin{bmatrix} \mathbf{B} & \mathbf{AB} & \mathbf{A}^2\mathbf{B} & \cdots & \mathbf{A}^{n-1}\mathbf{B} \end{bmatrix} \quad (10.62)$$

is nonsingular, where **P** is called the controllability matrix. Similarly, the system is observable if and only if the square matrix given by

$$\mathbf{Q} = \begin{bmatrix} \mathbf{C} \\ \mathbf{CA} \\ \mathbf{CA}^2 \\ \vdots \\ \mathbf{CA}^{n-1} \end{bmatrix} \quad (10.63)$$

is nonsingular, where **Q** is called the observability matrix.

Example 10.21: Controllability and Observability

Determine the controllability and observability for the second-order systems given in Example 10.20.

Solution

For the system in Part (a), the controllability matrix **P** is

$$\mathbf{P} = \begin{bmatrix} \mathbf{B} & \mathbf{AB} \end{bmatrix} = \begin{bmatrix} 2 & -8 \\ 0 & 0 \end{bmatrix},$$

Introduction to Feedback Control Systems

which is singular. Thus, the system is uncontrollable. The observability matrix $\mathbf{Q}$ is

$$\mathbf{Q} = \begin{bmatrix} \mathbf{C} \\ \mathbf{CA} \end{bmatrix} = \begin{bmatrix} 1 & 3 \\ -4 & -3 \end{bmatrix},$$

which is nonsingular because the determinant is 9, which is nonzero. Thus, the system is observable.

For the system in Part (b), the controllability matrix $\mathbf{P}$ is

$$\mathbf{P} = \begin{bmatrix} \mathbf{B} & \mathbf{AB} \end{bmatrix} = \begin{bmatrix} 2 & -8 \\ 1 & -1 \end{bmatrix},$$

which is nonsingular because the determinant is 6, which is nonzero. Thus, the system is controllable. The observability matrix $\mathbf{Q}$ is

$$\mathbf{Q} = \begin{bmatrix} \mathbf{C} \\ \mathbf{CA} \end{bmatrix} = \begin{bmatrix} 1 & 0 \\ -4 & 0 \end{bmatrix},$$

which is singular. Thus, the system is unobservable.

Note that the two systems in Example 10.20 have the same transfer function representation but different properties of controllability and observability. This implies that controllability and observability are functions of the state of the system and cannot be determined from a transfer function.

10.7.2 Control Design for Full-State Feedback

Consider an nth-order dynamic system given by Equation 10.60. If all of the states are measurable, then they can be fed back and used for computing the control input

$$u = -\mathbf{K}\mathbf{x} = -\begin{bmatrix} k_1 & k_2 & \cdots & k_n \end{bmatrix} \begin{Bmatrix} x_1 \\ x_2 \\ \vdots \\ x_n \end{Bmatrix}, \quad (10.64)$$

where $\mathbf{K}$ is the feedback gain matrix. The control law defined by Equation 10.64 is called full-state feedback. Figure 10.77 shows the block diagram of a closed-loop system with full-state feedback. Substituting Equation 10.64 into Equation 10.60 gives the state equation of the closed-loop system, that is,

$$\dot{\mathbf{x}} = (\mathbf{A} - \mathbf{B}\mathbf{K})\mathbf{x}. \quad (10.65)$$

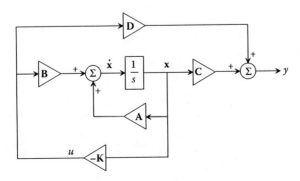

FIGURE 10.77 Block diagram of a closed-loop system with full-state feedback.

Note that the closed-loop poles are the eigenvalues of the matrix $\mathbf{A} - \mathbf{BK}$ and the closed-loop characteristic equation is

$$|s\mathbf{I} - (\mathbf{A} - \mathbf{BK})| = 0, \tag{10.66}$$

The left-hand side is an nth-order polynomial in s containing the gains $k_1, k_2, \ldots, k_n$. The pole placement method can be used to find the feedback gain matrix $\mathbf{K}$.

As discussed in Section 10.2, the performance of a controlled system is associated with the closed-loop poles. If a feedback gain matrix $\mathbf{K}$ is determined based on desired pole locations, then the closed-loop system with the feedback control law $u = -\mathbf{Kx}$ will achieve the desired performance. This is the basic idea of pole placement. Assume that the desired locations of the closed-loop poles are $s_1, s_2, \ldots,$ and s_n. Note that poles of a system are the roots of the characteristic equation of the system. Thus, the characteristic equation corresponding to the selected pole locations is

$$(s - s_1)(s - s_2)\ldots(s - s_n) = 0, \tag{10.67}$$

which is essentially the same as the closed-loop characteristic equation given by Equation 10.66,

$$|s\mathbf{I} - (\mathbf{A} - \mathbf{BK})| = (s - s_1)(s - s_2)\ldots(s - s_n). \tag{10.68}$$

Equating the coefficients of like powers of s on both sides yields the values of the gains $k_1, k_2, \ldots,$ and k_n.

Example 10.22: Full-State Feedback Control Design

Consider the DC-motor-driven cart discussed in Example 10.12, where a PD controller was designed to achieve the requirements: overshoot $M_p < 20\%$ and rise time $t_r < 0.1$ s.

a. Find a full-state feedback controller such that the closed-loop system achieves the same requirements.
b. Use MATLAB to find the control gain matrix $\mathbf{K}$.

Solution

a. The transfer function of the cart is given by

$$G(s) = \frac{Y(s)}{U(s)} = \frac{3.778}{s^2 + 16.883s},$$

where the output y is the position of the cart and the input u is the voltage applied to the DC motor. Using the position and the velocity as the state variables yields

$$\dot{\mathbf{x}} = \begin{bmatrix} 0 & 1 \\ 0 & -16.883 \end{bmatrix}\mathbf{x} + \begin{bmatrix} 0 \\ 3.778 \end{bmatrix}u,$$

$$y = \begin{bmatrix} 1 & 0 \end{bmatrix}\mathbf{x}.$$

For a second-order system, the gain matrix $\mathbf{K}$ is a 1×2 matrix and

$$\mathbf{K} = \begin{bmatrix} k_1 & k_2 \end{bmatrix}.$$

Applying Equation 10.66 yields the theoretical characteristic polynomial

$$|s\mathbf{I} - \mathbf{A} + \mathbf{BK}| = \left| \begin{bmatrix} s & -1 \\ 0 & s + 16.883 \end{bmatrix} + \begin{bmatrix} 0 \\ 3.778 \end{bmatrix} \begin{bmatrix} k_1 & k_2 \end{bmatrix} \right|$$

$$= \begin{vmatrix} s & -1 \\ 3.778k_1 & s + 16.883 + 3.778k_2 \end{vmatrix}$$

$$= s^2 + (16.883 + 3.778k_2)s + 3.778k_1.$$

Introduction to Feedback Control Systems

The time-domain specifications, $M_p < 20\%$ and $t_r < 0.1\,\text{s}$, indicate that $\zeta > 0.46$ and $\omega_n > 18\,\text{rad/s}$. Choose the same values of ω_n and ζ as in Example 10.12, where $\omega_n = 18.5\,\text{rad/s}$ and $\zeta = 0.5$. The desired closed-loop poles are located at $p_{1,2} = -9.25 \pm 16.02\text{j}$. Applying Equation 10.67 gives the desired characteristic polynomial

$$(s + 9.25 - 16.02\text{j})(s + 9.25 + 16.02\text{j}) = s^2 + 18.5s + 342.20.$$

Equating the two characteristic polynomials, we have

$$16.883 + 3.778 k_2 = 18.5,$$
$$3.778 k_1 = 342.20,$$

which gives $\mathbf{K} = [90.58 \quad 0.43]$. So the full-state feedback controller is

$$u = -\mathbf{Kx} = -90.58 x_1 - 0.43 x_2.$$

b. ◢ The following is the MATLAB session used to compute a full-state feedback control gain matrix $\mathbf{K}$ using pole placement design.

```
>> A = [0 1; 0 -16.883];
>> B = [0; 3.778];
>> p = [-9.25+16.02j -9.25-16.02j];
>> K = place(A,B,p);
```

The command `place` returns the gain matrix $\mathbf{K}$, for which the full-state feedback $u = -\mathbf{Kx}$ places the closed-loop poles at the desired locations.

It is interesting to see that the values of the gains k_1 and k_2 are the same as the values of the proportional and derivative gains k_P and k_D found in Example 10.12. This is because of the way we selected the states. In this example, $u = -90.58 y - 0.43 \dot{y}$, and in Example 10.12, $u = 90.58(r-y) + 0.43 d(r-y)/dt$. If the reference signal r is zero, the two controllers will end up with the same expression. This implies that the closed-loop system with full-state feedback is a regulation system. For tracking control, the control law $u = -\mathbf{Kx}$ needs to be modified and more details can be found in control texts. Note that if a different set of state variables is selected, the corresponding value of the gain matrix $\mathbf{K}$ will be different. The reader can solve Problem 5 in Problem Set 10.7 to verify this conclusion.

The full-state feedback control method requires that all state variables are measured. However, this is usually not a practical assumption. To make a full-state feedback controller practically implementable, an estimator or observer can be designed to compute an estimate of the state variables based on the measurements of the system. Then, the control law calculations are based on the estimated state rather than the actual state.

PROBLEM SET 10.7

1. For the system shown in Figure 10.78, derive the state-space equations using the state variables indicated. Make sure to give the $\mathbf{A}$, $\mathbf{B}$, $\mathbf{C}$, and D matrices. Solve for the poles of the system.
2. Repeat Problem 1 for the system shown in Figure 10.79.
3. Determine the controllability and observability for each of the following systems.

   a. $\begin{Bmatrix} \dot{x}_1 \\ \dot{x}_2 \\ \dot{x}_3 \end{Bmatrix} = \begin{bmatrix} -5 & -3 & 0 \\ 2 & 0 & 0 \\ 0 & 1 & 0 \end{bmatrix} \begin{Bmatrix} x_1 \\ x_2 \\ x_3 \end{Bmatrix} + \begin{bmatrix} 2 \\ 0 \\ 0 \end{bmatrix} u, \quad y = \begin{bmatrix} 0 & 1 & 6 \end{bmatrix} \begin{Bmatrix} x_1 \\ x_2 \\ x_3 \end{Bmatrix}$

   b. $\begin{Bmatrix} \dot{x}_1 \\ \dot{x}_2 \\ \dot{x}_3 \end{Bmatrix} = \begin{bmatrix} -1 & -1 & -2 \\ 4 & 0 & 0 \\ 0 & 0 & 2 \end{bmatrix} \begin{Bmatrix} x_1 \\ x_2 \\ x_3 \end{Bmatrix} + \begin{bmatrix} 2 \\ 0 \\ 0 \end{bmatrix} u, \quad y = \begin{bmatrix} 2 & 2 & 1 \end{bmatrix} \begin{Bmatrix} x_1 \\ x_2 \\ x_3 \end{Bmatrix}$

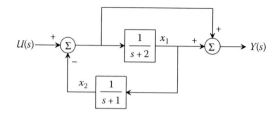

FIGURE 10.78 Problem 1.

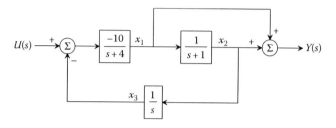

FIGURE 10.79 Problem 2.

c. $\begin{Bmatrix} \dot{x}_1 \\ \dot{x}_2 \\ \dot{x}_3 \end{Bmatrix} = \begin{bmatrix} 1 & 0 & 0 \\ 0 & -2 & 0 \\ 0 & 7 & 6 \end{bmatrix} \begin{Bmatrix} x_1 \\ x_2 \\ x_3 \end{Bmatrix} + \begin{bmatrix} 1 \\ 1 \\ 1 \end{bmatrix} u, \quad y = \begin{bmatrix} 1 & 1 & 0 \end{bmatrix} \begin{Bmatrix} x_1 \\ x_2 \\ x_3 \end{Bmatrix}$

d. $\begin{Bmatrix} \dot{x}_1 \\ \dot{x}_2 \\ \dot{x}_3 \end{Bmatrix} = \begin{bmatrix} 1 & 0 & 0 \\ 0 & 1 & 0 \\ 0 & 4 & 3 \end{bmatrix} \begin{Bmatrix} x_1 \\ x_2 \\ x_3 \end{Bmatrix} + \begin{bmatrix} 0 \\ 0 \\ 1 \end{bmatrix} u, \quad y = \begin{bmatrix} 1 & 1 & 1 \end{bmatrix} \begin{Bmatrix} x_1 \\ x_2 \\ x_3 \end{Bmatrix}$

4. Consider the two-degree-of-freedom mass–spring system as shown in Figure 10.80, where two masses are to be controlled by two equal and opposite forces f. The equations of motion of the system are given by

$$m\ddot{x}_1 + 2kx_1 - kx_2 = f,$$
$$m\ddot{x}_2 - kx_1 + 2kx_2 = -f.$$

Show that the system is uncontrollable. Using the concept of mode discussed in Section 9.4, associate a physical meaning with the controllable and uncontrollable modes.

5. Reconsider Example 10.22. Find the controllable canonical form (i.e., the controller canonical form in MATLAB) for the plant transfer function. Design a full-state feedback controller that places the closed-loop poles at the same locations as in the example.

6. A regulation system has a plant with the transfer function

$$G(s) = \frac{Y(s)}{U(s)} = \frac{-2}{s^3 + 3s^2 + 5s + 5}.$$

a. Transform the plant transfer function into the state-space form with the state vector $\mathbf{x} = \begin{bmatrix} y & \dot{y} & \ddot{y} \end{bmatrix}^T$.

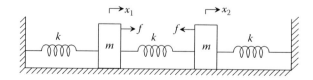

FIGURE 10.80 Problem 4.

b. Determine the state-feedback gain matrix **K** such that the closed-loop poles are located at $p_{1,2} = -2 \pm 2j$ and $p_3 = -5$.
c. ◀ Verify the result in Part (b) by using the MATLAB command `place`.

7. Consider the system

$$\begin{Bmatrix} \dot{x}_1 \\ \dot{x}_2 \end{Bmatrix} = \begin{bmatrix} 0 & 1 \\ -2.5 & -1 \end{bmatrix} \begin{Bmatrix} x_1 \\ x_2 \end{Bmatrix} + \begin{bmatrix} 0 \\ 1 \end{bmatrix} u, \quad y = \begin{bmatrix} 1 & 0 \end{bmatrix} \begin{Bmatrix} x_1 \\ x_2 \end{Bmatrix}.$$

a. Design a state-feedback controller so that the closed-loop poles have a damping ratio $\zeta = 0.707$ and a natural frequency $\omega_n = 3$ rad/s.
b. ◀ Verify the result in Part (a) by using the MATLAB command `place`.

8. Consider the system

$$G(s) = \frac{Y(s)}{U(s)} = \frac{1}{(s+2)(s+3)(s+4)}.$$

a. Design a state-feedback controller so that the closed-loop response has an overshoot of less than 5% and a rise time under 0.5 s. Set one of the closed-loop poles at -10.
b. ◀ Using MATLAB and/or Simulink, verify that your design meets the specifications.

10.8 SUMMARY

This chapter presented an introduction to feedback control systems. The essential components of a feedback control include a system we want to control, a controller we need to design, an actuator used to drive the controlled system, and a sensor used to measure the system output. Generally, the controlled system and the actuator are intimately connected, and they can be combined as one component called the plant. Unlike open-loop control, the output signal of the plant in a feedback control system is measured and fed back for use in computing the control signal. Compared with open-loop control, feedback can be used to stabilize unstable systems, reduce steady-state errors to disturbances, improve reference tracking performance, and reduce sensitivity to parameter variations.

Stability and performance are two important subjects in control. A linear time-invariant system is said to be stable if all the poles have negative real parts and is unstable otherwise. In terms of the pole locations in the s-plane, the imaginary axis is the stability boundary between the stable left-half s-plane and the unstable right-half s-plane. It is not an easy task to solve for the poles of a high-order linear system by hand. Routh's stability criterion is a method of obtaining information about pole locations without solving for the poles. A system is stable if and only if all the elements in the first column of the Routh array are positive.

The locations of poles in the s-plane are also associated with performance quantities, which are rise time t_r, overshoot M_p, peak time t_p, and settling time t_s in the time domain, and bandwidth ω_{BW} and resonant peak M_r in the frequency domain. For a second-order system with poles at $-\zeta\omega_n \pm j\omega_n\sqrt{1-\zeta^2}$, the correspondences between the system parameters and the time-domain specifications are given by

$$t_r \approx \frac{1.8}{\omega_n}, \quad M_p = e^{-\pi\zeta/\sqrt{1-\zeta^2}}, \quad t_p = \frac{\pi}{\omega_d}, \quad t_s \approx \frac{3.9}{\zeta\omega_n} \quad (2\% \text{ settling time}).$$

Compared with the time-domain performance specifications, the resonant peak M_r is similar to the overshoot M_p, both of which are related to the damping ratio ζ, while the bandwidth ω_{BW} is similar to the rise time t_r, both of which are related to the natural frequency ω_n.

The PID controller is a generic feedback control structure widely used in industries. It is described by the equation

$$\frac{U(s)}{E(s)} = k_p + \frac{k_I}{s} + k_D s.$$

In general, a larger proportional gain k_p results in a faster response and a smaller steady-state error. However, an excessively large proportional gain k_p leads to lightly damped oscillations and even instability. A larger integral gain k_I reduces steady-state errors more quickly, but reduces damping and leads to a larger overshoot. A larger derivative control decreases the overshoot, but slows down the speed of response. The PID gains can be tuned using the reaction curve method and the ultimate sensitivity method developed by Ziegler and Nichols.

Three different methods, including root locus, Bode plot, and state-space techniques, were introduced in this chapter for stability analysis and proportional feedback control design. The root locus and Bode plot work with graphs obtained from open-loop transfer functions, whereas the state-space method works directly with mathematical models in state-space form.

For a negative feedback system with $KL(s)$ as the open-loop transfer function, a root locus is a graph of the closed-loop poles or the roots of the closed-loop characteristic equation

$$1 + KL(s) = 0,$$

with respect to the control gain K. The rules for sketching a root locus are presented in Section 10.5. Using the root locus technique, it is very easy to determine the stability of a closed-loop system when the proportional gain K varies from 0 to ∞. For a particular value of K, the closed-loop system is stable if and only if all of the poles are in the left-half s-plane.

The Bode plot is a graph of the frequency response function, using a linear scale for magnitude (in dB) and phase (in degrees) and a logarithmic scale for frequency (in rad/s). For a frequency response function in the form of

$$KG(j\omega) = K \frac{(j\omega - z_1)(j\omega - z_2) \cdots (j\omega - z_m)}{(j\omega - p_1)(j\omega - p_2) \cdots (j\omega - p_n)}$$

the Bode plot can be easily drawn by hand using the rules described in Section 10.6. The stability of the corresponding closed-loop system can be determined by the GM and PM, both of which can be determined directly by inspecting the open-loop Bode plot.

For a dynamic system described in state-space form, the stability can be studied from the eigenvalues of matrix $\mathbf{A}$. If all the states are measurable, a full-state feedback controller given by

$$u = -\mathbf{K}\mathbf{x}$$

can be designed to improve stability and performance. The feedback gain matrix $\mathbf{K}$ can be determined using the pole placement method. Closed-loop poles are selected depending on the desired transient response. Equating the theoretical and desired closed-loop characteristic polynomials

$$|s\mathbf{I} - (\mathbf{A} - \mathbf{B}\mathbf{K})| = (s - s_1)(s - s_2) \cdots (s - s_n)$$

yields the values of the elements $k_1, k_2, \ldots,$ and k_n of the gain matrix $\mathbf{K}$.

REVIEW PROBLEMS

1. Consider the feedback control system as shown in Figure 10.81. Determine the range of K for closed-loop stability.

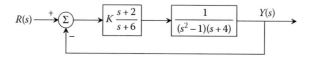

FIGURE 10.81 Problem 1.

Introduction to Feedback Control Systems

2. Consider the feedback control system shown in Figure 10.84.
   a. Design a PD controller such that the closed-loop poles are at $p_{1,2} = -1 \pm \sqrt{3}j$.
   b. Estimate the rise time, overshoot, peak time, and 1% settling time for the unit-step response of the closed-loop system.
   c. ◀ Use MATLAB to plot the unit-step response of the closed-loop system. Verify the answers obtained in Part (b).
3. Consider the feedback control system as shown in Figure 10.82.
   a. Assuming $C(s) = k_p$, determine the value of the proportional gain that makes the closed-loop system marginally stable. Find the frequency of the sustained oscillation.
   b. Using the gain and the frequency obtained in Part (a), apply the ultimate sensitivity method of Ziegler–Nichols tuning rules to design a PID controller.
   c. ◀ Plot the unit-step response of the resulting closed-loop system. Find the values of the rise time t_r, overshoot M_p, peak time t_p, and 1% settling time t_s.
4. Consider a unity negative feedback system with the open-loop transfer function

$$KG(s) = \frac{K}{s(s+1)(s+5)}.$$

   a. ◀ Use MATLAB to draw the Bode plots for $K = 1$. Determine the range of K for which the closed-loop system will be stable.
   b. Determine the range of K for closed-loop stability by sketching the root locus.
   c. Using Routh's criterion, determine the range of K for closed-loop stability.
5. ◀ Consider the feedback control system as shown in Figure 10.83.
   a. Determine the value of the gain K such that the undamped natural frequency ω_n and the damping ratio ζ of the dominant closed-loop poles are around 2 rad/s and 0.5, respectively.
   b. Determine the values of all closed-loop poles.
   c. Plot the unit-step response of the resulting closed-loop system. Find the values of the rise time t_r, overshoot M_p, peak time t_p, and 1% settling time t_s.

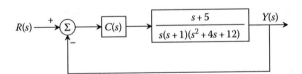

FIGURE 10.82 Problem 3.

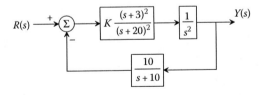

FIGURE 10.83 Problem 5.

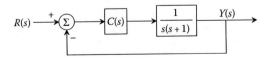

FIGURE 10.84 Problem 2.

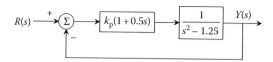

FIGURE 10.85 Problem 6.

6. Consider the unity negative feedback system with a PD controller as shown in Figure 10.85.
   a. Determine the value of the proportional gain k_p such that the damping ratio of the closed-loop system is 0.7.
   b. What is the GM of the system if k_p is set to the value obtained in Part (a)? Answer this question without creating the Bode plots.
   c. Verify your answer in Part (b) by creating the Bode plots using MATLAB.
7. Consider the system

$$\begin{Bmatrix} \dot{x}_1 \\ \dot{x}_2 \end{Bmatrix} = \begin{bmatrix} 0 & 1 \\ -2 & -3 \end{bmatrix} \begin{Bmatrix} x_1 \\ x_2 \end{Bmatrix} + \begin{bmatrix} 0 \\ 1 \end{bmatrix} u, \quad y = \begin{bmatrix} 1 & 0 \end{bmatrix} \begin{Bmatrix} x_1 \\ x_2 \end{Bmatrix}.$$

   a. Design a state-feedback controller so that the closed-loop unit-step response has an overshoot of less than 5% and a peak time under 1.5 s.
   b. Verify your result in Part (a) with MATLAB.
8. Consider the cart–inverted-pendulum system shown in Figure 5.79. Assume that the mass of the cart is 0.8 kg, the mass of the pendulum is 0.2 kg, and the length of the pendulum is 0.6 m.
   a. Determine the poles of the system. Is it stable or unstable?
   b. Design a full-state feedback controller such that the closed-loop poles are located at $p_{1,2} = -2.90 \pm 2.15j$, $p_3 = -10$, and $p_4 = -20$.
   c. Assume that the initial angle of the inverted pendulum is $5°$ away from the vertical reference line. Using the state feedback gain matrix **K** obtained in Part (b), examine the response of the closed-loop system by constructing a Simulink block diagram.

Bibliography

1. Beer, F.P., Johnston, E.R. Jr., and Cornwell, P.J., *Vector Mechanics for Engineers: Dynamics*. 9th ed., McGraw-Hill, New York, NY, 2009.
2. Beer, F.P., Johnston, E.R. Jr., Dewolf, J., and Mazurek, D., *Mechanics of Materials*. 5th ed., McGraw-Hill, New York, NY, 2008.
3. Cengel, Y.A., *Heat Transfer: A Practical Approach*. 2nd ed., McGraw-Hill, New York, NY, 2002.
4. Chabay, R. and Sherwood, B., *Matter & Interactions*. 2nd ed., Wiley, Hoboken, NJ, 2007.
5. Craig, J.J., *Introduction to Robotics: Mechanics and Control*. 3rd ed., Prentice Hall, Upper Saddle River, NJ, 2005.
6. Esfandiari, R., *Applied Mathematics for Engineers*, 4th ed., Atlantis, Los Angeles, CA, 2007.
7. Ewins, D.J., *Modal Testing, Theory, Practice, and Application*. 2nd ed., Research Studies Press, Baldock, Hertfordshire, England, 2001.
8. Floyd, T.L., *Electric Circuit Fundamentals*. 7th ed., Prentice Hall, Upper Saddle River, NJ, 2006.
9. Franklin, G.F., Powell, J.D., and Emami-Naeini, A., *Feedback Control of Dynamic Systems*. 5th ed., Prentice Hall, Upper Saddle River, NJ, 2006.
10. Hibbeler, R.C., *Engineering Mechanics: Dynamics*. 12th ed., Prentice Hall, Upper Saddle River, NJ, 2010.
11. Hibbeler, R.C., *Mechanics of Materials*. 7th ed., Prentice Hall, Upper Saddle River, NJ, 2008.
12. Inman, D.J., *Engineering Vibration*. 3rd ed., Prentice Hall, Upper Saddle River, NJ, 2007.
13. Meirovitch, L., *Fundamentals of Vibrations*. McGraw-Hill, New York, 2001.
14. Nilsson, J.W., *Electric Circuits*. 8th ed., Prentice Hall, Upper Saddle River, NJ, 2007.
15. Ogata, K., *Modern Control Engineering*. 4th ed., Prentice Hall, Upper Saddle River, NJ, 2002.
16. Palm, W.J. III, *Mechanical Vibration*. Wiley, Hoboken, NJ, 2007.
17. Spong, M.W., Hutchinson, S., and Vidyasagar, M., *Robot Modeling and Control*. Wiley, Hoboken, NJ, 2005.
18. Street, R.L., Watters, G.Z., and Vennard, J.K., *Elementary Fluid Mechanics*. 7th ed., Wiley, Hoboken, NJ, 1996.

Appendix A

TABLE A.1
The International System (SI) of Units

Basic Units

Quantity	Unit Name	Symbol or Unit
Length	Meter	m
Mass	Kilogram	kg
Time	Second	s
Electric current	Ampere	A
Voltage	Volts	V
Temperature	Kelvin	K
Amount of substance	Mole	mol
Solid angle	Radian	rad

SI Derived Units

Quantity	Unit Name	Symbol or Unit
(Circular) frequency (ω)		rad/s
Acceleration (a)		m/s^2
Angular acceleration (α)		rad/s^2
Angular velocity (ω)		rad/s
Area (A)		m^2
Area moment of inertia (I)		m^4
Density, mass density (ρ)		kg/m^3
Dynamic viscosity (b, c)		N·s/m
Electric capacitance (C)	Farad (F)	A·s/V
Electric charge (q)	Coulomb (C)	A·s
Electric resistance (R)	Ohm (Ω)	V/A
Force (f)	Newton (N)	kg·m/s^2
Frequency	Hertz (Hz)	1/s
Inductance (L)	Henry (H)	V·s/A
Magnetic flux	Weber (Wb)	V·s
Magnetic flux density	Telsa (T)	Wb/m^2
Mass moment of inertia (I)		kg·m^2
Power	Watt (W)	J/s
Pressure, mechanical stress	Pascal (Pa)	N/m^2
Specific heat (c, c_p, c_v)		J/(kg·K)
Specific volume (v)		m^3/kg
Thermal conductivity (k)		W/(s·m·K)
Velocity, speed (v)		m/s
Volume (V)		m^3
Wave length (λ)		1/m
Work (W), energy (E), heat (Q)	Joule (J)	N·m

TABLE A.2
Conversion Factors

Density	$1 \text{ g/cm}^3 = 62.43 \text{ lb}_m/\text{ft}^3$
Energy	$1 \text{ cal} = 4.184 \text{ J}$
Force	$1 \text{ lb}_f = 4.45 \text{ N}$
Length	$1 \text{ in} = 2.54 \text{ cm}$
	$1 \text{ ft} = 0.3048 \text{ m}$
	$1 \text{ mi} = 1609 \text{ m} = 5280 \text{ ft}$
Mass	$1 \text{ lb}_m = 0.4536 \text{ kg} = 16 \text{ oz}$
	$1 \text{ slug} = 32.174 \text{ lb}_m$
	$1 \text{ ton} = 2000 \text{ lb}_m$
Power	$1 \text{ W} = 3.413 \text{ Btu/hr}$
Pressure	$1 \text{ atm} = 1.0132 \times 10^5 \text{ Pa}$
Temperature	$°C = (°F - 32)/1.8$
	$°F = °C(1.8) + 32$
	$K = °C + 273.16$
	$°R = °F + 459.69$
	$1 K = 1.8 °R$
Thermal conductivity	$1 \text{ W}/(\text{m} \cdot °C) = 0.5778 \text{ Btu}/(\text{h} \cdot \text{ft} \cdot °F)$
Volume	$1 \text{ liter (L)} = 1000 \text{ cm}^3 = 0.0353 \text{ ft}^3 = 1.0564 \text{ quart}$
	$1 \text{ ft}^3 = 28.316 \text{ L}$
	$1 \text{ gal} = 3.785 \text{ L} = 4 \text{ quarts}$
	$1 \text{ quart} = 2 \text{ pints} = 67.2 \text{ in.}^3 = 0.9466 \text{ L}$
	$1 \text{ pint} = 16 \text{ oz}$

Appendix B: Useful Formulas

Trigonometric Expansions

Sum to Product

$\sin(a \pm b) = \sin a \cos b \pm \cos a \sin b$

$\cos(a \pm b) = \cos a \cos b \mp \sin a \sin b$

$\tan(a \pm b) = \dfrac{\tan a \pm \tan b}{1 \mp \tan a \tan b}$

Product to Sum

$\sin a \sin b = \dfrac{1}{2}[\cos(a-b) - \cos(a+b)]$

$\cos a \cos b = \dfrac{1}{2}[\cos(a+b) + \cos(a-b)]$

$\sin a \cos b = \dfrac{1}{2}[\sin(a+b) + \sin(a-b)]$

$\cos a \sin b = \dfrac{1}{2}[\sin(a+b) - \sin(a-b)]$

Double-Angle and Half-Angle Formulas

Double-Angle Formulas

$\sin^2 a = \dfrac{1}{2}(1 - \cos 2a)$

$\cos^2 a = \dfrac{1}{2}(1 + \cos 2a)$

$\sin 2a = 2 \sin a \cos a$

$\cos 2a = 2\cos^2 a - 1$

Half-Angle Formulas

$\cos \dfrac{1}{2}a = \sqrt{\dfrac{1}{2}(1 + \cos a)}$

$\sin \dfrac{1}{2}a = \sqrt{\dfrac{1}{2}(1 - \cos a)}$

Hyperbolic Functions

$\sinh a = \dfrac{1}{2}(e^a - e^{-a})$

$\cosh a = \dfrac{1}{2}(e^a + e^{-a})$

$\cosh^2 a - \sinh^2 a = 1$

$\sinh(a \pm b) = \sinh a \cosh b \pm \cosh a \sinh b$

$\cosh(a \pm b) = \cosh a \cosh b \pm \sinh a \sinh b$

$\sinh^2 a = \dfrac{1}{2}(\cosh 2a - 1)$

$\cosh^2 a = \dfrac{1}{2}(1 + \cosh 2a)$

Integration

$$\int x^n \, dx = \frac{x^{n+1}}{n+1} + c \quad (n \neq -1)$$

$$\int e^x \, dx = e^x + c$$

$$\int e^{ax} \, dx = \frac{1}{a} e^{ax} + c$$

$$\int e^{g(x)} g'(x) \, dx = e^{g(x)} + c$$

$$\int \frac{1}{x} \, dx = \ln|x| + c$$

$$\int \frac{g'(x)}{g(x)} \, dx = \ln|g(x)| + c$$

$$\int \sin x \, dx = -\cos x + c$$

$$\int \cos x \, dx = \sin x + c$$

$$\int \tan x \, dx = -\ln|\cos x| + c = \ln|\sec x| + c$$

$$\int \cot x \, dx = \ln|\sin x| + c = -\ln|\csc x| + c$$

$$\int \sec x \, dx = \ln|\sec x + \tan x| + c$$

$$\int \csc x \, dx = \ln|\csc x - \cot x| + c$$

$$\int \ln x \, dx = x \ln x - x + c$$

$$\int \frac{1}{a^2 + x^2} \, dx = \frac{1}{a} \tan^{-1} \frac{x}{a} + c$$

$$\int \frac{1}{a^2 - x^2} \, dx = \frac{1}{2a} \ln \left| \frac{x+a}{x-a} \right| + c = \frac{1}{a} \tanh^{-1} \frac{x}{a}$$

$$\int \frac{1}{\sqrt{a^2 - x^2}} \, dx = \sin^{-1} \frac{x}{a} + c$$

$$\int \frac{1}{\sqrt{a^2 + x^2}} \, dx = \sinh^{-1} \frac{x}{a} + c$$

$$\int e^{ax} \sin bx \, dx = \frac{1}{a^2 + b^2} e^{ax} (a \sin bx - b \cos bx) + c$$

$$\int e^{ax} \cos bx \, dx = \frac{1}{a^2 + b^2} e^{ax} (a \cos bx + b \sin bx) + c$$

Index

A

Accelerometer, 349
$A\cos\omega t + B\sin\omega t$ expressed as $D\sin(\omega t + \phi)$, 30–31
Adjoint matrix, 63–65
Algebraic eigenvalue problem, 337
Algebraic multiplicity (AM) of eigenvalue, 70, 71
Angular acceleration vector, 128
Angular position vector, 128
Angular velocity vector, 128
Armature-controlled motors, 217–221

B

Bandwidth, 369
Biot number, 258, 259, 266
Block diagonal matrix, 60, 61
Block diagram representation, of dynamic system, 103
    diagram construction, 111–113
    operations
        closed-loop (feedback) system, 105–107
        integration, 105
        parallel combination of blocks, 104–105
        series combination of blocks, 103–104
        summing junction, 103
    reduction techniques
        branch point moving, 107
        Mason's rule, 109
        summing junction moving, 107–108
Block triangular matrix, 60, 61
Bode diagram, 291–297, 401–413
    analysis using, 407–409
    of basic feedback system, 401–407
    control design using, 409–411
    of first-order systems, 292–294
    plotting in MATLAB®, 292
    of second-order systems, 295–297

C

Capacitors, 193–196
Cart-inverted-pendulum system, 169
Characteristic equation, 337, 414
Characteristic length, of solid object, 259
Closed-loop (feedback) system, 105–107
    negative feedback system, 106
Closed-loop transfer function (CLTF), 106–107
Complex analysis, 19
    complex numbers, in polar form, 22–25
        complex algebra, 23–24
        division using, 24
        integer powers, 25
        roots, 25
    complex numbers, in rectangular form, 19–21
        complex conjugate, 21
        magnitude of, 20–21
    functions, 25–26
    variables, 25
Complex plane, 19–20
Conduction, 255
Configurational form, of dynamic system, 79–80
    second-order matrix form, 80–82
Conjugates, 21
Controller canonical form, 97
Convection, 255
Convolution method, for inverse Laplace transformation, 45–46
Coordinate reference, 141–144
Coulomb damping, 130, 316–319
Cramer's rule, 61–63
Current, 189
Current–force relation, 216
Current–voltage relation. *See* Voltage–current relation

D

D'Alembert's principle, 145–146
Damper elements, 130–131
Damping ratio, 275
DC motor modeling, 215–216
Defective matrices, 73
Determinant of matrix, 58–59
Diagonal matrices, 57
Differential equations, 27
    linear, first-order differential equations, 27–28
    ordinary differential equations (ODEs), 27
    partial differential equations (PDEs), 27
    second-order differential equations with constant coefficients, 28–31
Differentiation, in MATLAB, 6–7
Dirac delta. *See* Unit-impulse (Dirac delta) function
Dominant pole, 394
Double pendulum, 173–176
Dynamic system model representation. *See* System model representation

E

edit command, MATLAB, 1–2
Effective moment, 155
Eigenvalue problem, 336–339
Eigenvalue problem, in matrices, 68
    properties, 69–70
        algebraic multiplicity (AM), 70, 71

Eigenvalue problem, in matrices (*continued*)
    generalized eigenverctors, 71–72
    geometric multiplicity (GM), 70, 71
  similarity transformations, 72
    defective matrices, 73
  solution, 68
Electrical elements, 189–190
  capacitors, 193–196
  inductors, 192–193
  resistors, 190–192
Electric circuits
  Kirchhoff's current law, 198–200
  Kirchhoff's voltage law, 196–198
  loop method, 202–204
  node method, 201–202
  state variables of circuits, 204–206
Electric elements, impedances of, 225
Electromechanical systems, 214–215
  armature-controlled motors, 217–221
  elemental relations of, 215–217
  field-controlled motors, 222–223
Elemental relations of, 215–217
Elementary row operations (EROs), in matrices, 57–58
Energy method, 171–176
Equivalence, 131–134
  parallel resistors, 191
  parallel springs, 132
  series springs, 132–133
  of cantilever beam, 133–134
EROs. *See* Elementary row operations (EROs), in matrices
Error signal, 360
Euler's formula, 22
`ezplot` command, MATLAB, 9

F

Feedback control systems, 359
  basic concepts and terminologies, 359–362
  benefits of, 370–380
    disturbance rejection, 373–375
    reference tracking, 375–377
    sensitivity to parameter variations, 377–379
    stabilization, 370–373
  Bode plot, 401–413
    analysis using, 407–409
    of basic feedback system, 401–407
    control design using, 409–411
  frequency-domain performance specifications, 368–369
  full-state feedback, 414–421
    control design for, 417–419
    state-space equations analysis, 414–417
  proportional–integral–derivative control, 380–389
    PID control, 383–385
    proportional control, 380–382
    proportional–integral control, 382–383
    Ziegler–Nichols tuning of PID controllers, 385–386
  root locus, 389–401
    analysis using, 394–396
    of basic feedback system, 389–394
    control design using, 396–398
  stability of linear time-invariant systems, 362–365
  time-domain performance specifications, 365–368

Field-controlled motors, 222–223
Final-value theorem (FVT), 47–48
First law of thermodynamics, 253–254
First-order systems, transient response, 270–273
  free response, 270
  ramp response, 272
  step response, 271–272
Force–current relation. *See* Current–force relation
Forced vibration, 320–327
  half-power bandwidth, 321–323
  harmonic base excitation, 325–327
  rotating unbalance, 323–325
Force equations
Fourier's law, 255
Free-body diagram, 139–141
Free response, 270
  of second-order systems, 275–277
    initial response in MATLAB, 276–277
Free vibration, 313–320
  Coulomb damping, 316–319
  logarithmic decrement, 313–316
Frequency equation. *See* Characteristic equation
Frequency response, 287–288
  Bode diagram, 291–297
    of first-order systems, 292–294
    plotting in MATLAB, 292
    of second-order systems, 295–297
  of stable, linear systems, 288–291
    first-order systems, 289–290
    second-order systems, 290–291
Full matrices, 57
Full-state feedback, 414–421
  control design for, 417–419
  state-space equations analysis, 414–417
FVT. *See* Final-value theorem (FVT)

G

Gain margin (GM), 407
Gear–train systems, 178–181
  single-degree-of-freedom, 179–180
  single-link robot arm, 180–181
General solution, differential equations, 27, 31
Generalized coordinates, 79
Geometric multiplicity (GM) of eigenvalue, 70, 71
Ground, in node method, 201

H

Half-power bandwidth, in forced vibration, 321–323
Harmonic base excitation, in forced vibration, 325–327
Heat, 254. *See also* Thermal system
Heat flow rate, 254
Hooke's law, 129
Hydraulic capacitance, 241–243, 265
Hydraulic resistance, 243–244

I

Ideal gases, 235–236
Identity matrix, 57
Impedance methods

electric elements, impedances of, 225
  mechanical impedances, 228–229
  series and parallel impedances, 225–228
Inductors, 192–193
Initial-value problem (IVP), 27, 46–47
Initial-value theorem (IVT), 48–49
Inline function, MATLAB, 3
Input–output (I/O) equation, 91–92
  to state-space form, 96–99
Integration, in MATLAB, 7
I/O equation. *See* Input–output (I/O) equation
Irreducible polynomial, 41
Isentropic process (reversible adiabatic process), ideal gas law, 236
Isobaric (constant-pressure) process, ideal gas law, 236
Isochoric (constant-volume) process, ideal gas law, 236
Isothermal (constant-temperature) process, ideal gas law, 236
Iterative calculations, in MATLAB, 4
IVP. *See* Initial-value problem (IVP)
IVT. *See* Initial-value theorem (IVT)

J

Jordan matrix, 73

K

Kinetic energy, 128, 129, 172
Kirchhoff's current law, 198–200
Kirchhoff's voltage law, 196–198

L

Lagrange's equations, 173
Laplace transformation, 31–32. *See also* Transfer functions
  of derivatives, 40
  differentiation of, 34–35
  final-value theorem (FVT), 47–48
  frequently used transformations, 33–34
  initial-value theorem (IVT), 48–49
  of integrals, 41
  integration of, 35
  inverse, 32, 41
    convolution method, 45–46
    partial fraction method, 41–45
  IVP solution, 46–47
  linearity, 32
  operations, 32
  periodic functions, 39–40
  poles, 32
    simple poles, 32
  shifted translation, 36–37
  unit-impulse (Dirac delta) function, 38–39
  unit-impulse and unit-step functions, relation between, 39
  unit-pulse function, 38
  unit-ramp function, 37–38
  unit-step function, 35–37
Law of heat conductance. *See* Fourier's law
Lead/log compensator, 398
Linear, first-order differential equations, 27–28
Linearization of system model, 116
  of nonlinear function, 116–117
  of nonlinear model, 117–121
    operating point, 118
    procedure, 118–120
  small-angle linearization, 120–121
Liquid-level systems, 241
  hydraulic capacitance, 241–243
  hydraulic resistance, 243–244
  modeling of, 244–250
Logarithmic decrement, in free vibration, 313–316, 353
Loop method, 202–204
Lower-triangular matrices, 57

M

Mason's rule, 109
  general case, 110–111
  special case, 109–110
Mass elements, 127–129
Massless junctions, 146
  two-degree-of-freedom, 146–147
Mass moment of inertia, 156–160
  of common geometric shapes, 157
Mass–spring–damper system
  single-degree-of-freedom, 139–141
  single-degree-of-freedom, rotational, 155
  two-degree-of-freedom, rotational, 155–156
MATLAB
  analytical expressions plotting, 9–10
  command prompt, 1
  data points plotting, 7–9
  differentiation, 6–7
  functions, defining and evaluating, 3–4
  integration, 7
  iterative calculations, 4
  matrices and vectors, 4–6
  partial fractions method in, 44–45
  screen capture of session, 2
  Simulink®. *See* Simulink
  user-defined functions, 1–2
Matrices, 55–56
  addition of, 74
  block diagonal matrix, 60, 61
  block triangular matrix, 60, 61
  Cramer's rule, 61–63
  determinant of, 58–59
    properties, 59
  diagonal matrices, 57
  eigenvalue problem. *See* Eigenvalue problem, in matrices
  elementary row operations (EROs), 57–58
  full matrices, 57
  homogeneous system, 63
  identity matrix, 57
  inverse, 63
    and adjoint matrix, 63–65
    properties, 65
  lower-triangular matrices, 57
  in MATLAB, 4–6
  product of, 55, 56
  rank of, 57, 59
  row-echelon form of REF(), 57
  skew-symmetric matrices, 57
  square matrices, 55
  symmetric matrices, 57

Matrices (*continued*)
　transpose of, 55
　upper-triangular matrices, 57
Mechanical impedances, 228–229
Mechanical systems, 127
　gear–train systems, 178–181
　mechanical elements, 127
　　damper elements, 130–131
　　equivalence, 131–134
　　mass elements, 127–129
　　spring elements, 129–130
　mixed systems (translational and rotational), 167
　　energy method, 171–176
　　force and moment equations, 167–171
　rotational systems, 152–153
　　general moment equation, 153–154
　　mass moment of inertia, 156–160
　　modeling of rigid bodies in plane motion, 154–156
　　pure rolling motion, 160–162
　translational systems, 138
　　coordinate reference, 141–144
　　D'Alembert's principle, 145–146
　　free-body diagram, 139–141
　　massless junctions, 146–147
　　Newton's second law, 138–139
　　static equilibrium position, 141–144
MIMO system. *See* Multi-input–multi-output (MIMO) system
Mixed mechanical systems (translational and rotational), 167
　energy method, 171–176
　force and moment equations, 167–171
Modal analysis, of vibrations, 336–347
　eigenvalue problem, 336–339
　orthogonality of modes, 340–342
　response
　　to harmonic excitations, 344–346
　　to initial excitations, 342–344
Modal forces, 345
Moment equations, 153–154, 167–171
Multi-input–multi-output (MIMO) system, 91. *See also* Single-input–single-output (SISO) system

N

Negative feedback system, 106, 361. *See also* Feedback control system
Newton's second law, 138–139
Node method, 198, 201–202
Nonlinear systems, 305–309
　RK4 method, 306
　　running in MATLAB, 307–309
Normalized amplitude, 291

O

ODEs. *See* Ordinary differential equations (ODEs)
Ohm's law, 190
Op-amp circuit, 213
Op-amp integrator, 211–212
Op-amp multiplier, 210–211
Operating point, of system model, 116, 118

Operational amplifiers, 209–214
Ordinary differential equations (ODEs), 27
　second-order equations, 29
Orthogonality, of vibrational modes, 340–342
Output equation, 85–86

P

Parallel impedances, 225–228
Partial differential equations (PDEs), 27
Partial fraction method, for inverse Laplace function, 41–43
　in MATLAB, 44–45
Particular solution, differential equations, 27
PDEs. *See* partial differential equations (PDEs)
Pendulum, 145–146
　cart-inverted-pendulum system, 169
　double pendulum, 173–176
Pendulum-bob system, 158–160
Periodic functions, 39–40
Phase (argument), of complex number, 22
Phase margin (PM), 407
PID control. *See* Proportional–integral–derivative (PID) control
piezoelectric transducer, 349
Plant, feedback control system, 359, 421
plot command, MATLAB, 7
Pneumatic capacitance, 236–237, 265
Pneumatic systems, 235
　ideal gases, 235–236
　modeling of, 237–240
　pneumatic capacitance, 236–237
Poles, in Laplace transformation, 32
Polytropic process, ideal gas law, 236
Positive feedback system, 361 *See also* Feedback control system
Potential energy, 128, 129, 130, 172
Power, 190
Proportional control, 380–382
Proportional–integral control, 382–383
Proportional–integral–derivative (PID) control, 380–389
　proportional control, 380–382
　proportional–integral control, 382–383
　Ziegler–Nichols tuning of, 385–386
Pulley system, 168–169
　energy method, 172–173
Pure rolling motion, 160–162

Q

Quarter-car model, two-degree-of-freedom, 142–144

R

Radiation, 255
Ramp function. *See* Unit-ramp function
Ramp response of first-order systems, 272
　steady-state error, 273
Regulator, 361
Resistors, 190–192
Resonant peak, 369
Right-hand rule, angular vector direction, 128
Rigid bodies modeling, in plane motion, 154–156
RK4 method, nonlinear systems, 306
　running in MATLAB, 307–309

Index

RLC circuit, 227
    parallel, 199–200
    series, 197–198
Root locus, 389–401
    analysis using, 394–396
    of basic feedback system, 389–394
    control design using, 396–398
Rotating unbalance, in forced vibration, 323–325
Rotational mechanical systems, 152–153
    general moment equation, 153–154
    mass moment of inertia, 156–160
    modeling of rigid bodies in plane motion, 154–156
    pure rolling motion, 160–162

S

Second-order differential equations with constant coefficients, 28–31
    homogeneous equation, 28–29
    particular solution, 29–31
        $A\cos\omega t + B\sin\omega t$ expressed as $D\sin(\omega t + \phi)$, 30–31
    recommended solutions, 29
Second-order matrix form, of dynamic system, 80–82
Second-order systems, transient response, 274–275
    to arbitrary inputs, 281–285
    free response, 275–277
    impulse response, 278
        in MATLAB, 278–279
    step response, 279–280
        in MATLAB, 280–281
Series impedances, 225–228
Servo system. *See* Tracking system
Shifted translation, unit-step function, 36–37
Simulink, 10
    block library, 10–11
    new model building, 12–13
    simulation, 13–15
Single-input–single-output (SISO) system, 91. *See also* Multi-input–multi-output (MIMO) system
Single-link robot arm, 180–181
    driven by DC motor, 219–221
Single-tank liquid-level system, with pump, 246–247
SISO. *See* Single-input–single-output (SISO) system
Skew-symmetric matrices, 57
Small-angle linearization, 120–121
Spring elements, 129–130
Square matrices, 55
State equation solution, 300–304
    formal solution of, 300–303
        in MATLAB, 302–303
        matrix exponential, 300–302
    via Laplace transformation, 303–304
    via state-transition matrix, 304–305
State-space form, of dynamic system, 82
    state equation, 83–85
        decoupling, 88–90
    state-variable equation, 83
    state variables, 82–83
    to transfer function, 99–101
State variables of circuits, 204–206
Static deflection, 320
Static equilibrium position, 141–144
Structural damping, 130

subs command, MATLAB, 4
Symmetric matrices, 57
System model representation
    block diagram representation, 103
        diagram construction, 111–113
        operations, 103–107
        reduction techniques, 107–111
    configurational form, 79–80
        second-order matrix form, 80–82
    input–output (I/O) equation, 91–92
        to state-space form, 96–99
    linearization, 116
        of nonlinear function, 116–117
        of nonlinear model, 117–121
    multi-input–multi-output (MIMO) system, 91
    operating point, 116, 118
    single-input–single-output (SISO) system, 91
    state-space form, 82, 86–88
        output equation, 85–86
        state equation, 83–85
        state equation, decoupling, 88–90
        state-variable equation, 83
        state variables, 82–83
        to transfer function, 99–101
    transfer functions, 93–94
System response, 269
    frequency response, 287–288
        Bode diagram, 291–297
        of stable, linear systems, 288–291
    of nonlinear systems, 305–309
        RK4 method, 306–309
    state equation solution, 300–304
        formal solution of, 300–303
        via Laplace transformation, 303–304
        via state-transition matrix, 304–305
    transient response, of first-order systems, 270–273
        free response, 270
        ramp response, 272
        step response, 271–272
    transient response, of second-order systems, 274–275
        to arbitrary inputs, 281–285
        free response, 275–277
        impulse response, 278
        step response, 279–280
    types of, 269

T

Temperature dynamics
    of heated object, 259–260
    of house with heater, 260–263
Thermal capacitance, 254, 265
Thermal resistance, 255–258
Thermal systems, 253
    first law of thermodynamics, 253–254
    modeling, 258–263
    thermal capacitance, 254
    thermal resistance, 255–258
Third root of 1, 25–26
Torque, 128, 131, 216, 222, 230
Tracking system, 361
Transfer functions, 93–94

Transient response
  to arbitrary inputs, 281–285
  of first-order systems, 270–273
    free response, 270
    ramp response, 272
    step response, 271–272
  of second-order systems, 274–275
    free response, 275–277
    impulse response, 278
    impulse response, in MATLAB, 278–279
    step response, 279–280
    step response, in MATLAB, 280–281
Translational mechanical systems, 138
  D'Alembert's principle, 145–146
  free-body diagram, 139–141
  massless junctions, 146–147
  Newton's second law, 138–139
  static equilibrium position and coordinate reference, 141–144
Transmissibility, 326
Two-tank liquid-level system, 247–250

U

Ultimate gain, feedback control system, 386
Ultimate sensitive method, 386
Unit-impulse (Dirac delta) function, 38–39
Unit-impulse and unit-step functions, relation between, 39
Unit-pulse function, 38
Unit-ramp function, 37–38
Unit-step function, 35–37
Upper-triangular matrices, 57

V

Vectors, in MATLAB, 4–6
Vibrations, 313
  forced vibration, 320–327
    half-power bandwidth, 321–323
    harmonic base excitation, 325–327
    rotating unbalance, 323–325
  free vibration, 313–320
    Coulomb damping, 316–319
    logarithmic decrement, 313–316
  modal analysis, 336–347
    eigenvalue problem, 336–339
    orthogonality of modes, 340–342
    response to harmonic excitations, 344–346
    response to initial excitations, 342–344
  suppressions of, 329–335
    vibration absorbers, 331–335
    vibration isolators, 329–331
  vibration measurement and analysis, 347–352
    system identification, 349–350
    vibration measurement, 348–349
Viscous damping, 130
Voltage, 189
Voltage–current relation
  for capacitor, 193
  for inductor, 192
  for resistor, 190

W

Work, 254

Z

Zeros, 363
Ziegler–Nichols tuning of PID controllers, 385–386